Quantum Electrochemistry

By John O'M. Bockris
Texas A & M University

and Shahed U. M. Khan
University of Bonn, West Germany

Kinetic electrochemistry at interfaces has become increasingly quantal in its orientation
in recent times. This volume responds to the
need for a comprehensive guide to quantum
aspects of electrochemistry both at interfaces and in solution.

The emphasis of this volume is on the molecular model approach in electrode kinetics,
although the contributions made from the
continuum dielectric viewpoint are also recognized. Among the topics covered are
electric double layers in metals, time-dependent perturbation theory, long-range energy
transfer in condensed media, proton transfer
in solution and at interfaces, and photoelectrochemical kinetics. This breakthrough
volume will serve as both a text for graduate
students and an interdisciplinary guide for
the practicing electrochemist.

Quantum
Electrochemistry

Quantum Electrochemistry

John O'M. Bockris
Texas A & M University
College Station, Texas

and

Shahed U. M. Khan
University of Bonn
Bonn, West Germany

Springer Science+Business Media, LLC

Library of Congress Cataloging in Publication Data

Bockris, John O'M
  Quantum electrochemistry.

  Includes bibliographical references and index.
  1. Electrochemistry. 2. Quantum chemistry. I. Khan, S. U. M., joint author. II. Title.
QD553.B634   541'.37   78-11167
ISBN 978-1-4684-2495-9

ISBN 978-1-4684-2495-9   ISBN 978-1-4684-2493-5 (eBook)
DOI 10.1007/978-1-4684-2493-5

© 1979 Springer Science+Business Media New York
Originally published by PLenum Press, New York 1979
Softcover reprint of the hardcover 1st edition 1979

Preface

The origin of this book lies in a time before one of the authors (J. O'M. B.) left the University of Pennsylvania bound for the Flinders University. His collaboration with Dennis Matthews at the University of Pennsylvania had contributed a singular experimental datum to the quantum theory of electrode processes: the variation of the separation factor with potential, which could only be interpreted in terms of a quantum theory of electrode kinetics. The authors came together as a result of graduate work of one of them (S. U. M. K.) on the quantum mechanics and photoaspects of electrode processes, and this book was written during a postdoctoral fellowship held by him at the Flinders University.

Having stated the book's origin, it is worthwhile stating the rationalizations the authors had for writing it. Historically, quantization in electrochemistry began very early (1931) in the applications of the quantum theory to chemistry. (See the historical table on pages xviii–xix.) There was thereafter a cessation of work on the quantum theory in electrochemistry until a continuum dielectric viewpoint, based on Born's equation for solvation energy, began to be developed in the 1950s and snowballed during the 1960s.

This book describes the other side of the penny from that development. It is the presentation of the molecular model approach in electrode kinetics, albeit with the complexity of the system recognized. Of course, the contributions made from the continuum viewpoint, part of which is always found to be useful, are described. We are attempting here to make up for the lack of quantization in earlier electrochemical studies and to recognize that there are those to whom the continuum approach, developed in such mathematical detail, remains tenuously connected to model concepts of electrodes and solution current in most study groups.

A second reason arises for writing a book of this kind. During the 1960s it was the fashion in the physical and inorganic divisions of chemistry departments to turn from the teaching of chemistry as a broad field with complex systems described phenomenologically and to treat particularly those systems more amenable to quantum chemical approaches, using data obtained spectroscopically. The good that this attempt at rigor brought had also negative side effects, including the spread of the feeling that chemistry was less in contact with what are so often described as *relevant* areas in chemistry. Parallel with the narrowing of scope of university chemistry went a loss of students, which might have occurred to a lesser degree had contact with more practical things (the attractive aspect of chemistry, compared with physics) been maintained. The attempt to deal, albeit mostly at a fairly crude level, with electrode processes in terms of quantum mechanics may be one lead to reintroducing into more chemistry departments the treatment of subjects that seem to have been squeezed out of the curriculum because they were too complex for exact quantum mechanics. Thus, there is no doubt of the *relevance* of electrode kinetics, e.g., in fuel cells, etc.

For whom is this book meant? It is certainly for advanced students in chemistry departments. However, it will appeal to a wider audience, for the practicing electrochemist comes from an interdisciplinary group of subjects that involves physics, chemistry, engineering, biology, energy studies, and occasionally further fields, such as geology and soil science. The book will be of use to those who wish to understand the connection of the usual phenomenological electrochemical considerations with more basic quantum chemistry.

One of the difficulties of writing a book like this is to know how much background to give. The book is a compromise between a monograph, appealing to workers in a specialized field, and a text that can be understood by any physical scientist. We have tried sections on our colleagues, and here particularly we have to acknowledge the help of John McCann and Darryl Rose, graduate students studying quantum electrochemistry at the Flinders University. When puzzlement was expressed, we have developed the material so that they and the other graduate students who helped us could understand it. Thus, the prerequisites of the book are an elementary knowledge of quantum mechanics and undergraduate physicochemical background. The necessary background of phenomenological electrode process chemistry (where it exceeds that available in undergraduate texts) is given in this book.

Our colleague, Errol McCoy, at Flinders, has been helpful in discussions of quantum mechanical points. Reg Cahill, of the Flinders Physics Group, has refereed the chapter on time-dependent perturbation theory (Chapter 4). Dr. U. Gösele (Max Planck Institute, Stuttgart) has given us the benefit of his opinion concerning the Forster theory of energy transfer in liquids. Dr. L. Kevan (Wayne State) has clarified details in his publications on solvated electrons. Dr. W. Schmickler (Bonn), with whom one of the authors (S. U. M. K.) is now in active collaboration, has been extremely forthcoming in frank criticisms and discussions of several chapters. Dr. Jens Ullstrup (University of Denmark) has contributed, in letters and discussions, to our understanding of the continuum viewpoint. Professor St. G. Christov (Bulgarian Academy of Science) has corrected misimpressions we have had of continuum concepts. Professor J. Walter Schulze (Berlin) has made a number of helpful contributions to our interpretations of tunneling through oxides at interfaces. Dr. P. Wright (University of Melbourne) has rendered mathematical help in preparing some of the appendixes.

We are very grateful to Plenum Publishing Corporation for their patience in waiting for this book a far longer time than at first anticipated, and to Mrs. D. Hampton for her forbearance and acumen in disinterring the authors' meaning and transferring it from manuscript to typescript.

Texas A & M University   John O'M. Bockris
University of Bonn   Shahed U. M. Khan

Contents

4.   Time-Dependent Perturbation Theory

5.   Long-Range Radiationless Energy Transfer in Condensed Media

6. Mechanisms of Activation

7. The Continuum Theory

8. Interfacial Electron Tunneling

9. Proton Transfer in Solution

10. Proton Transfer at Interfaces

12. Photoelectrochemical Kinetics

13. Quantum Electrode Kinetics

Quantum
Electrochemistry

Dates of Some First Contributions to Quantum Electrochemical Kinetics

	Molecular theory			Continuum theory			
Decade	Name and date	Country of work performance	Essence of contribution	Decade	Name and date	Country of work performance	Essence of contribution
1930–1940	Oliphant and Moon, 1930	U.K.	First formulation of conditions for neutralization of gaseous ion at electrode	1930–1940			No significant developments
	Gurney, 1931	U.K.	Seminal contribution formulating quantum mechanical transitions at electrode solution interface: $i \propto \int\int P_T N_E f(E)\, dE\, dx$				
	Fowler, 1931	U.K.	Quantum theory of light-stimulated electron emission at metal–vacuum interface				
	Bawn and Ogden, 1934	U.K.	First quantal attempt to explain separation factor				
	Horiuti and Polanyi, 1935	U.K.	Qualitative theory of effect of bonding of radicals to electrode; electrocatalysis				
	Butler, 1936	U.K.	Metal–hydrogen bonding added to Gurney formulation				
1940–1950	No significant developments			1940–1950			No significant developments
1950–1960	Gerischer, 1960	Germany	Reformulation of Gurney		Libby, 1952	U.S.	"Franck–Condon transition" in electrode processes; Born equation introduced

Period	Reference	Country	Description
	Weiss, 1954	U.K.	Born–Landau equation applied to electric kinetics; optical dielectric constant and activation energy in terms of ε_{op}
	Marcus, 1956	U.S.	Reorganization of solvent before electron transfer
	Zuolinski, 1956	U.S.	Tunneling through Libby-type barrier
	George and Griffith, 1959	U.K.	Bond stretching in redox reactions related to energy of activation
1960–1970	Christov, 1964	Bulgaria	Calculation of isotopic ratios
1960–1970	Levich and Dogonadze, 1960	U.S.S.R.	Hamiltonians formulated in terms of dielectric polarization theory
	Levich, 1970	U.S.S.R.	Polaron model for activation of electrode reactions
1970–1977	Schulze and Vetter, 1972	Germany	Kinetics through films
1970–1977	Bockris, Khan, and Matthews, 1974	Australia	Theoretical calculations from continuum theory inconsistent with experimental data unless bond stretching stressed
	Schmickler, 1976	Germany	Modelistic theory of reorganization energy
	Bockris and Sen, 1975	Australia	Frequency of large fluctuation of libratory energy on central ion too small to explain orders of magnitudes of currents observed
	Christov, 1977	Bulgaria	LDK proton discharge theory inconsistent with facts
	Khan, Bockris, and Wright, 1977	Australia	Time-dependent perturbation theory calculation
	Bockris, Khan, and Uosaki, 1977	Australia	Theory of light effects on hydrogen evolution on metals and semiconductors solution
	Schmidt, 1976	U.S.	Liquid drop model for inner sphere
	Schmickler, 1977	Germany	Resonance tunneling at electrodes

Electric Double Layers at Metals

1.1. Structure of the Double Layer

1.1.1. The Diffuse Layer

Gouy[1] and Chapman[2] gave equations for the excess net charge in a solution per unit area of an electrode as a function of the metal–solution potential difference, assuming only electrical interaction with the electrode, *and for point ions*. The corresponding equation that arises for the diffuse-layer capacitance is

$$C_{2-b} = \left(\frac{z^2 e_0^2 n\varepsilon}{2\pi kT}\right)^{1/2} \cosh \frac{z e_0 \phi_{2-b}}{2kT} \tag{1.1}$$

ϕ_{2-b} is the potential of the layer of ions in the outer Helmholtz plane (OHP) with respect to that of the solution, in which the concentration is n ions/cm^3.

This model of the diffuse layer is better used in conjunction with one for the inner layer. It has been validated for dilute solutions,[3] but there are doubts about its validity for higher concentrations. Thus, linear expansion of the potential-containing terms, used in the Debye–Hückel theory, cannot then be used. In the absence of such expansion, the principle of superposition of interaction energies from many ions to a central ion is no longer applicable.†

Stillinger and Kirkwood[4] applied correlation functions to calculate the situation in the diffuse layer in concentrated solutions; as expected, they found results agreeing with the Gouy–Chapman theory in the linear region. When $\phi_{2-b}F/RT > 1$, ionic layers of alternate sign, arising as a result of short-range repulsions, are indicated.

† This does not—indeed, cannot—prevent the application of zeta-potential concepts throughout electrochemistry, as, for example, in the treatment of the reduction of AgBr by Hoffmann and Billings.[157]

Levich, Kir'yanov, and Krylov[5] discussed the Stillinger and Kirkwood correlation function analysis. The result at moderate bulk concentrations is that there is a slightly faster fall in potential into the solution (see Buff and Stillinger[6]) than given by the original simple analysis (see Fig. 1.1).

The ability to apply confidently diffuse-layer theory is important. The values of contact adsorption calculated from thermodynamic data depend on this. The testing of the models due to Stillinger and Levich by comparing their predictions with experiment has not yet been done.

There are plenty of indications, of course, that the present double-layer treatments are approximations. The fact that the diffuse-layer theory may be grossly approximate at the higher concentrations, where its applicability is most needed, is not a good omen. Another approximation that has recently been criticized is the divisibility of the compact and diffuse layers into two capacitors in series. Cooper and Harrison[7] suggest that this may be justified only at low concentrations. That the division is an approximate modelistic interpretation is clear. What is as yet unclear is what is a less approximate modelistic interpretation of the obviously physically present compact and diffuse region in the double layer.

1.1.2. Stern–Grahame–Devanathan Model

Stern[8] combined the Helmholtz model with that of Gouy and suggested there were two kinds of adsorbed ions. In one, as Gouy had suggested, the ions remain separated from the electrode. In a second, ions touch the electrode ("contact adsorption") and may be present on it in an amount greater than that required for balancing the excess charge on the electronically conducting phase. The countercharge to make the total ionic charge in the double layer equal and opposite to that in the metal phase is

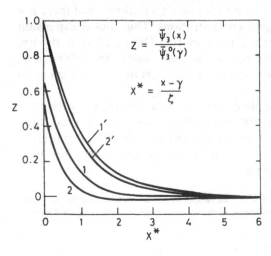

$$Z = \frac{\bar{\Psi}_3(x)}{\bar{\Psi}_3^0(\gamma)}$$

$$x^* = \frac{x - \gamma}{\zeta}$$

Fig. 1.1. The diffuse layer in concentrated solutions: (1) $5\,M$ electrolyte concentration; (2) $6\,M$ electrolyte concentration; (1',2') curves corresponding to the Gouy–Stern theory.

made up by an excess of ions in the diffuse layer, of charge opposite to ions in contact with the electrode.

Grahame[9] suggested that cations were adsorbed only in the first (Gouy) way, not in contact with the electrode. By using diffuse-layer theory to calculate ϕ_{2-b} from Γ_+ for a series of conventional electrode potentials, and Grahame's assumption, the diffuse-layer anionic charge can be obtained. One uses ϕ_{2-b} in the Gouy–Chapman expression for q_{diffuse} and thereby separates it from the thermodynamic total anionic charge, leaving over the charge that is contact adsorbed.

One doubtful point is Grahame's assumption. To this a number of recent contributions have been made that tend to suggest that the assumption is only correct at low concentrations.[10] A second doubtful point is the use of diffuse-layer theory in concentrated solutions, and the third is the assumption usually made that the specific surface excess of water is zero.

Devanathan[11] contributed equations that related relevant quantities in Stern's model together with, in particular, the capacitance, to q_{contact}, the specifically adsorbed charge.

1.1.3. Water Structure in the Double Layer

Lange and Miscenko[12] formulated the equation for the Galvani potential difference in terms of the difference of two outer potentials and the difference of two surface potentials,

$$\Delta\phi = \Delta\psi + \Delta\chi \qquad (1.2)$$

where $\Delta\psi$ is the Volta potential difference at an interface and $\Delta\chi$ represents the sum of the electron overlap and dipole potential differences. Bockris and Potter[13] were the first to apply such a model quantitatively to the double layer in an attempt to rationalize some anomalous pH data in electrode kinetics.

Macdonald[14] (1954) formulated a double-layer model in which adsorbed water molecules were present, but he did not at first account for their rotation or contribution and capacitance. In 1962, Macdonald and Barlow[15] pointed out that dielectric unsaturation near the potential of zero charge (p.z.c.) might be the reason for the capacitance hump. In the same year, a comprehensive paper on the double layer was published by Mott and Watts-Tobin.[16] They utilized a two-state model for water and showed that the water dipoles contributed an orientation polarization term to the dielectric constant in the double layer. This became important near to the p.z.c. and was in fact the cause of the capacitance hump there.

Bockris, Devanathan, and Muller[17] (1963) developed the two-state model for water in a different way from that which had been done by Mott and Watts-Tobin. They utilized the Lange and Miscenko equation [Eq. (1.2)] and expressed $\Delta\chi$ as a function of the two-state water model, but maintained the dielectric constant in the expression for $\Delta\chi$ as the saturated value (6 for water) at all potentials. The differential of $\Delta\psi$ with respect to q_m thus gives the reciprocal of the double-layer capacity as a function of the contribution from the water capacity, which only has the possibility of being significant near the point at which $N\uparrow \simeq N\downarrow$ (in a first approximation, near the p.z.c., but not necessarily there if one state of the water molecule differs from the other), thus accounting for the effect of dielectric unsaturation on the double-layer properties. Allowance for the water–water interaction energy has a very significant effect on the numerical results of a two-state water model. It reduces $\Delta\chi$ and increases C_{dipole}, which as a consequence has a reduced effect on the hump. The Mott–Watts-Tobin formulation was a gaslike zeroth approximation.

Adsorbed water in the double-layer properties is a focus of modern interest,[18] for its properties have wide-ranging effects.[19] The double-layer

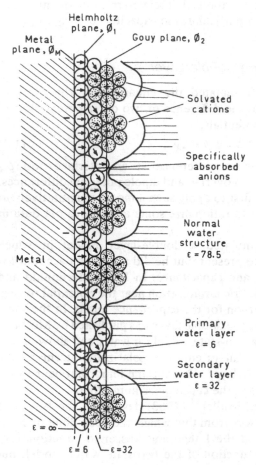

Helmholtz plane, $\emptyset_1$
Gouy plane, $\emptyset_2$
Metal plane, $\emptyset_M$

Solvated cations

Specifically absorbed anions

Normal water structure
$\varepsilon = 78.5$

Metal

Primary water layer
$\varepsilon = 6$

Secondary water layer
$\varepsilon = 32$

$\varepsilon = \infty$

$\varepsilon = 6$
$\varepsilon = 32$

Fig. 1.2. The Bockris, Devanathan, and Muller two-state model of the double layer.

structure with addition of an oriented water molecule [i.e., the Bockris–Devanathan–Muller (BDM) model] is shown in Fig. 1.2.[17] These matters have been recently reviewed by Reeves.[20] The most recent model for water at the interface is the three-state water model of Bockris and Habib.[21]

1.2. Methods of Investigation

1.2.1. Objectives of a Double-Layer Study

The objectives of a double-layer study are the evaluation of the intermolecular forces in detail such that the structural situation can be unambiguously calculated. There are two basic experimental dependencies that have to be established (for a series of ions at a number of interfaces) as a basis to the theoretical development:

(a) the evaluation of the coverage–concentration relation at constant charge, where coverage refers to the contact-adsorbed ions, and

(b) the evaluation of the contact-adsorption-potential situation at constant concentration.

1.2.2. Classical Electrocapillary Curve

The equations that pertain to this curve were deduced, but with ambiguities, by Gibbs.[22] Koenig[23] gives a substantial discussion. The derivation by Parsons and Devanathan[24] is more satisfactory. Objections have been raised to the measurement of surface tension by the capillary-rise method. Indicative results are given by Parsons and Zobel;[25] and Parsons and Trasatti[3] call their surface tension measurements "useless." They suggest that errors are due to the sticking of the mercury column. Kovac[26] studied this problem, resetting the reading at each potential by vibrating the mercury column several times. Bockris, Muller, Wroblowa, and Kovac[27] compared the results of electrocapillary measurements with those derived from capacitance measurements in aqueous HCl. There was agreement between the electrode charge and other double-layer properties derived separately in this system from surface tension and capacitance measurements, respectively. The degree of agreement was ±2 dyn cm^{-1}. In the region of 3 M concentrations, $\gamma_{capacitance} - \gamma_{direct} = \Delta\gamma$, discrepancies were significant. They did not arise from sporadic effects such as sticking of the mercury in a capillary. Any suggestions of an artifact as reason for the discrepancy seem difficult to make consistent with the fact that, when plotted against electrode charge, the discrepancy in $\Delta\gamma$ disappears (Fig. 1.3, curve B). It was therefore suggested by Bockris, Muller, Wroblowa, and Kovac[27] (but cf. Payne[28] and Bockris et al.[29]) that the discrepancy

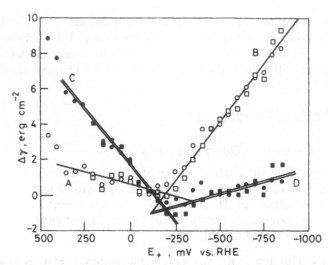

Fig. 1.3. Effect of the choice of the potential of zero charge upon discrepancies in the system Hg–0.01 N HCl. (A), (B), $E_{+zc} = -133$ mV ($E_{-zc} = -642$ mV); (C), (D), $E_{+zc} = -83$ mV ($E_{-zc} = -592$ mV); circles: Devanathan's values[164]; squares: Kovac's values[26] as compared to values calculated by integration of the measured differential capacities using either of the E_{zc} given. The difference in slopes between the lines of the anodic and cathodic sides is equivalent in either case to 1.6 μC cm^{-2}, or a discrepancy of $\Delta C = -10$ μF cm^{-2} over a region of 160 mV near the p.z.c.

arises because at constant charge there is no variation of $\Delta\chi$, and at constant potential such variation may occur. In ac measurements, the amplitude of the potential variation would cause a change in $\Delta\chi$ and hence in the apparent $\Delta\gamma$–potential relation.

If the surface tension of the solid–solution interface could be determined, advantages of the electrocapillary approach[30] (direct determination of electrode charge p.z.c., and Gibbs surface excess in the same type of measurement) could be obtained for solids. Beck was the first to make a measurement of the change of surface tension at the solid–solution interface as a function of potential[31] (Fig. 1.4). He extended a wire under a weight, the wire being sufficiently thin so that the surface tension affected its extension. Fredlein, Damjanovic, and Bockris[32] were the first to obtain an electrocapillary curve at a solid–solution interface using a gold electrode in 0.1 M KCl and also with platinum in 10^{-3} M H$_2$SO$_4$.

1.2.3. Capacitance Method

This method arises from the equation

$$C = \frac{\partial^2 \gamma}{\partial \Delta\phi^2} \tag{1.3}$$

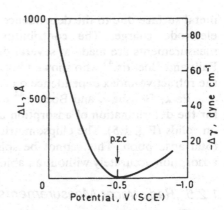

Fig. 1.4. "Electrocapillary curve" for gold ribbon in 0.1 *M* KCl (sweep speed = 0.25 V sec^{-1}).

Upon integration twice, one obtains the surface tension, the variation of which with concentration yields the surface excess. One of the needed integration constants is the p.z.c. and the other, the interfacial tension at the p.z.c. The establishment of p.z.c.'s (see Section 1.3) is now easier. The γ_{max} can be arbitrarily taken as zero when only the *change* of interfacial tension from that of some arbitrary situation is required, as is so for the evaluation of double-layer parameters.

Bridge design has been discussed recently by Schubert[33] and by Armstrong, Race, and Thirsk.[34] There are three difficulties with the application of capacitance methods to double-layer electrocapillary studies:

(a) The electrical equivalence model for the double layer may not be a pure capacitance,[35-37] although it is usually so presumed.

(b) The effects of roughness are difficult to eliminate.[38] The corresponding ac response has been examined by de Levie.[37] Roughness effects are important for frequencies at which $Dt/b^2 \ll 1$, where D is the diffusion coefficient in solution, t the time of the pulse or reciprocal of frequency, and b the surface roughness height.

(c) Pseudocapacitance. If there is a sufficiently high rate constant for a charge transfer reaction before the rate-determining step, the effect of injecting electrons into an electrode is not only to charge the double layer, but to transfer charge across the double layer. A pseudocapacitance arises. It can be eliminated by examining the dependence on frequency. This difficulty is more marked for catalytically active solid electrodes than for mercury.[39,40]

1.2.4. Ellipsometry

The first use of ellipsometry to determine adsorption in the double layer was published by Chiu and Genshaw.[41] Hansen[42] suggested that part of the changes observed were caused by changes in refractive index of the metal surface due to the dependence of the concentration of electrons on electrode charge. The contributions can easily be distinguished if measurements are made at several different wavelengths, as was done by Paik and Bockris,[43] who showed how one could subtract the error due to the refractive-index dependence on charge.

Paik, Genshaw, and Bockris[44] developed the ellipsometric approach for the determination of adsorption at interfaces and obtained much data on solids (Fig. 1.5). The ellipsometric method is convenient but measures total adsorption. This cannot be split up into diffuse layer and contact adsorption accurately without an ability to relate potential to charge.

1.2.5. Radiotracer Measurements

The first measurements using tracers were made by Aniansson.[45] Apparatuses were built simultaneously by Schwabe[46] and by Blomgren and Bockris.[47] The principle involves measurement of the activity through a thin foil from a solution. Calibration is simple.

Many measurements of adsorption of ions on solid electrodes that involve the radiotracer method have been carried out.[48] If suitable radioactive atoms exist for the systems considered, both cationic and anionic measurements are made and the p.z.c. can be measured. The degree of surface coverage is often linear with the surface concentration. Adsorption often follows a logarithmic law.[49] It can reach θ values much higher than those on mercury.

Extensive measurements of organic compound adsorption by radio-tracer methods have been made[50,51,52] (Fig. 1.6). One counts atom cores.

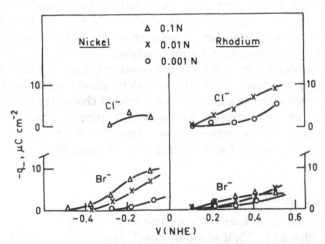

Fig. 1.5. Anion adsorption on nickel and rhodium electrodes.

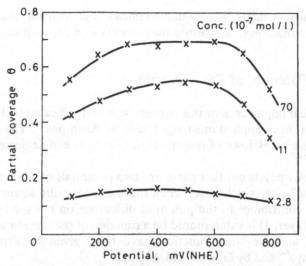

Fig. 1.6. Coverage–potential relationship 1 N H_2SO_4 at 50°C.[50]

No information concerning the radicals present arises *directly* from radio-tracer measurements. However, indirect conclusions arise, e.g., the entropy in the adsorption reaction can be calculated and related to the size of the particular radical being adsorbed.[52] Combination of radiotracer studies with potential sweep studies[53] can give rise to the determination of a radical.

1.2.6. Heat of Adsorption

The only present determinations are those of Parsons.[54]

1.2.7. Time Dependence

Work done before 1967 is reviewed in Gileadi's book.[55] In the theory developed,[56] the adsorption is supposed rate determined by diffusion; so not much information arises for the double layer. One obtains an equation of the form

$$\Gamma_t = \Gamma_0(1 - e^{-DAt/K\delta}) \tag{1.4}$$

where K is the equilibrium constant between surface and solution.

Indication that some adsorption rates are determined by a chemical step is given in the work of Bockris *et al.*[57] for various hydrocarbons on platinum. Sedova *et al.*[58] also found such behavior for methanol on platinum.

A distribution of adsorption rates is helpful interpretively. Unusual phenomena accompany pyridine adsorption[59] and can be interpreted in terms of a high attractive interaction coefficient between adsorbed pyridine molecules. Armstrong[60] suggested that clumps form, as with oxides.

1.3. The Potential of Zero Charge

The great importance of this concept was first indicated by Frumkin,[61] though it has been applied most vigorously by Antropov.[62] The associated subject of the breakdown of potentials in cells has been neglected by most workers.[63,64,65]

Antropov points out that there are two potentials of zero charge, one an "absolute" one, which arises when there is no specific adsorption; only the solvent contributes to the potential difference on the solution side of the double layer. This value should be a function of the metal and relations between it and the work function have been given by Frumkin and Gorodetskaya[66] and by Gileadi and Argade.[67]

1.3.1. Nature of the Potential of Zero Charge

The nature of the p.z.c. is discussed by Antropov,[62] who analyzes the various contributions to the cell potential. Stress upon a "ϕ scale of potential" was made by Antropov in 1950. There is a discussion of the nature of the p.z.c. by Bockris and Argade.[68]

The cell potential can be expressed in terms of the Galvani potential differences across the various interphases as

$$E = {}^{m_1}\Delta^s\phi_{test} - {}^{m_2}\Delta^s\phi_{ref} + {}^{m_2}\Delta^{m_1'}\phi \tag{1.5}$$

where the $\Delta\phi$ values are the Galvani potential differences across the various interfaces; superscripts m_1 and m_1', m_2 and s represent metals m_1 and m_2 and solution, respectively. Let the charge density on the test electrode be zero. Thus,

$$E_{q=0}^{m_1} = {}^{m_1}\Delta^s\phi_{q=0} - {}^{m_1}\Delta^s\phi_{ref} + {}^{m_2}\Delta^{m_1'}\phi \tag{1.6}$$

As the two metals in contact are in electronic equilibrium, one has

$$\bar{\mu}_e^{m_1'} = \bar{\mu}_e^{m_2} \tag{1.7}$$

and hence,

$$\phi^{m_2} - \phi^{m_1'} = (\mu_e^{m_2} - \mu_e^{m_1})/F \tag{1.8}$$

where $\bar{\mu}_e$ and μ_e are the electrochemical and the chemical potentials of the electron, respectively. Also

$$^{m_1}\Delta^s\phi_{q=0} = {}^{m_1}\Delta g_{q=0} \tag{1.9}$$

since the Volta potential difference at the metal–solution interface is zero.

$^{m_1}\Delta^s g$ is the surface potential difference across the metal–solution interface at zero charge. It can be expressed in terms of the surface potentials of the free phases, i.e., in contact with vacuum, and an interaction term that takes into account the changes in the orientation of the solvent dipoles and the electron overlap at the metal surface when the two phases are brought in contact with each other. Thus

$$^{m_1}\Delta^s g_{q=0} = \chi^{m_1} - \chi^s + \delta\chi \tag{1.10}$$

From Eqs. (1.6)–(1.10),

$$E_{q=0}^{m_1} = \chi^{m_1} - \chi^s + \delta\chi + [(\mu_e^{m_2} - \mu_e^{m_1})/F] - {}^{m_2}\Delta^s\phi_{ref} \tag{1.11}$$

It may be noted at zero charge that[69]

$$-\mu_e^{m_1} + F_\chi^{m_1} = -\alpha^{m_1} = \Phi^{m_1} \tag{1.12}$$

where α is the real potential of the electron in metal m_1 and Φ^{m_1} is its work function. Therefore, from Eqs. (1.10) and (1.11) one obtains by dropping superscript m_1 and putting $m_2 = $ ref,

$$E_{q=0} = (\Phi/F) - \chi^s + \delta\chi - {}^{ref}\Delta^s\phi + (\mu_e^{ref}/F) \tag{1.13}$$

Thus, the potential of zero charge measured with respect to a reference electrode is dependent upon the work function of the metal. As Frumkin[70] pointed out, the linearity of the relation between Φ and $E_{q=0}$ will depend upon whether the reorientation of the *solvent dipoles or the change in the distribution of the electron overlap is independent of the nature of the metal*. In short, if $\partial\delta\chi/\partial\Phi$ is a constant, the $E_{q=0}$ versus Φ plot will be linear.

1.3.2. Importance of the Potential of Zero Charge

Conventional potentials are complex, potentials of zero charge indicate the sign of the charge on the metal at conventional potentials. The increasing use of potentials of zero charge as potentials in terms of which modern electrochemical ideas can be expressed awaits mainly an increase in the accurate determination of p.z.c.'s.

1.3.3. Methods of Determination of the Potential of Zero Charge

These methods are discussed by Gileadi and Argade[67] and by Damaskin, Petrii, and Batrakov.[71]

1.3.4. For Liquid Metals

The surface-tension and the capacitance methods are sufficient.

1.3.5. For Solids

The difficulties have been stated in Section 1.2.[31,32] The major point depends upon the elimination of the difficulties and artifacts arising from the heterogeneous properties of solid surfaces. A high degree of polishing, smooth surfaces, and evaporated films are required, if results free of artifacts are to be obtained. Radioactive determinations have been used by Balashowa.[48]

Friction was first used to determine p.z.c.'s in the work of Rehbinder and Wenstrom,[72] Bockris and Parry-Jones,[73] Bockris and Argade,[68] and Bockris and Sen[74] (Fig. 1.7).

A recent method is Gokhshtein's.[75] An electrode is attached to a piezoelectric oscillator. Operation of this oscillator from an outside source causes the electrode (a flat plate in the surface of the solution) to oscillate. The electrode is placed horizontally in the surface of a solution, so that at an intermediate surface tension of those available to the interface concerned, much of the surface does make contact with the solution. The electrode is then subject to a potential sweep. The voltage amplitude of this sweep remains constant (at, say, 0.1 V), and the mean potential of the sweep is made to change across the potential range in which the double-layer phenomena are examined. Consideration of an electrocapillary curve shows that this means that there is a changing amplitude of $\Delta\gamma$ as the potential oscillations are moved over a range from the extreme anodic side

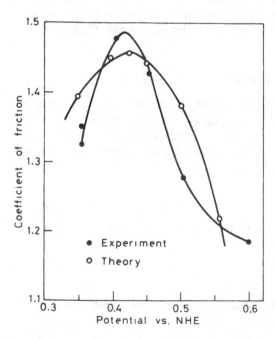

Fig. 1.7. Coefficient of friction versus potential for platinum on platinum in $HClO_4$.

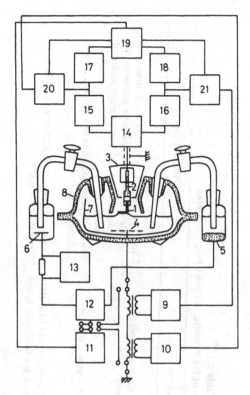

Fig. 1.8. Device for measuring the surface tension oscillations on solid electrodes at two frequencies simultaneously, (1) overmeniscus electrode, (2) piezoelement, (3) screen, (4) auxiliary electrode, (5) reference electrode, (6) polarizing electrode, (7) quartz vessel, (8) glass cell, (9 and 10) sinusoidal voltage generators, (11) linear voltage generator, (12) compensator, (13) current amplifier, (14), broad-band preamplifier, (15 and 16) selective amplifiers, (17 and 18) rectifiers, (19) multichannel oscillographic recorder, (20 and 21) blocks for phase measurements.

($\Delta\gamma$ small) through the period of steep rise of the curve ($\Delta\gamma$ great) through the p.z.c. region ($\Delta\gamma$ minimal) and on toward the extremely negative side ($\Delta\gamma$ small). If one can measure the change of surface tension quantitatively during this regime, a plot of its variation as a function of the range of the mid potential of the potential sweep means that there the minimum of this $\Delta\gamma$ occurs at the p.z.c. (Fig. 1.8). The better methods for solids are summarized in a table by Argade[76] (Table 1.1).

1.3.6. Numerical Values of the Absolute Potential Differences at an Interface

Attempts have been made to derive these values. Many earlier attempts were not consistent with allowances for surface potentials caused by adsorbed material, particularly water dipoles.

The theoretical calculation of p.z.c. is well described by Antropov.[62] It would lead to the determination of the absolute potential difference, so long as one knew the contact potential difference between two metals, and can make an allowance for a contribution due to $\Delta\chi$. This was attempted by Bockris and Argade.[68]

Table 1.1[155]

Various Methods of Determination of the Potential of Zero Charge on Metals

Method	Principle	Positive aspects	Negative aspects
Surface tension methods contact angle	Maximum in the contact angle versus potential	Basis similar to electrocapillary measurements	Usual objection of formation liquid phase under bubble improbable in aqueous solutions, but magnitude small and hence variation difficult to observe
capillary rise	A minimum in the height of the capillary rise versus potential	Basis similar to surface tension methods	Possible lack of resolution in a position determination
tension vibration measurement	Alternating current superimposed on dc applied to a flexed-beam electrode and vibrations measured by a piezo-element; amplitude proportional to $q \, \Delta E \cos \omega$	Possibility of measurement of surface free energy of solid–solution interface	At present, amplitude used to give sufficient sensitivity is 0.1 V
Change of surface area at constant potential (the immersion method)	Charging current proportional to the electrode capacitance and dV/dt	Simplicity of measurement; applied to many systems; possibly several means of exposing fresh metal surfaces	In presence of a Faradaic process, method inapplicable
open-circuit scrape	The potential/time decay curve comes to a plateau at zero charge	Has been used to study ionic specific adsorption on silver and gold	Scraping probably increases active sites and hence interference due to Faradaic processes

Method	Basis	Limitations
Capacitance measurement	A minimum in the C_{dl}/E plots in dilute solutions since diffuse layer capacitance is smallest at zero charge	Applicable only in dilute solutions; pseudocapacitance gives difficulty
	Applied extensively to many solids; measurement of double-layer capacitance can lead to q_m versus E relationship and electrocapillary thermodynamics; in principle, effects of adsorption and pseudocapacitance can be eliminated by working at high enough frequency; one of the most suitable methods	
Ionic adsorption	Tracer cationic and anionic charges are equal at zero charge	Inapplicable if adsorption of a third ionic species other than the radiotracer species is involved; availability suitable tracers questionable
	Applied to solid metals	
Organic adsorption	Γ_{org} versus E curves intersect at p.z.c. at various electrolyte concentrations while the concentration of the neutral organic compound is constant	Absence of specific adsorption of ions of the electrolyte necessary. The contribution of salting out effect must be small; measurement of Γ_{org} on solid metals of limited accuracy
	P.z.c. determination in presence of adsorption of neutral compounds possible	
Friction		
oscillating Herbert pendulum	The friction between the fulcrum of the pendulum and the metal is maximum at p.z.c.	Indirect way of measuring friction leads to confusion in interpretation whether friction or hardness is underlying physical property; experiments cannot easily be carried out under controlled conditions
	Easy applicability to any solid	
static friction	Static friction between two surfaces is at a maximum at zero charge	Friction depends upon the mechanical properties of the metal, e.g., plastic yield, hence variation of friction with mechanical properties limits accuracy
	Simple measurements; quantitative interpretation recently given; applicable to many systems	

continued overleaf

Table 1.1 (continued)

Method	Principle	Positive aspects	Negative aspects
Ultrasonic methods ultrasonic potential	Ultrasonic potential is maximum at zero charge	Not clear	Speculatively suggested; no data; difficult apparatus
dispersion of the electrode	Rate of weight loss is at a minimum at p.z.c.	Applied to few systems; possibly applicable to oxides	Method depends on the mechanical properties of the metal; weight loss measurements may be in error due to formation of oxide
Repulsion of diffuse double layers	Repulsion of ions in diffuse layer is at a minimum at p.z.c.	Applied to several systems; possibly applicable to "insulators"	Not applicable $>10^{-2}$ M; apparatus complex; not applicable in presence of specific adsorption

From the $E_{q=0}$ versus Φ relationship, it is possible to estimate the Galvani potential difference across the reference electrode–solution interphase. It can be seen from Eq. (1.13) that $^{\text{ref}}\Delta^s\phi$ can be determined provided one has the knowledge of the chemical potential of the electron in the reference electrode and the surface potential of the solution–gas interface and the interaction term $\delta\chi$. Hence, to calculate $^{\text{ref}}\Delta^s\phi$, one must know the chemical potential of an electron in platinum, since platinum is usually used in conjunction with the hydrogen electrode. The chemical potential of the electron (approximately the Fermi energy) for platinum is not known accurately. However, one knows fairly accurately the Fermi energy of sodium, for which the free electron gas model has been shown to be applicable.[77] The Fermi energy calculated on the basis of this model is. -3.2 eV. In order to estimate the absolute potential of the reference electrode, we will use the values of $E_{q=0}$, Φ, and $\delta\chi$ for mercury since these are reliable. We will refer the value of the potential of zero charge of mercury with respect to a standard reversible Na/Na^+ electrode, and hence from Eq. (1.13), one has

$$E_{q=0}^{\text{Hg}} = (\phi^{\text{Hg}}/F) - \chi^s + \delta\chi^{\text{Hg}} - {}^{\text{Na}}\Delta^s\phi_{\text{ref}} + (\mu_e^{\text{Na}}/F) \qquad (1.14)$$

and using $\phi^{\text{Hg}}/F = 4.51$ V and assuming the estimated $\chi^s = 0.2$ V,

$$\delta\chi^{\text{Hg}} = 0.26 \text{ V}, \qquad \mu_e^{\text{Na}}/F = -3.2 \text{ V}, \qquad (1.15)$$

$$^{\text{Na}}\Delta^s\phi_{\text{ref}} = 4.51 - 0.2 + 0.26 - 3.2 - 2.51 = -1.24 \text{ V} \qquad (1.16)$$

Thus, the estimated absolute potential across a Na/Na^+ standard reversible electrode is -1.24 V.

In a similar fashion, the absolute Galvani potential difference across the standard reversible hydrogen–nickel electrode can be estimated. Thus, one uses the values of $\delta\chi$, Φ, and $E_{q=0}$ for mercury, $\chi^s = 0.20$ V, the Fermi energy of nickel -5.5 eV;[77] then $^m\Delta^s\phi_{\text{NHE}}$ (on nickel) is found to be about -0.74 V. It may be observed that the value $(-^{\text{ref}}\Delta^s\phi + \mu_e^{\text{ref}})$ calculated by using the values of mercury for the quantities in Eq. (1.13) is -4.76 V. This may be compared with the value calculated from the $\sum \Delta G$,[68] which is -4.78 V.

The ideas briefly laid out here have been confirmed by Trasatti,[78] who, after rightly modifying the value chosen for the individual proton heat by Bockris and Argade,[68] has more or less taken the approach to himself.

Some further application of the concepts of Bockris and Argade[68] has been made by Bockris and Habib. They have shown it possible to calculate each component of the absolute Galvani potential[79]—not only the total value but also the contribution of the dipole potential difference at the p.z.c., and also the electron overlap potential difference. However, the accuracy of these calculations depends on the ability to make good calculations of the Fermi energy.

1.4. Forces in Contact Adsorption

Contact adsorbed ions used to be regarded as involving a covalent bond. This began to seem unlikely when it was realized that tetraalkyl-ammonium cations, Cs^+, and Tl^+ also adsorbed specifically on the surface. This would not be possible in terms of molecular orbital theory.

Andersen and Bockris[80] calculated the free energy change of transfer of ions from the outer Helmholtz plane (OHP) to the inner Helmholtz plane (IHP). They assumed that contact adsorption results from a net between diminution of free energy due to partial dehydration of the ion during adsorption and dispersion and image interaction of the ion with the metal. Modelistic hypotheses are in Table 1.2. Results are shown in Table 1.3. They are consistent with contact adsorption of halides and with an abnormally low adsorption of F^-. Bode[81] reexamined the model of Andersen and Bockris and confirmed its consistence with many of the facts.

A development of the concepts of Andersen and Bockris has been made by Barclay and Caja,[82] who have taken into account the possibilities arising when filled orbitals in the metal surface atoms can overlap vacant orbitals in organic adsorbents (Table 1.4). Thus, the much higher coverages of anions found by Paik and Bockris[44] on some solid metals would be

Table 1.2[80]

Constants Used in Evaluating Metal–Ion Interactions, and the Resultant Potential-Energy Terms[a]

Ion	Na^+	K^+	Cs^+	F^-	Cl^-	Br^-	I^-
r_{xl} (Å)[160]	0.98	1.33	1.67	1.33	1.81	1.96	2.19
n[161]	8.08	8.41	13.0	8.80	8.12	8.72	9.49
α ($cm^3 \times 10^{24}$)[162]	0.24	1.00	2.40	0.81	2.98	4.24	6.45
χ ($cm^3 \times 10^{29}$)[163]	−1.01	−2.42	−5.83	−1.56	−4.02	−5.74	−8.40
r_e(calc) (Å)	1.36	1.56	2.17	1.57	1.72	1.92	2.14
U_{image} (kcal/mol)	−56.7	−49.9	−36.6	−49.7	−45.6	−41.0	−37.0
U_{disp}(kcal/mol)	−7.0	−16.5	−14.8	−12.0	−28.9	−29.8	−31.9
U_{rep}(kcal/mol)	14.6	18.2	7.9	14.6	25.2	21.6	19.7
U_{net}(kcal/mol)	−49.1	—	−43.5	—	−49.3	−49.2	−49.2

^{a} $W_a = 15.4 \times 10^{-12}$ ergs, $N = 4.08 \times 10^{22}$ cm^{-3}, $\alpha_{Hg} = 5.05 \times 10^{-24}$ cm^3,[158] $\chi_{Hg} = -6.33 \times 10^{-29}$ cm^3.[159]

Table 1.3[80]
The Thermodynamics of Contact Adsorption

Ion	Water–electrode interactions			Ion–electrode interactions			Ion–water interaction			Total ΔG
	ΔH	ΔS	ΔG	ΔH	ΔS	ΔG	ΔH	ΔS	ΔG	
Na^+	28.1	3.6	27.0	−49.1	1.5	−49.6	39.5	11.2	36.1	13.5
K^+	21.1	6.7	19.1	−48.2	2.0	−48.8	33.9	5.2	32.3	2.6
Cs^+	20.9	10.7	17.7	−43.5	3.3	−44.5	17.7	−12.2	21.4	−5.4
F^-	38.1	6.7	36.1	−47.1	1.4	−47.5	35.0	6.4	33.1	21.7
Cl^-	22.0	12.4	18.3	−49.3	2.1	−49.9	17.7	−16.6	22.7	−8.9
Br^-	21.6	14.5	17.2	−49.2	2.8	−50.0	14.7	−22.4	21.4	−11.4
I^-	21.9	18.0	16.5	−49.2	3.6	−50.3	10.9	−32.6	20.7	−13.1

Table 1.4[82]
Adsorption of Anions as a Function of Solvation and Orbital Energy

Solvation	Orbital energy	Example	Adsorptivity
Low	High	I^-	High
High	Low	F^-	Low
High	High	OH^-	Low
High	High	$S_2O_3^{2-}$	High
Low	Low	ClO_4^-	Low

intermediable in such an orbital-oriented fashion. It is just this quantum-oriented viewpoint[83] that has been absent from double-layer interpretations.

1.5. Isotherms

There are three isotherms—Bell, Levine, and Calvert[84]; Macdonald and Barlow[15]; and Bockris, Devanathan, and Muller.[17] The three are similar in form.† They involve treatments of interactions between the contact-adsorbed ions. If one attempts to account for adsorption in terms of repulsion without ion imaging, the repulsion is too high. A solution is introduced by imaging the ions with the metal. Grahame[9] pointed out that there could also be imaging (and multiple imaging) with the solution.

In Levine, Bell, and Calvert (LBC)[84] and Macdonald and Barlow (MB),[15] multiple imaging in solution and metal has been developed to yield isotherms. However, an oversimple model—infinite imaging in solution and metal, assuming the dielectric discontinuity between the OHP and the solution was *sharp*—was used, instead of the diffuse and gradual change of dielectric constant that really exists.

BDM used the single-imaging approximation (within the metal). Their isotherm is easy to express and is (at $q_m = 0$):

$$\ln \frac{\theta}{(1 - p\theta)^p} = -A\theta^{3/2} + \text{const} \tag{1.17}$$

† Parsons[85] developed a number of isotherms (Volmer, Amagat, etc.) at an early stage, and the general orientation of his approach was to develop "equations of state" for the adsorbed film, in terms of the virial. One of his approaches[86] led to the first prediction of a maximum on capacitance–charge relations from ion repulsion. Such work was less quantitative in relation to experiment than approaches sketched later.

That of Levine, Bell, and Calvert (and Macdonald and Barlow) is similar, except that θ appears to the first power because of the overallowance for multiple imaging.

A test of these isotherms has been carried out by Wroblowa and Muller,[87] by Levine,[88] and particularly by Bockris and Habib.[89] One of the parameters in the theory is p, the ratio of the cation to water areas. Taking p as 1 (and also higher numbers), Wroblowa and Muller demonstrate a greater consistency of the single-imaging model by calculating the Essin–Markov effect, calculating the position of the hump, and comparing various theoretically calculable coefficients with experiment. Quite good agreement† was obtained if $p = 1$. Taking $p = 2$, Levine[88] found he could bring the multiple-imaging model into greater agreement with experiment, but hump characteristics and the possibility of calculating the q_m at which they occur were not investigated.

Bockris and Bonciocat[90] showed that the Bell, Levine, and Calvert isotherm that arises from the multiple-imaging calculations does not give rise to the observed shape of the capacitance–charge relation, namely that which has both a maximum and a minimum. The single-imaging model gives results in better accord with experiment than the model with imaging calculated on the idea that there is a sharp dielectric discontinuity at the OHP to diffuse double-layer planes. Some multiple imaging must occur. Correspondingly, the single-imaging model gives a C versus q_m curve in agreement with the widely observed shape; the multiple-imaging model does not. The single-imaging model gives a hump that depends on the anion, and the multiple-imaging model does not.

The future of isotherm work depends on getting a realistic multiple-imaging calculation, taking into account the *gradual* dielectric discontinuity over several angstroms out into the solution. If the change is sufficiently gradual, the effect of multiple imaging will be lost.

A quantitative estimate of the degree of multiple imaging was made by Bockris and Habib,[92] who used Booth's theory to establish the ε_x versus x relation near the electrode. They then calculated the primary, secondary, tertiary, quatenary, etc. image interaction, assuming a stepped dielectric constant–distance relation, and were then able to examine the effect of the height of the steps (degree of discontinuity) on the contribution of the multiple imaging to the net repulsive interaction between ions in the double layer. They found that a smooth dielectric-constant–distance plot implied a negligible image energy outside the compact double layer.

---

† There are some residual discrepancies with the BDM model. For example, Baugh and Parsons,[91] using data for PF_6^-, found that the model predicts a direction of the Essin–Markov coefficient incorrect in sign.

1.6. Dielectric Constants in the Double Layer

The variation of the dielectric constant of materials with field strength was first calculated in terms of a continuum model by Webb.[93] Grahame[9] extrapolated the results of Malsch to calculate the dielectric constant in the diffuse layer. Malsch[94] and Conway et al.[95] used Webb's formalism to obtain ε in the diffuse layer and also in the IHP. They found, remarkably, that although change to a high dielectric constant took place over a few angstroms (rather than the abrupt change, the assumption of which had led to the overestimation of the imaging effect), the effect on the calculated diffuse layer change was negligible.

An explicit theory for ε as a function of E was formulated by Booth,[96] who applied the Onsager and Kirkwood theories of polar dielectrics to high field strengths. He confirmed that saturation effects set in at field strengths of the order of those in the IHP. His equation for the dielectric constant at a field strength E is given by

$$\varepsilon = n^2 + \frac{\alpha\pi N_0(n^2+2)\mu_w}{E} L\left(\frac{\beta\mu_w(n^2+2)E}{kT}\right) \qquad (1.18)$$

where n is the optical refractive index, N_0 the number of molecules per unit volume, μ_w the dipole moment of the water molecule, E the field strength, and T the absolute temperature. $L(x)$ is the usual Langevin function, and α and β are numerical factors that have the following values:

(a) *Onsager method*: $\alpha = \frac{4}{3}$, $\beta = \frac{1}{2}$
(b) *Kirkwood method*: $\alpha = 28/3(73)^{1/2}$, $\beta = (73)^{1/2}/6$

From these formulas it is shown that the reduction of the dielectric constant due to the saturation effect is of importance for fields greater than 10^6 V cm^{-1} (Fig. 1.9).

Corresponding to this, Smyth and his co-workers[97] determined the dielectric constant of water under conditions where the frequency would be too great for orientation polarization to play a part. This work is the origin of the value of 6 suggested by BDM for the inner-layer contact adsorbed water molecules. Miller[98] concluded, by a quite different argument, that it was between 7 and 8. Correspondingly, Lawrence and Parsons[99] concluded from the interpretation of the adsorption of sulpholane at mercury that 7 was the correct value.

More recent work on double-layer properties taking into account the dielectric discontinuity near the electrode has been carried out by Kiryanov, Kuznetsov, and Damaskin,[100] but *does not provide a molecular model*, nor separate calculations, of the dielectric-constant–distance relation. Such is much needed in respect to image calculations.

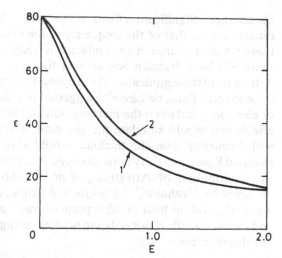

Fig. 1.9. The dielectric constant ε of water at 300°C as a function of the field strength E (in $10^7 \, V \, cm^{-1}$); (1) Booth,[96] (2) Webb.[93]

1.7. Relaxation of Solvent Origin in the Double Layer

Measurements of capacitance and the (seldom mentioned) double layer resistance on the mercury–solution interface led to the confirmation of characteristic variations of the resistance with frequency. These were interpreted by Bockris, Mehl, Conway, and Young[101] in terms of Maxwell displacement currents across the double layer; the capacitance would be pure, a capacitance in parallel with a resistance (and this independent of Faradaic currents). A review of other relevant work is available.[102]

One reason why dielectric displacement currents may be important is that the concept is consistent with a two-position water dipole model. If their dipole moments were not exactly orthogonal to the electrode surface, they would oscillate in the ac field. Correspondingly, if the frequencies were distributed (Cole and Cole[103]), one would obtain the type of relationship observed; i.e., a significant number of water molecules would have relaxation times in the range of the experimental frequencies. Important evidence for the type of suggestion made came from the work of D. C. Grahame,[104] who stated that the balancing resistance in the bridge was dependent on frequency—linear with log frequency.

The model for dielectric relaxation in the double layer was developed by Bockris, Gileadi, and Muller.[35,36] In it, the water molecules that give rise to $\Delta\chi$, the possibility of a capacitance hump, and the potential effect on organic adsorption are seen as existing at an angle to the electrode. Hence, as the ac field of the electrode causes them to vary in position, and hence in component of potential, they introduce a component that is equivalent to an admittance, i.e., a resistance in parallel with the pure double-layer capacitance. Significant effects occur when the relaxation times of the dipoles is near that of the frequency of the alternating current imposed. If there is a distribution of such relaxation frequencies (cf. Cole and Cole[103]), there will be a fraction having τ in this region. However, artifacts may detract from the significance of the apparent facts of the observed frequency dependence. Thus, de Levie[105] suggested that there may be creeping effects of electrolyte between the mercury wall and the glass, and these could give effects that would simulate the situation of apparent resistance variation with frequency. The film thickness would have to be 0.4 μm in a capillary radius of 8 μm, and this seems too large. The situation is made less certain by the contribution of Armstrong *et al.*,[106] who could not find variations observed by Grahame,[107] Bockris and Conway,[108] and others. The effect would depend on heat of adsorption of water molecules on the electrode, and this would affect not only the binding strength, but also the distribution of relaxation times, etc.

1.8. Double-Layer Properties as a Function of a Potential-Dependent Dipole Term

Elementary models of solvent structure at electrodes (Lange and Miscenko,[12], Bockris and Potter,[13] and Mott and Watts-Tobin[16]) have been referred to. The BDM version has been used in attempting to find out how much the intermolecular potential between the dipoles affects the contribution to the double layer, i.e., the $\Delta\chi$. The capacitance hump has been the subject of two interpretations. The first, mathematically formulated by Mott and Watts-Tobin,[16] suggested that the capacitance hump might be a result of dielectric unsaturation near the p.z.c., so that the rapid change in the number of dipoles from one state to another, as a result of a change of charge on the electrode, would give rise to a significant extra capacitance. A different explanation was mathematically formulated by Bockris, Devanathan, and Muller[17]—the capacitance hump was a result of the growing repulsion between ions as they increase in their population of the double layer. This would cause an inflection on the relation of $q_{contact}$ to q_m. BDM obtained excellent agreement between the metal charge (calculated from their model) at which capacitance humps should arise and the experimentally determined q_m for the hump maxima. Interesting contributions were here made by Payne,[28] who showed that in some organic solvents two capacitance humps were observed.

These alternative models have been clarified by Bockris and Habib,[21,109] who were able to explain the hump in terms of anion repulsion (see above) but were also able to explain why it is that, at potentials more positive than that of the capacitance hump, the double-layer capacitance once more begins to climb. It is consistent with a model first proposed by Blomgren and Bockris,[110] which takes into account dispersive attraction between ions on electrodes.†

Damaskin et al.[71] suggest that the solvent orientation model for capacitance humps is basically correct, but the solvent molecules that orient are attached to ions. Then, the (so undesirable from the viewpoint of a hump–water model) relation between the characteristic of the capacitance humps and the anions present could be qualitatively explained.

The evidence for an anion model for humps seems strong for aqueous solutions. Authors who advocate exclusively a water orientation do not take into account the water–water interactions that BDM showed reduced the water capacitance. Perhaps, in some systems, both models are applicable. The solvent orientation theory may be applicable at least when there are two humps.[28]

Numerical values of $\Delta\chi$ involve difficulties for the original BDM model. If the interactions between the water molecules are taken as sufficiently strong so that the water humps disappear, the $\Delta\chi$ potential is 1 V during a q_m change of some 20 μC. This is too much for the consistency of the Tafel slope over 1 V, as often observed. But, if the interaction potentials are reduced so that the $\Delta\chi$ value is reasonable, e.g., 100 mV over 20 μC, the solvent hump returns. This discrepancy can be alleviated[90] if the dispersion energy of interaction between the waters is taken into account. A more sophisticated interaction calculation is needed. The form of water on the surface certainly needs investigation. For example, Damaskin regards it as in groups, but Lawrence and Parsons,[99] from a study of sulfolane adsorption from a water–methanol mixture get best agreement with monomer water.

Some of these difficulties have been removed by a treatment of Bockris and Habib[21] known as the three-state water model. This model is basically similar to that of BDM with the difference that the number of "active" water dipoles on the surface is reduced by about three-quarters because of the tendency of a number of waters to form polymers, in which state they are bound into bodies (mostly dimers) that do not contribute to the χ_{dipole}. The value of $\Delta\chi$ is hence reduced to reasonably small values. Bockris and Habib's model originates with the unpublished work of Bockris and Law,[111] who measured the surface tension of the $Hg–H_2O_{vapor}$ interface at a number of temperatures so that they could determine the entropy of the adsorbed water. It corresponded to that of a dimer layer.

---

† Any doubt as to the real existence of an orientative water layer was allayed by Bockris and Habib,[109] who used Hills and Payne's measurements of the dependence of water-molecule entropy to calculate the water-molecule entropy as a function of electrode charge. The experimental results were in excellent agreement with the theoretical entropy–charge relation for a three-state model (in respect to the charge of the maximum, also a two-state one).

1.9. Adsorption of Undissociated Organic Molecules

Organic molecules often adsorb parabolically with electrode charge. Adsorption of undissociated molecules sometimes occurs. However, the undissociated character of the adsorbed entity in double-layer studies has been assumed too readily by most authors. It does sometimes apply in the adsorption of organic compounds, e.g., alcohols, aromatics, benzene. Current still passes; the amount of material on the electrode surface is larger in the molecular form than that of any radicals. The phenomenology of adsorption has been illustrated copiously in Frumkin and Damaskin,[112] Damaskin et al.,[113] and by Blomgren, Bockris, and Jesch.[114]

Russian data has been rationalized by various continuum treatments (Damaskin and Frumkin[115]) in which a batch of adjustable parameters are calibrated for each new system.† The free-energy change on adsorption is the work of charging the double layer caused by a change of dielectric constant and double layer thickness when the organic substance is introduced. The isotherm allows for interactions between the organic molecules (the exponential $a\theta$ term), but in its original form it neglects other interactions, particularly those concerned with the displacement of water, namely, water–water interactions, water–organic interactions, etc. The Frumkin isotherm may be represented as

$$\frac{\theta}{1-\theta} = Bc\, e^{2a\theta} \qquad (1.19)$$

It is of interest because it has been used to represent a large amount of data.†

Other isotherms have been suggested. It is sometimes unclear whether they apply to organic or ionic adsorption. Many do not account for intermolecular forces. Almost all neglect the application of the Flory–Huggins statistics. For example, the Blomgren and Bockris[116] isotherm was concerned with organic–organic interactions, because organic *ions* were being examined. The Blomgren and Bockris isotherm was the first to take into account water displacement during adsorption and the first in which intermolecular forces both of repulsion (coulombic) and attraction (dispersion) had been explicitly expressed. In Fig. 1.10 are shown the results of a plot indicating the changeover from one to the other upon increase of coverage.

Study of the adsorption of ethylene, benzene, and naphthalene on platinum showed that here *the largest component* on the surface was the

---

† M. A. Habib[156] showed that the relation of θ_{org} to electrode potential could be as well represented by $\theta = AV + BV^2 + CV^3 + DV^4 + \cdots$ with the same number of adjustable parameters as were needed by continuum theories.

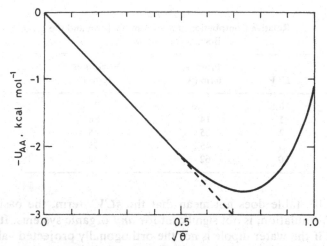

Fig. 1.10. Theoretical values of interaction energies for aromatic amines as a function of $\theta^{1/2}$. Parameters: $\varepsilon_s = 10$, $E_{ion} = 10$ eV, $S = 60$ Å.

unchanged molecule. Bockris, Green, and Swinkels[117] measured many cases in which (surprisingly) the organic adsorption overrides the formation of oxygen and hydrogen, which would be expected at some potentials, were no organic radicals to be present.

Competition for surface sites between organic and water molecules is an important consideration here. There is a small and positive entropy and enthalpy of adsorption for hydrocarbons on platinum. The low heat of adsorption from solution (compared with the high heat from the gas phase) can be interpreted in terms of the displacement of water molecules from the surface.[117] Is there equilibrium between solution and electrode? What is the $\theta_{undissociated}$ compared with θ values of radicals taking part in a reaction of oxidation of the adsorbed molecule?

A dominating effect of electrical water interactions with the metal exists at low coverages. Damaskin[118] attacked the BDM model, after that of BGM (which accounts for electrode–organic interactions) was published, for the neglect of organic–electrode interactions. Although Bockris, Gileadi, and Muller[119] account for these, they still neglect organic–organic interactions,† though see Bockris, Green, and Swinkels.[117] Damaskin was concerned by the neglect of the $\frac{1}{2}CV^2$ term in the molecular model of organic adsorption. The importance of the $\frac{1}{2}CV^2$ term compared with the water displacement term was examined by Gileadi[120] (Table 1.5).

---

† The need to account for them has also been expressed in an analysis of the adsorption of some C_4 compounds on mercury by Dukiewitz, Garnish, and Parsons.[121]

Table 1.5
Relative Contribution of the Frumkin Term and the
Bockris *et al.* Term[120]

E/V	Frumkin term (%)	Bockris *et al.* term (%)
0.5	7.6	92.4
1	14	86
2	25	75
5	45	55
10	62	38

Gileadi's table does not mean that the $\frac{1}{2}CV^2$ term, the basis of the Frumkin formulation, is not significant for *any* organic systems. It may be significant if the water dipole is not the orthogonally projected value used in the BDM and BGM models. The water positions on the surface are probably angled, or the water is polymerized, as in the three state water model of Bockris and Habib.[21]

Damaskin also stated that "the BDM theories disagree with thermodynamics," but this was shown not to be a solid statement by Gileadi.[120] He pointed out that eventually the Frumkin formulation, though thermodynamic in a mode of deduction, did indeed itself involve a model—though tacitly—so that when one compares the prediction of equations based upon the Frumkin approach with those based upon the molecular models of Bockris *et al.*, one is eventually comparing model with model (though Frumkin *et al.* have not stated what their model is). Lack of identity between the predictions of the different models must be expected.[121]

An advantage of the Bockris formulations for organic adsorption is their molecular level[122]; the degree of model admitted by Frumkin (the solvent layer as a jellium) is nugatory. Conversely, the Bockris models are too simple, and the first one (BDM) is an approximation restricted to an organic substance that has its polar group in the diffuse layer. At high coverages, the organic–organic interactions may be stronger than the water–water interactions, whereas in BDM, only the water–water and electrode–water interactions are involved. Frumkin's approach, conversely, accounts only for the organic–organic interaction. Needed is work along explicitly molecular model lines at a higher degree of mathematical sophistication, rather than the Frumkin approach, which is thermodynamic in treatment method and unfortunately calibrates *all* its parameters from experiment.†

† Further in the Frumkin isotherm [Eq. (1.19)], the constants *a* are sometimes found to change with θ so that the *form* is also of limited applicability.

Gradually, the concern with the overstress on continuum considerations is being left behind in double-layer work† before one has got rid of the approach completely in kinetics (Chapter 7). Thus, one aspect that such work has obscured is the need to allow for a proper statistics in consideration of adsorption. This was indicated in passing for the first time by Levine, Bell, and Calvert (Flory–Huggins approach),[84] and an alternative treatment is given by Swinkels *et al.*[123] (thermodynamics of displacement of several water molecules by one adsorbent). Correspondingly, entropy changes have been used to deduce the type of radicals on the surface by Heiland, Gileadi, and Bockris[50] (cf. also Conway and Gordon,[124] Barradas and Herman[125]).

1.10. Radicals on Electrodes

This subject hovers between the field of the double layer and that of electrode kinetics. A classical article on it by Conway and Gileadi exists.[40] Bockris, Gileadi, and Stoner[127] maintained an electrode at constant potential while observing the current produced by introducing an organic substance into solution. In the undissociated case, the current is calculable. It is less than that in which the molecule is dissociating and injecting charge into the electrode.

Few methods exist for identifying radicals on surfaces. The first was devised by Armstrong and Butler,[128] who determined the amount of hydrogen on platinum. This work was repeated by Will and Knorr[129a] and Breiter.[129b] Devanathan, Bockris, and Mehl[130] were the first to determine hydrogen on nonnoble metals (the metal dissolves during the anodic transient) by a galvanostatic double-pulse method; see also Mehl and Bockris.[131] Blomgren and Bockris[47] introduced radiotracer methods for looking at organic radicals on surfaces. Radiotracer and sweep methods were compared by Gileadi *et al.*[53] (Fig. 1.11). At relatively cathodic potentials the results (for benzene on platinum) were coincident, but at more positive potentials distinct deviations occurred due to the supposition of the sweep method that the number of electrons used per particle in producing current was that for the oxidation of an undissociated adsorbed *molecule*. Comparison of sweep and radiotracer results suggests the dominant radical present and sometimes may allow its identification.

Radiotracer methods have been developed by Podlovchenko *et al.*[132] He used labeled alcohol, and after adsorption of the alcohol from an aqueous solution, the solution was replaced by KOH into which the

---

† But residual struggles between various subviews of continuum theories (which do have unstated models hiding behind them) exist.[126]

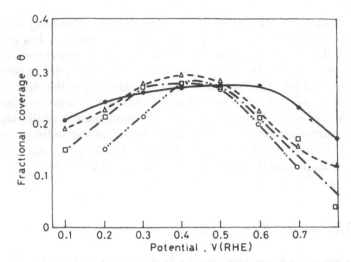

Fig. 1.11. Potential dependence of electrosorption of benzene—comparison of results obtained by four techniques. 0.50 V (RHE); 50°C. ●, Radiotracer method, 1.1 μM benzene; Δ, linear potential sweep method, 1.4 μM; ○, potentiostatic method, 1.1 μM; □, galvanostatic method, 1.3 μM.

chemisorbed residues were electrochemically oxidized. The quantity of electricity required for oxidation of the adsorbate was compared with the amount of carbonate formed. The number of electrons to form a CO_2 molecule was 2.08 ± 0.2. Thus the radical was HCO, but Biegler and Koch,[133] who used pulse methods to look at the adsorbed radicals, concluded that CO was dominant. Podlovchenko and Gorgonova[134] and Bagottskii and Vasil'ev[135] seem to confirm that the radical is HCO.

Some confusion (Brummer[136]) has existed about radicals that block the reaction sites and radicals that are present as intermediates in an organic oxidation reaction. Schuldiner and Piersma[137] have clarified the matter in the dehydrogenation of formic acid when the acid molecule and formate ion interact together to block the surface and stop further dehydrogenation. COOH seems to be the adsorbed radical when formic acid in solution is brought into contact with platinum.[138]

Methods for the study of organic radicals on surfaces ripe for development are transient ellipsometry (with relaxation times below the order of milliseconds)[139] and the use of this method with variable wavelengths to build up an identificatory ellipsometric spectroscopy. Multichannel counting with several radioactive isotopes has not yet been applied, but exchange problems are troublesome.

The analysis of organic reactions in which radicals taking part in the reaction also exist in solution is an ideal application for ESR. Humffray[140] states that many organic reactions involve radical stages in solution. In a book by Harrison and Thirsk,[141] called anomalously *A Guide to Electrode Kinetics*, it is the in-the-solution diffusion aspects of reactions to occur later at electrodes that are stressed, while the mechanism of the reaction at or on the electrode is relatively understressed. Electron microscopy at $\sim10^6$ V acceleration could be used to look at radicals through thin layers of solution.

1.11. Double Layer on Solids

Only a tiny fraction of work done on the double layer has been devoted to interfaces involving solid metals. The limited number of methods of looking at solid–solution interfaces has been given (Table 1.1). Two main result-bearing lines have so far been produced. The first is the thermodynamic approach of Frumkin, Petrii, and Damaskin,[142] in which the problem of the electrocapillary thermodynamics appropriate to adsorption on solids has been studied. In essence, the equations deduced apply to a situation in which a contact-adsorbed ion is no longer regarded as a unit charge but in which some degree of charge transfer has occurred. A study by Frumkin, Petrii, Kossaya, Entina, and Topolev[143] of the Essin–Markov effect in ion adsorption on platinum was a forerunner of this formulation, for it was found that, for certain situations, the Essin–Markov effect disappeared in a way consistent only with a degree of charge transfer on adsorption.

Another result of importance has been the ability to obtain, quickly and easily, total ionic adsorption data at solid–solution interfaces by ellipsometry. A number of results are shown in Fig. 1.12. It is seen that, particularly on gold and silver, the maximum adsorbed amounts (counting one ion adsorbed as one charge) are much higher than in the corresponding case for mercury—in confirmation of the inferences from the work of Paik *et al.*[44] and consistent with the approach of Barclay and Caja.[144]

One of the goals of modern electrochemistry is to gather the same degree of information about various solid–solution interfaces as has been obtained at the mercury–solution interface. This *can* be obtained from the capacitance measurements described, but a more direct way would be that of Fredlein, Damjanovic, and Bockris,[32] who measured an electrocapillary curve on a solid (platinum) for the first time. Direct thermodynamic information can be got therefrom about the total Gibbs surface excess, independently of the ambiguities of the interpretation of capacitance measurements (which may contain contributions from the nonpolarizable character of the interface). Gokhshtein's approach[75] may be of great importance in this area.

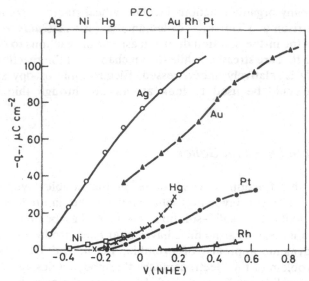

Fig. 1.12. Adsorption of Br^- ions from 10^{-2} M Br^- solutions on different metals.[44]

1.12. Oxygen on Electrodes

This area has recently emerged from a decade of controversy. The dependence of oxygen coverage (from a solution containing a given partial presence of O_2) on the electron structure of the metal has long been known.[145]

The controversy has been about the *kind* of layer formed by oxygen on noble metals (for platinum, above 0.9 V). Below this potential, workers agree that oxygen is adsorbed on the surface. Above it, three views existed:

(a) Oxygen remains "adsorbed" at monolayer concentration.
(b) Oxygen becomes dermasorbed.
(c) Oxygen forms an "oxide" in the sense of Damjanovic.[146]

The origin of the concept that an oxide film is formed at about 0.9 V on platinum was the work of Genshaw *et al.*,[147] who observed a sudden increase in ellipsometric intensity at this potential. Later workers[148] showed that ellipsometry could detect material at lower potentials, but this was interpreted by Genshaw and Bockris[149] in terms of sensitivity to ion adsorption. With increasing sensitivity of the ellipsometer and the introduction by Paik and Bockris[150] of the complete solution to the ellipsometric equations involving Δ, ψ and the reflectance parameters, this matter has been resolved. Figure 1.13 shows the results of Kim, Paik, and

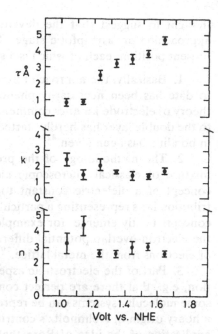

Fig. 1.13. The calculated values of the thickness τ, the real and imaginary parts of the refractive index of the film n and k at various potentials.

Bockris[151] that the film on platinum increases to about two monolayers, *while the refractive index remains constant.* Thus, a new entity is formed at 0.9 V, and remains of constant kind while its thickness is increasing. The material differs ellipsometrically from that which registers itself ellipsometrically at potentials negative to 0.9 V in a way that suggests that an adsorbed layer has been succeeded by one that has elements of a lattice.

Correspondingly, on iron electrodes, Brusic, Wroblowa, and Bockris[152] found that, as the iron electrode was made increasingly positive, oxygen was at first adsorbed, probably as OH^- ions, and thereafter there was diffusion of the OH^- to growth centers until the average coverage of the electrode was unity ($\theta = 1$) at the potential corresponding to the peak of the current-potential curve ("the passivation potential"). At more positive potentials, there is a qualitative change in the ellipsometric behavior that indicates a lattice oxide, and this grows, eventually, to ~30–40 Å.

1.13. The Near Future in the Development of the Model of the Interface

This book concerns many aspects of attempts to approach electrochemical situations quantally. However, most of the material does involve interfaces. The summary of the models given here should not be thought to suggest that the development of the double-layer model is approaching an asymptotic stage. The following are weaknesses in the present position, each of which is a starting point for a new approach:

1. Basically, the approach to the theory of the double layer structure to date has been necessarily phenomenological. Although the quantum theory of electrode kinetics commenced in the 1930s, a quantum approach to the double layer has hardly started, although some discussion of orbitals in bonding has been given.[82]

2. The methodology of the present models presents a fairly heady mixture of classical microscopic electrostatics involving the continuum concept of a dielectric constant together with the use of classical distribution laws representing a particular model. On top of this, quasiquantal concepts tacitly emerge, for example, Bockris and Habib's calculation of the electron-overlap potential difference, which implies the partial escape of electrons from the metal lattice.[79]

3. Part of the electrostatic aspects of the model involve hypostatization, e.g., that there are perfect condensers present. The hypothesis that some molecular systems can be represented as if they were condensers has a hoary origin in Helmholtz's contribution[153] and has blazed lustily in the application of the idea of Born that an ion can be regarded as a hollow conducting sphere; this hypothesis was much applied by Marcus and by Levich (see Chapter 7). But the degree of reality in regarding an ion as a conducting hollow sphere or a time-average layer of charge in solution as a conducting condenser plate is clearly open to betterment.† The modelistic assumption that the double layer can be regarded as two condensers in series or that the quadrupole layer of some kind of mixed species of water turn out to be reasonably represented by a dipole layer in series with other capacitative elements is all best-model play and is certainly only a shot at trying to represent reality in a manner simple enough to be communicated to others and set down in formulatable equations.

4. A long-known Achilles heel of double-layer studies is the use of Gouy's theory to represent the diffuse-layer charge. Here, the troubles extend beyond the simplification implied by the condenser model of the diffuse layer. Gouy's theory was the forerunner of the Debye–Hückel theory of electrolytes, the archetype of successful physical chemistry in electrochemistry, and at the same time the most disproven model (in

---

† But the condenser hypothesis is by no means an entirely leaky one. In sufficiently dilute solutions, there is excellent quantitative agreement between the calculated capacitance (which is all diffuse layer when the concentration is low enough) and the Gouy-derived equations. This agreement is similar to that of the Debye–Hückel zeroth approximation in representing the activity coefficients at 0.001 *M*. It suggests the *basis* to models at higher concentrations.

application to real solutions) in electrochemistry. All the devastating criticisms of the Debye–Hückel model (in application to solutions above 0.01 M) apply to Gouy's equation, which is nonetheless used at 0.1 M and upward to compute the diffuse charge—upon which depends the separability of the contact from the total Gibbs surface excess.

5. The concept of partial charge transfer[154] makes some quantities in the double layer numerically suspect.

6. On the side of the experimental data, there are two substantial deficiencies. One is the indirectness of the evaluations. There is as yet nothing analogous to the electron microscope† for looking at surfaces in solution. Correspondingly, mercury interfaces have provided 90% of the data, with an expected overemphasis on the adsorption of undissociated molecules.

References

1. G. Gouy, *J. Phys.* **9**, 457 (1910).
2. D. L. Chapman, *Phil. Mag.* **25**, 475 (1913).
3. R. Parsons and S. Trasatti, *Trans. Faraday Soc.* **65**, 3314 (1969).
4. F. H. Stillinger and J. G. Kirkwood, *J. Chem. Phys.* **33**, 1282 (1960).
5. V. G. Levich, V. A. Kir'yanov, and V. S. Krylov, *Dokl. Akad. Nauk. SSSR* **135**, 1425 (1960); V. S. Krylov, *Electrochim. Acta* **9**, 1247 (1964).
6. F. P. Buff and F. H. Stillinger, *J. Chem. Phys.* **39**, 1911 (1963).
7. I. L. Cooper and J. A. Harrison, *Electrochim. Acta* **22**, 519 (1977).
8. O. Stern, *Z. Elektrochem.* **30**, 508 (1924).
9. D. C. Grahame, *Chem. Rev.* **41**, 441 (1947).
10. P. Holmquist, *J. Electroanal. Chem.* **78**, 341 (1977).
11. M. A. V. Devanathan, *Trans. Faraday Soc.* **50**, 373 (1954).
12. E. Lange and L. Miscenko, *Z. Phys. Chem.* **149**, 1 (1930).
13. J. O'M. Bockris and E. C. Potter, *J. Chem. Phys.* **20**, 614 (1952).
14. J. R. Macdonald, *J. Chem. Phys.* **22**, 1857 (1954).
15. J. R. Macdonald and C. A. Barlow, *J. Chem. Phys.* **36**, 3062 (1962).
16. N. F. Mott and R. J. Watts-Tobin, *Electrochim. Acta* **4**, 79 (1961).
17. J. O'M. Bockris, M. A. V. Devanathan, and K. Muller, *Proc. Roy. Soc.* **A274**, 55 (1963).
18. T. Erdey-Gruz, *Kinetics of Electrode Processes*, Hilger, London (1973).
19. G. J. Hills, *J. Phys. Chem.* **73**, 3591 (1969).
20. R. Reeves, *Modern Aspects of Electrochemistry*, Volume 9 (B. E. Conway and J. O'M. Bockris, eds.), Plenum Press, New York (1973).
21. J. O'M. Bockris and M. A. Habib, *Acta Electrochim.* **22**, 41 (1977).
22. J. Gibbs, *The Collected Works*, Longmans, Green, New York (1928).
23. F. Koenig, *J. Phys. Chem.* **38**, 339 (1934).
24. R. Parsons and M. A. V. Devanathan, *Trans. Faraday Soc.* **49**, 404 (1963).
25. R. Parsons and F. G. R. Zobel, *J. Electroanal. Chem.* **9**, 2978 (1965).
26. Z. Kovac, Ph.D. Thesis, University of Pennsylvania, Philadelphia (1964).

---

† Electron microscopes capable of this exist but are so expensive that they have not yet been used for the purpose.

27. J. O'M. Bockris, K. Muller, H. Wroblowa, and Z. Kovac, *J. Electroanal. Chem.* **10**, 416 (1965).
28. R. Payne, *J. Electroanalyt. Chem.* **15**, 95 (1967).
29. J. O'M. Bockris, H. Wroblowa, K. Muller, and Z. Kovac, *J. Electroanal. Chem.* **15**, 101 (1967).
30. L. G. M. Gordon, J. Halpern, and B. E. Conway, *J. Electroanal. Chem.* **21**, 3 (1969).
31. T. R. Beck, *J. Phys. Chem.* **73**, 466 (1969).
32. R. Fredlein, A. Damjanovic, and J. O'M. Bockris, *Surface Science* **25**, 261 (1971).
33. F. Schubert, *Acta Electrochim.* **16**, 42 (1971).
34. R. Armstrong, W. P. Race, and H. R. Thirsk, *Acta Electrochimica* **13**, 215 (1968).
35. J. O'M. Bockris, E. Gileadi, and K. Muller, *J. Chem. Phys.* **44**, 1445 (1966).
36. J. O'M. Bockris, E. Gileadi, and K. Muller, *J. Chem. Phys.* **47**, 250 (1967).
37. R. de Levie, *Advances in Electrochemistry*, Volume 6 (P. Delahay and C. W. Tobias, eds.), p. 281, Interscience, New York (1967).
38. K. J. Vetter, *Electrochemical Kinetics*, Academic Press, New York (1967).
39. J. O'M. Bockris and H. Kita, *J. Electrochem. Soc.* **4**, 108 (1961).
40. B. E. Conway and E. Gileadi, *Modern Aspects of Electrochemistry*, Volume 3 (J. O'M. Bockris and B. E. Conway, eds.), Plenum Press, New York (1964).
41. Y. C. Chiu and M. Genshaw, *J. Phys. Chem.* **51**, 3148 (1969).
42. W. N. Hansen, *Anal. Chem.* **39**, 105 (1967).
43. W. K. Paik and J. O'M. Bockris, *Surface Science* **25**, 61 (1970).
44. W. K. Paik, M. Genshaw, and J. O'M. Bockris, *J. Phys. Chem.* **74**, 4266 (1970).
45. G. Aniansson, *J. Phys. Chem.* **55**, 1286 (1951).
46. K. Schwabe, *Chem. Tech. (Berlin)* **10**, 469 (1958).
47. E. A. Blomgren and J. O'M. Bockris, *Nature* **186**, 305 (1960).
48. N. A. Balshowa, *Elektrokhim.* **4**, 871 (1968).
49. Y. Veber, D. Pirtshhalava, Y. V. Vasil'ev, and V. S. Bagotsky, *Elektrokhim.* **5**, 1037 (1969).
50. W. Heiland, E. Gileadi, and J. O'M. Bockris, *J. Phys. Chem.* **70**, 1207 (1966).
51. B. T. Rubin, E. Gileadi, and J. O'M. Bockris, *J. Phys. Chem.* **69**, 3335 (1965).
52. H. Urbach, private communication (1971).
53. E. Gileadi, L. Duic, and J. O'M. Bockris, *Acta Electrochim.* **13**, 1915 (1968).
54. R. Parsons, "The determination of heats of adsorption of iodide on mercury," *Proc. 2nd Int. Conf. Surface Activity* **3**, 45 (1952).
55. E. Gileadi, *Electrosorption*, Chapter 3, Plenum Press, New York (1967).
56. P. Delahay and C. T. Fike, *J. Am. Chem. Soc.* **80**, 2628 (1958).
57. J. O'M. Bockris, H. Wroblowa, E. Gileadi, and B. J. Piersma, *Trans. Faraday Soc.* **61**, 2531 (1965).
58. S. S. Sedova, Y. V. Vasil'ev, and V. S. Bagotskii, *Elektrokhim.* **4**, 1221 (1968).
59. L. Griest, *Trans. Symp. Electrode Processes, Phila., 1959*, p. 294 (1961).
60. R. D. Armstrong, *J. Electroanal. Chem.* **20**, 168 (1968).
61. A. N. Frumkin, referred to on p. 265 of L. Antropov, *Theoretical Electrochemistry*, Mir, Moscow (1972).
62. L. Antropov, *Theoretical Electrochemistry*, Mir, Moscow (1972).
63. J. O'M. Bockris, *Energy Conversion* **10**, 41 (1970).
64. J. O'M. Bockris, E. Gileadi, and G. Stoner, *J. Electrochem. Soc.* **113**, 585 (1966).
65. J. O'M. Bockris, *J. Electroanal. Chem.* **34**, 1, 201 (1972).
66. A. N. Frumkin and A. Gorodetskaya, *Z. Physik. Chem.* **136**, 215, 415 (1928); *Zhur. Fiz. Khim.* **5**, 240 (1934).
67. E. Gileadi and S. Argade, *Electrosorption* (E. Gileadi, ed.), pp. 94–96, Plenum Press, New York (1967).

68. J. O'M. Bockris and S. Argade, *J. Chem. Phys.* **49**, 5133 (1968).
69. R. Parsons, in *Modern Aspects of Electrochemistry*, Volume 1, Chapter 3 (J. O'M. Bockris, ed.), Butterworths Publications, London (1954).
70. A. N. Frumkin, *Svensk. Kem. Tidskr.* **77**, 300 (1965).
71. B. B. Damaskin, O. A. Petrii, and V. V. Batrakov, *Adsorption of Organic Compounds on Electrodes*, Chapter 9, Plenum Press, New York (1971).
72. P. A. Rehbinder and E. K. Wenstrom, *Acta Physiochim.* **19**, 36 (1944).
73. J. O'M. Bockris and R. Parry-Jones, *Nature* **171**, 930 (1953).
74. J. O'M. Bockris and R. K. Sen, *Surface Science* **30**, 237 (1972).
75. A. Ya. Gokhshtein, *Electrokhim.* **2**, 1318 (1966); **7**, 3 (1971); *Electrochim. Acta* **15**, 219 (1970); *Dokl. Akad. Nauk. SSSR* **200**, 620 (1971).
76. J. O'M. Bockris, E. Gileadi, and S. Argade, *Electrochim. Acta* **14**, 1259 (1969).
77. A. H. Wilson, *The Theory of Metals*, Cambridge Univ. Press, Cambridge, England (1958).
78. S. Trasatti, *J. Electroanal. Interfacial Chem.* **52**, 313 (1974).
79. J. O'M. Bockris and M. A. Habib, *J. Electroanal. Chem.* **68**, 367 (1976).
80. T. Andersen and J. O'M. Bockris, *Electrochim. Acta* **9**, 347 (1964).
81. D. D. Bode, *J. Phys. Chem.* **76**, 2915 (1972); *Advan. Chem. Phys.* **21**, 362 (1971).
82. D. J. Barclay and J. Caja, *Croat. Chem. Acta* **43**, 221 (1971).
83. G. Blyholder, in *Modern Aspects of Electrochemistry*, Volume 7 (J. O'M. Bockris and B. E. Conway, eds.), Plenum Press, New York (1971).
84. S. Levine, G. M. Bell, and D. Calvert, *Can. J. Chem.* **40**, 518 (1952).
85. R. Parsons, *Trans. Faraday Soc.* **51**, 1518 (1955).
86. R. Parsons, *J. Electroanal. Chem.* **5**, 397 (1963); *Rev. Pure Appl. Chem. (Aust.)* **18**, 91 (1968).
87. H. Wroblowa and K. Muller, *J. Phys. Chem.* **73**, 3528 (1969).
88. S. Levine, *J. Colloid Interface Sci.* **37**, 619 (1971).
89. J. O'M. Bockris and M. A. Habib, *Z. Phys. Chem.*, *N.F.* **98**, 43 (1975).
90. R. K. Sen, Ph.D. Thesis, University of Pennsylvania, Philadelphia (1972).
91. L. M. Baugh and R. Parsons, *J. Electroanal. Chem.* **40**, 407 (1972).
92. J. O'M. Bockris and M. A. Habib, *J. Res. Inst. Catal., Hokkaido Univ.* **23**, 47 (1975).
93. H. Webb, *J. Amer. Chem. Soc.* **48**, 2589 (1926).
94. M. Malsch, *Ann. Physik.* **84**, 841 (1927); *Phys. Z.* **29**, 770 (1928).
95. B. E. Conway, J. O'M. Bockris, and I. A. Ammar, *Trans. Faraday Soc.* **47**, 756 (1959).
96. F. Booth, *J. Chem. Phys.* **19**, 391 (1951).
97. R. W. Rampolla, R. C. Miller, and C. P. Smyth, *J. Chem. Phys.* **30**, 566 (1959).
98. I. R. Miller, *Acta Electrochim.* **9**, 1453 (1964).
99. J. Lawrence and R. Parsons, *Trans. Faraday Soc.* **64**, 751 (1968).
100. V. A. Kiranov, A. Kuznetsov, and B. B. Damaskin, *Elektrokhim.* **3**, 12 (1967).
101. J. O'M. Bockris, W. Mehl, B. E. Conway, and L. Young, *J. Chem. Phys.* **25**, 776 (1956).
102. K. Muller, Ph.D. Thesis, University of Pennsylvania, Philadelphia (1965).
103. K. S. Cole and R. H. Cole, *J. Chem. Phys.* **9**, 341 (1941).
104. D. C. Grahame, private communication (1955).
105. R. de Levie, *J. Chem. Phys.* **47**, 2509 (1967).
106. R. D. Armstrong, W. P. Race, and H. R. Thirsk, *J. Electroanal. Chem.* **14**, 143 (1967).
107. D. C. Grahame, *J. Amer. Chem. Soc.* **76**, 4819 (1954).
108. J. O'M. Bockris and B. E. Conway, *J. Chem. Phys.* **28**, 707 (1958).
109. J. O'M. Bockris and M. A. Habib, *J. Electroanal. Chem.* **65**, 473 (1975).
110. E. Blomgren and J. O'M. Bockris, *J. Phys. Chem.* **63**, 1475 (1959).
111. J. T. Law, Ph.D. Thesis, London University, London (1952).

112. A. N. Frumkin and B. B. Damaskin, *Modern Aspects of Electrochemistry*, Volume 3, Chapter 3 (J. O'M. Bockris and B. E. Conway, eds.), Plenum Press, New York (1964).
113. B. B. Damaskin, O. A. Petrii, and V. V. Battrokov, *Adsorption of Organic Compounds on Electrodes*, Plenum Press, New York (1971).
114. E. A. Blomgren, J. O'M. Bockris, and C. Jesch, *J. Phys. Chem.* **75**, 2000 (1961).
115. B. B. Damaskin and A. N. Frumkin, *Reactions of Molecules at Electrodes*, (N. S. Hush, ed.), p. 1, Wiley (Interscience), London (1971).
116. E. A. Blomgren and J. O'M. Bockris, *J. Phys. Chem.* **63**, 1475 (1959).
117. J. O'M. Bockris, M. Green, and D. A. J. Swinkels, *J. Electrochem. Soc.* **111**, 743 (1964); J. O'M. Bockris and D. A. J. Swinkels, *J. Electrochem. Soc.* **111**, 736 (1964).
118. B. B. Damaskin, *J. Electroanal. Chem.* **23**, 431 (1969).
119. J. O'M. Bockris, E. Gileadi, and K. Muller, *Electrochim. Acta* **12**, 1301 (1967).
120. E. Gileadi, *J. Electroanal. Chem.* **30**, 123 (1971).
121. E. Dukiewitz, J. D. Garnish, and R. Parsons, *J. Electroanal. Chem.* **16**, 505 (1968).
122. R. Payne, *J. Electroanal. Chem.* **41**, 277 (1973).
123. M. Green, D. A. J. Swinkels, and J. O'M. Bockris, *Rev. Sci. Instr.* **33**, 18 (1962).
124. B. E. Conway and J. Gordon, *J. Phys. Chem.* **73**, 3609 (1969).
125. R. Barradas and R. Herman, *J. Phys. Chem.* **73**, 3619 (1969).
126. B. B. Damaskin, A. N. Frumkin, and A. Chizou, *J. Electroanal. Chem.* **28**, 93 (1970).
127. J. O'M. Bockris, E. Gileadi, and G. Stoner, *J. Phys. Chem.* **73**, 427 (1969).
128. R. Armstrong and J. A. V. Butler, *Trans. Faraday Soc.* **29**, 1261 (1933).
129a. F. G. Will and C. A. Knorr, *Z. Elektrochem.* **64**, 258 (1960).
129b. M. W. Breiter, *Electrochim. Acta* **8**, 925 (1963).
130. M. A. V. Devanathan, J. O'M. Bockris, and W. Mehl, *J. Electroanal. Chem.* **1**, 143 (1960).
131. W. Mehl and J. O'M. Bockris, *J. Chem. Phys.* **27**, 817 (1957).
132. B. I. Podlovchenko, V. F. Stenin, and V. E. Kazarinov, *The Double Layer and Adsorption on Solid Electrodes*, p. 121, Izd. Tartusk. Gos. Univ., Tartu, USSR (1968).
133. T. Biegler and D. F. A. Koch, *J. Electrochem. Soc.* **114**, 904 (1967).
134. B. I. Podlovchenko and E. P. Gorgonova, *Dokl. Akad. Nauk SSSR* **156**, 673 (1964).
135. V. S. Bagotskii and Yu. B. Vasil'ev, *Electrochim. Acta* **11**, 1439 (1966); V. S. Bagotskii and Yu. B. Vasil'ev, *Advances in the Electrochemistry of Organic Compounds*, p. 38, Izd. Nauka, Moscow (1966).
136. S. B. Brummer, *Elektrokhim.* **4**, 243 (1968).
137. S. Schuldiner and B. J. Piersma, *J. Am. Chem. Soc.* **74**, 2823 (1970).
138. M. Shisandaram, Y. B. Vasil'ev, and V. S. Bagotskii, *Elektrokhim.* **3**, 193 (1967).
139. B. Cahan, private communication (1978).
140. A. A. Humffray, in *Modern Aspects of Electrochemistry*, Volume 8, p. 106 (J. O'M. Bockris and B. E. Conway, ed.), Plenum Press, New York (1972).
141. G. Harrison and R. Thirsk, *A Guide to Electrode Kinetics*, Academic Press, New York (1972).
142. A. N. Frumkin, O. A. Petrii, and B. B. Damaskin, *J. Electroanal. Chem.* **27**, 81 (1970).
143. A. N. Frumkin, O. A. Petry, A. Kossaya, V. Entina, and V. Topolev, *J. Electroanal. Chem.* **16**, 175 (1968).
144. D. J. Barclay and J. Caja, *Croat. Chem. Acta* **43**, 221 (1971).
145. J. O'M. Bockris, A. Damjanovic, and M. L. B. Rao, *J. Phys. Chem.* **67**, 2508 (1963).
146. A. Damjanovic, in *Modern Aspects of Electrochemistry*, Volume 5 (J. O'M. Bockris and B. E. Conway, eds.), Plenum Press, New York (1969).
147. M. Genshaw, A. K. N. Reddy, and J. O'M. Bockris, *J. Chem. Phys.* **48**, 671 (1968).
148. R. Greef, *J. Chem. Phys.* **51**, 3148 (1969).
149. M. Genshaw and J. O'M. Bockris, *J. Chem. Phys.* **51**, 3149 (1969).

150. W. Paik and J. O'M. Bockris, *Surface Science* **28**, 61 (1971).
151. S. Kim, W. Paik, and J. O'M. Bockris, *Surface Science* **33**, 617 (1972).
152. V. Brusic, H. Wroblowa, and J. O'M. Bockris, *J. Phys. Chem.* **75**, 2823 (1971).
153. H. L. von Helmholtz, *Wied. Ann* **7**, 337 (1879).
154. For a comprehensive discussion, see M. A. Habib, *Comprehensive Treatise in Electrochemistry* (J. O'M. Bockris, B. E. Conway, and E. Yeager, eds.), Plenum Press, New York (1979).
155. J. O'M. Bockris, S. Argade, and E. Gileadi, *Electrochim. Acta* **14**, 1259 (1969).
156. M. A. Habib, Ph.D. Thesis, Flinders University of South Australia (1976).
157. A. Hoffmann and B. Billings, *J. Electroanal. Chem.* **77**, 97 (1977).
158. E. A. Moelwyn Hughes, *Physical Chemistry* (2nd edition), p. 392, Pergamon, London (1961).
159. *Handbook of Chemistry and Physics* (35th edition), p. 2387, Chemical Rubber Publishing Co., Cleveland (1953).
160. J. A. A. Ketelaar, *Chemical Constitution*, p. 28, Elsevier, Amsterdam (1958).
161. W. Paik and J. O'M. Bockris, *Surface Science* **28**, 557 (1971).
162. V. Brusic, H. Wroblowa, and J. O'M. Bockris, *J. Phys. Chem.* **75**, 91 (1971).
163. E. C. Stoner, *Magnetism*, p. 38, Methuen, London (1948).
164. M. A. V. Devanathan, Ph.D. Thesis, University of London,. London (1951).

2

Electrode Kinetics

2.1. Nature of Electrochemical Reactions

Electrochemical reactions are those involving a net transfer of charge in the overall reaction. They are usually interfacial; the charge transfer occurs between an electronically conducting face and an ionically conducting one. Electrochemical reactions differ from chemical ones in two ways.

(a) Thermodynamically, they do not give up or take in their heat of reaction. The heat given up or taken in during an ideal reversible electrochemical reaction is $T\Delta S$. The free energy of the reaction is all available, at limitingly low rates, as *electrical* energy. In practice, electrochemical reactions at significant rates are associated with a heat change which differs from $T\Delta S$ because of the existence of overpotential.

(b) The reactants in electrochemical reactions (e.g., C_2H_4, O_2) do not, in principle, need to be spatially near each other; they collide, respectively, with spatially separated electronic conductors.

Some heterogeneous reactions in condensed phases not obviously electrochemical probably occur by means of electrochemical mechanisms.[2] It is possible that gaseous heterogeneous reactions involve electrochemical steps (some of them depend markedly on the electronic conductivity of the substrate or upon intensive drying, which may alter the surface conductivity).[3] Drawing conclusions based on the work of Cope[4] and Mandel,[5] it has been suggested[6,7] that some biochemical reactions may be electrochemical in the same way. Electrochemical reactions occur spontaneously within a great range of happenings, e.g., in biological energetics,[8] engineering,[9] metallurgy,[10] and chemistry, pure and applied. The recognition of electrodic processes from the presence of electrochemical reactions in nature, except for those occurring in corrosion, dates from the 1960s. Academically, there are two types of studies of electrochemical reactions ("electrodics") going at this time.

(a) *Microscopic study.*[11] The field is related to surface chemistry and to heterogeneous chemical kinetics. The reaction is on the surface of an electron donor and acceptor: the structure of the double layer formed between this and the solution affects the reaction rate.† Electrochemical reactions involve an electron transfer across the interphase, but the heterogeneous chemical reactions that occur between adsorbed entities on the surface and the homogeneous partial reactions coupled to these electrochemical transfers in the solution are part of the overall electrochemical reaction. Indeed, the rate-determining step in the overall sequence need not be the electron transfer itself.

(b) *Macroscopic study.*[12] The double layer and the electrode surface are regarded as black-box entities. Attention is focused upon mass transfer aspects of the processes in circumstances when it has a significant effect upon the reaction rate. Such an approach is used particularly in electroanalytical chemistry (most of which depends on mass transport kinetics) and electrochemical engineering, where the practical importance of mass transfer may sometimes outweigh the importance of the electron transfer step.

2.2. Overpotential

When an electronic conductor is brought into contact with an ionic conducting phase, there is a change in the energy of the surface electron levels in the electronic conductor, because charge transfer between the phases takes place until equilibrium of the charge transfer reaction between solution and electrode occurs. A certain value of the surface electron level, different from that in the electronic conductor in contact with a vacuum, characterizes the situation of the conductor in solution, and is characteristic of the cell in which this particular conductor in contact with this particular solution finds itself. The potential difference registered by a potentiometer when it opposes the electrical driving force of the cell—so that the electrochemical reaction rate in the cell becomes negligible ("the reversible cell potential")—is characteristic of the overall reaction in the cell, and not of the metal–solution interfaces in the cell. A surface electron

---

† The term "rate" (r) in heterogeneous kinetics is given in moles per unit area and time. The electrochemical terminology deals in coulombs per unit time and area (i), but it is sometimes expressed as moles per square centimeter per second. The relation between the two is

$$r = i/nF$$

where n is the number of reactions per one act of the overall reaction and F is the Faraday.

level belongs to "the reversible potential,"[2,13,14] and this is associated with a double-layer potential difference. At this potential, $i = 0$. For a finite i, the energy level in the surface electrons in the electronic conductor is displaced, so that net electron flow in one direction (for example, from the metal to acceptors in the solution) occurs. It is this displacement that is the overpotential for the rate and reaction concerned.[15]

Thus, overpotential is the shift of the Fermi level from that pertaining when the reaction concerned is at equilibrium at the interface concerned to that needed to provoke a current density or net reaction rate.

Phenomenologically, overpotential is the difference between the electric potential (with respect to some reference electrode) of a certain electrode at which is occurring a given current density (i.e., a reaction rate) and the electric potential on the same scale corresponding to equilibrium for the interfacial electron transfer reaction concerned (of course, at the relevant concentration and temperature). Overpotential is a ubiquitous quantity. It is relevant to phenomena as diverse as biochemical reactions associated with digestion to the breakdown of the steering column of a bus due to hydrogen embrittlement.[16] The overpotential for the same reaction situation (nature of overall reaction, substrate, and same reaction rate) is different when the reaction is occurring spontaneously and when under the pressure of an outside electrical driving force. The difference is in sign, but it may also be in magnitude (Fig. 2.1).

Until the 1950s, there was confusion concerning the nature of overpotential. Thus, from Erdey-Gruz and Volmer,[17] the relation between i and η is

$$i = A \{\exp(-\alpha F\eta/RT) - \exp[(1-\alpha)F\eta/RT]\} \qquad (2.1)$$

where A is a factor not dependent on i or η. Thus, unless $|\eta| > 0$, $i = 0$. Until the general understanding of this equation took place, the fact that electrode reactions would only occur at significant rates when the potential was removed from equilibrium value was usually discussed in terms of artifacts that differed for each reaction.[16]

The basic overpotential is that defined above. However, reaction steps other than that of interfacial electron transfer may determine the rate of the overall reaction and the electron transfer. The displacement of the Fermi level is still the displacement that controls the electron flow, but it is determined predominantly by the rate constant of the nonelectrochemical rate-determining reaction, rather than by the properties of the interphasial electron transfer. Then, overpotential is often prefixed by a descriptive adjective. One refers to chemical overpotential (a rate-determining surface chemical, solution, or reaction or phase overpotential when phase transformations are rate determining in the overall reaction[18]).

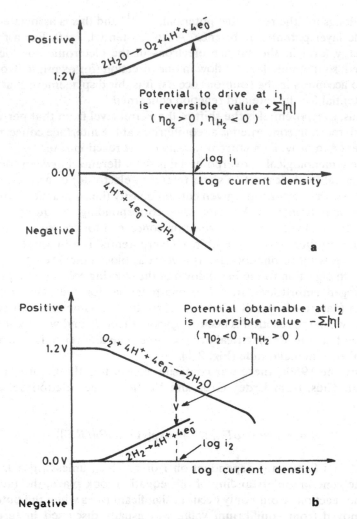

Fig. 2.1. (a) The two current (i.e., rate)–potential relations in an externally-driven hydrogen to oxygen cell (e.g., an electrochemical reactor). The energy needed to make the cell function at a given rate is *more* than that indicated from thermodynamics. (b) The two current (i.e., rate)–potential relations in a spontaneously acting cell (e.g., a fuel cell). The amount of energy available is *less* than that indicated by thermodynamics.

2.3. Rate as a Function of Overpotential

In chemical kinetics, the central measurement is rate as a function of temperature and the central law is the Arrhenius relation. In electrode kinetics, the Arrhenius relation is also applicable, but greater in importance is a relation that has a form similar to that of the Arrhenius relation but is an Arrhenius relation in which the heat of activation is potential dependent. The relation has the form:

$$i = A \exp(-\Delta H^{ox}/RT) \exp(-\alpha\eta F/RT) \qquad (2.2)$$

but is often written in the form:

$$i = i_0 \exp(-\alpha\eta F/RT) \qquad (2.3)$$

The corresponding equation in which $\log i$ is written as linear with η is called Tafel's equation.[19,20]

The Tafel equation is applicable to electrochemical reactions if conditions under which they are occurring are simple. Thus, they must be arranged so that there is *negligible interference with the rate of the interphasial electron transfer by transport processes in solution*; it will not be applicable if the surface changes significantly during a change of potential or with time (as, e.g., in the formation of oxide films during dissolution).

Finally, a typical reason for an occasionally reported nonapplicability of Tafel's law arises because some techniques of measuring electrochemical reactions involve a potential that is changed at a rate above the relaxation rate of the radical concentrations corresponding to each potential or so that diffusion affects the observed reaction rate.[21] Conversely, the Tafel equation has been shown to be applicable for complex reactions, as, e.g., in the oxidation of hydrocarbons.[22]

At high current densities (but under conditions when mass transport is not rate determining), there may occur a limiting current (electron transfer controlled), i.e., $di/d\eta \to 0$. This was first demonstrated experimentally by Despic and Bockris[23] (cf. Parsons and Passeron,[24] who worked later with redox couples).

There is evidence for the "perfection" of the linearity of the Tafel relation over some ten orders of magnitude for the hydrogen evolution reaction (her) and the oxygen evolution reaction (oer). Linear observations for the hydrogen evolution reaction on mercury follow from the work of Bockris and Azzam[25] and Nurnberg[26] from 10^{-10} A cm^{-2} to $10^{2.5}$ A cm^{-2}.[27] However, "perfect" linear relations between η and $\log i$ are obtained over equally long ranges for oxygen evolution.[28,29]

There are two straight lines in some Tafel relations. Such two-sloped Tafels were first observed systematically by Bockris and Conway[30] and Bockris, Ammar, and Huq.[31] They can sometimes be discussed in terms of the change of electrokinetic potential with electrode potential.[31] The general relation of rate to overpotential is shown in Fig. 2.2.

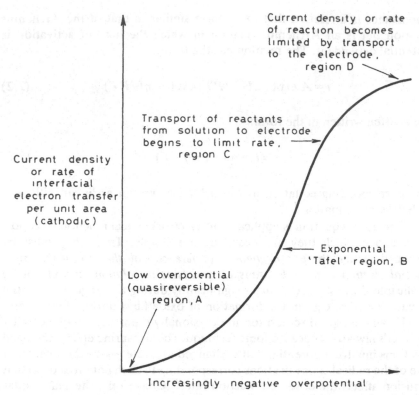

Fig. 2.2. The current–overpotential relation in general. All relations of current to potential have a linear region near to the reversible potential, have an exponential (or "Tafel") region, and finally become limited at sufficiently high rates by mass transport to or from the interface.

2.4. Exchange Current Density

At the equilibrium potential, $\eta = 0$, $i = 0$, and therefore

$$\vec{i} = \overleftarrow{i} \tag{2.4}$$

where $\vec{i}$ is the anodic and $\overleftarrow{i}$ the cathodic current density at $\eta = 0$. These i values are the exchange current density, the rate of an electrochemical reaction at equilibrium, in either direction. The term was first used by Dolin, Erschler, and Frumkin.[32]

The numerical values vary from around 10^{-1} down to 10^{-16} A cm^{-2} (around 10^{-6} to 10^{-21} mole cm^{-2} sec^{-1}). What area is referred to when one quotes an exchange current density? Phenomenologically, it is the unit area of apparent geometric surface of the electrode substrate, and results are given in this form.

One *should* refer to exchange current density to a real surface area. This is often two or three times greater than the apparent one, though with certain electrode preparations, it may become much more than this. Devanathan and Selvaratnam[33] refer the exchange current density to the free surface area, i.e., to the surface without radicals. But this is often not known.

There are many methods of obtaining exchange current densities; as the value gets smaller (e.g., $<10^{-2}$ A cm^{-2}), the determination comes easier, but interaction with competing reactions gets larger. With very fast electrode reactions ($>10^{-2}$ A cm^{-2}), mass transport effects cause difficulties. The following are some important methods of determination:

(a) The extrapolation of a Tafel relation to the value of zero η.[34-37]

(b) From a plot of i_0 against c_i: This method was first used by Vetter.[38] Details are given in the literature; in particular, see L. Antropov.[18]

In these methods, it is important to allow for the effect of the electrokinetic (zeta) potential. In earlier evaluations of Tafel lines and extrapolations therefrom, the variation of ζ with potential was neglected. It should be taken into account in deducing data from Tafel lines. This was first implied by Bockris and Conway,[30] but shown explicitly by Bockris, Ammar, and Huq[31] (cf. Antropov[18]).

(c) Determination from the rate of exchange of isotopes on electrodes.[39]

(d) From measurements of ac bridges, it is possible to isolate the Faradaic resistance and to plot this against $1/\omega^{1/2}$. The method was first suggested by Dolin, Erschler and Frumkin.[32] The method was rediscovered by Randles in 1947.[40]

(e) From transients. The observations of galvanostatic and potentiostatic transients (also potentiodynamic transients) can be made to yield exchange-current density values. The analysis is often complex if there is a significant surface reaction, a significant surface coverage of radicals on the electrode, etc. Much work has been done to fructify this method.[12] Measurements in the nanosecond range were first carried out by Blomgren, Inman, and Bockris.[41] Among difficulties of the analysis of the transient methods is the possibility of rate-determining adsorption on the electrode[12] and the possibility, at present ambiguous, concerning an alleged variation of the double-layer impedance with frequency.[42]

(f) From Faradaic rectification. The method is limited to situations where there is a difference in the transfer coefficients of anodic and cathodic reactions. The theory is not satisfactory.[43,44] The limiting factor in determinations for low values of electrode kinetic parameters is competition from reaction with impurities and the time taken to charge the double layer when the current density is low. The limiting factor for high values of reaction rate is mass transport.[45] Double-layer charging may interfere. If it is not possible to charge the double layer before mass transport control interferes strongly, determination of the electrode kinetic parameters becomes difficult.[41]

2.5. Rate Constants

In chemical kinetics, rate constants are quoted. In electrochemical kinetics, exchange current density (or exchange rate) is the most quoted quantity; it is difficult to ascertain the concentration of radicals that connect up exchange current density to rate constant. Thus,

$$i_0 = \frac{n}{\nu} F \Pi a_i \bar{k} \tag{2.5}$$

where i is a reactant and $\bar{k}$ the rate constant with which it is associated. In simple cases, e.g., redox reactions, there is no difficulty in relating i_0 to rate constants.

2.6. The Symmetry Factor

In one-step electrode reactions, one may relate an electrochemical rate constant to its "chemical" counterpart (rate constant at the reversible potential) by

$$\bar{k} = k_0 \exp(\pm \beta \eta F / RT) \tag{2.6}$$

The coefficient β is called the symmetry factor; it is near to $\frac{1}{2}$ in value but may have somewhat different values for forward and backward reactions. β is related to the gradients of the potential energy–distance profile for the representative points of reactant and products. If the slope of the relation near the intersection points is γ for the product and α for the reactant,

$$\beta = \frac{\tan \gamma}{\tan \alpha + \tan \gamma} \tag{2.7}$$

A geometric proof of this relation was first published by Bockris.[46] The relation was later reobtained by Marcus,[47] without reference to the previous work.

The meaning of β can be developed; e.g., according to Bockris and Matthews,[48] β is a coefficient controlling the transfer of electrical to chemical energy. Similar to this concept is that of Dogonadze,[49] in which β is a measure of the probability of the occupation of an orbital in the metal.† Very rudimentary attempts to calculate values of β absolutely have been made.[51-54]

2.7. The Transfer Coefficient

The coefficient β concerns the "pure charge transfer" step. When results of the measurements of α from a series of electrochemical reactions are considered, $d \log i/d\eta$ is often observed (in the absence of interfering mass transport) to have simple integral values related to F/RT. These may differ substantially from the value of β, seldom outside the range 0.4–0.6. Theoretical analysis indicates that the value of $d \log i/d\eta$ for consecutive reactions is related to β but contains phenomenological factors associated with the stoichiometry of the reactions (e.g., the number of times the rate-determining step is to occur from one act of the overall reaction) that are not involved in the rate-determining step. The relation between the α and β was first derived by Parsons in 1951[52] (cf. Bockris[46]). Practical realizations of some of the terms were difficult; an alternative definition was given by Matthews and Bockris[55] and is

$$\alpha = \frac{s}{\nu} + \beta r \qquad (2.8)$$

There are difficulties of communication in matters connected with β and α, because there is not yet unity among authors concerning the terminology. These questions have been discussed by Nagy and Bockris.[56] Many authors write electrochemical equations that involve the relation $\alpha = \beta z F$, so that the exponential part of the electrochemical rate constant is found in the literature as

$$\vec{k} = k_0 \exp(\beta z F/RT) \qquad (2.9)$$

where z is the overall number of electrons in the reaction. In reality it should be

$$\vec{k} = k_0 \exp(\alpha F \eta/RT) \qquad (2.10)$$

† Bockris and Reddy[50] point out that the concept of β becomes meaningless in electron transfer reactions in which there is no rearrangement of molecular species associated with the electron transfer, e.g., in electron transfer between two solid phases.

where

$$\alpha = \frac{s}{\nu} + \beta r \qquad (2.11)$$

and s is the number of electrons passing before the rate-determining step and r the number of electrons in the rate-determining step.

May we assume that there is *one* electron that always takes part in an act of an electrochemical charge transfer? That this should be so is indicated by theoretical calculations of heats of activation of electrochemical reactions.[57,58] If one electron is involved, then calculated heats of activation do fall in the requisite region of 0.5–1 eV, and if two electrons take part, the calculated value is increased quite outside the experimental region. However, note that Miles and Gerischer[59] have claimed experimental evidence of a two-electron rate-determining reaction.

2.8. Stoichiometric Number

The stoichiometric number is defined as the number of times the rate-determining step occurs for one act of the rate-determining reaction. It was introduced in electrode kinetics by Horiuti and Ikusima,[60] who published the first proof. The first measurements were made by Potter.[61] The stoichiometric number has been used less than one might have expected, because some authors have pointed out that it becomes a meaningless concept if there is not a clear rate-determining step.[62] The stoichiometric number will be meaningful and will have a simple integral value if there is such a step.

Situations in which the stoichiometric number concept is useful have been made by Makrides[63] and Conway and Salomon.[64] Recent applications have been carried out by Enyo.[39] The determination of α occasionally uniquely indicates a mechanism,[65] and the determination of stoichiometric number may also occasionally indicate uniquely a reaction (cf. Bockris and Potter[37]). More often, it is one of several factors that reduce the possibilities in reaction mechanisms. The situation is discussed by Antropov,[18] and also by Erdey-Gruz.[66] It is important to stress the value of stoichiometric number determinations in mechanism evaluation.

2.9. Stoichiometric Factors

These are the electrochemical analogs of order of reaction. They may be defined, following Parsons[67] and Vetter,[68] as

$$\left(\frac{\partial \ln \vec{i}}{\partial \ln a_Q} \right)_{\Delta\phi,\, T,\, a_{p},\, \ldots} = q^{1} \qquad (2.12)$$

where q^1 is the stoichiometric factor of the entity Q, and a_i is the activity of the species i in solution.

The stoichiometric factors may be related to the overall numbers of electrons involved in the reaction so long as the stoichiometric number is known. Thus, if in an overall reaction the number of particles occurring at certain species P is p, then the difference of the stoichiometric factors referring to P in the forward and backward reactions (namely, p' and p'', respectively) are related by equations of the type $p' - p'' = p/\nu$.[69,70]

If sufficient information were available on the stoichiometric factors for all the reactants and products in electrochemical reaction, then these combined with information on α, β, ν and the total number of electrons in the overall reaction could lead to a unique evaluation of the rate-determining step. This method of determining electrochemical reactions (first implied by Parsons[67]) is seldom practical, and auxiliary methods have to be used (see Bockris and Srinivasan[71]).

2.10. Rate as a Function of Temperature

Although a principal relation of electrode kinetics is $i = A\,e^{B\eta}$, the Arrhenius relation in chemical kinetics applies also to electrochemical kinetics at constant overpotential. (Earlier, it was not understood that the reference electrode must be at the same temperature as the working electrode.) Discussions of the temperature dependence of electrochemical reactions are rare.[72-74] Earlier, it was thought that determination of heats of activation could give mechanistic information (Gorbachev, as quoted in Antropov[75]). However, apart from an indication of rate-determining diffusion (more easily obtained by agitation), the method is nugatory.

In the ideal presentation of the determination of the heat of activation, it is assumed that the transfer coefficient is independent of temperature. Then,

$$i_\eta = A \exp(-E_{(\eta)}/RT) \tag{2.13}$$

If the experiment is arranged so that a potentiostat keeps η constant at a given value, measurement of variations of current as the temperature is varied gives rise to the heat of activation for the given η. Determinations over a series of $E_{(\eta)}$ values allow extrapolation of E_η to E^x at $\eta = 0$. Alternatively, one determines Tafel relationships at varied temperatures and plots $\log i_0$ against $1/T$. The heat of activation at the reversible potential is the quoted value.[76]

Greater attention would be paid to these determinations if their interpretation were less ambiguous. The first difficulty was pointed out by Baxendale.[77] Differentiation of the relation between $\log i$ and $1/T$

involves that of the reversible potential. A relation that may be deduced
with this background was given by Bockris[78] and is

$$\left[\frac{\partial \ln i_0}{\partial(1/T)}\right]_{\beta,\,\gamma,\,\delta,\,a_i} = -\frac{\Delta \bar{H}^{ox} + \delta - [(\beta + \gamma)/\nu]\Delta H_R^0}{R} \qquad (2.14)$$

where ΔH_R^0 is the heat of reaction, δ is a contribution arising from potential (similar to the electrokinetic potential), γ is a coefficient similar to β,
and the other symbols have their usual significance.

Thus, the "heat of activation" at the reversible potential involves a
heat of reaction and is not purely a heat of activation.

Another difficulty is the fact that the absolute potential difference at
the reversible potential has in the past not been known, so that when one
compares the heats of activation of a given reaction upon various substrates, one is only observing the effects (for bonding reactions) of, e.g.,
different bond strengths of various radicals, but also the effects of different
galvanic potential differences at the interface,[79] even though the reversible
potential on arbitrary reference scales is the same.

A third difficulty in the evaluation of heats of activation is the
temperature dependence of the transfer coefficient. It has been assumed in
the above discussion that $d\alpha/dT = 0$. However, Bockris, Parsons, and
Rosenberg[80] showed that this was sometimes not the case. Detailed
studies[73,74] over a wide range of temperatures develop this difficulty. Complication arose from the variations of coverage and anion concentration with
temperature, but these have less effect than that of β.

Thus, there are many complications in evaluating the significance of
heats of activation, except in simple reactions such as redox reactions,
where β may be temperature independent for a sufficient stretch of
temperature.[240]

2.11. Comparative Reaction Rates of Isotopic Reactions

The comparison of reaction rates of the same reaction, except that
atoms of different isotopic weight (nearly always hydrogen) have been
substituted, is a well-known part of chemical kinetics. Hitherto, the
determinations have always been made with the HER, using deuterium,[39]
tritium,[81,82] or oxygen-18,[83] or with some organic reactions.[84] The method
has been used to attempt to trace organic electrode mechanisms. One
utilizes various hypotheses of rate-determining steps and calculates
according to them what the different ratios of $i_{\text{isotope 1}}/i_{\text{isotope 2}}$ would be.
There are two determining factors in the theoretical discussion—the zero-
point energy conditions involved and the degree of quantum mechanical

character of the reactions. This is appreciated by considering qualitatively the two rate-determining steps:

$$H^+ + e \rightarrow MH \tag{2.15}$$

$$H^+ + H_M + e \rightarrow H_2 \tag{2.16}$$

In the second step, there are two zero-point energy conditions to be accounted for on the reactant side, and there is only one in the first. Mechanisms involving the second rate-determining step should have a much greater difference in the isotopic reaction ratio than those involving the first. The use of tritium is preferable in the hydrogen studies, because the quantum effects are greater. Thus, there are great differences in the isotopic separation factors predicted between the various reaction mechanisms for the hydrogen evolution reaction. Ambiguities of the theoretical calculations are relatively small, even using LEP methods,[85] and less using BEBO methods.[86]

Isotopic rate studies represent one of the more fruitful methods of examining the mechanisms in organic electrochemical kinetics.[84]

2.12. Chemical Surface Reactions

It is possible that a chemical surface reaction, one of a set in a sequence, is rate-determining and will affect and then determine the value of the "electrochemical" overpotential (cf. Bockris and Reddy[87]).

Many organic electrochemical reactions are rate-determined by a chemical surface reaction that follows an electrochemical reaction in the sequence of reactions giving the overall one. Most organic oxidation reactions follow such a path (Bagottskii and Vasil'ev[88]). This conclusion is only reached if one takes into account the correct isotherm for the situation, and the Tempkin isotherm has often been thus used.

Thus, neglect of the adsorptive stage in organoelectrochemical methods (cf. Bard[89] and Thirsk and Harrison[90]) eliminates what is often the most important (i.e. rate-determining) situation in electrochemical kinetics.

One sees here the importance of the stress laid by Antropov concerning the potential of zero charge and organic electrochemical reactions. This potential is a determining potential for the parabolic relationship between surface coverage and electrode charge (or potential)[91]; such a view can be compared with the structural relations of the adsorption of organic molecules upon potential by Bockris, Gileadi, and Muller.[92] The possibility of a direct dependence of the kinetics of electrode reactions upon the potential of zero charge, first suggested by Bockris and Potter,[35] is here of relevance.

A rate-determining chemical surface reaction within an electrochemical equation may give to a chemically determined limiting current.

2.13. Consecutive Reaction Rates

The electrode kinetic formulation of consecutive reaction kinetics is simple if there is a *sufficiently* rate-determining single step. An example is that of the electrodeposition of copper, rate-determined by the reduction of the cupric to cuprous ion in the double layer.[93] The cathodic slope should be $2RT/F$ and the anodic slope $2RT/3F$ as observed. The Tafel slope does seem here (as rarely) to have unique mechanistic significance.

A less approximate treatment of the consecutive kinetics for the hydrogen reaction has been given by Mauser and Bockris.[94] It yields a full, complex series of equations (see Table 2.1).

Clear solutions to the consecutive kinetics of electrochemical reactions have been reached for many reactions, among which may be quoted the oxidation of ethylene on platinum,[22] where part of the reaction mechanism is given†:

$$C_2H \rightleftharpoons C_2H_{4\text{ ads}} \tag{2.17a}$$

$$H_2O \rightarrow OH_{ads} + H^+ + e \tag{2.17b}$$

$$C_2H_4 + OH_{ads} \rightleftharpoons \text{organic radicals} \tag{2.17c}$$

The electrodeposition of iron[95] takes place by a perhaps somewhat unexpected mechanism of the type

$$Fe^{2+} + OH^- \rightleftharpoons FeOH^+ \tag{2.18a}$$

$$FeOH^+ + e \rightarrow FeOH_{ads} \tag{2.18b}$$

$$FeOH + OH^+ + e \rightarrow Fe + 2OH^- \tag{2.18c}$$

This kind of mechanism has been very well authenticated by Lorenz, Fischer, *et al.*[96–98] and the same obtains for nickel and cobalt.[99]

The mechanism of oxygen reduction[100] in acid solution on iridium takes place by a mechanism of the type

$$O_2 + 2M \rightleftharpoons 2MO \tag{2.19a}$$

$$MO + H^+ + e^- \rightarrow MOH \tag{2.19b}$$

$$MOH + H^+ + e^- \rightleftharpoons M + H_2O \tag{2.19c}$$

---

† It is noteworthy that the phenomenological deduction of mechanism is being joined by theoretical evaluation along molecular orbital lines.[250]

Table 2.1
The Electrochemical Desorption Reaction for Hydrogen Molecules to Hydrogen Ions[a]

Relations among the constants	Tafel equation	Coverage	$\left(\dfrac{\partial \eta}{\partial \ln \rho_{H_2}}\right)_4$
$\dfrac{\vec{k}_{2,R}}{k'}C_{H^+} > \dfrac{\vec{k}_{-1,R}}{k'} > \vec{k}_{-2,R\rho H_2}$ $> \vec{k}_{1,R}C_{H^+}$	$i = -2F\vec{k}_{-1,R}\dfrac{\vec{k}_{-2,R\rho H_2}}{\vec{k}_{2,R}C_{H^+}}$	$(C_H)_0 \to 0$	$-\dfrac{RT}{2F}\dfrac{\beta}{2-\beta}$

and

| $\dfrac{\vec{k}_{2,R}}{k'}C_{H^+} > \vec{k}_{-2,R\rho H_2} > \dfrac{\vec{k}_{-1,R}}{k'}$ $> \vec{k}_{1,R}C_{H^+}$ | $\exp\left[(3-\beta)\eta\dfrac{F}{RT}\right]$ | $(C_H) \to 0$ | |
| $\vec{k}_{-2,R\rho H_2} > \dfrac{\vec{k}_{2,R}}{k'}C_{H^+} > \vec{k}_{1,R}C_{H^+}$ $> \dfrac{\vec{k}_{-1,R}}{k'}$ | $i = -2F\dfrac{\vec{k}_{-1,R}}{k'}$ | $(C_H)_0 \to \dfrac{1}{k'}$ | $\dfrac{RT}{2F}$ |

and

| $\vec{k}_{-2,R\rho H_2} > \vec{k}_{1,R}C_{H^+} > \dfrac{\vec{k}_{2,R}}{k'}C_{H^+}$ $> \dfrac{\vec{k}_{-1,R}}{k'}$ | $\exp\left[(1-\beta)\eta\dfrac{F}{RT}\right]$ | $(C_H) \to \dfrac{1}{k'}$ | |
| $\dfrac{\vec{k}_{-1,R}}{k'} > \dfrac{\vec{k}_{2,R}}{k'}C_{H^+}\vec{k}_{1,R}C_{H^+} > \vec{k}_{-2,R\rho H_2}$ | $i = -2F\vec{k}_{-2,R\rho H_2}$ | $(C_H)_0 \to 0$ | $-\dfrac{RT}{2F}\dfrac{1+\beta}{1-\beta}$ |

and

| $\dfrac{\vec{k}_{-1,R}}{k'} > \vec{k}_{1,R}C_{H^+} > \dfrac{\vec{k}_{2,R}}{k'}C_{H^+}$ $> \vec{k}_{-2,R\rho H_2}$ | $\exp\left[(1-\beta)\eta\dfrac{F}{RT}\right]$ | $(C_H) \to 0$ | |
| $\vec{k}_{1,R}C_{H^+} > \dfrac{\vec{k}_{-1,R}}{k'} > \vec{k}_{-2,R\rho H_2} > \dfrac{\vec{k}_{2,R}C_{H^+}}{k'}$ | $i = -2F\vec{k}_{-2,R}\dfrac{\vec{k}_{-1,R\rho H_2}}{\vec{k}_{1,R}k'C_{H^+}}$ | $(C_H)_0 \to \dfrac{1}{k'}$ | $-\dfrac{RT}{2F}\dfrac{\beta}{2-\beta}$ |

and

| $\vec{k}_{1,R}C_{H^+} > \vec{k}_{-2,R\rho H_2} > \dfrac{\vec{k}_{-1,R}}{k'} > \dfrac{\vec{k}_{2,R}C_{H^+}}{k'}$ | $\exp\left[(3-\beta)\eta\dfrac{F}{RT}\right]$ | $(C_H) \to \dfrac{1}{k'}$ | |

[a] $\vec{k}_1$ and $\vec{k}_{-1}$ are forward and backward rate constants of the discharge reactions $H^+ + e \to H_{ads}$, $H_{ads} \to H^+ + e_0^-$, respectively. $\vec{k}_2$ and $\vec{k}_{-2}$ are the forward and backward rate constants of $H^+ + H_{ads} + e_0^- \to H_2$ and $H_2 \to H^+ + H_{ads} + e_0^-$, respectively.

Complications may occur by the absence of a clear rate-determining step.[101] The predicted relations depend upon the isotherm assumed.[102] An example is given by Johnson, Wroblowa, and Bockris.[103]

Diffusion complicates the situation, as has been treated in a detailed way by Hurd[104] and by Barnart,[105] the time dependence has been treated by Barnart and Johnson.[106]

2.14. Chemical Homogeneous Reactions

Many reactions that are overall electrochemical reactions do have a homogeneous rate-determining step, which may or may not involve an electron transfer. A good example can be quoted by the work of Brdicka,[107] the overall reaction being the reduction of formaldehyde to methanol. The formaldehyde is present in aqueous solutions in a hydrated state, and it turns out that the dehydration must take place before the reaction can occur; this is the rate-determining step. Thus,

$$CH_2O \cdot H_2O \rightarrow CH_2O + H_2O \qquad (2.20)$$

occurs, whereafter the formaldehyde is rapidly converted to methanol by $CH_2O + 2e + 2H^+ \rightarrow CH_2OH$. Such a reaction will give a limiting current, stirring independent. Chemical reactions appear to limit the cathodic production of nitric acid from nitrous acid. The steps are

$$2HNO_2 + 2H^+ \rightarrow 2NO^+ + 2H_2O \qquad (2.21)$$

$$2NO^+ + 2e \rightarrow 2NO \qquad (2.22)$$

$$HNO_3 + 2NO + H_2O \rightarrow 3HNO_3 \qquad (2.23)$$

The rate-determining step is (2.23).

Many equations involving electrochemical reactions coupled to chemical reactions in solution have been described in the diffusion-orientated book by Thirsk and Harrison[108] and are discussed in the polarographic literature.

2.15. Effect of Mass Transport on Electrochemical Reactions

Before the Erdey-Gruz and Volmer equation, much electrochemical kinetics was interpreted in terms of transport control. A vast literature concerns such treatments and is represented in modern times in the book by Thirsk and Harrison.[12] The stress in earlier years on mass transport control was augmented by the large size of the polarographic literature. A general relation that takes into account diffusion and activation control has been given by Bockris[109] and by Barnart.[105]

A large literature also exists on conventional transport, and a well-known book by Levich[110] gives a presentation of a field that has only indirect and operational relevance to the interphasially centred considerations of this book (a brief account is given by Antropov[111]). At sufficiently high current densities, all electrochemical processes become transport controlled (Fig. 2.2).

2.16. Electrode Kinetics as a Function of the Double-Layer Structure

The first equations of electrode kinetics[17] assumed that the concentration of reacting ions in the double layer was proportional to that in the bulk of the solution, and a seminal contribution to electrode kinetics was made in a paper of Frumkin in 1933.[112] Thus, the concentration of reactants to which electrons could be transferred would be related to that in the bulk through a Boltzmannian term involving the potential difference of the layer nearest to the electrode, which differs from that in the bulk by the electrokinetic, or Gouy, or zeta potential.

The electrokinetic potential is dependent both on concentration and on potential. A crude approximation gives $d\zeta/dV = 0$, but these relations assume that one is far from the potential of zero charge. It was shown for the first time by Bockris and Conway[30] (cf. Bockris, Ammar and Huq[31]) that the value of the Tafel slope was affected by the double-layer structure if potential range is sufficiently near the potential of zero charge. Parsons[113] elaborated these relationships for the effect of the double-layer structure on electrode kinetics, without reference to the previous literature.[114] Correspondingly, he took into account the effects of specific adsorption on the electrode kinetic area.

The concept of the presence of potential-dependent water molecules in the electrode[35,37,115,116] has given rise to an interesting concept by Gileadi and Stoner.[117] Reactants have to displace water molecules to become effective on electrodes. This introduces significant effects for large molecules and for charges far removed from the p.z.c.[247] The adsorption isotherm is correspondingly affected and a detailed model is given by Gileadi[118] (cf. Frumkin[119]). Schulze[120] has provided a quantitative basis for much of this with his study of water molecule layers on platinum. Their displacement as a prelude to a number of reactions takes about

23 kcal mole^{-1} (cf. Blomgren and Bockris,[121] who were the first to account in an isotherm for the displacement of water).

It is necessary to know these complications, because they may interfere with simpler ones and lead to an incorrect interpretation of results. For quantum electrochemistry, double-layer effects (as with diffusion) are disturbing complexities in the attempt to obtain basic relations between the rate of charge transfer and the corresponding quantum mechanics.

2.17. Reaction Rates and Isotherms

In basic papers in primitive electrode kinetics, it was assumed that a Langmuir isotherm is always applicable[35,122,123]: the heat of adsorption is independent of coverage. The more general isotherm is

$$\frac{\theta}{1-\theta} = Kc \, \exp[f(\theta)]/RT \tag{2.24}$$

where $f(\theta)$ represents some relation between the heat of adsorption and coverage, the simplest one, and one often chosen, being

$$\Delta H_\theta = \Delta H_0 - A\theta \tag{2.25}$$

where A is a constant.

Such relations have importance for electrode kinetics, because they alter the current–potential relation. This was first shown by Tempkin,[124] e.g., for the hydrogen evolution reaction, with a rate-determining combination reaction; the application of the Langmuir isotherm, with $\theta \ll 1$, is well known to give a relation of the type

$$i = i_0 \exp(2VF/RT) \tag{2.26}$$

but with a Tempkin isotherm (and $0.2 < \theta < 0.8$)

$$i = i_0 \exp(VF/2RT) \tag{2.27}$$

It is essential, therefore, for any basic electrode kinetic mechanism investigation to start off with an isotherm investigation.

There is much Russian work on the relation between electrode kinetic formulations and the isotherm (cf., e.g., Vol'kkovich, Vasil'ev, and Baggotsky[125]). The formation of the isotherm has been specifically considered by Bockris and Swinkels[126] and by Dhar, Conway, and Joshi.[127] The application of such equations to electrode kinetic problems can be seen in Conway and Gileadi.[128] Evaluations involving organic compounds where the compound is able to be in equilibrium with the surface (otherwise the isotherm is not applicable) are given by Wroblowa, Piersma, and

Table 2.2
Forms of a Number of Adsorption Isotherms and the Corresponding Reaction Order Factors[a]

Type of isotherm	Form of isotherm	$\left(\dfrac{\partial \ln \theta}{\partial \ln C}\right)_E$
"Henry's law"	$K(C_R)_b = RT\theta\Gamma_m$	1
Volmer	$K(C_R)_b = RT\dfrac{\theta\Gamma_m}{1-b\theta}\exp\left(\dfrac{b\theta\Gamma_m}{1-b\theta}\right)$	$1+\dfrac{b\theta}{1-b\theta\Gamma_m}+\dfrac{b\theta\Gamma_m}{1-b\theta\Gamma_m}$
Langmuir	$K(C_R)_b = \dfrac{\Gamma}{\Gamma_m-\Gamma}=\dfrac{\theta}{1-\theta}$	$1-\theta$
Flory–Huggins	$K(C_R)_b = \dfrac{\theta}{x(1-\theta)^x}$	$\dfrac{1-\theta}{1+(x-1)\theta}$
Virial	$K(C_R)_b = RT\theta\Gamma_m\exp(-r\theta\Gamma_m)$	$\dfrac{1}{1-r\theta\Gamma_m}$
Frumkin	$K(C_R)_b = \dfrac{\theta}{1-\theta}\exp(-r\theta\Gamma_m)$	$\dfrac{1-\theta}{1-r\theta\Gamma_m+r\theta^2\Gamma_m}$
Temkin	$K(C_R)_b = \exp(-r\theta\Gamma_m)$	$-\dfrac{1}{r\Gamma_m\theta}$

[a] Γ is the surface concentration of adsorbate; Γ_m the surface concentration corresponding to a monolayer at the surface; $K = \exp(-G_{ads}^0/RT)$; b is the molecular area of adsorbate species; θ the degree of coverage of the electrode surface by adsorbate, given by Γ/Γ_m; X the ratio of relative sizes of adsorbate and solvent molecules in the Flory–Huggins type of isotherm; and r the interaction parameter characterizing two-dimensional attraction or repulsion in the adsorbed layer in the virial or Frumkin isotherm or r is a heterogeneity parameter in Temkin's isotherm—units of r depend on how coverage is expressed, i.e., as $\Gamma(=\theta\Gamma_m)$ or in terms of θ.

Bockris.[22,129,130] An excellent example of the application of isotherms in electrochemical kinetics is that of Rudd and Conway.[131] Their table showing the effect of choice of isotherm is given here as Table 2.2.

2.18. Transients (Sweeps)

Baars[132] and, independently, Bowden and Rideal,[133] carried out the first transient (10^{-6} sec) measurements in electrode kinetics in 1928, long before transient measurements, as in flash photolysis, were made in chemical kinetics.† Thus, from early times, it has been recognized that transient measurements were of the essence in electrode kinetic technique. Measurements in a very short period would give advantages because of the

† The Nobel committee apparently remained ignorant of this fact, however, and awarded a Nobel Prize to Norrish and Porter for flash photolysis, and one to Eigen for relaxation methods in chemistry—which was originated 20 years earlier by electrochemists.

negligible change of surface with time. The superior reproducibility of measurements of growing mercury drops, i.e., intrinsically transient situations, stressed this, and the work of Conway and Bockris in the late 1940s made the irreproducibility associated with impurities a further reason for making measurements as rapidly as was rational. Two groups of phenomena may be distinguished in transient behavior:

Case A

Diffusion in solution of reactant, intermediate, or product, exerts rate-controlling influence. Many cases are dealt with by the equations of Sand,[134] and adsorption on the surface, diffusion controlled from solution, was considered by Delahay and Berzins[135] and Blomgren and Bockris[121] (with Delahay and Fike[136] solving a more sophisticated case in the 1950s). Complex possibilities have been derived in recent times by Nicholson and Shain[137] and Saveant and Vianello[138,139] and mechanistic sequences involving diffusion in solution of reactant, intermediate, or product have been summarized by Thirsk and Harrison.[12] At sufficiently high current densities, or with sufficiently high rate constant, there is usually *some* effect of mass transport on the kinetics. This raises difficulties. Rates in the nanosecond range were first measured by Inman, Bockris, and Blomgren.[41] In molten salts, such methods are often necessary. The impedance approach is reviewed by Power,[140] Schuhmann,[141] and most fully by Epelboin and Keddam.[142]

Case B

Time dependence of current at constant potential arises from the change of radical concentration on the electrode surface. Transients involving these were early investigated by Armstrong and Butler,[143] Knorr and Will,[144] and Breiter.[145] The first measurements involving hydrogen desorption from nonnoble metals, where the metal dissolves at the same time as hydrogen is desorbed, were made by Devanathan, Mehl, and Bockris,[146] who devised a double-pulse method, the first pulse reflecting dissolution of hydrogen and metal and the second only of metal.

Potentiostatic sweep measurements are sometimes used to measure the amount of adsorbed material on noble-metal surfaces. However, they involve uncertainties, e.g., the assumption that the material is desorbing by a reaction that involves the number of electrons per molecule corresponding to the undissociated molecule in the adsorbed sites. Such measurements are only fruitful if connected with some auxiliary measurements (radio tracer) that give a check of the assumption.[147]

An analysis of the concentration of intermediates in metal deposition reactions was first carried out by Mehl and Bockris.[148] They obtained the concentration of the intermediates in the deposition of silver in a domain at which the overall deposition reaction was controlled by surface diffusion to crystal-building sites.

A difficulty exists with potentiodynamic sweep techniques that is lessened by the use of potentiostatic and galvanostatic measurements. Thus, if it is desired to measure the kinetics of a reaction, although the analysis of radicals during transient situations is sometimes of value, information on the rate-determining step in the *steady state* needs a transient regime that reflects the sequence of steps with relative rate constants, coverages, etc. as exist in the steady state. This can give false results, as shown by Gileadi, Stoner, and Bockris.[21] These workers found that the Tafel parameters they were apparently determining during the oxidation of ethylene were constant with change of sweep rates, because the situation was influenced by partial diffusion control. It was only by going to an *extremely* slow (i.e., effectively steady state) sweep rate, that they obtained measurements that correspond to the absence of effects by diffusion of the reactant molecule in solution to the electrode.

The interpretation of potentiodynamic measurements is sometimes regarded in an oversimple way. Thus, the sophisticated analyses of Nicholson and Shain[137] are of full value *only* for reaction situations in which *adsorbed* intermediates are in defined states, although this may not be known. In organic reactions, adsorbed intermediates play an important part and take part in rate-determining steps. Then the interpretation of potentiostatic transients is complex and cannot yet be carried out in a general sense. Thus, the theory of potentiodynamic transients has been analyzed by Srinivasan and Gileadi,[149] Hale,[150] and by Stonehart *et al.*[151] When the reaction involves more than two steps, an analytical mathematical interpretation may become impractical (cf. Bockris and Fredlein[152]).

Many applications of potentiodynamic experiments have been made to organic reactions, where electrocatalysis is the objective. Much of this work is lacking in relevance to knowledge of the rate-determining step *in the steady state* (for a change of rate-determining step is expected as the surface coverage changes and the attainment of a steady state in coverage may consume many minutes, although the sweep is often completed in a few seconds).

Experimental artifacts complicate the interpretation of some transient measurements. Thus, it is possible to find an umbrella effect on double-layer measurements; current lines, if impeded in reaching the electrode, introduce impedance that is frequently dependent and can be confused with a variation of capacitance with frequency. Correspondingly, de Levie[154] has drawn attention to the effect of layers of solution between the capillary and metal in capacitance measurements on mercury, although such effects may be avoided, as argued by Bockris *et al.*[155]

De Levie[156] has worked out the effect of the surface roughness of an electrode upon possible frequency variation. Thirsk and Harrison[12] have suggested that the time delay during the galvanostatic transient during the initial stages of deposition of a metal in the presence of a normal concentration of dislocations need not be surface diffusion, but could be spherical diffusion with deposition to edges and corners. There are theoretical reasons[58] with respect to heat of activation for this reaction which indicate that this argument may not always be applicable.

2.19. Electric Equivalent Circuits

The equivalence of electrical circuits to reactions occurring at electrodes was pointed out by Warburg.[157] The first circuit that took into account the impedance offered by the interphasial electron transfer was by Dolin, Erschler, and Frumkin[32,158] in 1940. (The method is usually credited to Randles.[40] Grahame[159] also made an analysis of the Warburg impedance in 1952.)

Equivalent circuits have to be used in electrochemical work with ac bridges.[160] Cases stressing diffusion but neglecting the impedance of the surface reaction have been given by Thirsk and Harrison.[161] The effect of neutral molecule adsorption on impedance has been developed by Lorenz.[162] The development of the impedance analysis is largely due to the pioneering work of Epelboin.[163] It is described by Thirsk and Harrison.[164]

2.20. Electrocatalysis

Although a large (10^{10} times) range of rate constants for one reaction on many substrates was known from Bowden and Rideal,[34] Grubb[165] in 1963 was the first to use the term "electrocatalysis." The exchange current density for a reaction as a function of metal is the criterion. This criterion[84] eliminates work function of the metal but allows a formulation of the relation between rate and the properties of the surface.

A treatment of electrocatalysis was made simultaneously by Conway and Bockris[166] and by Gerischer.[167] † The hydrogen evolution reaction depends upon the bond strengths of hydrogen to the various metals in the following way: When coverage is low and the rate-determining step that of proton transfer, the exchange-current density increased with increase of

---

† The concepts were identical but independently derived. The publications were submitted on 14 and 20 March 1956, respectively. Similar considerations were thereafter published by Krishtalik[172] in 1960 but without recognition of the earlier literature.

the metal–hydrogen bond strength; for most transition metals, with the rate-determining step as electrochemical desorption on a full surface, the i_0 value would decrease with increase of the metal–hydrogen bond strength.

There is no *chemical* electrocatalysis for nonbonding reactions, because there is no bonding. Correspondingly, Damjanovic, Bockris, and Mannan[76] found that the heat of activation for redox reactions does not depend on the metal for noble metals.

Hoare[170] has pointed out that there are electrocatalytic effects for $Fe^{3+} \xrightarrow{e} Fe^{2+}$ on Pt–OH layers. There are certainly changes of rate constant among the noble metals for redox reaction,[76] but the independence of the heat of activation on the substrate and the dependence of ϕ_{2-b} on the absolute Galvani potential difference mean that there will be a dependence of rate on the substrate due to double-layer effects.[76]

The hydrogen evolution reaction on a given metal depends on the Fermi level, chemical bonding, surface defects and internuclear distance. A summary of these dependencies is shown in Fig. 2.3.[171] This reaction, the most studied in electrochemical kinetics, is now undergoing some study of detailed points. Giles, Harrison, and Thirsk[168] have determined the rate constant on edges and corners—not of crystals of noble metals, but of those on mercury. Analogous studies are by Fleischmann and Grenness,[169] who find that the rate-determining step of hydrogen evolution on a ruthenium carbon catalyst is surface diffusion.[36] The hydrogen evolution reaction is the basis of much else, for example, stress corrosion cracking.[156a] Similar electrocatalytic effects exist for oxygen evolution[173,248,249] (Fig. 2.4).

The effect of crystal face for hydrogen does give significant effects.[118] The close-packed planes are several times more active than loose-packed ones.[242] Volcano relations have been established for the electrochemical oxidation of ethylene on noble metals and their alloys.[84]

Studies of electrocatalysis are still primitive. The preexponential factor has not been separated from the exponential factor. The exponential factor needed is the heat of activation,[73] for which fundamental difficulties exist in the measurement. One can only use i_0 as a criterion. That this needs further analysis is shown by Appleby,[173] who illustrates the variation of preexponential and exponential factors, whereas hitherto most variation has been thrown into the exponential heat of activation.

In using alloy surfaces,[234] it is not always the alloy itself that forms the surface, so that some difficulties arise in representing "mean bond strength" for an alloy in a plot, say, rate against a property of the alloy. Thus, with rhodium–platinum surfaces, an oxygen reduction, the surface behaved as though it was either a rhodium or platinum surface.[174] The hydrophilic and hydrophobic nature of the metal surface has much to do with the final rate.[241]

ACID		CATALYSIS	ALLOY SYSTEM
r.d.s. on alloys	r.d.s. on metals		
Proton discharge r.d.s. Rate gets faster going from Au to Pt due to increased M—H bond strength. Finally on Pt, atomic combination rate determining. Proton discharge r.d.s. for Au–Pd alloys also.	Au electro-chemical desorption Pt atomic combination	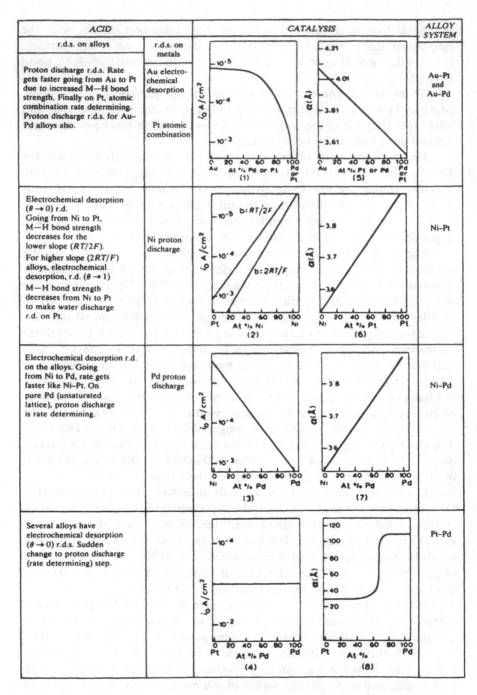	Au–Pt and Au–Pd
Electrochemical desorption ($\theta \rightarrow 0$) r.d. Going from Ni to Pt, M—H bond strength decreases for the lower slope $(RT/2F)$. For higher slope $(2RT/F)$ alloys, electrochemical desorption, r.d. ($\theta \rightarrow 1$) M—H bond strength decreases from Ni to Pt to make water discharge r.d. on Pt.	Ni proton discharge		Ni–Pt
Electrochemical desorption r.d. on the alloys. Going from Ni to Pd, rate gets faster like Ni–Pt. On pure Pd (unsaturated lattice), proton discharge is rate determining.	Pd proton discharge		Ni–Pd
Several alloys have electrochemical desorption ($\theta \rightarrow 0$) r.d.s. Sudden change to proton discharge (rate determining) step.			Pt–Pd

Fig. 2.3. A series of examples of electro-

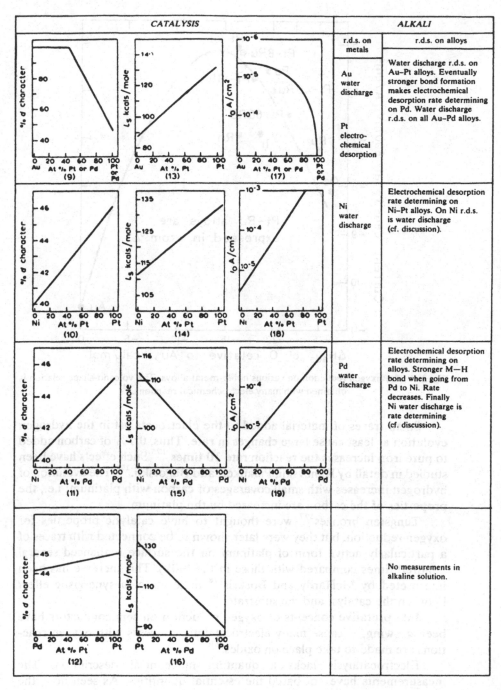

	CATALYSIS		*ALKALI*	
			r.d.s. on metals	**r.d.s. on alloys**
(9)	(13)	(17)	Au water discharge / Pt electrochemical desorption	Water discharge r.d.s. on Au–Pt alloys. Eventually stronger bond formation makes electrochemical desorption rate determining on Pd. Water discharge r.d.s. on all Au–Pd alloys.
(10)	(14)	(18)	Ni water discharge	Electrochemical desorption rate determining on Ni–Pt alloys. On Ni r.d.s. is water discharge (cf. discussion).
(11)	(15)	(19)	Pd water discharge	Electrochemical desorption rate determining on alloys. Stronger M—H bond when going from Pd to Ni. Rate decreases. Finally Ni water discharge is rate determining (cf. discussion).
(12)	(16)			No measurements in alkaline solution.

catalysis—the hydrogen evolution reaction.[171]

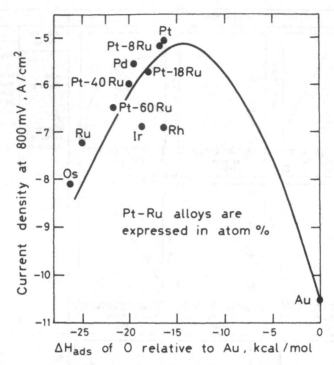

Fig. 2.4. The oxygen reduction on various noble-metal alloys. The volcano-shape relation is obtained with many electrochemical reactions.[230]

Small traces of material added to the electrocatalyst in the hydrogen evolution at least cause large changes in rate. Thus, 0.1% of carbon added to pure iron increases the reaction rate 10 times.[175] Such effects have been studied in detail by Kanevskii, Palanker, and Bagotskii.[176] The evolution of hydrogen increases with small coverages of carbon with platinum; i.e., the properties of the carbon are increased by the platinum.

Tungsten bronzes[177] were thought to have catalytic properties for oxygen reduction, but they were later shown to be connected with traces of a particularly active form of platinum on the surface (enhanced several hundred times compared with those in the bulk). This increase has been interpreted by McHardy and Bockris[178] in terms of a synergistic effect between the catalyst and the substrate.

Interpretative concepts of oxygen reduction on semiconductors have been growing, because many electrocatalytic concepts with oxygen reactions are made to take place on oxides.[236]

Electrocatalysis lacks a quantum mechanical description. The measurements have not bared the essential quantities. As seen now, the determining factors are the interplay of the equilibrium constants of steps that occur before the rate-determining step and the substrate–radical bond strengths of radicals occurring in the rate-determining step.[94] Conversely, the subject remains one of the most challenging and community-useful aspects of electrochemistry because of its relation to the economics of fuel cells.[221]

2.21. Mechanisms

Whether the determination of the mechanism of electrochemical reactions is possible depends on what is meant by "determination." If it means the evaluation of each of the consecutive steps past the rate-determining step, determination is possible in a practical sense only for simple reactions.[179]

If by mechanism determination is meant the evaluation of the rate-determining step, often with an indication of some aspects of the path, this can be done even for fairly complex organic reactions, because the rate-determining step there seems to occur within the first few steps of the reaction sequence. Some account of the methodology of a determination of electrochemical reaction mechanisms has been given by Bockris and Srinivasan.[181] A discussion of the extent to which the determination of a complex mechanism is possible is given by Despic.[182]

Determinations of reaction mechanisms are more difficult for fast reactions, where diffusion in solution interferes with the surface kinetics. This (arbitrary) criterion restricts attempts at mechanism determination to those of relatively slow reactions.

Many methods used in the determination of electrochemical reaction mechanisms are not electrodic, e.g., isotope ratio measurements. Reactions do not have a unique mechanism under change of ambient conditions. Nevertheless, gradually, generalizations become clear. "Tempkin" adsorption situations are prevalent with organic oxidation. Chemical rate-determining steps between adsorbed radicals, rather than those in solution, are important in organic oxidation; the combination of an adsorbed organic radical, arising from the chemisorptive and dissociated description of the entity in solution as the product of the discharge of water molecules, is often rate determining.

The unambiguous determination of an organic mechanism is often involved, because the part played by products of side reactions that are on the surface but do not take part in the reaction has often led to confusion.[183,184] The study of the surface radicals is of importance, but only insofar as it leads to knowledge of the rate-determining step. Occasionally, the all-radical approach leads to clarity in the latter goals, as with the work of Taylor, Pearce, and Brummer[185] on $HCOOH \rightarrow CO_2$.

The determination of all the rate constants (not only that of the rate-determining step) is becoming possible. Some approaches are discussed by Bewick and Thirsk.[186] The situation of the H_2O_2 intermediate in O_2 reduction has been evaluated—with intermediate rate constants—particularly by Damjanovic, Genshaw, and Bockris,[187] Hoare,[188] and Terasevich.[189] Bagotskii has developed such multirate constant determinations.[190]

2.22. Electrocrystallization

The first step toward modern work in the study of the kinetics of electrocrystallization was taken by Volmer,[191] who pointed out the necessity of the use of transients in metal-deposition kinetics because of the intrinsically changing nature of the substrate. There are two kinds of transients in metal deposition of M^{2+} onto M. In one kind, a maximum appears on the potential–time relation. Bockris and Razumney[192] gave a detailed account of such curves. They are due to the adsorption of impurities at growth sites. While they are frequent, overpotential grows abnormally, primarily because there is no possibility of growth sites accepting adions. When the overpotential conditions are such that the critical radius for the growth of a step has been surpassed, the step advances around the absorbed impurity.

Another characteristic phenomenon is the growth of spirals, predicted by Franck in 1949[193] and observed in electrodeposition by Steinberg in 1952.[194] Adsorption has a particularly great importance in determining electrocrystallization as opposed to any other electrochemical reaction. The rate of passage of steps across a surface and their height, together with many resultant crystal building phenomena, are dependent on bunching,[195] which arises from impurities that block the spreading of monatomic steps and make them bunch up to macrosteps about 1000 atoms high. One of the complications of examining the mechanisms of metal deposition arises from the fact that many metal cations are complexed in solution.

Budewski[196] and his colleagues[197,198] have published much beautiful work in an effort to establish Volmer's original theory of the rate of new phase formation and nucleation when a metal M is deposited upon some *other* metal M^1. The model was suggested before the discovery of dislocations and Franck's theory of the growth of crystals in terms of the rotation of screw dislocations to form spirals (it was first adduced as a mechanism for M^+ on M). Volmer's equation for rate-determining nucleation was first verified by Kaischew and Mustafchiew in 1966.[197] However, in actuality, metal nucleation has taken attention away from the more practical line of work in the field, the mechanism of the deposition of M upon

M under circumstances in which there are dislocations on the surface. The growth of a plane in terms of two-dimensional nucleation theorized by Volmer was shown to be possible by Budewski in 1966.[196]

The studies made by Kaischew, Budewski, *et al.*[196-198] of nucleation during deposition on a *foreign* substrate have been shown to have some relevance to real deposition if the surface is sufficiently covered by inhibitors. Then, nucleation of M on itself will begin as shown by Fischer and his co-workers.[199-201]

Correspondingly, Fleischmann and Thirsk and their collaborators have published sophisticated studies of crystal growth of various layers of oxides on metals, e.g., the growth of oxalates on mercury.[202] These contributions, particularly the methods used, will be worthily coupled at a future time with studies of the deposition of a metal on to itself, the case to which attention is of greatest use. Computer simulation studies have valuably been used by Fleischmann and his co-workers[203] in the study of nucleation phenomena in the deposition of a metal on a foreign substrate.†

2.23. Steps before Crystal Growth

A question in the deposition of M^+ upon M was whether the particles deposited at growth sites on the surface *directly* from solution ($M^+_{D.L.} \rightarrow M^+_{growth\ site}$) or whether the particles arise from solution and diffuse across the surface of the metal. This has been decided by transient studies at various overpotentials for new metals such as silver, copper, gallium, etc. When the overpotential is low, there tends to be a step that is rate-determining after the charge transfer between the OHP and the surface of the metal, and this is probably surface diffusion.[93,148,204,205] One of the determining aspects behind this conclusion lies in the calculations made by Conway and Bockris,[206] which suggest that the distortion of the solvation sheath at sites of landing other than the planar sites would give rise to a much enhanced heat of activation for the deposition step.[57] Two electron transfers seem to be energetically unlikely. The single electron transfer to planar surfaces—theoretically decided upon—is fully consistent with the later verification of the surface-diffusional path.[239]

It is possible[207] to calculate the adion concentration on the surface of dissolving and depositing metals under conditions in which the surface-diffusion step is rate determining as a function of distance from the step. Such studies indicate that, when surface diffusion is rate determining, deposition is *near* a growth site, though not *on* a growth site.

† These studies of deposition are accompanied by corresponding dissolution studies, e.g., those on iron, particularly by Beck-Nielson.[235]

The activity of the surface as a function of time is of interest; fresh (<0.1 seç old) surfaces have faster k values which become constant after 1 sec. The adions and their equilibration on crystal faces is a possible interpretation.[208]

The change in activity of a surface depends partly upon the effect of potential upon the equilibrium radius for rotation of growth spirals upon the surface.[180] Effectively, therefore, only a small number of growth sites are available for deposition when the potential is low, but this concentration increases as the potential is removed further from that of equilibrium, and in this way deviations from the normal Tafel behavior at low current densities are interpretable.

2.24. Crystal Growth

The first theory of crystallization which allowed absolute calculation of the rate is that of Barton and Bockris.[209] They measured the rate as a function of potential for the growth of dendrites on silver surfaces. The key point on the start of the growth of dendrites was the penetration of crystals through the diffusion layer associated with the substrate but with a radius that introduced spherical rather than linear diffusion. The crystal tip thus favored grew faster than other elements of crystal growth on the surface. Later, Diggle, Damjanovic, and Bockris[210] showed that the determining factor in the development of a tip was the rotation of spirals on the surface of the base model, itself potential dependent.

Knowledge of the mechanistic aspects of morphology is rudimentary.[242] The potential dependence of growth forms has a good phenomenology. It depends upon the adion concentration between growth steps, as a function of overpotential, the fraction of the steps that are active at a given overpotential, the surface concentration of different species in solution, anions and inorganic substances, and then the potential-dependent surface free energy of the elements on the surface.

A great effect upon the morphology of metal deposits is shown[243] by tiny traces of impurities on the surface, and these will make a deposit different if only present at a concentration of 10^{-8} mole/liter.[211]

The mechanism of leveling and brightening of surfaces has been worked out in recent years by Kardos[212] and others.[213] Leveling depends on an activation-controlled deposition in the presence of coverage of part of the surface by a leveling agent. The degree of coverage of this deposition is controlled by diffusion. Then surface coverage at recesses will be smaller than at peaks if the surface irregularities are small enough not to have an effect upon the outer profile of the diffusion layer. Thus, the primary current distribution will be changed in a way that causes electrochemical leveling. Mechanisms of bright electrocrystallization were reviewed by Bockris and Razumney.[214]

A fundamental advance in the treatment of inhibitors[215] has been made by Gileadi,[216] who has treated the effect of adsorbing phenol on the rate of oxidation of Br⁻ and I⁻ on a platinum electrode. He is able to interpret the observations by accounting for the effect of water dipoles on the surface and their change with potential, the first quantitative treatment of inhibition on electrode surfaces.[246]

2.25. Interphasial Charge Transfer in Engineering, Metallurgy, and Biology

Processes at electrodes have applicability to many branches of science and technology. Thus, in engineering, they give the molecular-level basis to the building of energy converters and storers.[1] For metallurgy, there are phenomena connected with corrosion and the stability of metals that are treated in electrode-kinetic terms.[10,217,218] Biochemical topics are beginning to receive discussion in the electrodic sense.[219,220] A basis for the electrochemical understanding of mechanisms involving ATP and ADP has been presented.[220] The interphasial transfer of charge from the material phase in the biochemical system (regarded as a semiconductor electrode) to ions in solution is an important aspect of biological processes.[5]

2.26. Techniques of Study

In complex electrochemical reactions, it is necessary to determine the number of electrons consumed in one act of the stoichiometric reaction; in simple reactions, this is often known. One must then measure the reaction rate at constant potential or overpotential as a function of the concentration of several of the reactants or products, respectively. If such studies can be made both in the cathodic and anodic direction, it is helpful.[233] A very widespread use of rotating disk electrodes is now made, not only to distinguish from activated steps but also to make distinction between reaction paths.[237]

Determination of the rate as a function of potential provides a method of determining the rate constant and the transfer coefficient, which can be auxiliary in the determination of the stoichiometric number.† Correspond-

---

† The contributions of differential and deviational aspects of electrochemical kinetics may contribute much in the future. In this respect, the treatments of noise given by Barker[222] is promising.

ingly, the determination of rate near the reversible potential allows one to obtain a measure of the stoichiometric number. These measurements are called steady-state measurements, although the technique of making them is to look at the transient behavior, usually at constant potential, from which the steady-state behavior is picked out. This avoids doubt-sowing contributions due to unstable surfaces.

Both potentiostatic and galvanostatic measurements can be helpful in establishing radical concentrations on the surface, which may help with elucidation of maxima.[153] Other transient techniques give information on radicals,[223] and hence this helps with the evolution of the reaction mechanism so long as other information needed is available. Directly it is clear that radicals *in solution* take part in electrode reactions, electron spin resonance becomes important and Bonnemay and Lamy[224] and Lamy and Malaterre[225] have made exploratory contributions. Thus, it is possible to study such matters as the nature of reaction sites on platinized carbon.[225]

Mechanism determination by means of separation factor studies is often determinative, particularly for reactions involving ^{238}H. Techniques that aid the study of mechanism involve optical techniques, (e.g., interference studies[226,211]) and ellipsometry,[245] which allows determinations of the characteristics of fractions of monolayers of films and adsorbed ions. This technique is helpful on metals that would dissolve under transient conditions. Optical studies which involve quantum efficiency calculations indicate energy level participation.[224] Mössbauer studies of thin films have recently become possible.[227]

Other techniques for looking at electrode surfaces are incipient, e.g., Raman spectra can be used for surface studies;[228] Auger and ESCA can be used for obtaining analysis of the electronic states of the surface.[229] The general techniques of dealing experimentally with electrodes, particularly those in which crystal growth is observed, have been dealt with by Damjanovic and Bockris,[230] McHardy,[231] and (at an instrumentation level) by Cahan.[232] Automation of a number of measurements, e.g., those of impedance, is under development.[248]

References

1. J. O'M. Bockris and S. Srinivasan, *Fuel Cells: Their Electrochemistry*, McGraw–Hill, New York (1970).
2. J. O'M Bockris, *Energy Conversion* **10**, 41 (1970).
3. J. Turkevich, private communication (1977).
4. F. W. Cope, *Arch. Biochem. Biophys.* **103**, 352 (1963).
5. L. Mandel, *Modern Aspects of Electrochemistry*, Volume 8 (J. O'M. Bockris and B. E. Conway, eds.), Plenum Press, New York (1972).
6. J. O'M. Bockris, *Nature* **224**, 775 (1969).

7. J. O'M. Bockris and D. Drazic, *Electrochemical Science*, Taylor and Francis, London (1972).
8. J. O'M. Bockris and S. Srinivasan, *Nature* 215, 197 (1967).
9. H. A. Liebhafsky and E. J. Cairns, *Fuel Cells and Fuel Cell Batteries*, John Wiley and Sons, New York, 1968.
10. C. Fontana and N. Green, *Corrosion Engineering*, McGraw–Hill, New York (1967).
11. T. Erdey-Gruz, *Kinetics of Electrode Processes*, Hilger, London (1972).
12. H. R. Thirsk and J. A. Harrison, *A Guide to the Study of Electrode Kinetics*, Academic Press, London/New York (1972).
13. E. Gileadi and G. Stoner, *J. Electroanal. Chem.* 36, 493 (1972).
14. J. O'M. Bockris, *J. Electroanal. Chem.* 36, 495 (1972).
15. E. K. Rideal, *Concepts in Catalysis*, Academic Press, New York (1969).
16. J. O'M. Bockris, *J. Chem. Educ.* 48, 352 (1971).
17. T. Erdey-Gruz and M. Volmer, *Z. Phys. Chem.* 150, 203 (1930).
18. L. Antropov, *Theoretical Electrochemistry*, pp. 343–386, Mir, Moscow (1972).
19. J. Tafel, *Z. Phys. Chem.* 50, 641 (1905).
20. K. Muller, *J. Res. Inst. Catalysis, Hokkaido Univ.* 14, 224 (1967).
21. E. Gileadi, G. Stoner, and J. O'M. Bockris, *J. Electrochem. Soc.* 113, 585 (1966).
22. H. Wroblowa, B. Piersma, and J. O'M. Bockris, *J. Electroanal. Chem.* 6, 40 (1963).
23. A. R. Despic and J. O'M. Bockris, *J. Chem. Phys.* 32, 389 (1960).
24. R. Parsons and E. Passeron, *J. Electroanal. Chem.* 12, 524 (1966).
25. J. O'M. Bockris and A. M. Azzam, *Trans. Faraday Soc.* 48, 145 (1952).
26. H. Nurnberg, *Studien Min. Modernen Technik, z. kinetik schneller Chem. u. Electrochem.*, chapter on *Schnitte von Protonen Ubergang Prozessen*, Verlag Chemie, Bonn (1968).
27. J. Appleby, J. O'M. Bockris, and R. K. Sen, *M.T.P. Volume on Electrochemistry*, Chapter 1, M.T.P., Butterworths, London (1973).
28. J. O'M. Bockris and A.K.M.S. Huq, *Proc. Roy. Soc.* A237, 277 (1956).
29. A. Damjanovic, A. Dey, and J. O'M. Bockris, *Electrochim. Acta* 11, 791 (1966).
30. J. O'M. Bockris and B. E. Conway, *Trans. Faraday Soc.* 48, 724 (1952).
31. J. O'M. Bockris, I. A. Ammar, and A.K.M.S. Huq, *J. Phys. Chem.* 61, 879, (1957).
32. P. Dolin, B. V. Erschler, and A. N. Frumkin, *Acta Physiochim. U.S.S.R.* 13, 779 (1940).
33. M.A.V. Devanathan and M. Selvaratnam, *Trans. Faraday Soc.* 56, 1820 (1960).
34. F. P. Bowden and E. Rideal, *Proc. Roy. Soc. (London)* 120A, 59 (1928).
35. J. O'M. Bockris and E. C. Potter, *J. Chem. Phys.* 20, 614 (1952).
36. J. O'M. Bockris, N. Pentland, and E. Sheldon, *J. Electrochem. Soc.* 104, 182 (1957).
37. J. O'M. Bockris and E. C. Potter, *J. Electrochem. Soc.* 99, 169 (1952).
38. K. Vetter, *Elektrochemische Kinetik*, p. 329, Springer-Verlag, Frankfurt (1961).
39. M. Enyo, *Modern Aspects of Electrochemistry* (J. O'M. Bockris and B. E. Conway, eds.), Volume 11, Chapter 6, Plenum Press, New York (1975).
40. J. E. B. Randles, *Discuss. Faraday Soc.* 1, 11 (1947).
41. E. A. Blomgren, D. Inman, and J. O'M. Bockris, *Rev. Sci. Inst.* 32, 11 (1961).
42. J. O'M. Bockris, E. Gileadi, and K. Muller, *J. Chem. Phys.* 47, 2510 (1967).
43. K. Doss and H. Agarwal, *Proc. Indian Acad. Sci.* 354, 45 (1952).
44a. G. C. Barker, *Trans. Symp. Electrode Proc., Philadelphia, 1959*, John Wiley and Sons, New York (1961).
44b. K. B. Oldham, *Trans. Faraday Soc.* 53, 50 (1957).
45. K. H. von Hamann, *Z. Elektrochem.* 71(6), 612 (1967).
46. J. O'M. Bockris, *Modern Aspects of Electrochemistry* (J. O'M. Bockris, ed.), Volume 1, Chapter 4, p. 236, Butterworths, London (1954).
47. R. A. Marcus, *J. Chem. Phys.* 43, 679 (1965).

48. J. O'M. Bockris and D. B. Matthews, *Proc. Roy. Soc.* **A292**, 479 (1966).
49. R. R. Dogonadze, in *Reactions of Molecules at Electrodes* (N. S. Hush, ed.), p. 135, Wiley, New York (1971).
50. J. O'M Bockris and A. K. N. Reddy, *Modern Electrochemistry*, Plenum Press, New York (1970).
51. A. Essin and K. Kosheurov, *Acta Physiochim. URSS* **16**, 169 (1942).
52. R. Parsons and J. O'M. Bockris, *Trans. Faraday Soc.* **47**, 914 (1959).
53. B. E. Conway and J. O'M. Bockris, *Can. J. Chem.* **35**, 1124 (1957).
54. J. O'M. Bockris, R. K. Sen, and B. E. Conway, *Nature* **240**, 143 (1972).
55. D. B. Matthews and J. O'M. Bockris, *Modern Aspects of Electrochemistry*, Volume 6, Chapter 4, Plenum Press, New York (1971).
56. Z. Nagy and J. O'M. Bockris, *J. Chem. Educ.* **50**, 839 (1973).
57. B. E. Conway and J. O'M. Bockris, *Proc. Roy. Soc.* **A248**, 1394 (1958).
58. B. E. Conway and J. O'M. Bockris, *Electrochim. Acta.* **3**, 340 (1961).
59. M. H. Miles and H. Gerischer, *J. Electrochem. Soc.* **118**, 837 (1971).
60. J. Horiuti and M. Ikusima, *Proc. Imper. Acad. (Tokyo)* **15**, 39 (1939).
61. E. C. Potter, Ph.D. Thesis, University of London, London (1950).
62. K. Vetter, *Elektrochemische Kinetik*, p. 206, Springer-Verlag, Frankfurt (1961).
63. A. C. Makrides, *J. Electrochem. Soc.* **104**, 677 (1957); **109**, 256 (1962).
64. B. E. Conway and M. Salomon, *Electrochim. Acta* **9**, 1599 (1964).
65. A. Damjanovic and A. Dey, *Electrochim. Acta* **11**, 791 (1966); *J. Electrochem. Soc.* **113**, 739 (1966).
66. T. Erdey-Gruz, *Kinetics of Electrode Processes*, p. 209, Hilger, London (1972).
67. R. Parsons, *Trans. Faraday. Soc.* **47**, 1332 (1951).
68. K. Vetter, *Z. Phys. Chem.* **194**, 284 (1949–50); **195**, 270 (1950); **195**, 337 (1950); **196**, 360 (1950); **199**, 285 (1952); **199**, 300 (1952); **199**, 22 (1952); *Z. Electrochem.* **55**, 121 (1951).
69. R. Parsons, private communication (1977).
70. R. Parsons, *Modern Aspects of Electrochemistry* (J. O'M. Bockris, ed.), Volume 1, Chapter 3, p. 103, Butterworths, London (1954).
71. J. O'M. Bockris and S. Srinivasan, *Fuel Cells: Their Electrochemistry*, pp. 469–488, McGraw–Hill Book Company, New York (1970).
72. J. N. Agar, *Discuss. Faraday Soc.* **1**, 81 (1947).
73. B. E. Conway and D. J. MacKinnon, *J. Electrochem. Soc.* **116**, 1665 (1969).
74. B. E. Conway, D. J. MacKinnon, and V. B. Tilak, *Trans. Faraday Soc.* **66**, 1203 (1970).
75. L. Antropov, *Theoretical Electrochemistry*, p. 394, Mir, Moscow (1972).
76. J. O'M. Bockris, R. J. Mannan, and A. Damjanovic, *J. Chem. Phys.* **48**, 1898 (1968).
77. H. Baxendale, *Discuss. Faraday Soc.* **1**, 46 (1947).
78. J. O'M. Bockris, *Modern Aspects of Electrochemistry* (J. O'M. Bockris, ed.), Volume 1, Chapter 4, Plenum Press, New York (1954).
79. M. Temkin, *Z. Phys. Khim.* **15**, 296 (1941).
80. J. O'M. Bockris, R. Parsons, and H. Rosenberg, *J. Chem. Phys.* **18**, 762 (1950).
81. J. O'M. Bockris, S. Srinivasan, and D. B. Matthews, *Discuss. Faraday Soc.* **39**, 239 (1965).
82. B. Dandapani and M. Fleischmann, *J. Electroanal. Chem.* **39**, 315 (1972); **39**, 323 (1972).
83. M. Salomon, *J. Electrochem. Soc.* **114**, 922 (1967).
84. A. T. Kuhn, H. Wroblowa, and J. O'M. Bockris, *Trans. Faraday Soc.* **63**, 1458 (1967).
85. J. O'M. Bockris and S. Srinivasan, *J. Electrochem. Soc.* **111**, 844 (1964); **111**, 853 (1964); **111**, 858 (1964).

86. J. O'M. Bockris and R. K. Sen, unpublished; work described by J. Appleby, J. O'M. Bockris, B. E. Conway, and R. K. Sen, in M.T.P., Vol. 6, Butterworths, London (1973).
87. J. O'M. Bockris and A. K. N. Reddy, *Modern Electrochemistry*, pp. 994–1002; Plenum Press, New York (1970).
88. V. S. Baggotskii and Y. V. Vasil'ev, *Electrochim. Acta* 9, 869 (1964); 12, 1323 (1967).
89. A. J. Bard and K. S. V. Santhanam, *Electroanalytical Chemistry*, Volume 4 (A. J. Bard, ed.), Dekker, New York (1970); D. H. Geske and A. J. Bard, *J. Phys. Chem.* 63, 1057 (1959); A. J. Bard, *Anal. Chem.* 40, 64R (1968); A. J. Bard and J. S. Mayell, *J. Phys. Chem.* 66, 2173 (1962); A. J. Bard and E. Solon, *J. Phys. Chem.* 67, 2326 (1967).
90. H. R. Thirsk and J. A. Harrison, *A Guide to the Study of Electrode Kinetics*, p. 103, Academic Press, New York (1972).
91. L. Antropov, *Theoretical Electrochemistry*, pp. 469–476, Mir, Moscow (1972).
92. J. O'M. Bockris, E. Gileadi, and K. Muller, *Electrochim. Acta*. 12, 1301 (1967).
93. J. O'M. Bockris and E. Mattsson, *Trans. Faraday Soc.* 55, 1586 (1959).
94. H. Mauser and J. O'M. Bockris, *Can. J. Chem.* 37, 475 (1959).
95. J. O'M Bockris, D. Drazic and A. Despic, *Electrochim. Acta* 4, 325 (1961).
96. F. Hilbert, Y. Miyoshi, G. Eichkorn, and W. J. Lorenz, *J. Electrochem. Soc.* 118, 1919 (1971).
97. G. Eickhorn, W. J. Lorenz, L. Albert, and H. Fischer, *Electrochim. Acta* 13, 183 (1968).
98. F. Hilbert, Y. Miyoshi, G. Eichkorn, and W. J. Lorenz, *J. Electrochem. Soc.* 118, 1927 (1971).
99. I. Epelboin and R. Wiart, *J. Electrochem. Soc.* 118, 1577 (1971).
100. J. O'M. Bockris and A. K. Reddy, *Modern Electrochemistry*, Plenum Press, New York (1973).
101 R. Parsons, *Trans. Faraday Soc.* 56, 1340 (1960).
102. B. E. Conway, *M.T.P. Volume on Electrochemistry* (J. O'M. Bockris, ed.) Volume 6, M.T.P., London (1972).
103. J. W. Johnson, H. Wroblowa, and J. O'M. Bockris, *Electrochim. Acta* 9, 639 (1964).
104. R. M. Hurd, *J. Electrochem. Soc.* 109, 327 (1962).
105. S. Barnart, *Electrochim. Acta* 11, 1531 (1966).
106. S. Barnart and C. A. Johnson, *J. Phys. Chem.* 71, 4430 (1967).
107. R. Brdicka, *Z. Elektrochem.* 64, 16 (1960); see also L. Antropov, *Theoretical Electrochemistry*, p. 339, Mir, Moscow (1972).
108. H. R. Thirsk and J. A. Harrison, *A Guide to the Study of Electrode Kinetics*, p.25, Academic Press, New York (1972).
109. J. O'M. Bockris, *Modern Aspects of Electrochemistry* (J. O'M. Bockris, ed.), Volume 1, Chapter 4, p. 190, Butterworths, London (1954).
110. V. G. Levich, *Physico-chemical Hydrodynamics*, Prentice-Hall, Englewood Cliffs, New Jersey (1962).
111. L. Antropov, *Theoretical Electrochemistry*, p. 333, Mir, Moscow (1972).
112. A. N. Frumkin, *Z. Phys. Chem.* 164A, 121 (1933).
113. R. Parsons, in *Advances in Electrochemistry and Electrochemical Engineering* (P. Delahay and C. W. Tobias, eds.), Interscience, New York, p. 176 (1961).
114. H. R. Thirsk and J. A. Harrison, *A Guide to the Study of Electrode Kinetics*, p. 14, Academic Press, New York (1972).
115. N. F. Mott and R. J. Watts-Tobin, *Electrochim. Acta* 4, 79 (1961).
116. J. O'M. Bockris, M. A. V. Devanathan, and K. Muller, *Proc. Roy. Soc.* A274, 55 (1963).
117. E. Gileadi and G. Stoner, *J. Electrochem. Soc.* 118, 1318 (1971).
118. E. Gileadi, *Israel J. Chem.* 9, 405 (1971).
119. A. N. Frumkin, *J. Res. Inst. of Catalysis Hokkaido Univ.* 15(1), 61 (1967).
120. J. W. Schultze, *J. Elektrochem.* 73(5), 483 (1969).

121. E. Blomgren and J. O'M. Bockris, *J. Phys. Chem.* **63**, 1475 (1959).
122. A. N. Frumkin, *Acta Physicochem. U.R.S.S.* **7**, 474 (1937); **18**, 23 (1943).
123. A. N. Frumkin, *Faraday Society Discussion on Electrode Processes*, p. 57 (1947).
124. M. Temkin, *Z. Phys. Khim.* **15**, 296 (1941).
125. Y. M. Vol'kkovich, Y. B. Vasil'ev, and N. S. Baggotskii, *Elektrokhim.* **5**, 1461 (1961).
126. J. O'M. Bockris and D. A. J. Swinkels, *J. Electrochem. Soc.* **111**, 736 (1964).
127. H. Dhar, B. E. Conway, and K. Joshi, *Electrochem. Acta* **18**, 789 (1973).
128. B. E. Conway and E. Gileadi, *Modern Aspects of Electrochemistry* (J. O'M. Bockris and B. E. Conway, eds.), Plenum Press, New York (1969).
129. J. O'M. Bockris, H. Wroblowa, E. Gileadi, and B. Piersma, *Trans. Faraday Soc.*, **61**, 2531 (1965).
130. H. Wroblowa, J. W. Johnson, and J. O'M. Bockris, *J. Electrochem. Soc.* **111**, 863 (1964).
131. B. E. Conway and E. J. Rudd, *Trans. Faraday Soc.* **67**, 440 (1971).
132. E. Baars, *Sitzber. ges. Beforder. Naturw. Marburg* **63**, 213 (1928).
133. F. P. Bowden and E. Rideal, *Proc. Roy. Soc.* **A120**, 80 (1928).
134. J. S. Sand, *Phil. Mag.* **1**, 45 (1900).
135. T. Berzins and P. Delahay, *J. Amer. Chem. Soc.* **75**, 55 (1953); **75**, 2486 (1953); **75**, 4205 (1953).
136. P. Delahay and C. T. Fike, *J. Am. Chem. Soc.* **80**, 2628 (1958).
137. R. S. Nicholson and I. Shain, *Anal. Chem.* **36**, 706 (1964).
138. J. M. Saveant and E. Vianello, *Electrochim. Acta* **12**, 629 (1967).
139. J. M. Saveant, *Electrochim. Acta* **12**, 99 (1967).
140. P. D. Power, *Soc. Electro-chem.* July 1969, p. 3.
141. D. Schuhmann, *J. Electroanal. Chem.* **17**, 45 (1968).
142. I. Epelboin and M. Keddam, *J. Electrochem. Soc.* **117**, 1052 (1970).
143. R. Armstrong and J. A. Butler, *Trans. Faraday Soc.* **29**, 1261 (1933).
144. F. G. Will and C. A. Knorr, *Z. Elek.* **64**, 258 (1960).
145. M. W. Breiter, *Electrochim. Acta* **8**, 925 (1963).
146. M. A. V. Devanathan, W. Mehl, and J. O'M. Bockris, *J. Electroanal. Chem.* **1**, 143 (1960).
147. E. Gileadi, L. Duic, and J. O'M. Bockris, *Electrochim. Acta* **13**, 1915 (1968).
148. W. Mehl and J. O'M. Bockris, *J. Chem. Phys.* **27**, 817 (1957); *Can. J. Chem.* **37**, 190 (1959).
149. S. Srinivasan and E. Gileadi, *Electrochim. Acta* **11**, 321 (1966).
150. J. M. Hale, *J. Electroanal. Chem.* **6**, 187 (1963); **8**, 332 (1964).
151. P. Stonehart, H. A. Kozlowska, and B. E. Conway, *Proc. Roy. Soc. (London)*, **A310**, 541 (1969).
152. J. O'M. Bockris and R. Fredlein, *Workbook of Electrochemistry*, p. 126, Problem 9, Plenum Press, New York (1973).
153. J. O'M. Bockris and G. Razumney, *J. Electroanal. Chem.* **46**, 185 (1973).
154. R. de Levie, *J. Chem. Phys.* **47**, 2509 (1967).
155. J. O'M. Bockris, E. Gileadi, and K. Muller, *J. Chem. Phys.* **47**, 2510 (1967).
156. R. de Levie, in *Advances in Electrochemistry* (P. Delahay and C. W. Tobias, eds.), Vol. 6, pp. 329–397, Interscience, New York, (1967).
156a. H. J. Flitt, R. W. Revie, and J. O'M. Bockris, *Austral. J. Corrosion* **1**, 4 (1976).
157. E. Warburg, *Ann. Physik.* **67**, 493 (1899).
158. P. Dolin and V. B. Erschler, *Acta Physicochem., U.R.S.S.* **13**, 747 (1940); *J. Phys. Chem. (U.S.S.R.)* **14**, 907 (1940).
159. D. C. Grahame, *J. Electrochem. Soc.* **99**, C370 (1952).
160. H. R. Thirsk and J. A. Harrison, *A Guide to the Study of Electrode Kinetics*, p. 75, Academic Press, New York (1972).

161. H. R. Thirsk and J. A. Harrison, *A Guide to the Study of Electrode Kinetics*, p. 53, Academic Press, New York (1972).

162. W. Lorenz, *Z. Elektrochem.* **62**, 192 (1958).

163. I. Epelboin, M. Keddam, and P. L. Morel, *Proc. Int. Congr. Metallic Corrosion, Moscow, 1966*, **3**, 110 (1966).

164. H. R. Thirsk and J. A. Harrison, *A Guide to the Study of Electrode Kinetics*, p. 73, Academic Press, New York (1972).

165. W. T. Grubb, *Low Temperature Hydrocarbons*, presented at the 17th Annual Power Sources Conference, Atlantic City, New Jersey (1963).

166. B. E. Conway and J. O'M. Bockris, *Nature* **48**, 178 (1956).

167. H. Gerischer, *Z. Phys. Chem.* **8**, 137 (1956).

168. R. D. Giles, J. A. Harrison, and H. R. Thirsk, *J. Electroanal. Chem.* **20**, 47 (1969).

169. M. Fleischmann and G. Grenness, *J. Chem. Soc. Faraday Trans.* **68**, 2305 (1972).

170. J. P. Hoare, *Electrochim. Acta* **17**, 1907 (1972).

171. A. Damjanovic, J. O'M. Bockris, and R. Mannan, *J. Electroanal. Chem.* **18**, 349 (1968).

172. L. I. Krishtalik, *Russ. J. Phys. Chem.* **34**, 53 (1960).

173. J. Appleby, *Modern Aspects of Electrochemistry* (J. O'M. Bockris and B. E. Conway, eds.), Volume 9, Plenum Press, New York (1973).

174. A. Damjanovic and J. O'M. Bockris, *Electrochim. Acta* **11**, 376 (1966).

175. J. O'M. Bockris and A. Drazic, *Electrochim. Acta* **7**, 293 (1962).

176. L. S. Kanevskii, V. Sh. Palanker, and V. S. Bagotskii, *Electrokhim.* **6**, 271 (1970).

177. A. Damjanovic, D. B. Sepa, and J. O'M. Bockris, *Electrochim. Acta* **12**, 746 (1967).

178. J. McHardy and J. O'M. Bockris, *J. Electrochem. Soc.* **120**, 53 (1973); **120**, 61 (1973).

179. T. Erdey-Gruz, *Kinetics of Electrode Processes*, p. 144, Hilger, London (1972).

180. H. Kita, M. Enyo, and J. O'M. Bockris, *Can. J. Chem.* **39**, 1670 (1961).

181. J. O'M. Bockris and S. Srinivasan, *Fuel Cells: Their Electrochemistry*, pp. 469–508, McGraw-Hill, New York (1970).

182. A. R. Despic, *Bull. Acad. Serbe Sci. Arts 1969, Classe Sci. Math. Nat. (Belgrade)* **46**(12), 79.

183. A. H. Taylor, S. Kirkland and S. B. Brummer, *Trans. Faraday Soc.* **67**(Pt. 3), 819 (1971).

184. M. W. Breiter, *J. Electroanal. Chem.* **19**, 131 (1968).

185. A. H. Taylor, R. D. Pearce, and S. B. Brummer, *Trans. Faraday Soc.* **67**(Pt. 3), 801 (1971).

186. B. Bewick and H. R. Thirsk, *Modern Aspects of Electrochemistry* (J. O'M. Bockris and B. E. Conway, eds.), Volume 5, Chapter 4, Plenum Press, New York, (1969).

187. A. Damjanovic, M. A. Genshaw, and J. O'M. Bockris, *J. Chem. Phys.* **45**, 4057 (1966).

188. J. P. Hoare, *Energy Conversion* **8**, 155 (1968).

189. M. R. Tarasevich, *Elektrokhim.* **5**, 713 (1969).

190. G. P. Samoilov, N. A. Shumilova, E. I. Khrushcheva, and V. S. Bagotskii, *Elektrokhim.* **4**, 1364 (1968).

191. M. Volmer, *Phys. Sowjet Union* **4**, 346 (1933).

192. J. O'M. Bockris and G. Razumney, *J. Electroanal. Chem.* **41**, 1 (1973).

193. F. C. Franck, *Discuss. Faraday Soc.* **5**, 48 (1949).

194. H. L. Steinberg, *Nature* **170**, 1119 (1952).

195. J. O'M. Bockris and G. Razumney, *Fundamental Aspects of Electrocrystallisation*, p. 100, Plenum Press, New York (1967).

196. E. Budewski, W. Bostanoff, T. Witanoff, Z. Stoinoff, A. Kotzewa, and R. Kaischew, *Electrochim. Acta* **11**, 1697 (1966).

197. R. Kaischew and Z. Mutaftschiew, *Electrochim. Acta* **10**, 643 (1965).

198. N. Pangarov and V. Velinov, *Electrochim. Acta* **11**, 1753 (1966).

199. D. Postl, G. Eickhorn and H. Fischer, *J. Physik. Chemie* **77**, 138 (1972); **77**, 149 (1972).
200. G. Eickhorn and H. Fischer, *Z. Physik. Chemie* **61**, 10 (1968).
201. G. Eickhorn, F. W. Schlitter, and H. Fischer, *Z. Physik. Chemie* **62**, 1 (1968).
202. R. D. Armstrong and M. Fleischmann, *Z. Phys. Chem.* **52**, 131 (1967).
203. J. W. Oldfield, Ph.D. Thesis, University of Newcastle-upon-Tyne, Newcastle-upon-Tyne, England, (1967).
204. J. O'M. Bockris and M. Enyo, *J. Electrochem. Soc.* **109**, 48 (1962).
205. J. O'M. Bockris and H. Kita, *J. Electrochem. Soc.* **108**, 4 (1961).
206. J. O'M. Bockris and B. E. Conway, *Electrochim. Acta* **3**, 340 (1961).
207. J. O'M. Bockris and A. Despic, *J. Chem. Phys.* **32**, 389 (1960).
208. J. O'M. Bockris and H. Kita, *J. Electrochem. Soc.* **109**, 928 (1962).
209. J. L. Barton and J. O'M. Bockris, *Proc. Roy. Soc.* **A266**, 485 (1962).
210. J. Diggle, A. Damjanovic, and J. O'M. Bockris, *J. Electrochem. Soc.* **117**, 1 (1970).
211. A. Damjanovic, M. Paunovic, and J. O'M. Bockris, *J. Electroanal. Chem.* **9**, 93 (1965).
212. O. Kardos, *Proc. Amer. Electroplaters' Soc.* **43**, 181 (1956).
213. R. Foulke and O. Kardos, *Proc. Amer. Electroplaters' Soc.* **43**, 172 (1956).
214. J. O'M. Bockris and G. Razumney, *Fundamental Aspects of Electrocrystallisation*, p. 134, Plenum Press, New York (1967).
215. M. Froment, G. Maurin, R. Schwarcz, and J. Thevenin, *Corrosion (France)*, **19**, 1 (1971).
216. E. Gileadi, *Coll. Czech. Chem. Commun.* **36**, 464 (1971).
217. J. O'M. Bockris and P. K. Subramanyan, *Corrosion Sci.* **10**, 435 (1970).
218. J. O'M. Bockris, M. Genshaw, and V. Brusic, *Electrochim. Acta* **16**, 1859 (1971).
219. L. Mandel, *Nature* **225**, 450 (1970).
220. D. Drazic, *M.T.P. Volume on Electrochemistry* (J. O'M. Bockris, ed.), M.T.P., London (1972).
221. J. O'M. Bockris, ed., *Environmental Chemistry*, Plenum Press, New York (1977).
222. G. C. Barker, *J. Electroanal. Chem.* **21**, 127 (1969).
223. J. O'M. Bockris and S. Srinivasan, *Fuel Cells: Their Electrochemistry*, p. 489, McGraw–Hill, New York (1970).
224. M. Bonnemay and C. Lamy, *J. Electroanal. Chem.* **32**, 183 (1971).
225. C. Lamy and P. Malaterre, *Surface Sci.* **22**, 325 (1970).
226. H. R. Thirsk and J. A. Harrison, *A Guide to the Study of Electrode Kinetics*, p. 146, Academic Press, New York (1972).
227. W. O'Grady and J. O'M. Bockris, *Surface Sci.* **38**, 249 (1973).
228. M. Fleischmann and A. Clarke, private communication (1977).
229. R. W. Revie, B. G. Baker and J. O'M. Bockris, *Surface Sci.* **52**, 664 (1975); *J. Electrochem. Soc.* **122**, 1460 (1975).
230. A. Damjanovic and J. O'M. Bockris, *Modern Aspects of Electrochemistry* (J. O'M. Bockris and B. E. Conway, eds.), Volume 3, Chapter 4, Plenum Press, New York (1964).
231. J. McHardy, Ph.D. Thesis, University of Pennsylvania, Philadelphia (1972).
232. B. Cahan, Ph.D. Thesis, University of Pennsylvania, Philadelphia (1968).
233. A. J. Appleby, *Modern Aspects of Electrochemistry*, Volume 9, p. 454 (1974).
234. Y. Takas and Y. Matsuda, *Electrochim. Acta* **21**, 133 (1976).
235. G. Beck-Nielson, *Electrochim. Acta* **21**, 634 (1976).
236. A. C. C. Tseung and S. Jasea, *Electrochim. Acta* **22**, 31 (1977).
237. R. D. Armstrong, T. Dickenson, and M. Reid, *Electrochim. Acta* **21**, 1143 (1976).
238. T. Matsushima and M. Enyo, *Electrochim. Acta* **21**, 1029 (1976).
239. S. Stucki, *J. Electroanal. Chem.* **80**, 375 (1977).
240. Y. Harima and S. Aoyagui, *J. Electroanal. Chem.* **81**, 47 (1977).

241. M. W. Breither, *J. Electroanal. Chem.* **81**, 285 (1977).
242. H. B. Sierra Alcaza and J. A. Harrison, *Electrochim. Acta* **22**, 627 (1977).
243. J. N. Jovicevic, D. M. Drazic, and A. R. Despic, *Electrochim. Acta* **22**, 589 (1977).
244. Z. C. Iwakura, K. Hirao, and H. Tamura, *Electrochim. Acta* **22**, 329 (1977).
245. See, for example, R. M. L. Manevich, E. B. Brick, and Y. M. Kolotyrkin, *Electrochim. Acta* **22**, 151 (1977).
246. R. Guidelli and M. L. Foresti, *J. Electroanal. Chem.* **77**, 73 (1977).
247. Z. Samec and J. Weber, *J. Electroanal. Chem.* **76**, 181 (1977).
248. R. D. Armstrong, M. F. Bell, and A. A. Metcalfe, *J. Electroanal. Chem.* **77**, 299 (1977).
249. N. Furuya and S. Motoo, *J. Electroanal. Chem.* **78**, 243 (1977).
250. F. A. Beland, S. O. Farwell, P. R. Callis, and R. D. Geer, *J. Electroanal. Chem.* **78**, 145 (1977).

3

Quanta and Surfaces

3.1. Introduction

Any consideration of the elementary act in electrochemical kinetics needs knowledge of the chemistry of the electronic (usually metal) phase and the ionic solution phase involved. In the latter (e.g., in a hydroxonium ion), the considerations are about the distribution and occupancy of energy levels in vibrational states of bonds and are well known to most physical chemists. The situation in the electronic conductor is entirely quantal, and quantal aspects of the solid state, particularly that of the surface structure, have to be understood by chemists interested in quantum electrochemistry. Much of the background material for this is presented in books on solid state physics. Among these is that written by Kittel[1] and simpler books such as that by Blakemore.[2]

Thus, consider what a physical chemist may need to understand to consider the absorption of photons by semiconductors and the corresponding emission of electrons to a proton in solution. He needs to know something of the spectroscopy of the solid as a function of the angle of incidence of the light beam, the density of states at various energy levels, the statistics of semiconductors, the transport characteristics of the charge carriers and their lifetimes, the surface states in the interface concerned, and the distribution of potential inside and outside the semiconductor.

Apart from these concepts and the quantum mechanical quantities they imply, such as the Hamiltonians of the surface functions, it is necessary to recognize a number of quantum "particles." Examples of these are the phonon, the plasmon, the polaron (Fig. 3.1), and the exciton. In particular, information concerned with the dynamics of surface movements, with adsorption energy, with spectra for adsorbed atoms, and states of adsorbed radicals are available in textbooks mostly at a phenomenological level, and the limited quantum chemistry that exists on these subjects is still in the original literature.

The electrochemist, therefore, has to step out into fields that have not been plowed, for the quantum work on surfaces that exists concerns that of

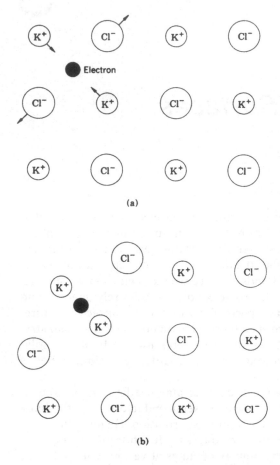

Fig. 3.1. The formation of a polaron. (a) A conduction electron is shown in a rigid lattice of an ionic crystal, KCl. The forces on the ions adjacent to the electron are shown. (b) The electron is shown in an elastic or deformable lattice. The electron plus the associated strain field is called a polaron. The displacement of the ions increases the effective inertia and, hence, the effective mass of the electron; in KCl, the mass is increased by a factor of 2.5 with respect to the band theory mass in a rigid lattice. In extreme situations, often with holes, the particle can become self-trapped (localized) in the lattice. In covalent crystals, the forces on the atoms from the electron are weaker than in ionic crystals, so that polaron deformations are relatively small.

a solid with a gas. Double-layer theory,[3] which has developed in terms of statistical thermodynamics to some degree of sophistication, has neglected the quantal aspects in this view, and the metal is treated for the most part as a continuum after the first few layers inside. Conversely, Oriani and Johnson,[78] Blyholder,[4] and Heiland[77] have all given accounts of metal surfaces oriented toward surface electrochemistry.

3.2. Quantum Particles

One of the concepts of quantum-oriented solid state science is that of "quantum particles." The evolution of concepts of radiation, earlier regarded as a wave motion, then as having corpuscular aspects, and then the meeting of these ideas with that of the electron as a particle and a wave motion is a part of the history of science. However, analysis of electron

movement in solids is rendered nugatory because of the many-body interactions concerned. It may be made tractable by considering a set of
independent fictitious bodies, the properties of which resemble those of the
real entities concerned.

One example of this is an electron passing through a solid. The mass of
the electron is known and unambiguous. However, it is preferable to let the
electron have an *effective* mass, differing from that which it is known to
have, thus to allow for the interaction of the electron with the nucleii it
passes. The introduction of the "fictitious body" (an electron with an
effective mass differing from the real mass but which is thereafter not
considered to interact) is an attempt to deal with the many-body system,
electron-many nuclei.

A good example is in the quanta of lattice vibrations, phonons. There
exist, in fact, complex motions of ions in solids in that each vibrator
interacts with its nearest neighbors, which in turn interact with the motion
of other ions throughout the crystal. The taking into account of the
complexity of this motion is helped by introducing the idea of an *elementary* excitation. A phonon is a "quantum particle" that represents the
oscillation of the atom in a solid lattice as though this oscillation were an
individual property of the atom itself and not of the lattice. It is a method
of dealing with the many-bodied problem. Thus, in this approach, we
regard the body as a set of individual quanta, each having an energy $\hbar\omega_K$,
together with a ground-state energy of $\frac{1}{2}\hbar\omega_K$. These are the phonons and,
although vibrating ions start off their conceptual life as particles interacting
in a complex way with all the other ions in the lattice, they are better
treated as particles that do not interact at all. They are pseudoparticles, the
imagined existence of which is useful in interpreting the properties of
solids.

Another fictitious particle is the hole in semiconductor theory. It is
electrons that are the real particles that move, although the fact that they
jump into holes and consequently make the hole seem to move makes
thinking of a hole as an entity a useful concept. Holes are spoken of as
though they were actual particles. In hydrodynamics, an empty hole turns
out, nonetheless, to have a mass. Thus, if we call the space vacated by an
electron movement a hole, its dynamics can only be calculated if we give it
a fictitional mass.

We have applied this kind of treatment to various other so-called
particles that do not actually exist any more than a hole really exists as an
entity unto itself, even though there is no actual real particle that does not
interact with the lattice. Nevertheless, it is useful to talk of such (only
conceptual) quantum "particles" and to picture them as though they existed to interpret some physical observations. In the following, some quantum particles used in electrochemical discussions will be considered.

3.2.1. The Phonon

The movements of a vibrator in a solid can be resolved into two types of motion—one is called acoustic and the other optical. In the acoustic mode, the motion is horizontal (see Fig. 3.2a). In the optical mode, the atom moves to and fro (Fig. 3.2b). Such motion has a frequency in the region of the so-called infrared spectrum and is therefore called optical.

The phonon illustrates well the concept of the quasiparticle. It contributes certain properties to a system. The actual real particle (for example, the vibrating atom) and the group of disturbed atoms around it are represented by the quasiparticle. The surrounding particles screen the effect of the force field of the original particle, and the quasiparticle, because of its weak or zero interaction with other quasiparticles, can then be treated as independent of them. The advantage of using particles, the properties of which are to an extent fictional, is that one may treat them as independent, individual entities, allotting to their fictional behavior, fictional parameters that differ from those of reality in such a way that they allow for the neglect of interaction.

3.2.2. The Plasmon

A plasma in energy science is a gas of electrons and ions and refers to temperatures sufficiently high so that electrons are removed from the nucleus. In a general sense, a plasma is a body consisting of charged particles the kinetic energy of which is greater than the potential energy of interaction between the particles. In a metal, the kinetic energy of the conduction electrons (a few electron volts) is much greater than the potential energy of interactions between the electrons and the metal ion cores. Hence, the condition just stated for a plasma applies to a metal and its electron gas.[6,7]

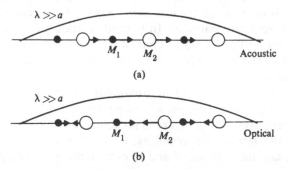

Fig. 3.2. (a) Atomic displacements in the acoustic mode. (b) Atomic displacements in the optical mode.

This electron gas has a special feature that concerns fluctuations in the density of electrons inside the metal. Why should there be any fluctuations? Let it be supposed that, for some reason, the electrons are moved from equilibrium. Then, when the density of the electrons is increased, they will repel each other and fall away from the dense region.

In the metal, one has to consider the attraction that the positive charges have on the electrons. Near an ion core, electrons tend to be attracted, the local density increases, repulsion between electrons occurs, they are repelled, but are again attracted to a nucleus, and so forth. Thus, there will be an oscillation back and forth of the electron density in its position with respect to the nucleus.

It is easy to get a simple expression for the electron plasma frequency, i.e., for the frequency with which the plasma oscillations occur. Following Elliott and Gibson,[5] one can consider the movement of an electron through a distance x. Because of interaction with neighboring positive charges, there will be a polarization P. By definition,

$$P = Ne_0 x$$

where N is the number of charged particles per unit volume and x is the displacement of an electron from a position at which zero local polarization is occurring. But

$$\text{field} = Ne_0 x / \varepsilon$$

if N equals the number of charged particles per unit volume. Therefore, force on a particle out of equilibrium is

$$\left(\frac{Ne_0 x}{\varepsilon}\right) e_0 = m_e \frac{d^2 x}{dt^2}$$

This is an equation of the simple harmonic motion and gives rise by the standard solution to a frequency ω, given by

$$\omega = \left(\frac{Ne_0^2}{m_e \varepsilon}\right)^{1/2}$$

This is the frequency of the density fluctuation of the electrons in the metal and is hence called the plasma frequency. The quantum corresponding to this is called the *plasmon*. A typical order of magnitude for a plasmon frequency would be 10^{15} Hz.

The experimental detection of plasma frequencies is made by shooting electrons of variable frequencies through a foil of the metal. Thus, when the electrons injected into the metal have a frequency in the region of the plasma frequency, they will be absorbed and show a fall in energy on passage through the foil.

3.2.3. The Polaron

The quantum called a "polaron" is important in quantum electrochemistry because it has been used in discussions by Levich, Dogonadze, and Kutznetsov.[8–11] These workers have regarded energy transfer from the surrounding system to ions, by means of which the latter obtain an energy of activation in electron interchange with electrons in metals, as occurring by means of polarons.

Just as a phonon concerns the vibrations of a solid lattice, so a polaron concerns the interaction energy that a local charge brings (by means of its electrostatic attractive and repulsive interactions) to the vibrations of the lattice. An electron in a crystal lattice interacts with ions or atoms of this lattice and deforms the lattice (electron–phonon coupling) (see Fig. 3.1). The electron moves rapidly in a lattice and the combination of the electron and the electrically polarized medium surrounding it is known as a "polaron" (cf. Kupper and Whitfield[12]). The concept of the polaron can be seen from the Fig. 3.1.

The polaron can be used to explain why the effective mass of an electron is usually larger than its real mass. It is because the ion cores are set in motion when the electron moves, and the electron can be pictured as dragging the ion with it. Another way of describing a polaron is that it is "an electron dressed up with a polarization field."

What of "wet polarons"? Thus, in a solution, the electron (which exists in an ion or an electrode) interacts with the solvent, and the electron–lattice interaction polaron of the solid state has been said by Levich and colleagues to have an analog in the electron–solvent interaction. This is a less clear concept because there are no quasifree electrons in a solvent as there are in the conduction band of a metal. The electron, which is regarded as giving rise to a polaron, is *in* an ion or a metal surface. The concept that a polaron could cause a transfer of electrical energy over a distance of hundreds of angstroms in a solution is discussed in Chapter 5.

3.2.4. The Exciton

The exciton is referred to in discussions of long-range energy transfer (Chapter 5) and consists of a bound electron–hole pair. Thus, the electron in a semiconducting solid interacts with a hole as a negative electron interacts with a positive nucleus, and the electron and hole move around their centers of gravity and migrate as a pair slowly through the crystal.

Another definition of an exciton ("Frenkel exciton"), more relevant to discussions of energy transfer, is the delocalized electronic excited state, where the excitation energy is transferred from one atom or molecule to another distant atom via electrostatic perturbation interactions with the neighboring atoms or molecules.

3.3. Electron Distribution in the Metal Electrode

There sometimes arises a need for explanation of the terms Fermi distribution law, density of states, and Fermi surface. The most well known of these three concepts is the Fermi distribution. The law governing it can be discussed in relation to the two other distribution laws that exist for gases, the Boltzmann and the Bose–Einstein laws.†

The physical meaning of these three laws is easy to describe. When systems can be looked at (e.g., for dilute gases) as existing without interaction between the atoms, then the statistical thermodynamic concepts give rise to the equation called the Boltzmann distribution law. The distinguishability of states is the distinguishability of a number of atoms with the same energy. Thus, the concept of cells that are empty or occupied by more than one particle does not come in for one atom, and the space in which it moves is one cell. On the other hand, in the Fermi–Dirac distribution law, dealing as it does with mutually repelling electrons, it is meaningful to refer to the probability of the occupancy of an energy state, which can be less than one. However, nothing is said when Fermi's law about *how many of the states* exist per unit volume of the system is concerned. Fermi's law gives the probability that each state, having a given energy, will be occupied. A corresponding consideration gives the number of these states that exist per unit volume.

3.3.1. Fermi Distribution Law

Thus, in the Fermi distribution law, the statistical thermodynamic discussion begins with the idea that because of repulsion between electrons there will be a number of empty states (in the Boltzmann law, every state is assumed to be occupied).

Carrying through these thoughts, the result is that the probability of having the particle (electron) occupy a state at energy E, is given by the equation (Fermi law)

$$P_E = \left[\exp\left(\frac{E - E_f}{kT}\right) + 1 \right]^{-1} \tag{3.1}$$

where P_E is this probability, E_f is the Fermi energy of the electron in the metal at which the probability of having the particle (i.e., the occupancy) is

---

† In the Bose–Einstein statistics, as distinct from the Boltzmann noninteractive statistics and the Fermi–Dirac (repulsion) statistics, the situation assumes that the particles will attract each other and thus *more than one particle* can exist in one state. But Fermi particles (e.g., electrons) must obey the Pauli exclusion principle, i.e., the occupancy of any allowed state is restricted to either one or zero electrons.

half. The Fermi energy can also be defined as the maximum kinetic energy of the conduction electrons in the metal when $T \to 0$.

3.3.2. Density of States

Fermi's law [Eq. (3.1)], gives the expression for the probability that a given state will be occupied. To compute the current density in quantum electrode kinetics, the number of electrons at a given energy (having an energy between E and $E + dE$) is needed. This is the product of the density of states that have the given energy (and that may or may not be occupied) times the probability of the occupancy of them by an electron (the probability given by the Fermi law).

The density of states $\rho(E)$ that exist (filled or not) having an energy E is given per unit volume by the equation

$$\rho(E) = \frac{1}{2\pi^2}\left(\frac{2m_e}{\hbar^2}\right)^{3/2} E^{1/2} \tag{3.2}$$

where m_e is the (actual) mass of electrons.

Thus, when we wish to express the number of electrons per unit volume that have an energy E and $E + dE$, we shall express it as

$$\rho(E)f(E)dE = \frac{1}{2\pi^2}\left(\frac{2m_e}{\hbar^2}\right)^{3/2}\frac{E^{1/2}\,dE}{\exp[(E-E_f)/kT]+1} \tag{3.3}$$

Thus, the number of electrons n per unit volume in *all* energy ranges (0 to ∞) is

$$n = \int_0^\infty \rho(E)f(E)\,dE = \frac{1}{2\pi^2}\left(\frac{2m_e}{\hbar^2}\right)^{3/2}\int_0^\infty \frac{E^{1/2}\,dE}{\exp[(E-E_f)/kT]+1} \tag{3.4}$$

This cannot be integrated analytically. The density of states, $\rho(E)$ and $\rho(E)f(E)$ as a function of electron energy is given in Fig. 3.3.

3.3.3. Fermi Surface

The Fermi surface does not refer to any actual physical surface but to a mentally conceived surface in momentum space or in wave number, i.e., k, space.

Thus, considering an alkali metal, most of its electrons will be bound to the nucleus, but the valency electrons break off and are shared not with the nucleus from which they begin in the gas phase before condensation but among all the nuclei. They are called conduction electrons—the ones

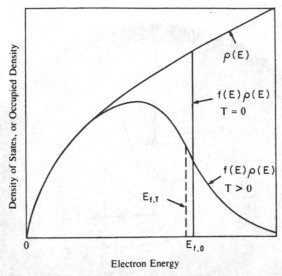

Fig. 3.3. The density of electronic states as a function of energy on the basis of the free electron model and the density of occupied states dictated by the Fermi–Dirac occupancy law. At a finite temperature, the Fermi energy moves very slightly below its position for $T = 0°K$. The effect shown here is an exaggerated one; the curve in the figure for $T > 0$ would with most metals require a temperature of thousands of degrees Kelvin.

that take part in the electrochemical reaction—and exist in Bloch states, which extend throughout the crystal. The energy of the conduction electrons can be expressed as

$$E = \frac{\hbar^2 k^2}{2m_e^*} \qquad (3.5)$$

where m_e^* is the effective mass of the electron and k is the wave number. This was originally defined as the reciprocal of the electron wavelength λ and is now usually defined as $k = 2\pi/\lambda$. The momentum of the conduction electrons is p.

Consideration of the equation just given, the equation for the energy of the conduction-band electrons, shows that if one plots a diagram in which the x, y, and z coordinates of real space are replaced by k_x, k_y, and k_z (corresponding to the electron velocity, or momentum components in x, y, and z direction) in k-space, one can construct spheres of different radius values of $k = (k_x^2 + k_y^2 + k_z^2)^{1/2}$ in k-space. A sphere in which the radius k value corresponds to $k_f = (2m_e^* E_f)^{1/2}/\hbar^2$ (i.e., the wave number corresponding to the Fermi energy E_f) is called the Fermi sphere. The surface of this sphere, having the constant value k_f or constant energy E_f is called the *Fermi surface* (Fig. 3.4) in k-space or in momentum space ($\hbar k = p$).

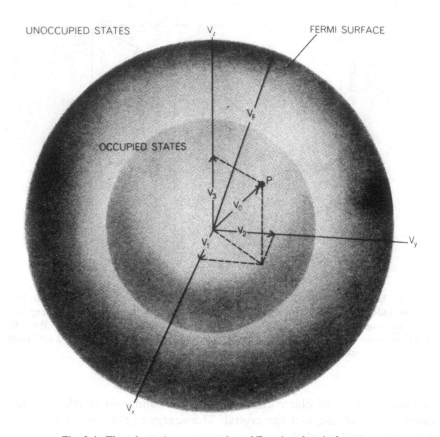

Fig. 3.4. The schematic representation of Fermi surface in k-space.

One can visualize the Fermi surface from its analogy in space. In real space, the k_x, k_y, and k_z are the x, y, and z coordinate distance, and k_f will correspond to a particular radial vector r_0. The importance of the Fermi surface is that the electrical properties of a metal are influenced by its shape, since the flow of an electric current is due to the change in the occupancy of states near the Fermi surface. Physically, the Fermi surface separates the unfilled orbitals in the metal from the filled orbital in momentum space at absolute zero.

The Fermi surface picture given above is an ideal one, and if the electron mass were the mass of a free electron, it would correspond to a perfect gas in a metal. Taking into account that the effective mass allows for the interaction for the quasi-nature of the "free electron," the situation is not as good as this, and the real Fermi surface differs from that ideal one given in Fig. 3.4. Thus, a Fermi space for copper appears as given in Fig. 3.5.

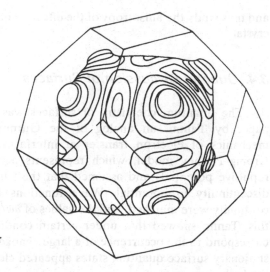

Fig. 3.5. The Fermi surface for copper.

3.3.4. Cyclotron Resonance

Cyclotron resonance is briefly described here because it provides one of the ways in which one can obtain the effective mass of the electron, as distinct from its real mass, in the absence of other interactions. Thus, in a cyclotron, a magnetic field is applied to electrons and exercises a force upon them of He_0v, where H is the strength of the magnetic field, e_0 the electron charge, and v the velocity of the electron. The centripetal force of the rotating body is given by mv^2/r, where v is the velocity of the body rotating under central force, r is the radius, and m is the mass. If the magnetic field is regarded as the central force, an electron will rotate in a circle under its influence, and the radius of this circle is then given by equating He_0v with mv^2/r. Thus, a body has a "cyclotron frequency," given by $\omega = He_0/m_e^*$.

In determining the effective mass of the electron (in a semiconductor, for example), some particles are introduced in the conduction band, and a magnetic field is switched on so that the particles begin to rotate due to the cyclotron effect and hence at the cyclotron frequency. Another electromagnetic signal of variable frequency is applied, and this signal will undergo resonance and will be absorbed when it reaches the cyclotron resonance frequency. Thus, knowing the cyclotron frequency and the magnetic field, the effective mass m_e^* of the electron in the particular system can be obtained from the above equation.

Sophisticated things can be done in this way: for example, one can change the orientation of a magnetic field relative to the crystalline axis and thus study the anisotropy of the effective mass with the direction of the crystal.[13]

3.4. Quantal Discussion of Surfaces

The quantum chemistry of surfaces was commenced by a seminal paper by Tamm, an analog of the Gurney paper for the quantum mechanics of electron transfer at interfaces.[14] Thus, Tamm took the Kronig–Penney model, which represents a linear array of δ function repulsive potentials and assumed that the surface was represented by a discontinuity in potential. Wave functions before and after the discontinuity were matched, and the values of $\partial \psi / \partial x$ were obtained. By doing this, Tamm showed that under certain conditions, which turned out to correspond to the occurrence of a large change of potential at the surface, stationary surface quantum states appeared clearly separate from those in the bulk.

Another contribution analogous to that of Tamm was by Shockley.[15] He showed by matching wave functions across the discontinuity at the surface that under certain conditions described by him as "the bulk bands being crossed," surface states would appear, but this time the condition is that the perturbation caused by the surface should be sufficiently small. Tamm states are therefore to be distinguished from Shockley states: both are surface states, and both are hypothetical because these "pure surface states" (which might be experienced in an extremely high vacuum in the presence of a completely clean solid) are usually superseded by "real surface states," states on the surface introduced by adsorption, in the electrochemical case of ions or molecules on the surface.

The research effort on semiconductor surfaces and surface state theory was accelerated when it was realized that these surfaces have technological importance in transistor technology. Shockley and Pearson[16] modulated the potential of a semiconductor free surface and showed that part of the space charged was mobile but that some was immobile, particularly when a secondary field was applied parallel to the surface. The immobile charge was thought of as due to the presence of trapped electrons in the surface, and these electronic states were called *surface states* and related to the oxygen absorbed upon the surface.

Koutecky[17] treated the quantum theory of localized states as ideal surfaces. From 1960, experimental investigations began to be made of these semiconductor surfaces by means of low-energy diffraction, and these works gave plentiful confirmation of the theoretical ideas introduced particularly by Tamm[14] and by Shockley.[15] Thus, when free (zero adsorbed content) surfaces were observed, they were seen to introduce some spots on the screen that were themselves unstable and temperature dependent. These spots represented not atoms but local perturbations in confirmation of the reality of Shockley states.

Quantum Models for Adsorbed Atoms

The experimental study of surface states[18-20] begins the modern period in quantum-oriented research on surfaces. In one of the approaches,[19] where it is attempted to treat the quantum mechanical interaction between an atom and the surface, some kind of bond, e.g., a Heitler–London interaction, is assumed, and one obtains, therefore, a Hamiltonian. The resulting solution involves the wave function for the adsorbed atoms or molecules, and from this it is possible to see to some extent what happens in the process of chemisorption.[21]

In one such approach (Anderson[22]), the states of the interacting metal atom system are constructed from the linear combination of wave functions of the unperturbed continuum of metal states $\psi_{k\sigma}$ and that of the discrete ground state of the atom $\psi_{a\sigma}$ with its eigenvalue. The eigenvalues of the metal are represented by $\varepsilon_{k\sigma}$ and that of the atom in the ground state by $\varepsilon_{a\sigma}$. The label k indicates the dynamic quantum numbers of the metal state (the wave number) and σ represents the spin of the electron at a given state. The coupling between the adatom and the metal is responsible for adsorption and is represented by the off-diagonal matrix element $V_{ak} = \langle \psi_{a\sigma} | H | \psi_{k\sigma} \rangle = \langle a | H | k \rangle$, in which H is the full Hamiltonian of the combined system.

In the Anderson model, electron–electron interactions between electrons of opposite spin when they are on the adatom are allowed for. This tends to raise the energy $\varepsilon_{a\sigma}$ in the atomic state. Thus, there is an energetic difference between a state with a single spin on the adatom and a state with a net charge that is the same but with both spin-up and spin-down electrons present. This effect, added to that of V_{ak}, gives rise to the lowering of energy, which is a characteristic of an adsorbed state.

The Hamiltonian that corresponds to this model is written in the second quantized notation[23] as[21]

$$H = \sum_{k\sigma} \varepsilon_{k\sigma} n_{k\sigma} + \sum_{\sigma} \varepsilon_{a\sigma} n_{a\sigma} + \sum_{k\sigma} (V_{ak} C_{a\sigma}^{+} C_{k\sigma} + V_{ak}^{*} C_{k\sigma}^{+} C_{a\sigma}) + U_{a\sigma} n_{a\sigma}$$

$$(3.6)$$

The diagonal matrix elements of the number operators, $n_{k\sigma}$ and $n_{a\sigma}$, gave the number of electrons in the states $\psi_{k\sigma}$ and $\psi_{a\sigma}$, respectively. The co-operators are Fermian creation and annihilation operators. In Eq. (3.6), the first two terms are unperturbed Hamiltonians belonging to the metal and the atom, respectively. The third term represents the atom–metal coupling, which causes the transfer of an electron between the metal and the atom and gives bonding and adsorption. The last term represents the intraatomic Coulomb repulsion between up- and down-spin electrons in the adatom. In the theory of Anderson,[22] the effective strength of the Coulombic repulsion between up and down electrons U was to be taken from a Coulomb integral involving electron states in the atom with opposite spin, which turned out to be too large even for the gas-phase treatment and is due to the fact that correlation effects between the movements of the electrons were not built into the wave functions of electrons in the atom.

Later, Schrieffer and Mattis[24] included correlation effects but found a reduction in the U and a corrected U of the Coulomb integral, compared with that which it was in the gas phase, without taking into account a correlation effect.

Another correction of the Coulomb integrals U is given by considering the screening of the Coulomb potential e_0^2/r_{12} by the free-conduction-band electrons of the solid.[25] The Fermi–Thomas screening is expressed as

$$\frac{e_0^2}{r_{12}} \exp(-k_{FT} r_{12}) \tag{3.7}$$

where k_{FT} is the Fermi–Thomas screening parameter, and r_{12} is the average electron–electron separation.

The model, which turns up from the Anderson calculations, is a quasilocalized virtual state.† One can see the situation for the noninteracting atom and metal in Fig. 3.6. A narrow d band centered at energy ε_d (taking the Fermi level energy as zero) is indicated by the dense spacing of the level. As the atom is brought to the surface, the V_{ak} and U terms shift both the valence level $\varepsilon_{a\sigma}$ and the affinity level below the vacuum potential. All metal states with energies approximately equal to that of the shifted atom state mix well with the atomic state. The discrete adatom state turns out to have been converted into a broadened virtual state and is shown in Fig. 3.6b.

A full description of the Anderson approach (a molecular orbital theory) is given by Gadzuk.[21] The main conclusion is that the molecular orbital rather than the valency-bond type of approach is more appropriate to the study of the adsorbed state.[26–28]

One of the difficulties of the situation concerns the expense of the computer time. It is acceptable to work up interactions of a given metal

---

† A virtual state is one having an energy that is simply a parameter of the system but does not represent a stationary state of the system.

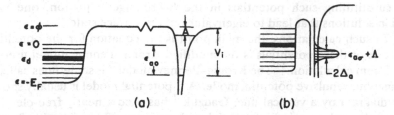

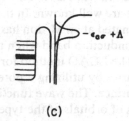

Fig. 3.6. (a) Schematic potential and energy-level diagram for noninteracting atom and metal. The occupied portion of the conduction band lies within the range $-E_F \leqslant \varepsilon \leqslant 0$. A narrow d band is centered at $\varepsilon = \varepsilon_j$. (b) Adsorption for which the broadened atomic virtual state lies below the Fermi level and is thus totally occupied. (c) Ionic adsorption for which the broadened valence lies above the Fermi level and is thus almost totally unoccupied.

with about 10 electrons, but as a further increase in number makes a difference, the complexities of the calculations give rise to difficulties in respect to the accuracy of computation.

3.5. Theory of Surface States

The theoretical calculation of surface states involves two basic approaches. One may be called the "potential method"; the other is the "LCAO method." In the potential method, a relation is considered between the potential in the electrode and the distance from the surface.

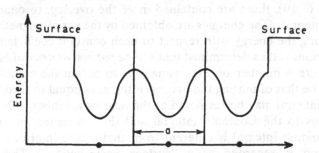

Fig. 3.7. Schematic representation of a one-dimensional monatomic lattice in the nearly-free-electron approximation, showing the lattice sites at potential minima and the lattice terminated at a potential maximum without distortion.

By substituting such potentials in the Schrödinger equation, one may obtain solutions that lead to eigenvalues of the surface state.

In such calculations, the main point is the equation for the potential, and one of these potentials is the modified Kronig–Penney model used in the Tamm calculation.[14] The Kronig–Penney model[29] is sometimes called a δ function, repulsive potential mode. A δ potential model is usually shown in a diagram by a vertical line. Gadzuk[21] has used a nearly free-electron approach, and a periodic potential with no edge effects is used. Such a potential is shown in Fig. 3.7. Inside a crystal, the Schrödinger equation reduces to what is called a Mathieu differential equation, the properties of which are well known. In the case of an ionic salt, a one-dimensional lattice terminated at an M^+ ion has an M-like surface state, which moves nearer to the conduction band when the ionicity increases.

The LCAO method for calculating surface states avoids the difficulties that arise by utilizing a potential relation for the electron in its relation to the surface. The wave function ψ of the finite crystal can be represented by a sum of orbitals of the type

$$\psi = \sum_r C_r \phi_r \qquad (3.8)$$

There is latitude in the choice of atomic orbitals. A hydrogenlike orbital is sometimes used. Bloch orbitals[30,31] have often been used in Schrödinger's equation. One substitutes the selected atomic orbitals in the Schrödinger wave equation

$$H\Psi = E\Psi \qquad (3.9)$$

where H is the Hamiltonian operator for the one electron situation and E is the orbital energy. Multiplying by Ψ and integrating, one obtains the mean energy of the state:

$$\langle E \rangle = \int \psi H \psi \, d\tau \Big/ \int |\psi|^2 \, d\tau \qquad (3.10)$$

In Eq. (3.10), there are contained in H the overlap, resonance, and Coulomb integrals. The energies are obtained by the variation method, i.e., by minimizing the energy with respect to each constant coefficient, and a set of equations is thus determined that *can be put best as determinants*. It is usual to make a number of approximations to obtain the solution. One path might be that of putting the overlap integral as equal to zero, and the resonance integral may be restricted to the nearest neighbors. One has to assign values to the Coulomb integral and the resonance integral. The surface Coulomb integral is different from that of the interior Coulomb integrals and is regarded as a "surface perturbation." The surface resonance integral is related to the interior resonance integral but has a "deformation." Solution of these equations gives the energy of the surface states. One of the difficulties of these approaches is that there has been no experimental verification of the calculations made. More experimental knowledge of surface states and their energies is highly desirable.

Surface states have been calculated not only at semiconductor surfaces, but also at metals as with calculations by Carruthers on aluminum.[32] Thus, Carruthers studied the electronic structure of a slab with 13 atomic layers, using reasonable potentials, though the self-consistent approach was not used. Thirty-nine slabs were calculated to obtain the difference between structures associated with thin films and those associated with the surface. A number of surface states were discovered by this theoretical approach. Some of these went 15 Å into the bulk.

A large number of calculations have been performed in respect to semiconductor surface states.[33] The nature of the states in the energy gap has often been discussed. Chemically, those present in the gas phase probably represent states associated with bonds broken at the surface. Appelbaum and Hamann[34] support the "dangling bond" hypothesis. It is possible to obtain detailed evidence of surface states by inelastic electron loss measurement, such as those of Rowe and Ibach.[35]

Typical of the calculations are those carried out by Appelbaum and Hamann[34] for the unreconstructed silicon (111) surface, and two different assumptions concerning the position of the last plane in the surface in respect to the bulk are made. A smooth charge density was assumed, and this was interpolated between a vacuum charge density supposed to be increased from the surface outward exponentially and the correct bulk density, so that the charge neutrality was maintained at the surface origin. The calculations yielded values near to those of the measured work function.

After physically reasonable potentials had thus been constructed, the Schrödinger equation was used to search the surface. In the case of silicon, this search was made at two symmetry points in a hexagonal zone. At the center of the zone face, the band of surface states found was 0.6–0.8 eV above the bottom of the band gap. Other bands of surface states have been found in silicon, some being 2–4 eV below the valency-band maximum.[36]

3.6. Surface Energy

The energy of a solid E_{total} can be written as

$$E_{total} = \varepsilon_v V + \varepsilon_s S \tag{3.11}$$

where V is the volume of the solid, S is its surface area, and ε_v and ε_s are the energies per unit volume of the solid and per unit area, respectively.

To define the surface energy of a crystal face, the bulk solid can be thought of as being divided along a plane parallel to that face formed by two new surfaces. The energy required for a separation of the surfaces per unit area is the surface energy. Most measurements of surface energy associated with metals have been made with liquids, for the reason that, then surface tension can easily be determined. Thus, the measurements do not apply to any particular crystallographic face. Impurities have a marked effect. For insulators and semiconductors, it is possible to form a new surface by cleavage, and one can obtain the free surface energy by measuring the work of cleavage calometrically.

There is a limited experimental knowledge of surface energy. One of the problems is the surface reconstruction and relaxation. Knowing the difficulty of the experimental situation, there have been some interesting quantum mechanical calculations of surface energies. The energy of the solid consists of the kinetic energy of the valence electron. The kinetic energy E_K can be written as

$$E_K = \sum_i \int_{z>z_B} \psi_i^*(\mathbf{r})(-\tfrac{1}{2}\nabla^2)\psi_i(\mathbf{r}) \, d^3\mathbf{r} \tag{3.12}$$

where the integration has been limited to a region to the right of a plane z_B within the bulk of the solid.

The electrostatic energy, E_{es}, of the valence electron is written simply as

$$E_{es} = \frac{1}{2} \int_{z>z_B} \rho_T(\mathbf{r}) V_{es}(\mathbf{r}) \, d^3\mathbf{r} - E_{es,c} \tag{3.13}$$

where V_{es} and ρ_T are the total electrostatic potential and charge density, respectively. The self-energies of the individual atomic cores $E_{es,c}$ have been subtracted. There is the exchange of correlation energies of valency electrons, and one of the equations used to express these is

$$E_{xc} = -0.738 \int_{z>z_B} \rho^v(\mathbf{r}) \left[1 + \frac{0.959}{1+12.57f(\mathbf{r})} \right] f(\mathbf{r}) \, d^3\mathbf{r} \tag{3.14}$$

where

$$f(\mathbf{r}) = [\rho(\mathbf{r})]^{1/3} \tag{3.15}$$

Finally, the energy of the core E_{core} is given by

$$E_{core} = \int_{z>z_B} V_{core}(\mathbf{r})\rho(\mathbf{r}) \, d^3\mathbf{r} \tag{3.16}$$

Some recent quantum mechanical calculations for the surface energy of metals have been carried out by Lang and Kohn.[37-39] They first calculated the electron charge density for jellium and found agreement for liquid metals, where the experimental values were sound. For high-density metals, the jellium surface energy is negative whereas it should be positive, and to rectify this, the discrete ion lattice was produced, which tends to stabilize the solid against spontaneous surface formation. Lang and Kohn found reasonable agreement between theory and experiment at all electron densities for the sp-bonded simple metals. Surface energies for ionic surfaces have been calculated.[40,41] Experimental values are higher than the theoretical ones, but it may be that the experimental ones are in error because the measurements are difficult and the calculations are generally performed over about half a dozen different crystal planes.

A general difficulty here is that the experiments are poor and there is not much test of the equations. No experimental work has been reported directly aimed at testing the quantum mechanical calculations of surfaces in contact with a solution, e.g., examining the theoretical potential dependence in quantum mechanical terms.

3.7. Quantum Mechanical Calculations of Adsorption Energy

To simplify the calculations, the value aimed at is the initial heat of adsorption at zero coverage, a measure of the binding area on a clean surface. One sums the total of the attractive and repulsive potentials as the atom approaches the surface. Calculations of this type were carried out for hydrogen and helium on nickel by Pollard,[42] who added the exchange energy to the dispersion energy and calculated the heat of absorption for the benzene–mercury system (Fig. 3.8). The experiment gives $-15.4 \, \text{kcal mole}^{-1}$ and theory $-10.2 \, \text{kcal mole}^{-1}$. The definition of the

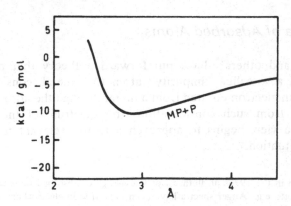

Fig. 3.8. The adsorption energy for a benzene–mercury system.

electron cloud of the adsorbate is important, and it will be strong for alkali metals, which have a low ionization potential. The adsorbing entity then ionizes at some distance from the metal and has a different polarizability and different characteristics from those of the free atom.

The LCAO method is used in calculations of adsorption heats. For example, Toya[43] has calculated quantum mechanically the adsorption energy of hydrogen on platinum. The wave functions at the surface is a linear combination of the wave functions for (1) the exchange configuration, (2) a neutral configuration corresponding to an electron transition with influence of adsorption on a higher free level, (3) two ionic configurations, M^+—H^- and M^-—H^+. The wave function of the free electrons was assumed to be plane waves. The calculated adsorption energy was -5 eV and the distance approached 0.8 Å.

Newns[44] has published calculations on the adsorption energy of hydrogen on titanium, chromium, nickel, and copper, using the self-consistent model. He has adopted what are called Anderson Hamiltonians.[22] Newns[44] obtains the adsorption energy near the experimental observation. In general, the calculations of the surface energy show an approach to reality, but only simple cases have as yet been studied.

These simple cases of quantum mechanical calculations of heats of adsorption are not impressive because of their introduction of empirical parameters to iterate to reality. However, they give insight into adsorption, which is relevant to the present purpose. It becomes a complicated many-electron phenomenon, and this makes an improvement difficult because of the general difficulty of many-electron problems in quantum mechanics. One of the things not well known is the value of the Fermi energy in effecting the shape of the Fermi surface. To illustrate the difficulty, we do not even know the exact position of surface atoms; for example, they may often be surface ions.

3.8. Spectra of Adsorbed Atoms

Gadzuk and others[45] have put forward a theory that relates the adsorption of a so-called "impurity" atom to a metal, considering the transition of an electron emitted from a metal through the adsorbed atom. They deduce from such considerations the spectrum of the adsorbed atoms. Such a view begins to approach a theory relevant to the electrochemical situation.†

---

† There has been in the 1970s an upthrust of knowledge of absorbed films through spectroscopic methods, e.g., Auger spectra have been measured in electrochemically relevant situations.[81]

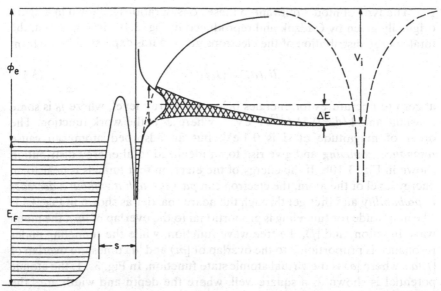

Fig. 3.9. Energy-level diagram relevant to a metal–surface impurity problem. The dashed curve is the ion–core potential for the particle at infinity. The solid curve is a schematic of the combined atomic and metal potential for the atom a distance s from the surface. The details of the potential between the atom and metal are only qualitatively depicted here. V_i represents the ionization energy of the isolated atom, ΔE is the shift of the energy level, and Γ is the natural broadening due to the atom's interaction with the solid. The "tornadolike" structure depicts this graphically.

The theory of resonance tunneling of field-emitted electrons through atoms adsorbed at metal surfaces was first treated by Duke and Alferieff.[46] It is at first a good idea to look at the model used by them, which is seen in Fig. 3.9. When an atom is adsorbed upon a metal, as shown in the figure, the original ground state level is shifted and made broader so that it forms a *virtual* state due to configuration interaction with the metallic states. The situation is similar to the making of real and virtual impurity states localized in a solid. The theoretical treatment of the atom–metal interaction in adsorption has emphasized the calculation of the change in energy and the width of the level represented by Γ in Fig. 3.9. If the numerical quantities corresponding to this change of energy and the width are known, the charge on the adatom–dipole moment and the atomic bonding energy can be calculated.[7,47,48] It was the virtue of the Duke and Alferieff work[46] to understand that the line shapes of the perturbed atomic spectra would be observable in field emission experiments on tunneling.†

---

† This work is discussed from the viewpoint of the theory of tunneling in Chapter 8.

The type of model relevant to these considerations is shown in a figure originally given by Gadzuk and reproduced in Fig. 3.10. In Fig. 3.10a, the total energy distribution of the electrons given out is expressed in the form

$$dj_0/dE = (J_0/d)\, e^{E/d} \tag{3.17}$$

at zero temperatures for energies below the Fermi level, where j_0 is some constant and $1/d = 0.51\phi^{1/2}/F$ eV^{-1}, where ϕ is the work function. The order of magnitude of d is 0.1 eV, but an adsorbed atom can cause *resonance tunneling* and give rise to an idealized total energy distribution shown in Fig. 3.10b. If the energy of the electron that tunnels is near to an energy level of the atom, the electron can *pass through it without a decrease in probability* and thus get through the nearer barrier as shown in Fig. 3.11. The amplitude for tunneling is proportional to the overlap of $|m\rangle$, the metal wave function, and $|f\rangle$, the free wave function, while the tunneling under resonance is proportional to the overlap of $|m\rangle$ and $|a\rangle$ times the overlap of $|f\rangle|a\rangle$, where $|a\rangle$ is the virtual atomic state function. In Fig. 3.11, the atomic potential is shown as a square well, where the depth and width are "the right size" for the atom and can produce bound states at the appropriate energy. The resonance tunneling gives rise to an enhancement factor $R(E)$,

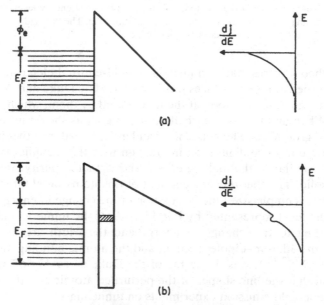

(a)

(b)

Fig. 3.10. (a) Model potential and total energy distribution for field emission from a metal. (b) Model potential and total energy distribution for resonance-tunneling field emission from a metal with a narrow-band adsorbate.

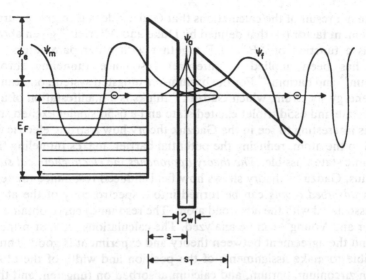

Fig. 3.11. Schematic model showing the idealized potentials relevant in resonance tunneling. The electron wave functions are: ψ_m, the unperturbed metal function; ψ_a, the localized virtual impurity function; and ψ_f, the emitted electron function that will be used in subsequent calculations.

and Gadzuk[45] expressed this as

$$R(E) = \frac{dj}{dE} \bigg/ \frac{dj_0}{dE} \qquad (3.18)$$

where dj/dE is the total energy distribution in the presence of an atom. Structure should appear in $R(E)$, which should reflect the virtual atomic state. The maximum in the enhancement factor for resonance tunneling can be 10^4.

A model first used by Iogansen[49] was applied by Duke and Alferieff to show the resonance tunneling (Fig. 3.11). The atomic potential is represented both by a square-well attractive core and the repulsive δ-function pseudopotential performing the role of orthogonalization to the core states.

Gadzuk[45] took the work of Duke and Alferieff[46] and analyzed it to show how the measured $R(E)$ curves can be analyzed to obtain values for ΔE and Γ. He reformulated the theory of resonance tunneling in a different way from that of Duke and Alferieff[46] and made a closer identification with other work that had been carried out on surface-impurity virtual states.[47,50-52] The Gadzuk theory is a hybridization of the Oppenheimer perturbation theory, collision theory, configuration interaction theory, the WKB approximation, and the surface impurity theory.

The net result of the calculations that Gadzuk does is to give a current enhancement factor (cf. that defined by Duke and Alferieff[46] given above), which is a function of ΔE and Γ together with other parameters. The theory has been applied to interpret resonance tunneling through zirconium[53] and barium.[54] For the barium-on-tungsten system, one can see a true energy spectrum, which Gadzuk[45] thinks is a manifestation of a $6s^2$ ground state and 6s5d triplet excited state and a 6s5d singlet excited state.

It is interesting to see in the Gadzuk theory how a strong electric field applied to the atom, reducing the potential barrier, makes tunneling from the atomic state possible. *The theory approaches the electrochemical situation.* Thus, Gadzuk's theory shows how field-induced resonance tunneling *through adsorbed atoms* can be turned into a spectroscopy of the atomic levels associated with the adsorbed atom. The resonance curve obtained by Plumber and Young[55] can be analyzed. The calculations are first-principle ones, and the agreement between theory and experiment is good. Thus, it is possible to make assignments of the position and width of the atomic levels in zirconium, barium, and calcium adsorbed on tungsten, and these are experimentally accessible in field-emission experiments.

There is a tremendous current enhancement resulting from resonance tunneling through adsorbed atoms, and this can explain anomalies in past field-emission studies. Thus, in the consideration of the microscopic level of atom–metal interactions, it is necessary to have a knowledge of ΔE and the width Γ of the atomic energy levels to calculate the adsorbate charge, work function, and binding energies. Resonance-tunneling spectroscopy gives information on the surface atomic levels. It is only after spectroscopic knowledge of these levels has been connected with theory that further programs here can be made.[56]

3.9. Further Work on the Quantum Mechanics of Adsorbed Species

Some other work has been done on the crystal LCAO calculations of adsorbed species, particularly that by Koutecky.[57] If the adsorbed atom has potential energy terms that are different enough from that of the crystal lattice, then a Tamm localized state may be produced.[19] But if the adsorbed atom does not produce a perturbation to give rise to Tamm states, Shockley states are produced instead. The localized states are not necessary for chemisorption, because the adsorption bond may involve delocalized states, as is seen often in chemisorption. A good account of the localized states in adsorbed hydrogen has been given, with electrochemical overtones, by Blyholder.[58]

3.9.1. Ionic Adsorption

Electronic interaction is taken as having only a small effect on the initial surface state of an adsorbent when ionic adsorption occurs.[80] It has been made out by Mark[59] for ionic adsorption on semiconductors. It is important to think of the electronic energy level of the adsorbate when an electron has been transferred to or from the band structure of the adsorbent. The potential energy level of an ion in free space must be ionization potential $-I$ or the electron affinity $-A$, depending upon whether an electron has been gained or lost. Near an ionic lattice, the electrostatic interaction with the lattice changes the potential. The lattice interaction can be obtained from the Madelung constant C_a for the ion when it occupies the lattice site one lattice distance a_0, above the surface. The value of C_a can be obtained from

$$C_a = C - C_s \tag{3.19}$$

where the Madelung constant is defined by

$$C = \sum_{i,j,k} (q_{i,j,k}/R_{i,j,k}) = \sum_{i,j,k} Q \tag{3.20}$$

The Madelung constant for an ion in the surface C_s is defined by

$$C_s = \sum_{i \ge o,j,k} Q \tag{3.21}$$

Here $q_{i,j,k}$ is the charge on an ion at an indexed position (i, j, k). $R_{i,j,k}$ is the distance from the origin, the *free space potential is corrected* by ΔV_a, where

$$\pm \Delta V_a = z_a C_a e_0/a_0 \tag{3.22}$$

with the plus sign for a position above a negative lattice site and the minus sign for a positive lattice site. The charge z_a for the lattice ion is usually fractional, as indicated in 1958 by the calculations of Conway and Bockris on metal ions undergoing electron transfer at interfaces.[79]

A potential energy diagram is shown in Fig. 3.12. The efficiency band for the adsorption is shown crosshatched on the left. The conduction band is shaded. The Fermi level and intrinsic surface states occur between the valence and the conduction bands, and the vacuum level is considered the zero potential. The electrophobic adsorbate is characterized by an ionization potential with a positive electron affinity. If the corrected potential on the surface $-I_d(s) = -I_d \pm \Delta V_a$ is above the Fermi level, the adsorbate will donate electrons to the surface. The electrophilic absorbate is characterized by a large ionization potential and the electron affinity on the surface; $-A_a(s) = -A_a + \Delta V_a$ lie below Fermi level and produce acceptor states.

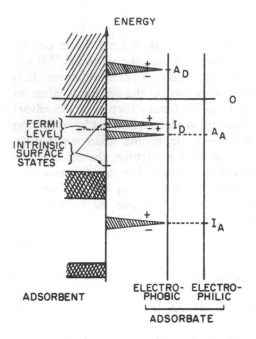

Fig. 3.12. Potential energy diagram for chemisorption on an ionic surface.

As the surface index of a crystal plane increases, Eq. (3.19) indicates that C_a and hence $|\Delta V_a|$ will increase. During crystal growth, the highest growth rate will be for the high-indexed planes. This will have, as Blyholder[58] points out, effects on the electrode processes.[80] If a component is present in the solution that will inhibit electron transfer when adsorbed, the adsorbed species will be greater in concentration on the high-index plane and the electron transfer will have to occur on the low-index plane, but if the adsorption of the species promotes the electron transfer, the reaction may proceed more readily on the high-index plane, because here the chemisorption process has occurred most readily.

3.9.2. Other Treatments of Adsorption

Blyholder and Coulson[60] have looked at a surface compound approximation. They found that small, finite models give the same surface states when treated in the Hückel approximation as calculations for semiinfinite crystals. A diatomic treatment has been tried by Eley[61] where, considering surface–atom as a diatomic molecule, the Pauling[62] approximation for the bond strength is used.

Treatments based on the consideration of charge transfer complexes have been used, particularly by Mulliken.[63] Mignolet,[64] Matson et al.[65] and Brodd[66] have treated this field as well.

Hückel LCAO calculations have been carried out, particularly by Blyholder.[67] Laforgue *et al.*[68] has been considering a Hückel π electron model for ethylene interacting with either groups of one to four atoms or a chain or rectangular array of up to 25 atoms. Dunken and Opitz[69-71] have studied the adsorption of CO, HCN, and C_2H_4 in the Hückel π-electron approximation.

Blyholder[4] has concentrated on the production of localized states, but the work indicates that even where the localized states are responsible for adsorption, contributions from the nonlocalized states can be important in determining the most favorable adsorption site.

Perturbation processes have been developed, particularly by Jansen,[72] but have not been successful in respect to hydrogen, because they were not able even to interpret its dissociation in adsorption upon metals.

Grimley[73,74] has looked at the application of perturbation theory using semiempirical approximations for adsorbed species. He examines the situation that results from the interaction of an adsorbed molecule and metal energy levels within the framework of one-electron orbital theory. Thus, interaction energy between CO molecular orbitals and the metal bond states is given by perturbation equation

$$E - H_0 = \sum_k [|V_{0K} - H_0 S_{0K}|^2 / (H_0 - H_k)] \tag{3.23}$$

which can be simplified to

$$E - H_0 = 4 \int_{E_V}^{E_F} [\omega_\pi(E)(E + A)] \, dE - 2 \int_{E_F}^{E_{max}} [\omega_\sigma(E)(E + I)] \, dE \tag{3.24}$$

where A and I are the electron affinity and the ionization potential, respectively, of, e.g., CO. On the right, the first term is for the electron transfer from metal levels E below the Fermi level E_F to the vacant 2π-level E_0 of CO. Electron transfer is represented by the right-hand term upon the field CO σ-level to the empty metal levels above the Fermi level. The interaction terms are defined by

$$\omega_\sigma(E) = \sum_k |V_{0k} - H_0 S_{0k}|^2 \delta(E - E_K) k > K_F \tag{3.25}$$

and

$$\omega_\pi(E) = \sum_k |V_{0k} - H_0 S_{0k}|^2 \delta(E - E_K) k > K_F \tag{3.26}$$

where V_{0k} is the interaction term between the ground state and level k of the metal and K_F represents the Fermi level.

The difficulty with this work is that the interaction energy represented by the terms $\omega_\sigma(E)$ and $\omega_\pi(E)$ are adjustable parameters and given "reasonable values." Since the density of states enters the expression for the probability of electron transfer from an electrode, it follows that a correlation between the adsorption energy and electron transfer rate should exist, as indeed is known to be so from a phenomenological approach (Conway and Bockris[75]). Further details are given by Blyholder.[58]

3.10. Concluding Remarks

The situation is difficult in respect to calculations of adsorption, because of the lack of comparisons between experimental values and work in which the answers have *not* been obtained by substituting arbitrary parameters. The extension to electrochemical situations must begin with interpretations at a quantum mechanical level of simple phenomenon, such as the specifically absorbed situation between, say, iodine and mercury (cf. Anderson and Bockris[76]), or the states of oxides in passive layers.[81]

References

1. C. Kittel, *Introduction to Solid State Physics* (4th edition), John Wiley and Sons, New York (1971).
2. J. S. Blakemore, *Solid State Physics*, W. B. Saunders, London (1969).
3. M. A. Habib and J. O'M. Bockris, *Comprehensive Treatise of Electrochemistry* (J. O'M. Bockris, ed.), Volume 4, Plenum Press, New York (1977).
4. G. Blyholder, *Modern Aspects of Electrochemistry* (J. O'M. Bockris and B. E. Conway, eds.), Volume 8, Plenum Press, New York (1973).
5. R. T. Elliott and A. F. Gibson, *Solid State Physics*, Macmillan, New York (1974).
6. W. K. Paik and J. O'M. Bockris, *Surface Sci.* **27**, 191 (1971).
7. P. M. Platzman and P. A. Wolf, *Solid State Physics* (H. Ehrenreich, F. Seitz, and D. Turnbull, eds.), Academic Press, New York (1973).
8. V. G. Levich, *Physical Chemistry, An Advanced Treatise* (H. Eyring, D. Henderson, and Y. Jost, eds.), Volume IXB, Academic Press, New York (1970).
9. R. R. Dogonadze, *Reactions at Electrodes* (N. S. Hush, ed.), Wiley Interscience, New York (1971).
10. R. R. Dogonadze, A. Kuznetsov, and V. G. Levich, *Electrochim. Acta* **14**, 1001 (1969).
11. J. O'M. Bockris and R. K. Sen, *Mol Phys.* **29**, 357 (1975).
12. C. G. Kupper and G. D. Whitfield, eds., *Polarons and Excitons*, Plenum Press, New York (1963).
13. M. A. Omar, *Elementary Solid State Physics*, p. 161, Addison–Wesley, London (1975).
14. I. Tamm, *Z. Physik* **76**, 849 (1932); *D. Phys. Z. Sowjet* **1**, 733 (1932).
15. W. Shockley, *Phys. Rev.* **56**, 317 (1939).
16. W. Shockley and G. Pearson, *Phys. Rev.* **74**, 232 (1948).

17. J. Koutecky, *Phys. Rev.* **108**, 13 (1957).
18. J. D. Levine and S. G. Davison, *Phys. Rev.* **174**, 911 (1968).
19. J. D. Levine, *Phys. Rev.* **171**, 701 (1968).
20. S. G. Davison and J. D. Levine, in *Solid State Physics* (H. Ehrenreich, F. Seitz, and D. Turnbull, eds.), Academic Press, New York, Vol. 25, p. 271 (1970).
21. J. W. Gadzuk, *Surface Physics of Materials* (J. M. Blakely, ed.), Academic Press, New York (1975).
22. P. W. Anderson, *Phys. Rev.* **124**, 41 (1961).
23. E. Merzbacker, *Quantum Mechanics*, John Wiley and Sons, New York (1970).
24. J. R. Schrieffer and D. C. Mattis, *Phys. Rev.* **140A**, 1412 (1965).
25. C. Herring, "Exchange interactions among itinerant electrons," in *Magnetism* (G. T. Rado and H. Suhl, eds.), Volume IV, Academic Press, New York (1966).
26. A. Bagchi and M. H. Cohen, *Phys. Rev.* **B9**, 4103 (1974).
27. W. Brendig and K. Schönhammer, *Z. Phys.* **267**, 201 (1974).
28. A. Madhukar, *Phys. Rev.* **8**, 448 (1973).
29. R. de L. Kronig and W. G. Penney, *Proc. Roy. Soc.* **A130**, 499 (1931).
30. M. Abramowitz and I. A. Stegun (eds.), *Handbook of Mathematical Functions*, Dover Publications, New York (1968).
31. F. Bloch, *Z. Phys.* **52**, 555 (1928).
32. E. Caruthers, L. Kleinman, and G. P. Alldredge, *Phys. Rev.* **8**, 4570 (1974).
33. S. G. Devisson and J. D. Levine, *Solid State Phys.* **25**, 37 (1970).
34. J. A. Appelbaum and D. R. Hamann, *Phys. Rev. Lett.* **31**, 106, 1106 (1973).
35. J. E. Rowe and H. Ibach, *Phys. Rev. Lett.* **31**, 103 (1973).
36. J. A. Appelbaum, *Surface Physics of Materials* (J. M. Blakely, ed.), Volume 1, Academic Press, New York (1975).
37. N. D. Lang and W. Kohn, *Phys. Rev.* **B1**, 4555 (1970).
38. N. D. Lang and W. Kohn, *Phys. Rev.* **3**, 1215 (1971).
39. N. D. Lang and W. Kohn, *Phys. Rev.* **B7**, 3549 (1973).
40. G. C. Benson, *J. Chem. Phys.* **39**, 302 (1963).
41. G. C. Benson and K. S. Yun, in *Solid–Gas Interface* (E. A. Flood, ed.), Arnold, London (1967).
42. W. G. Pollard, *Phys. Rev.* **60**, 578 (1941).
43. T. Toya, *J. Res. Inst. Catalysis Hokkaido Univ.* **6**, 308 (1958).
44. D. M. Newns, *Phys. Rev.* **178**, 1123 (1969).
45. J. W. Gadzuk, *Phys. Rev.* **1**, 2110 (1970).
46. C. B. Duke and M. F. Alferieff, *J. Chem. Phys.* **46**, 923 (1967).
47. A. J. Bennett and L. M. Falicov, *Phys. Rev.* **151**, 512 (1966).
48. L. Schmidt and R. Gomer, *J. Chem. Phys.* **45**, 1605 (1966).
49. L. V. Iogansen, *Zhur. Eksper. Teor. Fiz.* **45**, 207 (1963).
50. J. W. Gadzuk, *Surface Sci.* **6**, 133 (1967).
51. D. M. Edwards and D. M. Newns, *Phys. Lett.* **24A**, 236 (1967).
52. J. W. Gadzuk, *The Structure and Chemistry of Solid Surfaces* (G. A. Somorjai, ed.), John Wiley and Sons, New York (1969).
53. E. W. Plummer, J. W. Gadzuk, and R. D. Young, *Solid State Commun.* **7**, 487 (1969).
54. J. W. Gadzuk, E. W. Plummer, and R. D. Young, *Bull. Amer. Phys. Soc.* **11**, 399 (1969).
55. E. W. Plummer and R. D. Young, *Phys. Rev.* **B1**, 2088 (1970).
56. J. W. Gadzuk, *Surface Physics of Materials* (J. M. Blakely, ed.), Chapter 2, Academic Press, New York (1975).
57. J. Koutecky, *Advan. Chem. Phys.* **9**, 85 (1965).
58. G. Blyholder, *Modern Aspects of Electrochemistry* (J. O'M. Bockris and B. E. Conway, eds.), Plenum Press, New York (1972).

59. P. Mark, *Catal. Rev.* **1**, 165 (1968).
60. G. Blyholder and C. A. Coulson, *Trans. Faraday Soc.* **63**, 472 (1967).
61. D. D. Eley, *Discuss. Faraday Soc.* **8**, 34 (1950).
62. L. Pauling, *The Nature of the Chemical Bond*, Cornell University Press, Ithaca, New York (1939).
63. R. S. Mulliken, *J. Am. Chem. Soc.* **74**, 811 (1952).
64. C. P. Mignolet, *J. Chem. Phys.* **21**, 1298 (1953).
65. F. A. Matson, C. A. Makrides, and N. Hackerman, *J. Chem. Phys.* **22**, 1800 (1954).
66. R. J. Brodd, *J. Phys. Chem.* **62**, 54 (1958).
67. G. Blyholder, *J. Phys. Chem.* **68**, 2772 (1964).
68. A. Laforgue, J. Rousseau, and B. Imelick, *Advan. Chem. Phys.* **8**, 141 (1965).
69. H. H. Dunken and C. Opitz, *Z. Chem.* **6**, 234 (1966).
70. H. H. Dunken and C. Opitz, *Z. Phys. Chem.* **60**, 25 (1968).
71. H. Dunken, *Z. Chem.* **10**, 158 (1970).
72. L. Jansen, *Phys. Rev.* **162**, 63 (1967).
73. T. B. Grimley, *Molecular Processes on Solid Surfaces* (Drauglis, Gretz, and Jaffee, eds.), McGraw–Hill, New York (1969).
74. T. B. Grimley, *J. Vacuum Sci. Technol.* **8**, 31 (1971).
75. B. E. Conway and J. O'M. Bockris, *Nature* **178**, 488 (1956).
76. T. Anderson and J. O'M. Bockris, *Electrochim. Acta* **9**, 347 (1964).
77. W. Heiland, *Modern Aspects of Electrochemistry* (J. O'M. Bockris and B. E. Conway, eds.), Volume 11, Chapter 3, Plenum Press, New York (1975).
78. R. A. Oriani and C. A. Johnson, *Modern Aspects of Electrochemistry* (J. O'M. Bockris and B. E. Conway, eds.), Volume 5, Chapter 1, Plenum Press, New York (1968).
79. B. E. Conway and J. O'M. Bockris, *Proc. Roy. Soc.* **248A**, 394 (1958).
80. B. G. Baker, J. O'M. Bockris, and R. W. Revie, *J. Electrochem. Soc.* **122**, 1460 (1975).
81. B. G. Baker, J. O'M. Bockris, and R. W. Revie, *Surface Sci.* **52**, 664 (1975).

4

Time-Dependent Perturbation Theory

4.1. Introduction

The part of science that deals with the *rate* of change of concentration in reactions (chemical kinetics) was based initially upon a primitive phenomenological viewpoint. Thus, the concept that transitions between one compound or atom and another (e.g., $H_2 + I_2 \rightleftarrows 2HI$) is connected with *encounters* between molecules in the gas phase was current in the nineteenth century.[1] The first quantitative theory of the speed at which molecular change occurred arose from the kinetic theory of gases. In these early thoughts, the gas phase only was considered; no attention was paid to happenings on the surfaces nor to those in condensed phases.

The expression for the number of collisions between two particles of a different kind, N_A per unit volume of A and N_B per unit volume of B, is[2]

$$Z_{AB} = \pi N_A N_B r_{AB} \left(\frac{8 \pi k T}{\mu_{AB}} \right)^{1/2} \tag{4.1}$$

where μ_{AB} is the reduced mass of A and B and r_{AB} is the mean of molecular diameter.

Comparison between theory and experiment indicates that every collision is by no means fruitful.[3] The fraction of collisions that are fruitful is generally $<10^{-6}$. Successful collision may, therefore, be infrequent. The idea that molecular collisions are the prerequisites of chemical reactions is the most basic idea in reaction kinetics. However, it is an *insufficient* idea, and there must be at least one concept in addition to account for the great disparity between the collisional consideration and observed rate (see Section 6.4).

A second concept was introduced by Arrhenius[3]—only those collisions result in reaction in which the total energy of the pair is greater than some critical energy. At a crude level, molecules that have less than a certain velocity (and thus less than a certain kinetic energy), when they strike each other, do not strike with sufficient force (and therefore sufficient translational energy to be converted to vibrational states within the bonds) to cause an interchange of the bonding in the molecules. They merely glance off each other without effecting molecular change.

The time of these thoughts was subsequent to the publication in 1860 of the Maxwell law of the distribution of velocity.[4] Evidence became available from the work of Arrhenius[3] that, experimentally, the logarithm of the rate of reaction and the reciprocal of absolute temperature are linearly related. Thus, the rate of reaction V_R is

$$V_R = Z \exp(-E_a/kT) \qquad (4.2)$$

where Z is the collision number and E_a is the energy of activation of reaction.

The third great step on the way to the present quantal theory of reaction rates was the theory of the calculation of the absolute rate of chemical reactions associated with the name of Henry Eyring[5] (but see also Evans and Polanyi[6]). The molecules A and B not only collided with each other and reacted if they made the collision with sufficient total translational energy, but attention was focussed upon the colliding pair as an entity, and a concept of an "activated complex" became fruitful.

The basic expression in the theory of absolute reaction rates is

$$V_R = \frac{kT}{h} C^* \qquad (4.3)$$

where these symbols have their usual significance and C^* is the concentration of the activated complex. All activated complexes, independent of their molecular constitution, decompose at the rate kT/h per complex. Thus, if it is possible to calculate the value of C^*, the concentration of the activated complex, it becomes possible to calculate the absolute rate of the reaction concerned. This theory, which was the leading one in kinetics in the period 1936 until the end of the 1960s, was shown to give rise to calculations of rates in reasonable agreement with experiment for very simple reactions.[6]

How the concentration of the activated complex can be calculated in terms of the partition functions of the reactants and that of the activated complex is described in numerous books,[7,8] but we wish to comment here on one aspect of the theory. Thus, in the expression kT/h of Eq. (4.3), there occurs Planck's constant. This indicates that the Eyring theory has a quantal aspect. The transition state theory represents a stage on the way from theories of reaction kinetics that use only classical statistical mechanics and the kinetic theory of gases to models that are quantal.

Every reaction must be associated with energy changes. An essential process during collision is the transfer of energy from the high translational energies of the (two) reactants and its redistribution first to the vibrational states in the activated complex and finally to the bonds in the products. It is in consideration of such redistribution that the concepts of the theory of rates begin to change from those that are particlelike and classical to those that are wavelike and quantal. For if the essential act during the lifetime of the activated complex is the exchange of bonds and the transfer of energy, then we are dealing with the behavior of electrons undergoing a change from one energy state to another, and all matters connected with energy redistribution at a molecular level must be treated quantum mechanically.

4.1.1. Time-Dependent Perturbation Theory in Kinetics

A part of quantum mechanics that deals with rate processes where one molecular state is converted to another (e.g., spectroscopic transitions, scattering of particles, and electron transfer at an interface) is termed the *time-dependent perturbation theory*.[9] We will explain some of the terms in this branch of quantum chemistry and what they mean and then state what we shall do in the mathematical development.

Thus, the title of the field—the time-dependent perturbation theory—may be explained as follows. By means of the theory, one describes the *time dependence* of some molecular processes. However, it is not accurate to equate the term "time-dependent perturbation theory" and "rate of reaction." The term "rate of reaction" is a phenomenological one and thus refers to overall observable facts. The term "time-dependent perturbation theory" refers to one aspect of the quantal theory of reacting particles. For example, the theory can be used to find the probability of transition occurring between two states of equal or different energy and to find the probability that a molecular state will have a certain critical energy. The theory states nothing about the distribution of energy within the system, nor does it give information about the frequency of proximity of two molecular species. A more historically correct way to regard the meaning of the title "time-dependent perturbation theory" is to regard the "time dependence" as referring to that of *the perturbation* itself.

We may focus upon the type of time dependence we mean. Consider for example a C—C bond. Let it be existing when we first consider it in the lowest possible vibrational–rotational energy state, and the transition, the rate of which we wish to calculate, is the transition to a higher energy state of the same bond after receiving energy due to collision or from the arrival of some electromagnetic radiation (Fig. 4.1).

The disturbance of molecular bonds by the radiational field is the disturbance that is referred to as the perturbation in the idea "time-

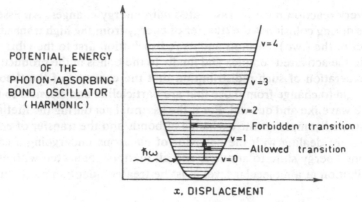

Fig. 4.1. Excitation of a diatomic bond vibration (harmonic) from ground state to first excited state due to perturbation by electromagnetic radiation of energy ω.

dependent perturbation theory." Thus, the electromagnetic field that effects the bonds and perturbs their energy is varying with time, so that we are dealing with the theory of the effects of a perturbation that is time dependent.

In electrode kinetics the fundamental act is the transfer of an electron to or from an electron conductor, from or to a bond in an entity in solution. Thus we are talking about the system, electron in a solid and acceptor (or donor) in solution. The electron is the particle that exists in two states, and it is the transfer of this particle between these two states (contrast the transfer of the vibrational energy state of the bond activated by photons) that we are discussing when we apply time-dependent perturbation theory to electrochemistry (Fig. 4.2). Thus, the time dependence in time-depen-

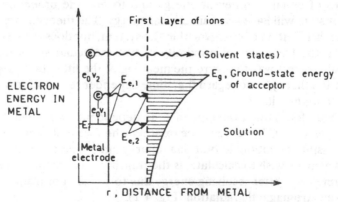

Fig. 4.2. Transition of electron from the energy state in the metal electrode $E_{e,1}$ to the equal energy state $E_{e,2}$ in the acceptor in solution at the interface.

dent perturbation theory in electrochemistry (applied to find the probability of transition of an electron) is the time the perturbation acts on the electron in the system.

Now the perturbation in time-dependent perturbation theory does not have to be itself varying with time; the time dependence may refer to the time during which the perturbation is applied. One would be the electric field between an electrode and an ion in contact with it. When an electron is on the surface of the electrode, it is influenced by the field existing between the surface of the electrode and the acceptor ions in solution. Its motion and energy are thus perturbed and it is this perturbation, itself constant with time, that is referred to in the time-dependent perturbation theory by means of which we deal with the rate of transition of electrons between two states. Thus, time-dependent perturbation theory can be applied to find the transition probability of the electron between the electrode and the solution. The perturbation itself does not have to be time dependent.[10]

In this chapter, we will discuss transitions that are radiationless and compare them with the well-known spectroscopic transitions. These latter are radiational, in the sense that a state m produced from n has in it an extra amount of energy (i.e., the excitation energy) from that of the state n, which energy may express itself as an electromagnetic radiation.

4.1.2. Radiationless Transition

In transitions in interfacial electrochemistry, no radiation is emitted or absorbed.† We assume that the electron in the initial state (the electron within the metal) has the same energy as the electron in the atom (i.e., the ion in solution after it has received an electron[11]; see Fig. 4.2). Thus§

$$E_{e,1} = E_{e,2} \qquad (4.4)$$

The calculation of an electrochemical reaction rate in terms of quantum theory can be regarded as having three separate parts. In the first, we calculate the probabilities of the existence of the electrons in the electronic energy state $E_{e,1}$ in the metal. This can be calculated from Fermi's distribution law and a knowledge of the density of states. The second part is the calculation of the probability of having the necessary partner energy state $E_{e,2}$ in solution. This may be calculated using time-dependent perturbation theory (see Section 6.11). The third part is the heart of the

---

† This is usually true and is the most basic postulate in this book. However, it is not always true, as, for example, in electrochemical luminescence.[12]
§ Historically, this condition was first stated by Ronald Gurney[11] in 1931.

quantal part of the calculation, that of the rate of transfer between these states of equal energy but *different constitution*. The typical situation in electrochemistry is the transition between an electronic state in a metal and an electronic state in an ion in solution.

4.2. General Background

Time-dependent perturbation theory is concerned with the time variation of the stationary state† of the unperturbed system under the action of a perturbation. Thus, time-dependent perturbation theory is designed for the formulation of the probability of a transition between two stationary states of two unperturbed systems, respectively. There are two distinct ways in which such transitions between states can occur.

One is that in which the system is influenced from outside in some way by a time-dependent perturbing force that changes the energy of the state and thus causes it to make a transition from one energy state to another. An example of this is the excitation of an electron in an atom to a higher state in energy by the absorption of a quantum of light (Fig. 4.3a).

The second way in which a transition can occur is that of a system passing between two states, each of equal energy, by the influence of a time-independent perturbing force§ acting for a certain (short) time. An example of this is the radiationless transition of an electron from the Fermi level in an electronically conducting system to an electronic state in an

---

† At first sight the idea of "time dependence of a stationary state," apparently meaning the variation of the probability of having a state with time, is a contradiction in terms. However, the concept of the "time dependence of a stationary state" has a quantum mechanical meaning in the following way. A "stationary state" in quantum mechanics means the state that is the result of the solution of the time-dependent Schrödinger equation in which the potential function is not time dependent. Thus, considering the lowest energy state of a hydrogen atom, we may calculate the equation for the state of the electron in the situation and obtain its stationary state. The physical analog is that of a string in which the oscillations have been set up in a stationary way or the bond oscillation of a diatomic molecule.

We can, however, take such a stationary state of the type $\psi(x, t) = \psi(x) e^{-iEt/\hbar}$ and, by perturbing it, persuade it to make some of the molecular entities of which it consists to undergo a transition to another state, which is also stationary in the quantum mechanical sense. As the many entities that make up the actual macroscopic meaning of the first stationary state gradually change as individuals to the second stationary state, we may speak about the time dependence of the stationary state (i.e., transition of a diatomic bond from its stationary ground state to a higher quantum state that is also stationary).

§ Note that the time independence is of the *force* acting upon the system, not of the *energy*. Thus, if a force acts upon a particle in a certain time it undergoes an energy change and this itself may depend upon time, but we still speak about a time independent perturbation. However, if the force varies with time, then the perturbation is said to be time dependent, and if the force does not vary with time, not time dependent.

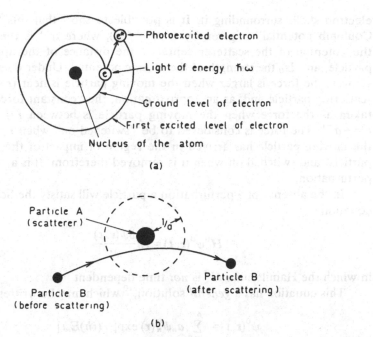

Fig. 4.3. (a) Excitation of an electron in an atom due to perturbation of light energy of quantum ω.   (b) Scattering of a particle B with the atom A (scatterer) during collisional perturbation.

acceptor ion in solution. The ion is to have the energy of one of its outer electronic states equal to that of the electron in the metal. The time-independent electric field between the metal and the ions in the double layer in solution provides a force on the electron during its transition from the metal to the solution and *this* is the origin of the energy that perturbs the electron and stimulates its transition.

Another example of a phenomenon of transition that involves a time-independent perturbing force is the scattering of a beam of particles by a fixed center of force such as an atomic nucleus, where the particles are deflected into another direction, sometimes with no change in their energy (elastic scattering, Fig. 4.3b).

Acceptance of this scattering process as one that is consistent with the definition of time independence in respect to the perturbation needs some sophistication in understanding. Thus, at first sight, the Coulombic force that acts between the charge centers of the two particles in scattering is connected to distance, and therefore, in respect to a moving particle approaching a charge center, the force should be in fact time dependent. An approximation is made (see below) in which one deals with the inter-action of the particle in terms of the screening of the nucleus by the electron shells surrounding it. It is possible to show that this "screened Coulomb potential" is given by $U_0(e^{-ar}/ar)$, where a^{-1} is the range of the potential of the scatterer center, r the distance of the approaching particle, and U_0 the strength of the atomic potential. Under these circumstances, the force is larger when the moving particle is near $(r < a^{-1})$ the scattering particle and, as an approximation, the "constant force" can be taken as the force when the moving particle is between $r = +a^{-1}$ and $r = -a^{-1}$. The force is considered to be "switched on" when $r = a^{-1}$, i.e., the moving particle has arrived in the region of impact of the scattering particle, and switched off when it is removed therefrom. It is a "constant" perturbation.

In the absence of a perturbation a particle will satisfy the Schrödinger equation

$$H^0\psi^0(\mathbf{r}, t) = i\hbar \frac{\partial \psi(\mathbf{r}, t)}{\partial t} \tag{4.5}$$

in which the Hamiltonian H^0 is *not* time dependent.

This equation has a *general* solution,[10] which may be written as

$$\psi^0(\mathbf{r}, t) = \sum_{n=0}^{\infty} a_n \psi_n^0(\mathbf{r}) \exp[-(i\hbar)E_n t] \tag{4.6}$$

where $\psi^0(\mathbf{r}, t)$ is the wave function before any perturbation (time dependent or not) is applied, the values a_n are *time-independent constants* and represent the amplitude of the waves; E_n is the energy of any unperturbed state n (not a particular chosen state).

Here, the $\psi^0(\mathbf{r}, t)$ is not the ψ of a definite state, such as ψ_3, or ψ_m, etc., but it is the wave function of the whole system with all its states. All the wave functions $\psi_n^0(\mathbf{r})$ satisfy the time-independent Schrödinger equation:

$$H^0(\mathbf{r})\psi_n^0(\mathbf{r}) = E_n\psi_n^0(\mathbf{r}) \tag{4.7}$$

Now, in the presence of a perturbation energy U^{pert}, the time-dependent Schrödinger equation has the form

$$(H^0 + U^{\text{pert}})\psi(\mathbf{r}, t) = i\hbar \frac{\partial \psi(\mathbf{r}, t)}{\partial t} \tag{4.8}$$

where U^{pert} may be either time dependent or time independent.[10]

Though we cannot solve Eq. (4.8) explicitly, the solution may be expressed as an expansion in terms of unperturbed wave functions in the following way:

$$\psi(\mathbf{r}, t) = \sum_n a_n(t) \psi_n^0(\mathbf{r}) \exp[-(i/\hbar)E_n t] \tag{4.9}$$

where now the coefficient a_n becomes time dependent as a result of the perturbation (whether this itself is time dependent or not), and ψ is the amplitude.

When Eq. (4.9) is substituted in Eq. (4.8), the result involves the sum of the ψ values of the various states:

$$\sum_n a_n(t) H^0 \psi_n^0(\mathbf{r}) \exp\left(-\frac{i}{\hbar} E_n t\right) + \sum_n a_n(t) U^{\text{pert}} \psi_n^0(\mathbf{r}) \exp\left(-\frac{i}{\hbar} E_n t\right)$$

$$= i\hbar \sum_n \frac{\partial a_n(t)}{\partial t} \psi_n^0(\mathbf{r}) \exp\left(-\frac{i}{\hbar} E_n t\right) + \sum_n a_n(t) \psi_n^0(\mathbf{r}) E_n \exp\left(-\frac{i}{\hbar} E_n t\right) \quad (4.10)$$

By the use of Eq. (4.7), one can write the first term of the left-hand side of Eq. (4.10) as $\sum_n a_n(t) E_n \psi_n^0(\mathbf{r}) \exp[-(i/\hbar)E_n t]$, which then cancels the second term of the right-hand side of Eq. (4.10), and thus Eq. (4.10) becomes

$$\sum_n a_n(t) U^{\text{pert}} \psi_n^0(\mathbf{r}) \exp\left(-\frac{i}{\hbar} E_n t\right) = i\hbar \sum_n \frac{\partial a_n(t)}{\partial t} \psi_n^0(\mathbf{r}) \exp\left(-\frac{i}{\hbar} E_n t\right) \quad (4.11)$$

or

$$i\hbar \sum_n \frac{\partial a_n(t)}{\partial t} \psi_n^0(\mathbf{r}) \exp\left(-\frac{i}{\hbar} E_n t\right) = \sum_n a_n(t) U^{\text{pert}} \psi_n^0(\mathbf{r}) \exp\left(-\frac{i}{\hbar} E_n t\right) \quad (4.12)$$

The aim of this discussion is to find an expression for the time dependence of the coefficient, a_n. Since we are interested in the rate of change of a state (say, the state l) to another state (say, the state m), we shall be interested in learning the rate of formation (or rate of change) of a_m, i.e., da_m/dt, from the state l. If we know the value of this differential coefficient, we can integrate it and obtain a_m at a time t, $a_m(t)$.

Why should we be interested in knowing $a_m(t)$? As will be established in Section 4.4, it is because the square of the quantity a_m [i.e., $|a_m(t)^2|$] represents the probability of transition from the stationary state l to the stationary state m.

In order to find da_m/dt, we use a mathematical device, namely, we multiply from the left both sides of Eq. (4.12) by the complex conjugate of a particular unperturbed wave function, i.e., $\psi_m^{0*}(\mathbf{r}, t) = \psi_m^{0*}(\mathbf{r}) \exp(iE_m t/\hbar)$ and integrate over all space. We obtain

$$i\hbar \sum_n \frac{\partial a_n(t)}{\partial t} \exp\left[\frac{-i(E_n - E_m)t}{\hbar}\right] \int \psi_m^{0*}(\mathbf{r}) \psi_n^0(\mathbf{r}) \, d\mathbf{r}$$

$$= \sum_n a_n(t) \exp\left[\frac{-i(E_n - E_m)t}{\hbar}\right] \int \psi_m^{0*}(\mathbf{r}) U^{\text{pert}} \psi_n^0(\mathbf{r}) \, d\mathbf{r} \quad (4.13)$$

The integral on the left-hand side of Eq. (4.13) becomes zero if m and n are different states (orthonormal condition), but if n and m are in fact the same states, it becomes unity. When $n = m$, $E_n = E_m$, and the left-hand side of Eq. (4.13) is simply

$$i\hbar \frac{\partial a_n(t)}{\partial t}$$

The summation ($n = 0, 1, 2, \ldots$) is also dropped, for n only is equal to m at *one* value of n.
Hence†

$$i\hbar \frac{da_m(t)}{dt} = \sum_n U_{mn}^{\text{pert}} a_n(t) \exp\left[\frac{-i(E_n - E_m)t}{\hbar}\right] \qquad (4.14)$$

where

$$U_{mn}^{\text{pert}} = \int \psi_m^{0*}(\mathbf{r}) U^{\text{pert}} \psi_n^0(\mathbf{r}) \, d\mathbf{r} \qquad (4.15)$$

is the matrix element of perturbation that connects an initial state, one of the values of n, say l, the ground state, with the final state m, the excited state of an atom or a molecule. Here, $d\mathbf{r}$ in Eq. (4.15) is $dx\,dy\,dz$ in Cartesian coordinate and $r^2 \sin\theta\,d\theta\,d\phi$ in spherical polar coordinate systems.

The rate of increase of the coefficient $a_m(t)$, da_m/dt, from its initial value of zero (when the whole system is in the state n) is [according to Eq. (4.14)] proportional to (a) the magnitude of the initial coefficient $a_n(t)$, (b) the matrix element of perturbation U_{mn}^{pert} between two states n and m, and (c) the oscillatory factor $\exp[-i(E_m - E_n)t/\hbar]$. Equation (4.14) also shows that at any finite time t, the rate of increase of the coefficient $a_m(t)$ out of an initial state n is zero when the perturbation energy U^{pert} is zero. In this circumstance, the transition matrix U_{mn}^{pert} will also be zero. Hence, the matrix element of the perturbation U_{mn}^{pert} between two states can be regarded as responsible for the building of $a_m(t)$, i.e., the state m being formed from an initial state l among the set of states n. For this reason, the matrix element U_{mn}^{pert} is called, in general, the *transition matrix*.

---

† It must be understood that in Eq. (4.14) E_n is not equal to E_m. Thus, in Eq. (4.13), we have obtained the situation resulting there, in which we have removed the summation sign before the term $da_m(t)/dt$. We did this by taking a special condition in which $E_n = E_m$ due to orthogonality. The left-hand side of Eq. (4.13) is therefore entirely self consistent. In it, the summation sign has been removed, and we are only considering one of the many values of E_n.

On the right-hand side of Eq. (4.13), and correspondingly Eq. (4.14), we do not remove the summation sign, and therefore we are considering all the values of E_n. There is nothing inconsistent in equating the left-hand side of Eq. (4.13) (one value of E_n only considered) with the right-hand side, where the summation sign involves the consideration of all the values of E_n.

4.3. An Interim Comment

It would be helpful if we could solve equations such as those of Eq. (4.14). The reason for this is that, although a_m, which we should get from such a solution, is not itself a wave function but only a coefficient of a wave function, it can be shown (and we shall show it in Section 4.4) that the product of a_m with its complex conjugate a_m^* gives in fact the transition probability, the computation of which is the objective in the quantum contribution to the theory of reaction rates.

Before we show how Eq. (4.14) can be solved, let us first *prove* that $a_m^* a_m$ does indeed represent the probability of a transition from one state (above) called l, one of the states in the series of n (where $n = 0, 1, 2, \ldots, l$) to a state m.

4.4. Probability of Transition

Let us consider the reaction

$$A \rightarrow B \qquad (4.16)$$

The rate of reaction†

$$V_R = \frac{dC_B}{dt} = pC_A \qquad (4.17)$$

when p is the *probability of transition from A to B per unit time* and C_A and C_B are the concentrations of A and B, respectively. Therefore,

$$p = \frac{1}{C_A} \frac{d}{dt}(C_B) \qquad (4.18)$$

Now, we find out the probability of transition P *during the time* the perturbation is turned on (not per unit time, for which the probability is p). We suppose that the perturbation is turned on at $t = 0$ and turned off after a short time at $t = \tau$ (perturbation time). Hence, the probability that a transition will occur during the time interval 0 to τ can be obtained by

---

† This expression (4.17) is not inconsistent with Eq. (4.3), since one can show that

$$V_R = \frac{kT}{h} C_A^* = \frac{kT}{h} C_A \exp\left(\frac{-E_a}{kT}\right) = PC_A \qquad (4.17a)$$

where $P \equiv (kT/h)\exp(-E_a/kT)$, i.e., probability per unit time.

integrating Eq. (4.18) from 0 to τ. Thus,† the total transition probability in time τ, P_T, is

$$P_T = p \int_0^\tau dt = \frac{1}{C_A} \int_0^\tau \frac{d}{dt}(C_B)\, dt \qquad (4.19)$$

Therefore,

$$P = p\tau = \frac{C_B}{C_A} \equiv \frac{P_B N}{P_A N} \qquad (4.20)$$

where P_A represents the probability of having the system in the initial state A and P_B represents the probability of having the system in the state B at time τ, and N is the total number of particles before reaction.

One of the ways of relating the wave functions of Eq. (4.20) to coefficients such as a_m and a_n begins with one general kind of expression for the *average* energy in quantum mechanics. The kind of energy to which we refer has the technical name, the *expectation value* of the energy, because it is the average value of the energy of the system counting the contributions from all the states that belong to the system.

This expectation value is defined by the equation

$$E = \int_{-\infty}^{+\infty} \psi(\text{energy operator})\psi\, d\tau \qquad (4.21)$$

where $d\tau$ is a volume element. What is the energy operator? Now, the general (i.e., including time dependence of the states) form of the Schrödinger equation is

$$H\psi = i\hbar \frac{d}{dt}(\psi) \qquad (4.22)$$

But, in a stationary state,

$$H\psi = E\psi \qquad (4.23)$$

Hence, using Eqs. (4.22) and (4.23),

$$\langle E \rangle = \int \psi^* H\psi\, d\tau$$

$$= \int \psi^* i\hbar \frac{d}{dt}(\psi)\, d\tau \qquad (4.24)$$

† The equations derive and involve the ratio of the concentrations of the two states at a time corresponding to the end of the period during which the perturbation is on, i.e., between 0 and τ sec. The ratio C_B/C_A is in no sense to be confused with an equilibrium ratio (for the concentrations do not refer to the state at which equilibrium has been reached).

or

$$\langle E \rangle = \int_{-\infty}^{+\infty} \psi^*(\mathbf{r}, t) i\hbar \frac{d}{dt} \psi(\mathbf{r}, t) \, d\tau \tag{4.25}$$

From Eqs. (4.9) and (4.25), one gets

$$\langle E \rangle = \int_{-\infty}^{\infty} \sum_{n=1}^{\infty} a_n^* \psi_n^*(\mathbf{r}) \exp\left(\frac{iE_n t}{\hbar}\right) i\hbar \sum_{n=1}^{\infty} a_n \psi_n(\mathbf{r}) \left(\frac{-iE_n}{\hbar}\right) \exp\left(\frac{-iE_n t}{\hbar}\right) d\tau$$

$$= \sum_{n=1}^{\infty} E_n a_n^* a_n \exp\left[\frac{-i(E_n - E_n)t}{\hbar}\right] \int_{-\infty}^{\infty} \psi_n^*(\mathbf{r}) \psi_n(\mathbf{r}) \, d\tau$$

$$= \sum_{n=1}^{\infty} E_n a_n^* a_n \tag{4.26}$$

(using the principle of orthogonality).

However, the general expression for the expectation value of the energy of a system, or the mean value, can be written as

$$\langle E \rangle = \sum_{n=1}^{\infty} E_n P_n \tag{4.27}$$

when P_n is the probability of the system at state n. Hence, by comparing Eqs. (4.26) and (4.27), we find that the probability of a certain state n can be written as

$$P_n \equiv a_n^* a_n = |a_n|^2 \tag{4.28}$$

Similarly, we can find the probability of certain state m as

$$P_m = a_m^* a_m = |a_m|^2 \tag{4.29}$$

Hence, $|a_m|^2$ represents the probability of the existence of the state m and thus,

$$\psi_m^* \psi_m = |a_m(t)|^2 \tag{4.30}$$

Now, using the Eq. (4.28) in Eq. (4.19), we get the transition probability[13] [see Eq. (4.19)]

$$P_T = \frac{|a_B(t)|^2}{|a_A(t)|^2} = |a_B(t)|^2 \tag{4.31}$$

since at $t = 0$ the probability of finding the system in the initial state A is unity, and thus $|a_A(t)|^2 \simeq 1$. Similarly, $P_T = |a_m(t)|^2/|a_n(t)|^2 = |a_m(t)|^2$, for in the initial state, $|a_n(t)|^2 = 1$.

Hence, we have proved the necessity for obtaining $a_m(t)$ by solving Eq. (4.14). If we can solve it, we shall know the transition probability. But the transition probability per second is the rate of transition from l to m, or, in general, from one state to another.

4.5. Golden Rule for Transition Rates

A complete mathematical solution to the general differential equation of Section 4.2 [Eq. (4.14)] is not easy to obtain owing to the presence of summations over all the various energy states on the right-hand side. To attempt a solution, it is necessary to make an approximation. Instead of taking into account the successive values of a_n, corresponding to the states 1, 2, 3, ... (e.g., $a_1 + a_2 + a_3 + a_l + \cdots$), we would take into account only that one state that contributes most to the sum,† e.g., a_l. As there is now (by this arbitrary approximation) only one state, or one value of a_n, we shall no longer need the summation sign.§ The coefficient (i.e., a_l) changes with time t after the perturbation is on. At $t = 0$, $a_l = 1$. As a further approximation, we take $a_l \simeq 1$ at $t \geq 0$, i.e., we consider *small* changes.

The aim of this is to obtain not a_l, the coefficient of the *initial* state, which we are subjecting to the approximations we have been discussing, but a_m, the corresponding coefficient at the *final* state. It is clear that, just before the perturbation is applied, at $t \leq 0$, the value of a_m must be zero and, for the short time we are to consider the system, it will then become finite and increase, after the perturbation has been switched on.¶

We obtain a_m from the new version of Eq. (4.14), namely, that which takes into account the approximation, and becomes

$$i\hbar \frac{da_m}{dt} = U_{ml}^{\text{pert}}(1) \exp\left[\frac{-i(E_l - E_m)t}{\hbar}\right] \tag{4.33}$$

where the (1) is the value of a_l under the approximation that we are looking at the situation under a time scale such that $a_l \simeq 1$.

---

† The question, of course, arises, Which state? This depends upon the problem and we have to exercise choice of the single state considered so that it *is* the greatest contributor. It is clear that the validity of the approximation depends on the system's having one such predominant state (e.g., the ground state), and the applicability of the approximation in practice depends on knowing what this state is.

§ There is no contradiction with the argument leading up to Eq. (4.13). On the left-hand side of Eq. (4.13), we are only considering *one* state—there is no approximation as yet. But on the right-hand side, we are considering a *sum* of states. This is the part we approximate. The physical meaning of the approximation is that we are considering no longer a transition from any state to any state but from *one* state to any state.

¶ Some might see a contradiction in the fact that a_m is *increasing* when the *parent* coefficient (i.e., the coefficient of the initial state a_l), is taken as remaining almost at unity for $t \geq 0$. This is the way of things in this approximation. It reminds us that one of the limitations of this time-dependent perturbation theory is that the time of the perturbation τ is small and is defined by the equation[10,14]

$$\tau \leq \frac{h}{\Delta E} \tag{4.32}$$

where τ is the time of perturbation and ΔE is the difference in energy between the initial and final state. If, for example, $\Delta E = 1$ eV, $\tau \leq 10^{-14}$ sec. So long as we are within the condition (4.32), it is admissible to take $a_l \simeq 1$ at $t > 0$.

Integrating Eq. (4.33) with respect to t we get

$$a_m(t) = \frac{1}{i\hbar} \int_0^t U_{ml}^{\text{pert}} \exp(-i\omega_{ml}t) \, dt \qquad m \neq l \tag{4.34}$$

where

$$U_{ml}^{\text{pert}} = \int \psi_m^* U^{\text{pert}} \psi_l \, dt \tag{4.35}$$

and

$$\omega_{lm} = \frac{E_l - E_m}{\hbar} \tag{4.36}$$

For a perturbation that is turned on at $t = 0$ and off at time t, Eq. (4.34) may be integrated to obtain a_m as a function of time. It is easy to show, thus, that

$$a_m(t) = \frac{U_{ml}^{\text{pert}}}{i\hbar} \int_0^t \exp(-i\omega_{lm}t) \, dt$$

$$= \frac{U_{ml}^{\text{pert}}}{\hbar} \frac{\exp(-i\omega_{lm}t) - 1}{\omega_{lm}} \tag{4.37}$$

The probability that the system, known to have been in the initial state l at $t = 0$, will be in the unperturbed final state m at time t is given [Eq. (4.31)] by $|a_m|^2$. Hence,†

$$|a_m(t)|^2 = \frac{|U_m^{\text{pert}}|^2}{\hbar^2 \omega_{lm}^2} \left| \exp(-i\omega_{lm}t) - 1 \right|^2$$

$$= \frac{|U_m^{\text{pert}}|^2}{\hbar^2 \omega_{lm}^2} \{[\exp(i\omega_{lm}t) - 1][\exp(-i\omega_{lm}t) - 1]\}$$

$$= \frac{|U_m^{\text{pert}}|^2}{\hbar^2 \omega_{lm}^2} [2 - \exp(i\omega_{lm}t) - \exp(-i\omega_{lm}t)]$$

$$= \frac{2|U_m^{\text{pert}}|^2}{\hbar^2 \omega_{lm}^2} [1 - \cos \omega_{lm}t] \tag{4.38}$$

† Why do we take the modulus of $a_m(t)$ and square it, and not just a_m? It is because we must have a mathematically *real* answer for the probability of existence of a state. Of course, if a_m before squaring does not contain i, squaring will always produce a real answer in the mathematical sense, i.e., one not containing a complex number. If we square a_m simply, and it does contain a complex, the answer will be negative, meaningless as far as probability is concerned. However, a *modulus* is the multiplicand of a quantity times its complex conjugate. Thus, suppose one has a quantity ia, then its complex conjugate is $-ia$, so that $ia(-ia) = +a^2$, and is therefore positive and real. Now if the quantity is a (not ia), the complex conjugate is a, and hence $a \cdot a = a^2$, positive and real.

Expression (4.38) is an oscillating function of time. When a *weak* perturbation is applied (and for what is meant by "weak," see below) the probability amplitude for transition is at a maximum when the unperturbed energies of the initial and final states are very close together.

This can be proved by expanding $\cos \omega_{lm} t$ in Eq. (4.38) as follows:

$$|a_m(t)|^2 = \frac{2|U_{lm}^{\text{pert}}|^2}{\hbar^2} \left(\frac{1 - \cos \omega_{lm} t}{\omega_{lm}^2} \right)$$

$$= \frac{2|U_{lm}^{\text{pert}}|^2}{\hbar^2} \left| \frac{1 - (1 - \omega_{lm}^2 t^2/2! + \omega_{lm}^4 t^4/4! \cdots)}{\omega_{lm}^2} \right|$$

$$= \frac{2|U_{lm}^{\text{pert}}|^2}{\hbar^2} \left(\frac{t^2}{2!} - \frac{\omega_{lm}^2 t^4}{4!} \cdots \right) \tag{4.39}$$

This equation indicates that as ω_{lm} becomes smaller, $|a_m(t)|^2$ becomes larger in number.

Neglecting the second term in brackets in Eq. (4.39) for $\omega_{lm} t \ll 1$,† one obtains[10]

$$|a_m(t)|^2 \simeq \frac{1}{\hbar^2} |U_{lm}^{\text{pert}}|^2 t^2 \tag{4.40}$$

where it is appropriate to consider t as the time *during which the perturbation is acting on the system*. The time t is a specific time, that of the short perturbational interaction, e.g., 10^{-14} sec. This gives transition probability from a *certain* state l to another *certain* state m, where $\omega_{lm} t \ll 1$.

4.5.1. A More Realistic Approach to the Calculation of the Probability of Transition

If we consider Eq. (4.40), it is seen that the transition probability depends on the time *squared*. Now at first, there seems nothing against this conclusion. But what is rate? The rate of change of one state to another is

---

† In Eq. (4.38), the exponential expression is an oscillatory function. In the case of a transfer under the Gurney condition in electrochemical processes, the value of $(E_m - E_l)$ is zero because of the radiationless transfer condition. In order to make sense of the equation, we assume that $E_m - E_l$ is subject to the uncertainty principle so that the $\Delta E = h/\Delta \tau$, or h/τ, where τ is the time of perturbation, which is small enough to have the first-order approximation valid, so that $\Delta \tau = h/\Delta E$, and ΔE is obtained from $\Delta p \, \Delta x = h$ when Δx is the uncertainty in the position of electron, which is roughly equal to the de Broglie wavelength of the outcoming electron. This is of the order of the thickness of the double layer, $\sim 5 \times 10^{-8}$ cm. Using this value of Δx, Δp is obtained and thus ΔE is found from $\Delta E = (\Delta p)^2/2m$. Thus with the use of this ΔE value one gets a perturbation time, $\tau \sim 10^{-15}$ sec (for the electrochemical case, this is of the order of the transition time of an electron across the double layer). In this time, one is always in the first phase of the oscillatory variation indicated by Eq. (4.38).

identical with the probability of transition per unit time. Hence, regarding Eq. (4.40), we see that, according to it, the rate R_t is

$$R_t = \frac{|a_m(t)|^2}{t} = \frac{1}{\hbar^2}|U_{ml}^{\text{pert}}|^2\frac{t^2}{t} \tag{4.41}$$

Thus, according to this first treatment, rate† is proportional to the time during which the perturbation acts on the system.

The transitions in real systems do not actually occur precisely from one state to one single state; i.e., the uncertainty principle does not allow us to make a measurement that refers to a transition from a given state l precisely to a given state m. Apart from the fact that the uncertainty principle forbids an exact definition, there are practical reasons why it is not possible to observe a transition as precise as that implied in the treatment we have given, e.g., lack of absolutely monochromatic light. For this reason, we shall now calculate the probability of transition from a state l to a series of states m (e.g., $m_1, m_2, \ldots$), all of which are near to each other.§

---

† Here again, we must guard against confusion. The rates we refer to in phenomenological matters are, of course, "real rates," i.e., they refer the direct result of observing, e.g., the production of one new molecular species from the interaction of two others. In referring here to rates in the theory of time-dependent perturbation, we are talking about a more restricted meaning of rate, namely, what the rate would be if the collision frequency were unity and the probability of existence of the energy state l were also unity. Of course, there are no real cases where this is so, but this does not mean that our discussions of time-dependent perturbation theory are not applicable to real cases, for we can easily use for the collision frequency and probability of existence of a matching quantum state in our expression for the special meaning of rate we have calculated here.

§ Thus, at the beginning of our considerations, we considered the rate of transition from a *series* of initial states a_1, a_2, a_3 to *one* final state [cf. Eq. (4.14)]. Because of the mathematical difficulties of solving (4.14), we then approximated to consider the probability of transition from *one* state l to *one* state m [Eq. (4.33)]. In making the change to this second, and better, calculation of the probability of transition, we consider now transition to several states from one initial state and the question obviously arises: Why can we do this when we have limited the initial state and accepted this limitation as a reasonable approximation?

What we are doing is a reasonable procedure for the following cause. In considering the *initial* state, and limiting it to the consideration of just one substate, we correspond to a frequently met reality, e.g., that a vibrating chemical bond is in fact predominantly in its ground state. It is therefore admissible to neglect contributions for the other state, for their concentration is often small.

However, even though we start from an initial state that can be simplified to be *almost* a single substate, the state *to which* transition is made is certainly likely to be *significantly* more than one substate. For example, in the excitation by light of an initial vibrational ground state, the bond is not activated *precisely* by the amount $h\nu$ received but to that amount plus or minus the uncertainty principle energy, $\Delta E = h/\tau'$, where τ' is the lifetime of the final state.

To find the total probability $P_{l \to m}$ that the particle has made a transition from the initial eigenstate l to some series of final eigenstates m, we write

$$P_{l \to m} \equiv \sum_{m=0}^{m=\infty} |a_m(t)|^2 \tag{4.42}$$

In order to replace the sum by an integral,† we assume that there are a large number of closely spaced (quasi-continuum) final eigenstates in the range ΔE ($\sim h/\Delta t$); the number of final eigenstates dN_m per energy interval dE_m ($\approx dE_m$) is the density of the final states, i.e., $\rho_m(E_m) = dN_m/dE_m$. Hence, we get from Eq. (4.42),

$$P_{l \to m} = \int |a_m(t)|^2 \, dN_m \tag{4.43}$$

Using Eq. (4.38), since $2(1 - \cos \omega_{lm} t) = 4 \sin^2 \frac{1}{2}\omega_{lm} t$, we get

$$P_{l \to m} = \frac{4}{\hbar^2} \int_{-\infty}^{\infty} |U_{ml}^{\mathrm{pert}}|^2 \left| \frac{\sin^2 \frac{1}{2}\omega_{lm} t}{\omega_{lm}^2} \right| \frac{dN_m}{dE_m} \, dE_m$$

$$= \frac{4}{\hbar^2} \int_{-\infty}^{\infty} |U_{ml}^{\mathrm{pert}}|^2 \left| \frac{\sin^2 \frac{1}{2}\omega_{lm} t}{\omega_{lm}^2} \right| \rho(E_m) \, dE_m \tag{4.44}$$

In respect to the limits of integration in Eq. (4.43), the value of the expression is found to have nearly all its contribution from a quite small interval of energy—in fact, the interval ΔE around the energy range of the E_m. It is mathematically convenient to continue to work with the above expression using the limits of integration $-\infty$ to $+\infty$, but when we come to deal with $\rho(E)$ and U_{ml}^{pert} and ask ourselves in what way they are functions of energy, it is a reasonable approximation to take them as constant, since the integrand in Eq. (4.44) is a sharply peaked function around E_m; i.e., most contributions come from the range ΔE. On this basis, we take§ $\rho(E)$ and U_{ml}^{pert} out of integral sign.

Hence, Eq. (4.44) can be written as

$$P_{l \to m} = \frac{4|U_{ml}^{\mathrm{pert}}|^2}{\hbar^2} \rho(E_m) \int_{-\infty}^{\infty} \frac{\sin^2 \omega_{lm} t}{\omega_{lm}^2} \, dE_{lm} \tag{4.45}$$

---

† The difficulty of introducing the integration sign in developing Eq. (4.43) is different here than it was in Eq. (4.13). In Eq. (4.13), the integration was introduced to make use of the orthogonality condition in order that we could eliminate the summation sign on the left-hand side of the equation. In (4.34), it is introduced to obviate the difficulty of the infinite summation. The first step is not an approximation and the second step *is* one.

§ It should not be forgotten that at l, the initial state is single and defined, but m is a series of states close together.

Now, since

$$E_{lm} = \hbar\omega_{lm}$$

$$dE_{l-m} = \hbar d\omega_{lm}$$

we can change the variable in Eq. (4.45) from E to ω_{lm} and get

$$P_{l\to m} = \frac{|U_{ml}^{\text{pert}}|^2}{\hbar^2}\hbar\rho(E_m)\int_{-\infty}^{\infty}\frac{\sin^2\frac{1}{2}\omega_{lm}t}{\omega_{lm}^2}\,d\omega_{lm}$$

$$= \frac{|U_m^{\text{pert}}|^2}{\hbar}\rho(E_m)t^2\int_{\infty}^{\infty}\frac{\sin^2\frac{1}{2}\omega_{lm}t}{\frac{1}{4}\omega_{lm}^2t^2}\,d\omega_{lm} \qquad (4.46)$$

Now, putting

$$\tfrac{1}{2}\omega_{lm}t = z$$

so that

$$d\omega_{lm} = 2dz/t \qquad (4.47)$$

then, with Eq. (4.47) in Eq. (4.46),

$$P_{l\to m} = \frac{2|U_{ml}^{\text{pert}}|^2}{\hbar}t\rho(E_m)\int_{-\infty}^{\infty}\frac{\sin^2 z}{z^2}\,dz$$

Now, as $\int_{\infty}^{\infty}(\sin^2 z)/z\,dz = \pi$, one obtains

$$P_{l\to m} = \frac{2\pi}{\hbar}|U_{ml}^{\text{pert}}|^2\rho(E_m)t \qquad (4.48)$$

Thus, by this second approach to the probability of transition, we have found a proportionality between probability and the time during which the perturbation effects the system t, not t^2, as in our first attempt [cf. Eq. (4.40)]. This makes, of course, a very substantial difference.

4.5.2. Calculation of Rate

The rate from the initial state l may be regarded as the transition probability per unit time and is, therefore,

$$R_t = \frac{dP_{l\to m}}{dt} = \frac{2\pi}{\hbar}|U_m^{\text{pert}}|^2\rho(E_m) \qquad (4.49)$$

Equation (4.49) is the famous *Fermi's golden rule*. It is widely used because of its general applicability in the field of kinetics, scattering problems, and other velocity calculations in quantum chemistry and physics.

4.6. Applicability of Time-Dependent Perturbation Theory (TDPT)

The time-dependent perturbation theory to find the transition probability or the rate of transition is applicable under two major approximations. The *first approximation* concerns the *time* condition for the applicability of the theory. This condition is that the time of perturbation τ will be small enough such that[14]

$$\tau \gtrsim \frac{h}{\Delta E} = \frac{2\pi}{\omega_{ml}} \tag{4.50}$$

where $\Delta E = E_m - E_l$.

The *second approximation* concerns the *perturbation energy* limitation and qualitatively we know that energy of the perturbation must be relatively smaller than the energy content of the system. How small must it be? One of the conditions for the maximum of the transition probability is that

$$\omega_{ml} \simeq 0 \tag{4.51}$$

Hence,

$$E_l \simeq E_m \tag{4.52}$$

This means that perturbation energy must be in this sense, therefore, so small that we can essay what this means by using the uncertainty principle:

$$\Delta E \, \Delta t \approx h \tag{4.53}$$

when ΔE will be our perturbation energy.

Hence, for the applicability of time-dependent perturbation theory

$$\Delta E \leq \frac{h}{\Delta t} \tag{4.54}$$

where $\Delta t \simeq \tau$ is the time of perturbation. Therefore,

$$\hbar \omega_{ml} \leq \frac{h}{\tau} \tag{4.55}$$

or

$$\hbar \omega_{ml} \tau \leq h \tag{4.56}$$

Hence,

$$\int_0^\tau \hbar \omega_{ml} \, dt \leq h \tag{4.57}$$

or in general,

$$\int_0^\tau U^{\text{pert}}\, dt \leqslant h \qquad (4.58)$$

Equation (4.58) will be the general condition of perturbation energy that will limit the applicability of the *first-order* time-dependent perturbation theory.

4.7. Example of the Applicabilities of TDPT

In this section we will discuss the applicability of conditions (4.50) and (4.58) for (a) the transition of electrons across an interface; (b) spectroscopic dipole transition, and (c) the energy transfer occurring during a collision between molecules.

(a) For the *electrochemical electron transfer case*, if we consider that the electron transfer time is the time of duration of the perturbation, τ, since when the perturbation gets hold of the initial state electron in the metal surface at $t = 0$, it undergoes transition in the next 10^{-16} sec† and thus the perturbation can act for only 10^{-16} sec on the outcoming surface electron. ΔE_{ml} is close to zero as is obtained from the uncertainty principle using the relation

$$\Delta p\, \Delta x \simeq h \qquad (4.59)$$

If the uncertainty of the position of the electron is the distance between the metal surface and the acceptor state, which is assumed to be in the first row of ions in solution, i.e., 5 Å, then $\Delta x = 5 \times 10^{-8}$ cm, and we get Δp from Eq. (4.59) and find $\Delta E \simeq 10^{-12}$ erg. Using Eq. (4.54), one finds $\Delta t \simeq \tau$ and

$$\tau \leqslant 2 \times 10^{-13}\, \text{sec} \qquad (4.60)$$

Thus, we notice that the restriction concerning the time of perturbation [cf. Eq. (4.50)] is satisfied in the case of the electrochemical transition if the transition time of electron ($\sim 10^{-16}$ sec) is considered as the duration of the perturbation.

† This figure is obtained as follows:

$$\text{Electron transfer time} = \frac{\text{distance between the electrode and acceptor ion}}{\text{velocity of Fermi-level electron}}$$

$$= \frac{5 \times 10^{-8}\, \text{cm}}{3 \times 10^8\, \text{cm sec}^{-1}} \simeq 10^{-16}\, \text{sec}$$

However, this calculation is an approximate one and is applicable for electron transfer over the barrier. More accurate calculation will involve the quantum mechanical calculation of finding the time delay of the electron in the barrier region.

(b)   In the *spectroscopic case*, say, the excitation of a vibrational bond from one energy state to the other, one finds that $\omega_{ml} \simeq 10^{13}$ sec^{-1}. Hence from Eq. (4.50) one finds the limit of perturbation is

$$\tau \leqslant 10^{-13} \text{ sec} \tag{4.61}$$

The increase in energy of a bond in acceptance of an absorbed photon is $h\nu_{\text{radiation}}$. At the midpoint of the vibration, $h\nu_{\text{radiation}} = \frac{1}{2}m(v^2 - v_0^2)$, where v and v_0 are the velocities of the atoms in the first excited and ground state, respectively. Thus, one knows v, and it is reasonable to estimate the time of excitation, that in which the perturbation is active, as the time for an atom in the bond to travel a distance equal to the increase in amplitude of vibration between the first excited and ground states. This is $\Delta x/v$. Taking Δx as 0.1 Å and calculating v from the above, one obtains (using the vibrational excitation of H_2 as the example) about 6×10^{-16} sec. Hence condition (4.50) is satisfied.

(c)   For the *energy transfer during collisional interaction* the energy of interaction is about $kT \approx 0.025$ eV at $T = 300°$K. Hence, $\Delta E \simeq 4 \times 10^{-14}$ ergs. Thus, one gets from Eq. (4.50) that the time of perturbation

$$\tau \leqslant \frac{h}{4 \times 10^{-14}} \approx 10^{-13} \text{ sec} \tag{4.62}$$

The time the colliding molecules remain in contact for an impact, i.e., the time during which the perturbation occurs due to collision, may be obtained in the following way. Let us consider that the kinetic energy during collision between two molecules is converted to the potential energy of interaction between them, and that this potential energy can be expressed in terms of Morse's equation,

$$U \simeq D(1 - e^{-as})^2 \tag{4.63}$$

where D is the dissociation energy, a is the Morse constant, and s is the displacement. Therefore,

$$U = Da^2 s^2 = p^2/2m \tag{4.64}$$

where p is the relative momentum of the colliding molecules.
Hence, from Eq. (4.64) one gets

$$s = \frac{p}{a(2mD)^{1/2}} \tag{4.65}$$

We know from Newton's law of motion that the displacement s at time τ can be expressed as

$$s = \frac{1}{2}f\tau^2 \tag{4.66}$$

for an initial velocity equal to zero, where f is the acceleration. The acceleration f is obtained as

$$f = -F/m$$

$$= \frac{1}{m}\frac{dU}{ds}$$

$$\simeq 2Da\,e^{-as}/m \tag{4.67}$$

Hence, from Eq. (4.66) we get

$$s = \frac{Da\,e^{-as}}{m}\tau^2 \tag{4.68}$$

Now putting the value of s from Eq. (4.65) into Eq. (4.68), one gets the time τ during which the displacement s occurred (i.e., the time of perturbation) as

$$\tau \simeq \frac{p^{1/2}m^{1/4}}{D^{3/4}a\,e^{-as/2}}$$

$$\simeq 10^{-16}\,\text{sec} \tag{4.69}$$

for $p = 10^{-22}$ erg sec cm^{-1}, $m = 10^{-27}$ g, $D = 3$ eV, $a = 10^8$ cm^{-1}, and $s = 10^{-11}$ cm.

Thus, comparing Eqs. (4.62) and (4.69) we find that condition (4.50) concerning the time of perturbation is satisfied in the case of transition occurring due to collisional perturbation.

4.8. Magnitude of the Perturbation

Now, let us discuss whether condition (4.58) regarding the limit of the strength of perturbation is satisfied for the above three cases.

(a) In the electrochemical electron transfer case, U^{pert} is given as $U^{\text{pert}} = e_0\mathbf{Xr}$. The integral (4.58) becomes

$$I = e_0\mathbf{X}\int_0^\tau \mathbf{r}\,dt = e_0\mathbf{Xr}\tau \tag{4.70}$$

Now, considering perturbation energy $e_0\mathbf{Xr}$ of the order of an electron volt (outer Helmholtz plane) and the time of perturbation $\tau \simeq 10^{-16}$ sec, we find that the integral in Eq. (4.70) becomes

$$I = 1.6 \times 10^{-28}\,\text{erg sec} \leqslant h \tag{4.71}$$

(b) For spectroscopic perturbations due to electromagnetic radiation, the expression for U^{pert} can be written as

$$U^{\text{pert}} = U^{\text{pert}}(\mathbf{r}) \cos \omega t \qquad (4.72)$$

Hence, the integral (4.58) becomes

$$I = \int_0^\tau U^{\text{pert}}(\mathbf{r}) \cos \omega t \, dt$$

$$= \frac{U^{\text{pert}}(\mathbf{r})}{\omega} \sin \omega \tau \qquad (4.73)$$

Now, considering the time-independent part of the perturbation interaction energy as the energy of the electromagnetic radiation that effects the transition from state l to state m having energy difference $\hbar \omega_{ml}$, $\hbar \omega_{\text{photon}} \simeq \hbar \omega_{ml} \simeq U^{\text{pert}}(\mathbf{r})$. Thus, we get $\omega \simeq \omega_{ml} \simeq 10^{13} \sec^{-1}$ (for a dipole transition in the infrared region). We have shown that the time of perturbation is $\sim 10^{-13}$ sec. Hence, we get from Eq. (4.73) that

$$I = \frac{\hbar \omega}{\omega} \sin(10^{13} \times 10^{-13})$$

$$= 0.84 \hbar$$

$$\leqslant \hbar \qquad (4.74)$$

(c) If one considers collision as a quick impact, i.e., the interaction is on when the colliding particles (atom or molecules) are in contact and off when they are apart, we may consider such perturbation as time independent. Hence the integral (4.58) becomes

$$I = U^{\text{pert}} \int_0^\tau dt$$

$$= U^{\text{pert}} \tau \qquad (4.75)$$

The collisional perturbation can be taken as the energy of activation, which, for chemical reactions, is of the order of 1 eV ($=1.6 \ 10^{-12}$ ergs atom^{-1}). Hence,

$$I = (1.6 \times 10^{-12}) \tau$$

But we have $\tau = 6 \times 10^{-16}$. Thus,

$$I = 1.6 \times 10^{-12} \times 6 \times 10^{-16} \simeq 9 \times 10^{-28} \text{ erg sec} \qquad (4.76)$$

and is $< \hbar$.

4.9. Relation of Time-Dependent Perturbation Theory to Reaction Kinetics

Now let us find in what way the transition probability or the transition rate expressions of TDPT can be linked with the experimental reaction rate. We know that one of the early applications of quantum mechanics to reaction kinetics is the use of the WKB tunneling probability expression to find the contribution of the rate from a system that tunneled through a potential energy barrier. Since the WKB method (see Section 8.5) is based on the drastic approximation that the potential energy is a very slowly varying function of distance in the region of the de Broglie wavelength, its use is doubtful in many cases. Moreover, the tunneling calculation[15] does not involve any specific states of the system in either the initial or final state apart from uncertainties in the barrier parameters. Also, no time is mentioned in the tunneling case.

The time-dependent perturbation theory is the more general and more sophisticated method of calculation of transition probability from initial state to final state where the states are well defined and the time development of the processes is also included.

The rate of first-order chemical reactions can be expressed in general as

$$V_R = \kappa Z C_A^0 \exp(-E_a/kT) \qquad (4.77)$$

where Z is the frequency of collision, and κ is the quantum mechanical transmission coefficient here equivalent to the transition probability. *It is a version of this factor that one finds from the TDPT.* The probability of activation of the reactants is given in terms of the Boltzmann factor.

But one can also use the time-dependent perturbation theory to find the probability that the system will go to some higher energy state from its ground state due to collision or photon interaction in the quantum mechanical sense. Thus, one may in general express Eq. (4.77) in terms of two probabilities, i.e., the probability of transition P_t (or the rate of transition R_t) and the probability of activation P_a. Thus

$$V_R = C_A^0 Z_C P_t P_a \qquad (4.78)$$

or

$$V_R = C_A^0 R_t P_a \qquad (4.79)$$

where P_t, R_t, and P_a can be obtained from the time-dependent perturbation theory.

4.10. Perturbations by Electromagnetic Radiation: Bohr's Resonance (Coherence) Condition

The electric or the magnetic vector of the incident electromagnetic radiation may interact with an atom or a molecule. The perturbation interaction due to electromagnetic radiation is given as[16]

$$U^{\text{pert}} = 2F^0(\mathbf{r}) \cos \omega t = F^0(\mathbf{r})(e^{i\omega t} + e^{-i\omega t}) \tag{4.80}$$

Using Eq. (4.34) we get

$$a_m(t) = \frac{1}{i\hbar} F_{ml}^0(\mathbf{r}) \int_0^t (e^{i\omega t} + e^{-i\omega t}) \exp(-i\omega_{ml}) \, dt$$

$$= -\frac{i}{\hbar} F_{ml}^0(\mathbf{r}) \int_0^t (e^{i\omega t} + e^{-i\omega t}) \exp(i\omega_{ml}t) \, dt$$

$$= -\frac{i}{\hbar} F_{ml}^0(\mathbf{r}) \int_0^t \{\exp[i(\omega + \omega_{lm})t] + \exp[-i(\omega - \omega_{ml})t]\} \, dt$$

$$= F_{ml}^0(\mathbf{r}) \left\{ \frac{1 - \exp[i(\omega + \omega_{ml})t]}{\hbar(\omega + \omega_{ml})} + \frac{1 - \exp[-i(\omega - \omega_{ml})t]}{\hbar(\omega - \omega_{ml})} \right\} \tag{4.81}$$

where we have used

$$\omega_{ml} = (E_m - E_l)/\hbar$$

$$= -(E_l - E_m)/\hbar = -\omega_{lm} \tag{4.82}$$

and the matrix element

$$F_{ml}(\mathbf{r}) = \langle \psi_m(\mathbf{r}) | F^0(\mathbf{r}) | \psi_l(\mathbf{r}) \rangle \tag{4.83}$$

where $\psi_l(\mathbf{r})$ and $\psi_m(\mathbf{r})$ are the radial wave functions of initial and final states, respectively.

It is important to note that the first term of Eq. (4.81) dominates the second term when $\omega \sim -\omega_{ml}$,† that is, when

$$\hbar\omega \simeq -\hbar\omega_{ml} = E_l - E_m \tag{4.84}$$

This corresponds to the physical process of *stimulated (induced) emission*.

---

† Here, one should not be confused by the negative sign on frequency. It is not the frequency that is negative, but the energy is of negative sign.

An incident photon of energy $\hbar\omega$ induces a transition from the state E_l to a lower energy state; i.e., $E_m < E_l$. This process has a maximum probability when $\hbar\omega = E_l - E_m$. In like manner, the second term of Eq. (4.81) dominates the first term when $\omega \sim \omega_{ml}$, that is, when

$$\hbar\omega \simeq \hbar\omega_{ml} = E_m - E_l \qquad (4.85)$$

This is recognized as the *resonance absorption*, in which the photon excites the system to a higher energy state $E_m > E_l$. Therefore, we can conclude that for values of ω that are not close to $\pm\omega_{ml}$, neither term of Eq. (4.81) becomes large, and $a_m(t)$ is negligibly small; i.e., there is found no absorption or emission. This is the *Bohr frequency (coherence) condition*—a system making a transition from one state to another emits or absorbs a photon with energy equal to the difference in energy of the two states [Eqs. (4.84) and (4.85)].

4.11. Constant Perturbation for Electron Transition at Interfaces: Gurney Condition of Radiationless Transfer of Electrons

The electron transfer at the interface involves a *non-time-dependent and nonoscillating* perturbing field of energy[17] (double-layer field)

$$U^{\text{pert}} = e_0\mathbf{X}\mathbf{r} \qquad (4.86)$$

where e_0 is the electronic charge U_{pert}, $\mathbf{X}$ is the field across the electrode–solution interface, and $\mathbf{r}$ is the distance across which the electron travels between the electrode and the acceptor.

Substitution of Eq. (4.86) in place of the perturbing energy U_{ml}^{pert} in Eq. (4.34) gives the probability of an electron transition to an available state across an interface as

$$P_t = |a_m(t)|^2 = \left| \frac{U_{ml}^{\text{pert}}}{i\hbar} \int_0^t \exp(i\omega_{lm}t) \, dt \right|^2 \qquad (4.87)$$

where the matrix element of transition is

$$U_{ml}^{\text{pert}} = \langle\psi_m|e_0\mathbf{X}\mathbf{r}|\psi_l\rangle \qquad (4.88)$$

where ψ_l represents the initial state of electrons in the metal in the partially filled level, ψ_m is the final state of electron in the acceptor in solution (or vice versa), and ω_{lm} is the frequency corresponding to the energy difference between the initial and the final state. After integration, Eq. (4.87) becomes

$$P_t = \left| \frac{U_{ml}^{pert}}{i\hbar} \left(\frac{\exp(i\omega_{ml}t) - 1}{i\omega_{ml}} \right) \right|^2 \tag{4.89}$$

On further simplification, Eq. (4.89) becomes

$$P_t = \frac{4|U_{ml}^{pert}|^2}{\hbar^2} \left(\frac{\sin^2 \frac{1}{2}\omega_{ml}t}{\omega_{ml}^2} \right) \tag{4.90}$$

Equation (4.89) can be expressed in terms of energy, since $E_m - E_l = \hbar\omega_{lm}$:

$$P_t = \frac{|U_{ml}^{pert}|^2}{\hbar^2} \left(\frac{t^2 \sin^2\{[(E_m - E)/2\hbar]t\}}{\{[(E_m - E_l)/2\hbar]\}^2 t^2} \right) \tag{4.91}$$

The plot of $(t^2 \sin^2 z)/z^2$ [cf. Eq. (4.91)] versus z shows a maximum as $z \to 0$, where $z = (E_m - E_l)t/2\hbar$. Thus, in contrast to the dipole transition, the transition probability of electron at the interface is a maximum when $E_m - E_l \simeq 0$ (or for $E_m \simeq E_l$). Equation (4.91) represents a proof[18] given of a suggestion made intuitively many years ago by Ronald Gurney[11] concerning the radiationless transition of electrons at the interface. He suggested that electrons would only cross from electrode to solution when $E_l = E_m$, and based upon this (to him obvious) condition the first paper[11] on the quantum mechanical theory of the rate of a reaction—that of electrons donated to protons at interfaces.

4.12. Types of Perturbation: Adiabatic and Nonadiabatic

Consider some molecular system, e.g., an HCl molecule. Let the molecule be perturbed in two sorts of ways. In the first way, called adiabatic, the perturbation (e.g., an electric field) is introduced slowly. What is meant is that the application is slow enough so that the molecule responds "completely" to the electric field, so that it enters another energy state. This kind of "adiabatic transition" is, therefore, a successful transition.

Now consider what may happen if we switch on and switch off the electric field very rapidly. The state of the system approaches the maximum or the lower level of Fig. 4.4 rapidly. It does not adjust to optimize the result of applying the field, remaining on the lower level curve. In fact, under the very fast application of the perturbing field, the system fails to go to the state II as in Fig. 4.4 and, as it were, misses the intersection (the optimum condition) and merely rises in its original form to a higher energy version thereof (state III). Thus, as shown in Fig. 4.4 the molecule goes

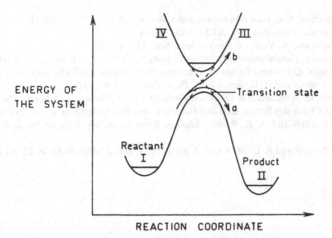

ENERGY OF THE SYSTEM

Fig. 4.4. Adiabatic (a) and nonadiabatic (b) motion in the region of closest approach of two potential energy curves.

temporarily to state III, thereafter relaxing, not to state II (a successful transition), but to state I—no final transition has occurred. This is a nonadiabatic happening.

What, then, is meant by the phrase "nonadiabatic *reaction*"? It might be thought that, if a perturbation is applied too rapidly, then there will be no reaction. The "nonadiabatic reaction" is a reaction in which *many* of the attempted transitions are not successful.

There is a term associated with these ideas. One of them is the "adiabatic approximation." In an *idealized* adiabatic situation, the perturbation is applied infinitely slowly. In the adiabatic approximation, it is recognized that many perturbations vary slowly with time, and then the variation with time is neglected.

References

1. M. Bodenstein, *Z. Phys. Chem.* **13**, 56 (1894); **22**, 1 (1897); **29**, 295 (1899).
2. C. N. Hinshelwood, *Kinetics of Chemical Change in Gaseous Systems*, Oxford University Press, Oxford, (1976).
3. S. Arrhenius, *Z. Phys. Chem.* **7**, 226 (1889).
4. J. C. Maxwell, *Phil. Mag.* **19**, 31 (1860).
5. H. Eyring, *J. Chem. Phys.* **3**, 107 (1935).
6. M. L. Evan and M. Polanyi, *Trans. Faraday Soc.* **31**, 775 (1935).
7. K. J. Laidler, *Theories of Absolute Reaction Rate*, McGraw–Hill, New York (1969).
8. S. Glasstone, K. J. Laidler, and H. Eyring, *The Theory of Rate Processes*, McGraw–Hill, New York (1941).
9. S. Gasiorowicz, *Quantum Physics*, John Wiley and Sons, New York (1974).

10. E. Marzbacher, *Quantum Mechanics,* John Wiley and Sons, New York (1970).
11. R. W. Gurney, *Proc. Roy. Soc.* **A134**, 166 (1931).
12. M. S. Waite and A. Vecht, *J. Electrochem. Soc.* **121**, 109 (1974).
13. R. M. Eisberg, *Fundamentals of Modern Physics,* John Wiley and Sons, New York (1967).
14. H. L. Strauss, *Quantum Mechanics,* Prentice–Hall, Englewood Cliffs, New Jersey (1968).
15. J. O'M. Bockris and R. K. Sen, *Chem. Phys. Lett.* **18**, 166 (1972).
16. C. R. Gatz, in *Introduction to Quantum Chemistry* (T. L. Brown, ed.), Merril Physical and Inorganic Chemistry Series, Bell and Howell Company, Columbus, Ohio (1971).
17. J. O'M. Bockris and A. K. Reddy, *Modern Electrochemistry,* Plenum Press, New York (1973).
18. J. O'M. Bockris and S. U. M. Khan, *J. Res. Inst. Catal. Hokkaido Univ.* **25**, 63 (1977).

5

Long-Range Radiationless Energy Transfer in Condensed Media

5.1. Introduction

One of the ideas introduced into electrode kinetics in recent years is the concept of the Russian school led at the time (1970) by Levich that activation energy in electrochemical reactions is not due to momentum transfer in collisions nor to electrostatic interactions with its nearest neighbors, but to the transfer of energy in the form of polarons over long distances, e.g., 10^3 Å.[1] It is important, therefore, to know the mechanism by which energy may be transferred over long distances, so that we may understand at a better level the concept of long-distance activation used by Levich,[1] Dogonadze,[2] and others.[3-6]

There are two ways by which the energy can be transferred from one molecule to another remote molecule in a condensed medium. One is energy transfer in the form of electromagnetic radiation, the other is transfer in a form that is not electromagnetic. The former is well known while the latter is of importance in explaining, e.g., observations of the deexcitation of excited molecules, which is radiationless but not connected with direct collisional transfer of momentum.[5] The main question is how is energy transferred in a nonradiative way over distances up to 1000 Å?

In this chapter we will point out facts that support the idea of the noncollisional long-range radiationless transfer of energy in a condensed medium and subsequently discuss some possible mechanisms.

5.2. Experimental Evidence for Radiationless Energy Transfer over Long Distances

Experimental studies of the transfer of excitation energy that does not involve *electromagnetic* radiation have been made in gases, liquid solutions, and solids between both like and unlike atoms and molecules. As early as 1923, the existence of "radiationless" transfer of energy was unequivocally demonstrated by Cario and Franck.[7] They illuminated a mixture of mercury and tellurium vapors with light ($\lambda = 2537$ Å) from a mercury-resonance lamp. This vapor emitted, in addition to the mercury resonance line, a number of tellurium lines, although in the absence of mercury, the tellurium vapor neither absorbed nor emitted radiation. This experiment indicates that it is the mercury vapor that absorbs the light, and transfers energy to tellurium in a nonradiative way, and this tellurium vapor becomes as a result excited and emits light characteristic of tellurium, which then loses energy in a radiative way.

An explanation of how the energy is transferred between mercury and tellurium would not be difficult to find in the gas phase, without the hypothesis of the transfer of energy over a distance, for in this kind of system (in contrast to the situation in condensed phases), there are frequent molecular collisions between the mercury (originally excited by the incident radiation) and the tellurium. Although it is not certain that the excitation of these latter atoms occurs by collisional mechanisms, it is an acceptable hypothesis, so that the emission of radiation by the tellurium atoms would not be regarded as proof of radiational energy transfer from the mercury atoms over a distance.

However, when similar phenomena of an apparent transfer of energy to a second molecule, which does not absorb the primary radiation, are found in liquids and solids, the explanation is more difficult, for here there are no collisions in the time domain concerned between the atoms that absorb the radiation and those that emit it. Such radiationless energy transfer in solution was first observed by Perrin.[8a] Forster[9] illuminated the mixture of tryptaflavin and rhodamine B by light of frequency that was absorbed only by rhodamine B. The expected fluorescence of rhodamine B was then found to be suppressed due to transfer of its excitation energy to tryptaflavine by a nonradiative mechanism and, in effect, the fluorescence of tryptaflavine was observed. By determining the effects of the concentration of donor and acceptor and of the viscosity of the solvent, Forster was able to demonstrate that the nonradiative transfer of excitation energy occurs efficiently at distances as great as 70 Å and therefore cannot be collisional. Had there been a *radiative* energy transfer, the rhodamine B would also have fluoresced.

Similarly, Bowen and Livingston[10] explain the fact that in liquid solutions containing two fluorescent solutes, where the emission spectrum of the first solute (A) overlaps the absorption spectrum of the second (B), the fluorescence of A is quenched, while the fluorescence of B is sensitized due to radiationless transfer of energy from excited molecules of A to the unexcited molecules of B. Thus, B becomes excited and emits fluorescence.

The measurement of the intensities of the fluorescence of each pair of hydrocarbons, such as chloroanthracene–perylene, chloroanthracene–rubrane, and cyanoanthracene–rubrane, indicates that radiationless transfer of excitation between the two solutes is more efficient than can be accounted for by a radiative process of emission and reabsorption. The data are consistent with the postulate that a nonradiative transfer of energy involves a bimolecular interaction between an excited molecule A and an unexcited molecule B. The rate of this interaction is not determined by diffusion, because it is independent of the viscosity of the solvent and at least ten times faster than a diffusion-controlled process. Nonradiative transfer of energy between solute molecules, such as *p*-terpheryl- and tetrapherylbutadiene in toluene, was also reported by Birks and Kuchela.[11]

The fluorescence[12] of pure anthracene is in the blue region. But a *green* fluorescence of anthracene is observed if trace amounts of naphthacene are present in solid solution with anthracene. The green fluorescence of anthracene is due to the fact that the blue fluorescence of anthracene is quenched by naphthacene in concentrations as little as 0.1% in solid solution. The naphthacene is therefore receiving energy from the anthracene. Because no radiation in the appropriate wavelength range is observed, the transfer from anthracene to naphthacene must be radiationless.

Interesting experiments were led by Kühn,[13] who studied radiationless energy transfer between monomolecular layers of dyes separated by a uniform inert substrate. In some cases, the evaporated film of a metal or semiconductor was used as a substrate. When the distance between dye layers was in the range of 30–100 Å, energy transfer across the intervening paraffin layer was observed, although no radiation could be detected.

The energy transfer from pyrene to perylene in poly-methylmethacrylate and polystyrene systems were studied by Mataga *et al.*[14] They reported a transfer distance 37 Å. Radiationless energy transfer between naphthalene and perdeuteronaphthalene mixed crystal were reported by several authors.[15–17]

Radiationless transfer of energy in another system was observed by Simpson,[18] illuminating one side of an aromatic hydrocarbon crystal that can absorb light and that was free from absorbed substance. The other side of this crystal was joined to a contiguous crystal containing impurity centers capable of trapping the excitons produced in the first crystal and emitting their characteristic fluorescence and thus acting as a detector, although no radiation characteristic of the top crystal was observed. The schematic diagram of this situation is given in Fig. 5.1.

The transfer distance of energy in crystalline tetracene was measured by Vaubel and Baessler[19] and found to be 580 ± 50 Å below 190°K, to decrease in value at higher temperature, and to approach a value of

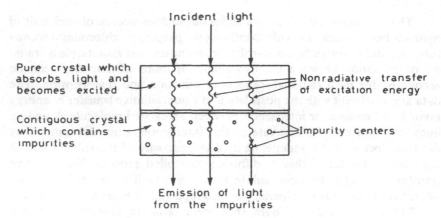

Fig. 5.1. A schematic representation of the nonradiative transfer of excitation energy from one pure crystal to a contiguous crystal that contains impurities.

$120 \pm 10 \,\text{Å}$ at room temperature. The transfer of electronic excitation energy in solid solutions of perylene in *N*-isopropylcarbazole has been reported by Klopffer,[20] who found the total displacement from the starting point of the exciton to be 1350 Å in its lifetime.

Energy transfer and the effect of traps in pure and mixed crystals were considered by Birks and Leite.[21] Host traps and nonradiative transfer are shown to be important in energy transfer in mixed anthracene–tetracene crystals.

Recently, Pantos *et al.*[22] investigated the long-range radiationless energy transfer in a doubly doped (xenon and benzene) solid neon. They observed that when this doubly doped neon is exposed to light of energy about 10.4 eV, absorption by xenon occurs and the energy is then transferred through neon in a nonradiative way and excites the benzene molecule. Subsequently, the excited benzene molecules give out a characteristic fluorescence. Energy transfer over long distances was also observed by several workers in rare gas crystals doped by another rare gas.[23–28] These experimental studies indicate, then, that there are some mechanisms by which energy is transferred from one center to another, separated by distances above 1000 Å, in a nonradiative way.

5.3. Mechanisms of Radiationless Long-Range Energy Transfer

There are several mechanisms available in the literature by which energy can be transferred in nonradiative ways. Some of these mechanisms are not of interest to us. For example, the radiationless electronic energy transfer by internal conversion and intersystem crossing are basically due to direct collisional interaction with the surrounding medium. We are interested in the mechanisms that can explain energy transfer but that are noncollisional and nonradiative and involve no particle transfer. Hence, we will consider a number of mechanisms according to which excitation energy between distant and different electronic systems may be transferred in a nonradiative way.

5.3.1. Resonance Transfer Mechanisms

Let us consider two atoms, one (D) the donor atom and the other (A) the acceptor atom. The two atoms are considered to be at a large distance from each other. In the energy transfer step in a nonradiative way from D to A, the system (A and D) undergoes transition from an initial state i to a final state f. Energy conservation for radiationless transfer requires that the probability $p(E)$ be large only if the energy difference between excited and ground state at D is within ΔE of A. This is why this type of energy transfer is called *resonance transfer*.

What is the mechanism? The cause of the interaction is the dispersion energy between atoms D and A. The probability of a transfer of energy is given by

$$p(E) = \frac{2\pi}{\hbar} |\langle \psi_f | U | \psi_i \rangle|^2 \rho(E) \tag{5.1}$$

and

$$U = -\frac{e_0^2}{\varepsilon |\mathbf{r}_1 - \mathbf{r}_2|} \tag{5.2}$$

where e_0 is the electronic charge, ε is the dielectric constant of the medium, and $|\mathbf{r}_1 - \mathbf{r}_2|$ represents the distance between D and A.

Using $\psi_i = \phi_D^*(\mathbf{r}_1)\phi_A(\mathbf{r}_2)$ and $\psi_f = \phi_D(\mathbf{r}_1)\phi_A^*(\mathbf{r}_2)$ and Eq. (5.2) in Eq. (5.1), one gets

$$p(E) = \frac{2\pi}{\hbar} \frac{e_0^2}{\varepsilon} \left[\iint \left| \phi_D(\mathbf{r}_1)\phi_A^*(\mathbf{r}_2) \frac{1}{|\mathbf{r}_1 - \mathbf{r}_2|} \phi_D^*(\mathbf{r}_1)\phi_A(\mathbf{r}_2) \right| \right]^2 d\mathbf{r}_1 \, d\mathbf{r}_2 \, \rho(E) \tag{5.3}$$

where $\mathbf{r}_1$ and $\mathbf{r}_2$ represent the position of electrons in donor and acceptor, respectively.

The terms $\phi_D(\mathbf{r}_1)\phi_A^*(\mathbf{r}_2)$ and $\phi_D^*(\mathbf{r}_1)\phi_A(\mathbf{r}_2)$ in Eq. (5.3) can be considered as the charge distribution near D and A atoms, respectively. If the transition at an atom D goes between excited p and ground s state, the charge distribution near D may be characterized by a dipole moment, μ_D. Similarly, the charge distribution near A can be characterized by the dipole moment μ_A. The matrix element can be written approximately as[29]

$$\langle \psi_j | U | \psi_i \rangle \approx \frac{\mu_D \mu_A}{\varepsilon R^3} \tag{5.4}$$

where R is the distance between D and A. Using Eq. (5.4) in Eq. (5.1), it can be shown that the probability of transition per unit volume is

$$p(E) = \frac{2\pi \mu_D^2 \mu_A^2}{\hbar \varepsilon^2 R^6} \rho(E) \tag{5.5}$$

Now we shall discuss the transition probability due to resonance transfer in terms of experimentally measurable quantities given by Forster[30,31] and later developed by Dexter.[32a]

Forster's Treatment[30,31]

The theory of long-range interactions in condensed medium that do not depend upon radiational transfer was given earlier by Perrin,[8a] who treated the quantum mechanics of the interaction of oscillators that have similar frequencies and related this to the transfer of activation energy over long distances. His contribution used the theory of Kalliemann and London.[8c] Forster[30,31] carried out a normal treatment of the quantum mechanical calculation of the transition probability of excitation energy. He also considered, like Perrin, a dipole–dipole interaction between donor and acceptor atoms and molecules where the donor molecule was excited by the acceptance of electromagnetic radiation and the acceptor had received excitation energy from the donor, although the donor did not have any apparent radiation. He applied the quantum mechanical treatment of electrodynamic interaction to fluorescence and its quenching due to the presence of an impurity having nearly the same resonance energy as that of the host molecule in a condensed medium.

Using the general treatment of time-dependent perturbation theory, *but finally inserting experimentally determined parameters*, Forster[30,31] (later, Dexter[32a]) gave the expression for the probability of radiationless transition of excitation energy as

$$p_{DA}(DD) = \frac{3\hbar^4 c^4 Q_A}{4\pi n^4 R^6 \tau_s} \left(\frac{X_c}{\varepsilon^{1/2} X_c} \right)^4 \int \frac{f_D(E) f_A(E)}{E^4} dE \tag{5.6}$$

Here $p_{DA}(DD)$ is the probability for the transfer of dipole–dipole inter-action, c is the velocity of light, n and ε are, respectively, the index of refraction and the dielectric constant of the specimen, and X and X_c are the dipole fields exerted in a vacuum and a condensed medium, respectively. $X/\varepsilon^{1/2}X_c$ has been, perhaps doubtfully, taken as unity. The function $f_D(E)$ is the distribution curve of the emission band of donor D and $f_A(E)$ is that of the absorption spectrum of acceptor A, where E is the energy either of emission or absorption. These two curves are normalized with respect to E. The fact that the principal contribution to the integral containing $f_D(E)f_A(E)$ in the above equation comes from the region where the overlap is large corresponds to the energy conservation requirement. The denominators in the integrands result from the normalization condition. Q_A is the integrated absorption cross section $\sigma(E)\,dE$ and is proportional to the area under $f_A(E)$. Therefore, Q_A is proportional to μ_A. The quantities can be expressed in terms of the average lifetimes τ_D and τ_A of the excited states at D and A.

The right-hand side of Eq. (5.6) contains only measurable quantities, the properties of the solute molecules. For example, if one takes $\frac{1}{3}$ eV for the integral of Eq. (5.6) and assumes that E varies slowly in the region where the integrand is large and is about 5 eV, we get, taking $n^4 = 6$ and $Q_A = (10^{-8}\text{ cm})^2$ eV,

$$p_{DA}(DD) \simeq (27/R)^6(1/\tau_s) \tag{5.7}$$

where τ_s is in units of 10^{-8} sec and R is in Å.

A difficulty of Forster's theory is in the magnitude of the probability of transfer. To cause an electrochemical reaction to occur, we need the ion concerned to receive energy of about 1 eV. However, if we substitute $R = 1000$ in the above equation and $\tau_s = 1$, one obtains $\sim 4 \times 10^{-10}$ for the probability of the transfer. The transfer of energy of the order of 1 eV would be obtained only by the cooperative interaction of some 10^{10} molecules. There are about 10^8 molecules in this volume. The probability that they would all transfer in the given time domain appears, however, to be low.

The mechanism of energy transfer over a long distance, as presented by Forster[30,31] and later by Dexter,[32a]† seems to lack experimental confirmation. It needs reworking with derived wave functions to see if it yields energy transfers that would correspond to those observed without the substitution of experimental results. A positive view of the Forster mechanisms is given by Bennett[32b] and a combination of the Forster mechanism with that of diffusion by Klein *et al.*[32c]

---

† Dexter[32a] introduced a mechanism of radiationless energy transfer when the particles are very near. But this mechanism is not relevant in this chapter where long-range energy transfer is considered.

5.3.2. The Exciton Theory

A more promising theory of radiationless transfer of energy over long distance in a crystal is provided by the exciton model. Electronic excitation in the process of light absorption by a crystal does not remain confined to a particular atom or molecule[33] but is distributed between all of them in the form of "excitation waves" similar to a phonon spectrum. The delocalized excited state is called a "Frenkel exciton."† Thus, one finds a difference between the normal excited state of an isolated atom or molecule that is not mobile and the important property of the Frenkel exciton, namely, its ability to hop from one atom (or molecule) to the other, the ease of hopping being dependent upon the coupling between the neighbors. Thus, it is possible to imagine an excitation wave traveling through the crystal, and in this way one may explain radiationless transfer of energy between two ends of a crystal separated by a long distance compared to the molecular dimensions. The exciton moves in a crystal until it becomes trapped by an impurity atom or molecule or is emitted back as fluorescence.[18]

How is the excitation energy handed on to a neighboring particle? This is possibly due to the perturbation of the interaction between the neighboring atom and its surroundings. Let us consider that in the initial state the molecule m is not excited while the neighboring molecule n is in an excited state f in the crystal.[28,35,36] This initial state can be represented by a wave function $\phi_m^0 \phi_n^f$. In the final state, the molecule m becomes excited to the fth state due to transfer of excitation from the molecule n as a result of the perturbation interaction $V_{\text{Coulombic}}$ between the molecules m and n. The final state can be represented by the wave function $\phi_n^0 \phi_m^f$. Thus, one may write that the amount of excitation energy transfer is proportional to the transition matrix M_{mn}, i.e.,

$$M_{mn} = \int \phi_n^{*0} \phi_m^{*f} V \phi_m^0 \phi_n^f \, d\tau \tag{5.8}$$

where M_{mn} is the matrix element of excitation transfer between molecules m and n. Thus, the excitation is transferred due to a perturbation interaction V_{mn} from m to n and then n to another neighboring molecule and so on. This, in turn, gives rise to the hopping motion of an exciton until it is trapped by an acceptor (trap) molecule.

---

† There is another kind of exciton. Excitation of an electron by light absorption produces an electron–hole pair in a semiconductor or in any molecular crystal. The electron and hole together form a bound state in which the two particles revolve about their common center of mass as an electron around a proton in a hydrogen atom. Such a state is also referred to as an exciton. Such excitons may be called Wannier excitons.[34] Energy transfer by such an exciton (pair of electron and hole) in the crystal involves particle movement and is not relevant in the present discussion.

One can determine the wave function of the exciton wave in a crystal of N atoms by solving the Schrödinger with a periodic potential. The solution of such an equation gives a wave function of the Bloch form, but for N particles, as

$$\psi_k = (N!)^{-1/2} \sum \psi_l \exp(i\mathbf{k} \cdot \mathbf{R}_l) \qquad (5.9)$$

where $\mathbf{k}$ is the wave vector of the excitation and $\mathbf{R}_l$ is the position vector of the lth excited atom; ψ_l is the wave function of N atoms, which contain among them *one* atom represented by a subscript l, which is excited and has a wave function ϕ_l^*. Then, ψ_l can be expressed as

$$\psi_l = \phi_1 \phi_2 \cdots \phi_{l-1} \phi_l^* \phi_{l+1} \cdots \phi_{N-1} \phi_n \qquad (5.10)$$

where $\phi_1, \phi_2, \ldots, \phi_N$ are the ground-state wave function of atoms $1, 2, \ldots, N$, and ϕ_l^* is the wave function of the excited atom of level l.

Exciton–Phonon Coupling

Recently, several treatments[37–41] have been put forward regarding the motion of excitons in a crystal due to exciton–phonon coupling. In the treatment of exciton motion in a molecular crystal by Gover and Silbey,[37] it was found that the migration process is similar to diffusion on a long time scale for all temperatures as long as the exciton–phonon interaction is considered. However, if there is no exciton–phonon coupling, the migration is wavelike. To determine the transport property of excitons, the authors considered the density matrix (which gives the density of excitons in a particular part of the crystal) in "clothed" exciton representation (where the "clothed" process corresponds to relaxation of the initially formed exciton by phonon emission).† In the Gover–Silbey treatment, a model system with one band of Frenkel excitons and several bands of optical phonons is considered (see Section 3.2). The Hamiltonian for the exciton–phonon system can be represented as[37]

$$H = H_{exciton} + H_{phonon} + H_{exciton-phonon} \qquad (5.11)$$

where the three terms in Eq. (5.11) are defined as

$$H_{exciton} = \sum_k \varepsilon(\mathbf{k}) a_k^+ a_k \qquad (5.12)$$

$$H_{phonon} = \sum \omega_{\lambda,\alpha} b_{\lambda,\alpha}^+ b_{\lambda,\alpha} \qquad (5.13)$$

† There is some similarity between the mechanism of this kind of energy transfer and that of the polaron mechanism. The difference lies in the lattice distortion by a charge that comes in the polaron mechanism.

$$H_{\text{exciton-phonon}} = N^{1/2} \sum_{\mathbf{k},\lambda,\alpha} f(\mathbf{k}, \lambda) a_{\mathbf{k},\lambda}^{+} a_{\mathbf{K}}(b_{\lambda,\alpha} + b_{\lambda,\alpha}^{+}) \qquad (5.14)$$

where $a_{\mathbf{k}}^{+}$, $a_{\mathbf{k}}$ are the creation and annihilation operators for an exciton wave vector $\mathbf{k}$. Similarly, $b_{\lambda,\alpha}^{+}$ and $b_{\lambda,\alpha}$ create and destroy a phonon wave vector λ in band α. Both the exciton and phonon operators are assumed to satisfy Bose–Einstein commutation relations. The term $\varepsilon(\mathbf{k})$ is the free-exciton energy and $\omega_{\lambda,\alpha}$ is the phonon frequency.

Abram and Silbey[39] investigated electronic energy transfer between molecules embedded in a host lattice. They also considered the exciton-phonon coupling and solved the density matrix of the electronic system in the absence of radiative decay. In the weak coupling limit, they found that the large transfer rate becomes a simple exponential function of time.

5.3.3. Energy Transfer by Polaron Motion

Here we will discuss an alternative long-range radiationless energy transfer mechanism via polaron motion in a crystal. A polaron (see Chapter 3, Section 3.2.3) can be regarded as a charge in a dielectric which deforms under the influence of the charge.

Consider that one of the electrons in an impurity ion is in an excited state (e.g., in 2p, 3d, etc.) and another impurity ion far away from the first is in the ground state. In the excited state, the electron has a different charge distribution than it has originally in the ground state. Thus, this excited-state electronic charge distribution in the first impurity ion interacts differently with the neighboring lattice ions in the crystal than it interacted when it was in ground state and produces an excited polaron. This interaction perturbs the lattice vibrations (phonons), which in turn perturb the electrons in the second impurity ion and excite it. Now the second impurity ion, after being excited, perturbs the surrounding lattice and produces an excited polaron. Thus, electronic excitation energy from the first excited impurity ion moves to the second unexcited one and so on, which in effect constitutes motion of the polaron. Hence, one can visualize the energy transfer in a nonradiative way via polaron motion in a crystal (Fig. 5.2). Thus, the basic process of energy transfer via polaron motion is that the energy is transferred from the excited ion, giving its energy to the lattice, and the second ion subsequently becomes excited by absorbing energy from the lattice, and so on.

So far our discussion has been limited to the transfer of energy from an excited ion to another unexcited one via polaron motion. Can one consider this mechanism of energy transfer to be applicable for the activation of an ion receiving energy from excited phonons surrounding it in a polar medium? Such a discussion is given in Chapter 6, where another mechanism of activation due to phonon–vibron coupling will also be outlined.

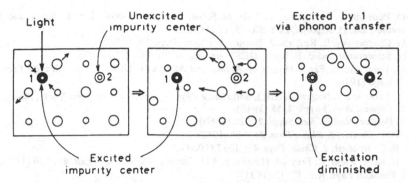

Fig. 5.2. A schematic representation of nonradiative energy transfer from an excited impurity center to another impurity center via phonon transfer. The polaron aspect enters when the excited impurity is a charge.

References

1. V. G. Levich, *Physical Chemistry, An Advanced Treatise* (H. Eyring, D. Henderson, and Y. Jost, eds.), Volume IX, p. 985, Academic Press, New York (1971).
2. R. D. Dogonadze, *Reactions of Molecules at Electrodes* (N. S. Hush, ed.), Wiley-Interscience, New York (1971).
3. R. A. Marcus, *J. Chem. Phys.* **43**, 679 (1965).
4. P. P. Schmidt, *J. Chem. Phys.* **58**, 4290 (1973).
5. W. Schmickler and W. Vielstich, *Electrochim. Acta* **18**, 883 (1973).
6. Yu. I. Kharkats and J. Ulstrup, *J. Electroanal. Chem.* **65**, 555 (1975).
7. G. Cario and J. Frank, *J. Physik.* **17**, 202 (1923).
8a. F. Perrin, *Ann. Phys.* **17**, 20 (1932).
8b. J. Perrin, *Deuxième Conseil de Chimie Solvay*, Bruxelles (1924).
8c. H. Kollemann and F. London, *Phys. Chem.* **132**, 207 (1929).
9. T. Forster, *Ann. Physik* **2**, 55 (1948).
10. E. J. Bowen and R. Livingston, *J. Amer. Chem. Soc.* **76**, 6300 (1954).
11. J. B. Birks and K. N. Kuchela, *Proc. Phys. Soc.* **77**, 1083 (1961).
12. E. J. Bowen, *J. Chem. Phys.* **13**, 306 (1945).
13. H. Kühn, *Naturwiss.* **54**, 429 (1967).
14. N. Mataga, H. Obashi, and T. Okada, *J. Phys. Chem.* **73**, 370 (1969).
15. H. K. Hong and G. W. Robinson, *J. Chem. Phys.* **54**, 1369 (1971).
16. V. L. Broude, A. V. Leidermann, and T. D. Tratas, *Sov. Phys. Solid State* **13**, 3058 (1972).
17. H. Port, D. Vogel, and H. C. Wolf, *Chem. Phys. Lett.* **34**, 23 (1975).
18. O. Simpson, *Proc. Roy. Soc. (London)* **A238**, 6102 (1957).
19. G. Vaubel and H. Baessler, *Mol. Cryst.* **12**, 47 (1970).
20. W. Klopffer, *J. Chem. Phys.* **50**, 1689 (1969).
21. J. B. Birks and M. S. S. C. P. Leite, *J. Phys.* **B3**, 513 (1970).
22. E. Pantos, S. S. Hasnain, and I. T. Steinberger, *Chem. Phys. Lett.* **46**, 395 (1977).
23. J. Jortner, *Vacuum Ultraviolet Radiation Physics* (E. E. Koch, R. Haensel, and C. Kunz, eds.), p. 263, Pergamon/Vieweg, London/Braunschweig (1974).
24. Z. Ophir, N. Schwentner, B. Raz, M. Skibowski, and J. Jortner, *J. Chem. Phys.* **63**, 1972 (1975).

25. D. Pudewill, F. J. Himpsel, V. Saile, N. Schwentner, M. Skibowski, E. E. Koch, and J. Jortner, *DESY Report*, DESY SR-75/17 (1975).
26. O. Chesnovsky, B. Raz, and J. Jortner, *J. Chem. Phys.* **59**, 5554 (1973).
27. N. Schwentner and E. E. Koch, *DESY Report*, DESY SR-76/01 (1976).
28. U. Fano and L. Fano, *Physics of Atoms and Molecules*, University of Chicago Press, Chicago (1970).
29. R. Kubo and T. Nagamiya (eds.), *Solid State Physics*, McGraw-Hill, New York (1969).
30. T. Forster, *Ann. Physik* **2**, 58 (1948).
31. T. Forster, *Radiat. Res. Suppl.* **2**, 326 (1960).
32a. D. L. Dexter, *J. Chem. Phys.* **21**, 836 (1953).
32b. R. G. Bennett, *J. Chem. Phys.* **41**, 3037 (1964).
32c. U. K. A. Klein, R. Frey, M. Hauser, and U. Goselle, *Chem. Phys. Lett.* **41**, 139 (1976).
33. J. Frenkel, *Phys. Rev.* **37**, 17 (1931).
34. G. H. Wannier, *Phys. Rev.* **52**, 191 (1937).
35. D. L. Dexter and R. S. Knox, *Excitons*, Wiley-Interscience, New York (1965).
36. A. S. Davydov, *Theory of Molecular Excitons*, Plenum Press, New York (1970).
37. M. Gover and R. Silbey, *J. Chem. Phys.* **52**, 2099 (1970); **54**, 4893 (1971); see also U. Göselle, *Chem. Phys. Lett.* **43**, 1 (1976).
38. V. Ern, A. Suna, Y. Tomkiewicz, P. Avakian, and R. P. Groff, *Phys. Rev.* **B5**, 3222 (1972); see also A. I. Burshstein, *Sov. JETP Phys.* **35**, 882 (1972).
39. I. I. Abram and R. Silbey, *J. Chem. Phys.* **63**, 2317 (1975).
40. D. Emin, *Advan. Phys.* **22**, 57 (1973).
41. A. Nakamura, *J. Chem. Phys.* **64**, 185 (1976); R. G. De Losh and W. J. C. Grant, *Phys. Rev.* **1**, 1754 (1970).

6

Mechanisms of Activation

6.1. Mechanism of Activation in the Gas Phase

It is well known[1-5] that molecules in gaseous media become activated in a direct collision between the translators. When a reference molecule undergoes collision with another molecule, the total kinetic energy of the two molecules is redistributed between them. The possibility exists for transfer between the translational kinetic energy of the colliding molecules and the energy stored in the vibrational, rotational, or electronic modes of the molecules.

If the translational energy per pair of molecules is particularly large, then significant parts of that energy may be converted into vibrational energy so that this energy for each of the two reacting molecules exceeds the point at which their potential energy–distance relationships intersect. If the vibrational energy is thus heightened in the reacting pair, the original location of the electron orbitals present in each molecule before collision may be changed so that the bonding of each atom is different after the collision from that before collision.

The necessary translational energy to provoke a reaction will depend upon several factors. What is the probability of the conversion of translational energy to vibrational energy? What is the degree to which the vibrational states have to be raised before the vibrational energies of the bonds that are about to interchange their electron orbitals become equal? The necessary translational energy of the two colliding particles will differ depending upon which two bonds have to be reformed and the new bonds that are to be formed.

The history of attempts to calculate the reaction path and the energetics of the reaction date from the early 1930s.[1-5] The course of such calculations, however, has been difficult, not only because of the difficulty of making potential energy surface calculations in situations other than those of the simplest reacting particles, but also because of the difficulty of taking into account the quantum mechanical probability of transition.

6.2. Conversion of Translational Energy during Collision

It is important to discuss the way energy conversion from the translational to the vibrational and rotational occurs as a preliminary to reaction due to molecular collision.

6.2.1. Translational Energy to Vibrational Energy

Translational–vibrational energy transfer between atoms and diatomic molecules in collision has been studied extensively in recent years by classical,[6–11] semiclassical,[12–29] and quantum mechanical theory.[30–37] The translational–vibrational energy transfer was first treated classically by Landau and Teller.[6] They gave an expression for the probability of energy transfer from a translating atom A to a vibrating molecule BC when the molecule was considered as a classical harmonic oscillator.

Kelley and Wolfsburg[8,9] calculated the amount of energy ΔE transferred for the collinear collisions (Fig. 6.1) between an atom and a diatomic molecule. The potential of the oscillator (diatomic molecule) was considered to be both harmonic and Morse-like in type. The interaction between colliding atom and the diatomic molecule was taken both as an exponential repulsion and as a Lennard-Jones potential. They found that the total energy transferred from translation to vibration ΔE decreases when the initial energy of the oscillator increases. A similar type of classical treatment was carried out by Rapp and Kassal.[10,11] Unfortunately, all classical treatments compare poorly with experiment.

In recent years, most work on translational–vibrational energy transfer is based on the semiclassical treatment. The semiclassical treatment of collisional energy transfer was originated by Zener.[12] The translational motion of the colliding atom is treated classically, and the molecule is considered to have quantized vibration. The situation considered in most of these treatments[13–24] is that of the collinear collision (Fig. 6.1), since mechanically this is the most profitable direction of collision for the translational energy transfer. Time-dependent perturbation theory may be applied to find the probability of such energy transfer.

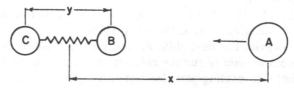

Fig. 6.1. Collinear collision of an atom A with a diatomic molecule BC. Impact is represented to be between atoms A and B.

Calvert and Amme[25] presented an approximate three-dimensional semiclassical treatment of the collisional interaction of an impinging atom with an harmonic oscillator (a molecule). The interaction between the colliding systems was considered in terms of a Morse-type potential. Shin[26] has utilized a semiclassical approach to examine the orientational effect on the value of the probability of vibrational excitation from the ground state to a higher vibrational quantum state during collision between an atom A and a molecule either of the type X—X or X—Y. Shin[27] also carried out the two-dimensional semiclassical calculation of the vibrational excitation probability for the argon atom and H_2 molecule collision. The probabilities of vibrational transitions from ground state to the first excited state, then $0 \rightarrow 2$ and $1 \rightarrow 2$ states, simultaneously occurring with the rotational changes have been calculated for the translational velocity range 10^5–10^6 cm sec^{-1} (Fig. 6.2). Shin found that simultaneous vibrational–rotational transition probability becomes a maximum in a velocity range near 10^5 cm sec^{-1} and decreases with increasing velocity. A three-dimensional semiclassical calculation for the collisional energy transfer to vibration was carried out by Wartell and Cross.[28]

Doll and Miller[29] calculated the transition probability for the simultaneous vibrational–rotational transition of H_2 molecules from the $0 \rightarrow 1$

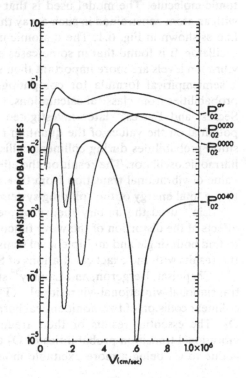

Fig. 6.2. Plots of the simultaneous and orientation-averaged transition probabilities for $0 \rightarrow 2$ at velocities below 10^6 cm sec^{-1}. P_{02}^{0040}, P_{02}^{0000}, P_{02}^{0020}, and P_{02}^{av} represent the transition probabilities from 00 to 40, 00 to 00, 00 to 20, and vibrational level and rotational transition from 00 to 40, 00 to 00, 00 to 20, and average, respectively.

vibrational level and from the $0 \rightarrow 2$ rotational levels due to collision with the helium atom. The calculations were carried out considering the H_2 molecule's vibrational potential as harmonic and also considering it as a Morse potential with inclusion of the effect of anharmonicity. They found for the total energy conservation of amount $2\hbar\omega$ and $3\hbar\omega$ (ω being the frequency of vibration) the transition probability became 1.57×10^{-7} and 8.40×10^{-5}, respectively. These results are comparable to the result 1.40×10^{-7} and 6.72×10^{-5}, respectively, of the exact quantum mechanical calculation by Secrest and Johnson.[30]

Closed-form expressions were obtained by Eastes and Doll[31] for the vibrational transition probabilities in a semiclassical formulation for the inelastic hard sphere collision of an atom with a harmonic oscillator in colinear interaction (Fig. 6.1). The result of transition probabilities from such expressions to different vibrational levels are found to be in good agreement with the numerically computed quantum mechanical results.[30]

There are rare works on pure quantum mechanical treatments of the translational–vibrational energy transfer during collisional interaction between an atom and a molecule or between molecules. In 1966, Secrest and Johnson[30] carried out an exact quantum mechanical calculation of the transition probabilities for the collinear collision of an atom with a diatomic molecule. The model used is that of a diatomic molecule colliding with an atom constrained in such a way that all three atoms lie on a straight line as shown in Fig. 6.1. The diatomic molecule is treated as a harmonic oscillator. It is found that in some cases even triple quantum jumps in the vibration levels are more important than single quantum jumps. They gave a semiempirical formula for computing quantum mechanical transition probabilities from classical calculations. Using the same type of treatment, Secrest and Eastes[32] later investigated the effect of an attractive well potential on the values of the quantum mechanical vibrational state transition probabilities during collinear collision of a particle with a diatomic harmonic oscillator. The result on the effect of square potential well in the value of vibrational transition from $0 \rightarrow 1$ is given in Fig. 6.3 with respect to the total energy of the colliding system.

Liu[33] used the unitary theory of inelastic scattering (which includes effects of the distortion of the wave function of the system due to collision) to find both single and multiple level transition probabilities and compared the results with the exact calculations of Secrest and Johnson.[30]

Chapuisat, Bergeron, and Launay[34] studied quantum mechanically the translational–vibrational–vibrational (TVV) energy transfer in the collinear collision of two nonidentical harmonic diatomic molecules N_2 and O_2. The essential results of their treatment are: (a) the translational–vibrational transition probabilities of O_2 are greater than those of N_2; this is due to O_2 being a more excitable molecule and being acted on by the

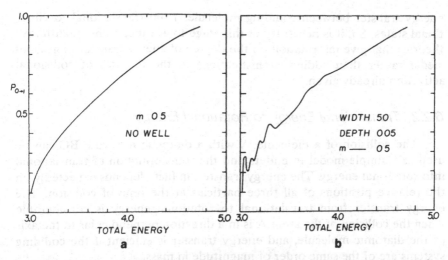

Fig. 6.3. Transition probability for a square barrier potential: (a) without a square attractive well, (b) with a square attractive well.

more excited field; (b) the translational–vibrational one-quantum jump in O_2 and vibrational–vibrational one quantum exchange between N_2 and O_2 are the two principal processes; (c) in the case of the translational–vibrational one-quantum jump in N_2, the mechanism is not unique according as O_2 is vibrationally excited or not prior to collision.

Halavee and Levine[35] made computations of quantal transition probabilities from classical trajectories for the collinear reactive $H + Cl_2$ system and compared it with the previous quantal results[36] and obtained good agreement. Bass[37] used a non collinear model for calculating the translational energy transfer between oxygen atom and NH_3 molecule and oxygen atom and CO_2 molecule. He used a time-dependent interaction to cause excitation in this treatment.

From the previous discussion one finds that most of the treatment used a collinear collisional model. But in reality molecules do not collide in such a well-fashioned manner. Most real situations have a rather complicated pattern of collisional impact, and energy transfer will be dependent on the angle of the axis of the molecule and the directional vector of collision. Bass[37] attempted to calculate the energy transfer probability where the collision is not simply a collinear one but is angle-dependent.

However, apart from the modelistic simplicity, most of the treatments have been concerned with the excitation probability of a diatomic oscillator due to collisional interaction without entering into the problem of practical interest, e.g., the determination of the probability of activation of gas molecules by collision during reaction, including numerical calculation of energy transfer between colliding molecules from translational to vibrational states. So, it is necessary at this stage to advance some quantitative theories that give information on the degree of energy transfer in gaseous media rather than adding sophistications to the theories of collisional activation already given.

6.2.2. Translational Energy to Rotational Energy

The collision of a molecule A with a diatomic molecule BC can be used as a simple model to understand the transformation of translational into rotational energy. The energy transfer, in fact, depends expectedly on the relative positions of all three particles at the time of collision. The energy transfer from translational to rotational energy is considerable when the collision of the atom A is in a direction perpendicular to the axis of the diatomic molecule, and energy transfer is efficient if the colliding systems are of the same order of magnitude in mass.

Recent investigations[38-42] on the collisional interaction and the rotational transition mainly deal with the scattering cross section without going into detail as to how much energy is transferred and what the probability is that excitation occurs from a particular rotational energy level to another higher level by collisional interaction during gaseous reactions.

Stallcope[43] used the impact parameter method and the sudden approximation to determine the total probability of inelastic translational to rotational transitions arising during a collision of an atom and a homonuclear diatomic molecule. The result of this treatment has been applied to collisions between an argon atom and an O_2 molecule. In this treatment, it is considered that the collisional energy is larger than the spacing of the rotational energy levels, i.e., the difference between two successive rotational levels. The interaction between the atom and molecule is considered due to electron exchange and overlap forces during collision. Stallcope[43] obtained an expression for the transition probability of rotational excitation using an exponential excitation of interaction. Figure 6.4 illustrates the dependence of total inelastic rotational transition probability at various impact parameters. The geometry of the interaction process is given in Fig. 6.5.

Pattergill[44] reported a semiclassical generalized phase-shift (GPS) treatment of rotational excitation during collision of the atom and N_2 molecule. He compared the results of transition probability of excitation to different rotational levels from the initial rotational level 10, based on GPS treatment, to that using the sudden approximation.

In the recent overviews of Nikitin[45] and Amme,[46] the vibrational and rotational excitation in gaseous collisions are discussed. However, it can be concluded that no theory is at present specific regarding details of the

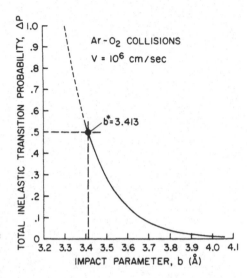

Fig. 6.4. The plot of the total inelastic transition probability versus impact parameter.

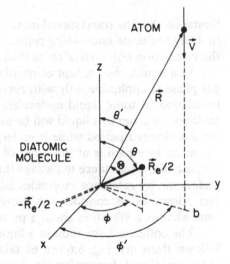

Fig. 6.5. Geometry of the collision of an atom with a diatomic molecule having relative velocity **V** and impact parameter *C*. The center of the coordinate system is located at the center of the diatomic molecule, and the direction of **V** is chosen to be parallel to the *z* axis.

energy transfer during collision and subsequent activation in gaseous reaction. It is desirable to *correlate the theory of energy transfer in the gas phase with the activational process in chemical reactions.*

6.3. Collisional Activation Model in Liquids

The collisional model for activation in liquid solutions is considered virtually to be that in the gas phase.[47–50] In a liquid medium, there is the

Fig. 6.6. Movement of a reference
molecule in a liquid and collision process.

limitation that the translational motion of liquid molecules is restricted and
that any molecule undergoing collision in a liquid is, in fact, influenced by
the interaction potential of more than one other molecule.[51]

In a liquid, the concept of translation in the sense understood in the
gas phase is applicable only with reservations. There may be a restrictive
translation of some liquid molecules, but for the most part, the average
molecule or atom in a liquid will be held in contact with another molecule
for a prolonged period while it undergoes a series of vibrations. Because
the mean free path is of the order of molecular diameter in a liquid and
because of the resistance to passage through other molecules, the collisions
undergone by reference molecules add up to a series of reflections. This
fact allows us to consider the movement of a molecule in the liquid
somewhat as a vibration about a particular point, as depicted in Fig. 6.6.

The collisional situation in a liquid can be pictured in more detail as
follows. Basically, Fig. 6.6 can be taken to represent the model. Thus, in
the model liquid, two molecules, A and B, are held in close contact. They
collide in a vibrational sort of way. Thus, A collides with B, and this
collision occurs vibrationally many thousands of times.

The average translator in a liquid has an energy in the region of kT.
The average energy of a vibrator for which $h\nu/kT \gg 1$ is much greater than
kT and the amount of energy transferable upon one collision is as much as
$h\nu$. Thus, one can understand qualitatively the possibility of activation by
collision in a liquid on a model of vibrators. One must only remember that
the Boltzmann fraction has to be replaced by a quantum fraction that
corresponds to the distribution of energy states in vibrators.

6.3.1. Expressions for the Free Energy of Activation in Liquid

George and Griffith[52] (1959) derived an expression for the free energy of activation for electron exchange between the redox ions, e.g.,

$$M^{2+}(H_2O)_6 + M^{3+}(H_2O)_6 \leftrightarrows M^{3+}(H_2O)_6 + M^{2+}(H_2O)_6 \qquad (6.1)$$

It was assumed as a first approximation that the force constants f for all the ion oxygen bonds are the same. These were expressed as f_r and f_0 for the $M^{2+}-O$ and $M^{3+}-O$ bond, respectively. The equilibrium bond lengths were considered to be q_0 for the $M^{3+}-O$ bond and $q_0 + \Delta q$ for the $M^{2+}-O$ bond; $q_0 + \alpha$ and $q_0 + \Delta q - \beta$ are the corresponding bond distances in the activated (transition) state. The vibrational energies (via harmonic approximation) of the n bonds [for the $M^{2+}(H_2O)_6$ complex $n = 6$] before electron transfer in the transition state is given by (Fig. 6.7):

$$F_i = \tfrac{1}{2}[nf_0(AC)^2 + nf_r(CB)^2]$$

$$= \tfrac{1}{2}[nf_0\alpha^2 + nf_r\beta^2] \qquad (6.2)$$

After electron transfer (final state) the energy becomes (Fig. 6.7):

$$F_f = \tfrac{1}{2}[nf_r(AB - AC)^2 + nf_0(AB - CB)^2]$$

$$= \tfrac{1}{2}[nf_r(\Delta q - \alpha)^2 + nf_0(\Delta q - \beta)^2] \qquad (6.3)$$

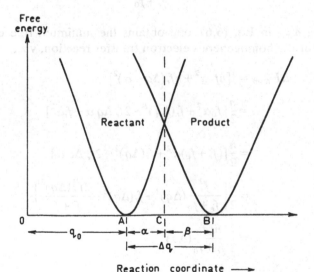

Fig. 6.7. The plot of free energy vs. reaction coordinate to represent the free energy of activation corresponding to the crossing points of reactant and product curves. A and B represent the equilibrium position of reactant and product oscillator, respectively, and C represents the position of the activated complex.

for the radiationless transition of an electron the total free energy must be conserved during the transition and therefore,

$$F_i = F_f \tag{6.4}$$

From Eqs. (6.2), (6.3), and (6.4) one finds that

$$\Delta q - \beta = \alpha \tag{6.5}$$

Using Eq. (6.5) in Eq. (6.3), one obtains the free energy of activation of the system $\Delta F_{hom}^{\ddagger}$ in homogeneous medium as

$$\Delta F_{hom}^{\ddagger} = \tfrac{1}{2}[nf_0\alpha^2 + nf_r(\Delta q - \alpha)^2] \tag{6.6}$$

Minimizing Eq. (6.6) with respect to α, one obtains

$$\frac{\partial F}{\partial \alpha} = \frac{\partial}{\partial \alpha}\left\{\frac{1}{2}[nf_0\alpha^2 + nf_r(\Delta q - \alpha)^2]\right\} = 0 \tag{6.7}$$

From Eq. (6.7), the minimum value of α is given by

$$\alpha_{min} = \frac{f_r}{f_r + f_0}\Delta q \tag{6.8}$$

Substituting α_{min} in Eq. (6.6), one obtains the minimum free energy of activation for the homogeneous electron transfer reaction, viz.,

$$\Delta F_{hom}^{\ddagger} = \tfrac{1}{2}[nf_0\alpha^2 + nf_r(\Delta q - \alpha)^2]$$

$$= \frac{n}{2}[f_0\alpha^2 + f_r(\Delta q)^2 - 2f_r\Delta q\,\alpha + f_r\alpha^2]$$

$$= \frac{n}{2}[(f_r + f_0)\alpha^2 + f_r(\Delta q)^2 - 2f_r\Delta q\,\alpha]$$

$$= \frac{n}{2}\left[\frac{f_r^2}{f_r + f_0}(\Delta q)^2 + f_r(\Delta q)^2 - \frac{2f_r^2(\Delta q)^2}{f_r + f_0}\right]$$

$$= \frac{1}{2}\frac{nf_r f_0}{f_r + f_0}(\Delta q)^2 \tag{6.9}$$

Since it is known[53] that the free energy of activation for the electrochemical redox reaction is half the amount required for the homogeneous reaction, one can calculate the free energy of activation from the stretching of the ion–solvent bond in the electrochemical case as

$$\Delta F^{\ddagger}_{\text{electrochemical}} = \frac{1}{2}\left[\frac{1}{2}\frac{nf_r f_0}{f_r+f_0}(\Delta q)^2\right] \tag{6.10}$$

Marcus[53] (1965) rederived the expression derived by George and Griffith[52] for the free energy of activation of ions in solution by considering the vibrational energy increase of an ion–solvent bond due to collisional interaction with the surrounding solvent molecules. He took the ion–solvent bond vibrations as harmonic, giving the expression for the free energy of activation from ion–solvent bond oscillations by the harmonic approximation as

$$\Delta F^{\ddagger}_{\text{hom}} = \frac{1}{2}\sum_{j=1}^{n} k_j(\Delta q_j)^2 \tag{6.11}$$

where k_j is the reduced force constant of the jth ion–solvent bond given by the expression[53]

$$k_j = \frac{f_j^r f_j^p}{f_j^r+f_j^p} \tag{6.12}$$

In Eq. (6.12), f_j^r and f_j^p are the force constants of the jth ion–solvent bond in its reactant and product states, respectively. In Eq. (6.11), Δq_j is the difference in the equilibrium values of the bond coordinates of the jth ion–solvent bond in its reactant and product states, i.e.,

$$\Delta q_j = q_j^r - q_j^p \tag{6.13}$$

where q_j^r and q_j^p are the equilibrium value of the bond coordinates of the ion in its reactant and product states, respectively (usually taken as the equilibrium bond length of the ion–solvent bond). The summation in Eq. (6.11) is over the number of ion–solvent coordinated bonds. Using Eq. (6.12) in Eq. (6.11), one obtains

$$\Delta F^{\ddagger}_{\text{hom}} = \frac{1}{2}\sum_{j=1}^{n}\frac{f_j^r f_j^p}{f_j^r+f_j^p}(\Delta q_j)^2 \tag{6.14}$$

6.3.2. Mechanism of Activation

Moelwyn-Hughes (1971) reviews the basic theories of reactions in solution from a mechanistic point of view in some ten pages (in a book[47] of five hundred pages on reactions in solution). Even this material, however, is oriented toward the resuscitation of equations from reactions in the gas phase. The "transition state theory," and "statistical theory," are typical descriptions in Moelwyn-Hughes' discussion of mechanisms of activation in solution.[47]

The heart of the matter is that these theories attempt (as in the earlier theories of liquids) to carry over theory from the gas phase to the solution. The principal difficulty of such theories is that they leave the heart of the matter empty by making the *implicit* assumption that there is a Maxwellian distribution of translators in solution. The assumption is that the classical Boltzmann distribution of heat energy among (implied) translators exists in solution so that the equations in the gas phase can be taken over entirely and the only thing that has to be modified is the collision number.

Correspondingly, most theories of collision in solution concentrate upon changes in the collision number that must be made between the gas and solution. For example, in the solution there is a cage effect, molecules being kept in contact with each other, constantly activating each other, and colliding at a higher frequency than in the gas phase. Such approaches, however, come into the difficulty that one cannot apply a simple Boltzmann continuous translational statistic to vibrators, and the tacit implication that a sufficient number of translators is always present is presumptuous.

In water, there *is* evidence for a small number of translators,[54a,b] i.e., translational motion in the classical way. The evidence has come out of the spectroscopic work on liquid water compared with water in the gas phase, the liquid water showing peaks that have the energy of translators. It may be that there are a number of free water molecules in solution that have a restricted translation.[70] However, to rely upon the existence of these translators present in small[54] concentration as the basis to the theory of the translator-oriented transfer of energy and activation in solution, and the implied justification for considering the situation to be gaslike, may not be satisfactory.

The theory given above by George and Griffith[52] (and those given by Marcus[53]) is formalistic. They present equations to describe a mechanism of bond stretching and a vibrationally energy-rich molecule that reacts by the same mechanism as that in the gas phase; the vibrational levels become energized and activated so that at the intersection of the two potential energy curves reaction can take place. Such theories miss the point: they imply that vibrational energy is distributed and that activated vibrational stages exist in liquids without stating how the thermal energy is transferred to the reactant in a liquid in the virtual absence of translators. The energy transfer mechanism is tacitly ignored.

There are some quantum mechanical theoretical treatments for the vibrational relaxation,[55,56] i.e., the dissipation of vibrational energy in a condensed medium. Unfortunately, there is no theory to consider the augmentation of vibrational–rotational energy with special reference to *activation* in condensed media prior to activation and realignment of bonds.

6.4. Need for Alternate Activation Mechanisms in Condensed Media

Collisional mechanisms in which translational energy is converted to vibrational energy (i.e., the gas phase type of activation mechanism) is uncertainly present in reactions in solutions. A point that seems at first to favor a translation to vibration mechanism is that there are reactions that have the same heat of activation in the gas phase as in solution. Thus, in the decomposition of chlorine monoxide, isomerization of pinene, and the decomposition of ozone catalyzed by chlorine,[57] the reaction products and activation energy are the same in the gas phase and in solution.

However, this evidence in tenuous. There may be only one path and rate-determining step available. Hence, if a different mechanism of activation occurs in the solution phase, compared with that in the gas phase, it may still give rise to the same value of the activation energy. The following facts suggest that one does need a model of activation in condensed media different from the collisional one.

(a) Reactions occur in molten salts. Here, there are no translators, so that the energy of particles in collision cannot be Maxwellianly distributed. Relevant data in this area arise from the work of Horne.[58] He carried out measurements on the rate of the ferrous–ferric electron exchange reaction at $-84°C$ in ice. The log rate constant versus $1/T$ plot was continuous for both liquid and solid phases, and this seems to indicate that the mechanism is the same.† In ice there cannot be translators; i.e., the distribution of energy must be sought in other forms of molecular arrangements. But if the rate constant in liquid solutions has a continuity with that in ice, there may be no translational mechanism of activation in the liquid either. Similar types of results were reported by Baulch[59] and also by Nitzan and Wahl.[60] Arvia and Videla[61] reported the kinetics of hydrogen evolution on a platinum electrode during the electrolysis of molten potassium bisulfate where they observed Tafel behavior. Several other workers[62,63] have observed Tafel-like behavior in molten salt systems. Thus, Chebotin et al.[64] observed Tafel behavior for oxygen evolution in both the ZrO_2 and Y_2O_3 solid electrolyte systems.

(b) Recently, Armstrong et al.[65] observed Tafel behavior for the Cu/Cu^+ charge transfer reaction in solid electrolyte systems at room temperature. These facts about the Tafel behavior of the electrochemical

† The continuity of the log rate $-1/T$ plots through the melting point is more significant than the identity of slopes implying the same energy of activation, for the former implies the same arrangement energy for the activated complex as well as the same magnitude for the heat of reaction.

current in molten salt and solid electrolyte systems suggest that it is not the collisional mechanism that is responsible for activation, since there are no translators in these media to produce exponentially distributed acceptor levels. As one observes Tafel behavior in aqueous solutions, in molten salts, and in solid electrolyte systems, it would be more reasonable to find an activation mechanism common to all three systems.† In aqueous solutions, free water molecules (translators) are limited in extent at room temperatures[54a,b]§ and this raises the question as to whether a collisional activation process is operative for charge-transfer reactions even in aqueous solution.

6.5. Activation Due to Continuum Solvent Polarization Fluctuation

The most thorough-going attempt that has been made to replace the normal theories of activations in liquids by a new theory has been made by the Russian school of electrochemists, the work originating under the names of Levich[66,67] and, in recent times, Dogonadze.[68,69] The origin of this work arose in the following way. One of the basic laws in electrochemistry is Tafel's law, and this equation has been tested thoroughly in certain instances over a change of electrode potential of more than 1 V. During this change, the line remains linear, and in particular, it is not *stepped* in any way.

The Russian workers originally saw this fact as evidence for a mechanism of activation avoiding translators. They reasoned in the following manner. The electrons in a metal emit to the solution and are received by, say, H_3O^+ ions. They must be received by the vibrational states of a given electronic level of the H_3O^+ ion. However, in the gas phase, these vibrational states would have an energy difference of the order of some $\frac{1}{2}$ eV.

Now, as an electrode potential is moved over a range of 1 eV itself, the Fermi level in the electrode from which the majority of electrons emit in cathodic reactions to ions in solution passes *continuously* over an energy difference equivalent to about two vibrational levels (for H_3O^+) insofar as the gas phase structure is concerned. In this case, the Tafel line would be stepped (Fig. 6.8a), and hence a mechanism of activation that would fill in the energy differences seemed necessary.

† However, Tafel behavior is observed due to exponentially distributed acceptor energy states. Hence, one needs to find out the mechanism by which such a distribution arises in the absence of translators in liquid and solid.
§ From spectral investigation, Stevenson[54a] found that the concentration of monomer is only 1.30% in a pure liquid water at 25.5°C. But earlier, Marchi and Eyring[54b] reported the concentration of monomer is about 5% in a liquid water at 25°C.

Fig. 6.8. (a) The log *i* versus potential plot that represents the expected current potential relation according to Levich's[67] view that energy levels of the ion–solvent bond oscillators in solution are far apart from one another. (b) A schematic representation of the electron transition situation across the electrode–solution interface. In the solution side, well-separated acceptor levels are shown according to Levich.[67] No acceptance of electron is shown in between two well-separated acceptor energy states.

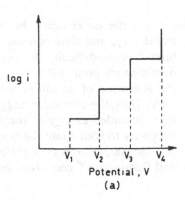

(a)

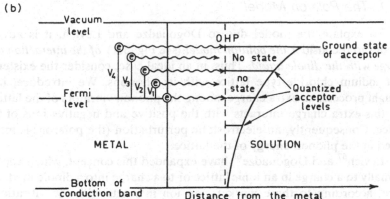

The Russian workers ignored the fact that the molecules in solution would have a smoothing out of their vibrational–rotational states into a virtual continuum (cf. the spectra of H_3O^+ ions in solution[70]). They thought that the acceptance of electrons in the cathodic reaction would be good when the Fermi level was equal to that of a given vibrational level, but between the vibrational levels there would be a forbidden zone and no current should be observed (Fig. 6.8b). There is no evidence in experimental Tafel lines that support this view.†

To face and overcome the difficulty that he thought existed, Levich[67] attempted to involve some type of activator that would behave classically (continuous energy distribution) and yet would not be a translator. He focussed upon the libratory movements of the water molecule because the frequency of libration has a magnitude 10^{11} Hz, i.e., is classical (compared with the vibrational frequency in the range of 10^{14} Hz).

† In the original paper by Gurney of 1931 (see Chapter 8), the continuum nature of the vibrational–rotational states in water was stressed, and it was indicated there that the smoothness of the Tafel line originated from them.

On the other hand, having got out of the difficulty of the missing translatory, and thus continuous, activation source, Levich had involved himself in the difficulty that, as he saw it, a great number of librators would be necessary (and, indeed, some kind of coordinated movement) to bring the activation of an ion in solution which has the order of magnitude of 1 eV. Thus, the average energy of a librator in electron volts is 0.001 eV, so that the added energy of something like one thousand librators would be necessary to contribute the needed 1 eV.† In order that the energy of some one thousand librators could be applied to activate a central ion, Levich[67] and Dogonadze[68] conceived of an *electrostatic* type of energy transfer.

6.5.1. The Polaron Model

To explain the model due to Dogonadze and Levich, it is advantageous to reconsider *the polaron model* (see Fig. 3.1) *of the interaction of a charge with the dipole liquid.* Thus, in an ideal case, consider the existence of a sodium chloride-type lattice with its phonons. We introduce, in a thought process, an extra charge among the phonon vibrators of the lattice, and this extra charge interacts with the positive and negative ions of the lattice. Consequently, an electrostatic perturbation (the polaron) is introduced in the phonon energy of the lattice.

Levich[67] and Dogonadze[68] have expanded this concept, which applies normally to a charge in an ionic lattice or to a charge into a dipole medium. Here, according to these workers, an ion interacts with the vibrational energy of the dipoles that constitute the fluid. These dipoles interact, after having received an interactional energy from the ions, back upon the ion, and thus change the energy of the ion itself.

In order to ascertain what is the maximum energy of the interaction of a single librator with the central ion, we may fall back upon the fact that this interaction cannot be more than the energy of the librator itself. A typical libratory energy is of the order of 0.001 eV per particle.

6.5.2. Frequency of a Fluctuation in the Energy of a Central Ion

To understand how it might be possible for such a small interaction energy of the librator with the central ion to give rise to the large 1 eV increase in energy association with the central ion, it has been suggested by Levich[66,67] that a "fluctuation" of the energy of interaction of the central ion with many (of the order of one thousand or more) librators could give rise to such an energy.

---

† It seems that the tail of the Maxwell–Boltzmann distribution of the classically behaving high-energy librators was not considered by Levich.

Levich's group[66-69] did not further develop this point, i.e., did not calculate the actual fluctuational *frequency* that would arise to allow the central ion to experience the full (coupled) interaction with the central ion and the thousand surrounding librators.†

Bockris and Sen[71,72] were the first to examine the frequency of similar fluctuational interaction. They treated the situation in terms of a Lorentzian *internal* field and showed that the probability that the fluctuational energy of the central ion due to the surrounding solvent dipoles would correspond to about 1 eV and was too low to account for the observed currents (except for $|\eta| > 1$ V). This model is presented in Section 7.6.1.

On the other hand, the theory of Bockris and Sen[71] was classical in the sense that it dealt with ion–dipole interactions and their libratory variations, and did not take into account the polaron component of the phonon energy and *its* fluctuation.

Levich and Dogonadze[66-69] assumed that these fluctuations would be sufficiently frequent. If they are, and because the model is a continuum one, the energy (mainly electron) of the ion will be continuously distributed, because it originates in librators for which $\hbar\omega/kT < 1$. Hence, the continuous nature of the Tafel line is expected. Thus, before further looking into such processes, one needs to calculate quantum mechanically the frequency of such fluctuations of energy, and investigate whether it is sufficiently large to correspond to the requirements of experiment.

Thus, to check the validity of the fluctuation model one needs to calculate the probability of energy fluctuations from the quantum mechanical viewpoint. The average energies $\langle E \rangle^2$ and $\langle E^2 \rangle$ to use in the probability of fluctuation expression

$$p_{\text{fluctuation}} = \frac{1}{(2\pi)^{1/2}} \exp\left(-\frac{(E - \langle E \rangle)^2}{2[\langle E^2 \rangle - \langle E \rangle^2]}\right) \qquad (6.15)$$

may be expressed as

$$\langle E \rangle^2 = |\langle \psi_1 | H_{\text{ion–librator}} | \psi_1 \rangle|^2 \qquad (6.16a)$$

and

$$\langle E^2 \rangle = |\langle \psi_1 | H^2_{\text{ion–librator}} | \psi_1 \rangle| \qquad (6.16b)$$

where ψ_1 is the wave function of the phonon and $H_{\text{ion–librator}}$ the ion–librator interaction Hamiltonian, a type of ion–phonon interaction

---

† Of course, the use of the number one thousand is only meant in a very global manner. It arises because the activation energy has to be of the order of 1 eV and the energy of an individual librator is of the order of 10^{-3} eV. In reality, all the librators of a solution can be regarded as contributing to the energy of activation of each individual ion.

Hamiltonian, that is given as[73]

$$H_{\text{ion–librator}} = \sum_\omega g(B_\omega + B^+_{-\omega})C^+_i C_i \qquad (6.17)$$

where

$$g_\omega = \frac{4\pi e_0^2}{\omega}\left(\frac{z^2 N}{2\Omega_p M}\right)^{1/2} \qquad (6.18)$$

and ω = phonon frequency, $\Omega_p = (4\pi z^2 e_0^2 N^2 / M)^{1/2}$, z = change on the ion undergoing activation, N = the number of ions per cubic centimeter, and M = mass of the librator. B^+_ω and B^-_ω are the phonon creation and annihilation operators, respectively, given by the expressions

$$B^+_\omega = \left(\frac{\omega}{2\hbar}\right)^{1/2} q + \left(\frac{\hbar}{2\omega}\right)^{1/2}\frac{\partial}{\partial q} \qquad (6.19)$$

$$B^-_\omega = \left(\frac{\omega}{2\hbar}\right)^{1/2} q - \left(\frac{\hbar}{2\omega}\right)^{1/2}\frac{\partial}{\partial q} \qquad (6.20)$$

where q is the coordinate distance. C^+_i and C_i are the creation and annihilation operator of the ion.

By determining $\langle E^2 \rangle$ and $\langle E \rangle^2$ from Eqs. (6.16a) and (6.16b), one could find the probability of a fluctuation using Eq. (6.15) and investigate the frequency of energy fluctuations on the basis of quantum mechanical calculations. The theory of such a calculation at the moment is demanding because of the restrictions of the many-body problem.

6.6. An Expression for the Free Energy of Activation from the Continuum Solvent Polarization Fluctuation Model

Levich[66,67] stated an expression for the free energy of activation for an electron transfer reaction in a polar solution. The expression is said to be derived from "the quantum theory." However, it is possible to obtain this same well-known expression classically with trivial modelistic assumptions. Considering the free energy system with two overlapping similar parabolic energy wells for reactions and products, one can formulate the relation between the free energy of activation and that of the so-called reorganization or repolarization energy E_s of Levich's continuum model. An illicit assumption has to be made that the force constant k for the reactant and product solvent system are the same. In the schematic diagram of Fig. 6.9, the parabolic curves show the change of polarization energy E_s when, e.g., Fe^{3+} becomes Fe^{2+}.

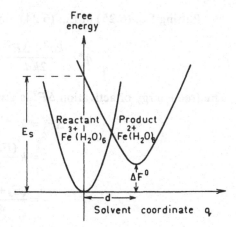

Fig. 6.9. The free energy versus solvent coordinate curve showing the relation between reorganization energy E_s, free energy of reaction ΔF^0, and free energy of activation $\Delta F^{\ddagger}$.

One can use the equation for the harmonic oscillator to represent the free energy of the reactant and product systems in terms of solvent coordinate q and force constant k in the following very simple way devised by Appleby et al.[74] This method obtains the often-quoted equation for free energy of activation in the continuum model.

$$F_R = kq^2 \tag{6.21}$$

for reactant system and

$$F_p = k(q-d)^2 + \Delta F^0 \tag{6.22}$$

for product system, where d is the displacement between reactant and product system corresponding to respective equilibrium positions. At the activated state, the two reactant and product system parabolas intersect where

$$F_R = F_p \tag{6.23}$$

Now using Eqs. (6.21) and (6.22) in Eq. (6.23), one can find out the solvent coordinate corresponding to the activated state q^* as

$$kq^{*2} = k(q^*-d)^2 + \Delta F^0$$

or

$$kd^2 - 2kdq^* + \Delta F^0 = 0 \tag{6.24}$$

Now from the definition, the repolarization energy is the energy of the product system when its solvent coordinate is still the same as that of the reactant state, i.e., where $q = 0$. Thus, putting $q = 0$ in Eq. (6.22), one gets

$$F = kd^2 = E_s \tag{6.25}$$

Putting Eq. (6.25) in Eq. (6.24), one obtains after rearrangement

$$q^* = \frac{kd^2 + \Delta F^0}{2kd} = \frac{(E_s + \Delta F^0)}{2kd} \tag{6.26}$$

The free energy of activation $\Delta F^{\ddagger}$ is given as

$$\Delta F^{\ddagger} = kq^{*2}$$

$$= k\frac{(E_s + \Delta F^0)^2}{4k^2 d^2}$$

$$= \frac{(E_s + \Delta F^0)^2}{4E_s} \tag{6.27}$$

In Eq. (6.27), ΔF^0 is the standard thermodynamic free energy change in the reaction. E_s is sometimes called (rather vaguely) the energy of reorganization. Its meaning can be better seen from Eq. (6.27). It is the energy of the product system $[Fe^{2+}(H_2O)_6$, say] when it is deformed to have the configuration of the reactant system.

One of the more controversial items confronting a student in this area is the expression for E_s. Bockris, Khan, and Matthews[75] found satisfactory agreement with experiment when they used Eq. (6.25) and equations similar to those of George and Griffith.[52] The implication, of course, is that the activation process involves activation of vibrational levels ("bond stretching"). However, the older workers derived a value for E_s along a thought process which *perhaps* went somewhat as follows:

The self-energy of an ion in vacuum is $(z_i e_0)^2/2r_i$. In solution at equilibrium, the electrostatic continuum expression for the energy of an ion neglecting the structure of the first layer is $-(z_i e_0)^2/2\varepsilon_s r_i$, where ε_s is the static dielectric constant. This energy is a free energy that represents the total polarization of the solvent around the ion. This is well known to include that polarization due to rearrangement and orientation ("reorganization") of the dipoles in an electric field; stretching of nuclei in an electric field and the stretching of the electron shells that occur with a short relaxation time, comparable with the time of electronic transfer, $(z_i e_0)^2/2r_i \varepsilon_{op}$, one removes this energy away from the total, and one will obtain, then, that due to reorganization. Hence

$$\frac{(z_i e_0)^2}{2r} \left| \frac{1}{\varepsilon_{op}} - \frac{1}{\varepsilon_s} \right|$$

is the energy of reorganization E_s.

Of course, such an evaluation of the reorganization energy avoids many questions and assumes no bond stretching energy is involved.

One of the parts of the model that is more difficult to accept is that "only the outer sphere solvation energy," i.e., the Born component, is to be taken into account. This has been described by some workers in this area as implying "no bond stretching." However, potential energy changes due to movement of the bond occur in activation. There must also certainly be some change of electrostatic interaction of dipoles in the *first* layer of water molecules near an ion, and the Born equation does not apply here. This is because the ion–dipole interaction depends on the charge in the ion, i.e., it is given as a first approximation by $ze_0\mu/\varepsilon r^2$. The neglect of the change of interaction with the dipoles neglects up to about half the interaction of the ion with the surrounding medium.

6.7. Expression for E_s

A neglect of the structure of the solvated ion has occurred in all the formulations of the continuum viewpoint. Thus, it might be controversial as to whether *stretching* of the inner part of the ion–solvent complex is associated with electron transfer. It is not possible, however, to deny that the inner part of the solvation shell, that associated with the ion–dipole inter-action, changes upon change of charge on the ion. However, this (a larger component than the Born outer solvation part) has not been explicitly considered in theories derived by Marcus and Levich.

The physical meaning of E_s is usually taken as the change of the Born solvation energy of the ion with the same organization of solvent as the charge on the ion changes.[66-69] A change in energy of the ion–solvent interaction takes place, therefore, only in the optical region. Thus, considering the change in energy of the ion, the expression for E_s in the optical frequency due to the Born component, the ion–inner sphere solvent dipole, and the ion–quadrupole interactions, one gets an expression for E_s as†

$$E_s(\text{electrode}) = (\Delta z)^2 e_0^2 \left(\frac{1}{2a} - \frac{1}{R}\right)\left(\frac{1}{\varepsilon_{op}} - \frac{1}{\varepsilon_s}\right) - n_s(\Delta z)^2 e_0^2 \left(\frac{\mu}{a^2} + \frac{Q}{a^3}\right)\left(\frac{1}{\varepsilon_{op}} - \frac{1}{\varepsilon_s}\right)$$

(6.28)

where Δz is the change in charge on the ion before and after electron

---

† However, induced and multipole interactions have not been taken into account, so that the present expression is simply an advance to a better approximation than that of Born.

transfer. The radius of the ion is a, and μ is the dipole moment. R is double the distance between the center of the ion in the outer Helmholtz plane and the electrode. Q is the quadrupole moment of the solvent and n_s is the solvation primary number.

6.8. Formulation of Activation Energy

Weiss[79] was the first to apply the Born equation and the Frank–Condon principle to obtain the energy of reorientation of dipoles on electron transfer in terms of ε_{op} and ε_s. However, Weiss regarded this energy in a Libby-like manner as forming part of the barrier to transfer. Marcus[53] was the first to use such ideas for the formulation of a reorganization energy before transfer. The following derivation of E_s owes much to that of Hush[80,81] and Schmickler.[82]

The energy in the initial and final states are, respectively,

$$F_{op,i}(z_i) + F_{s,i}(z_i)$$

and

$$F_{op,f}(z_i + \Delta z) + F_{s,f}(z_i + \Delta z)$$

where Δz is the change in charge.

The energy of the activated state can be written as

$$\tfrac{1}{2}[F_{op,i}(z_1) + F_{op,f}(z_i + \Delta z)] + F_s(z_i + \tfrac{1}{2}\Delta z_i)$$

Hence

$$\Delta F^{0\ddagger} = \tfrac{1}{2}[F_{op,i}(z_i) + F_{op,f}(z_i + \Delta z)] + F_s(z_i + \tfrac{1}{2}\Delta z_i) - F_{op,i}(z_i) - F_s(z_i) \quad (6.29)$$

For an isotopic reaction,

$$F_{op,i}(z_i) + F_s(z_i) = F_{op,f}(z_i + \Delta z_i) + F_s(z_i + \Delta z_i) \quad (6.30)$$

Substituting $F_{op,f}(z_i + \Delta z_i)$ from Eq. (6.30) into Eq. (6.29),

$$\Delta F^{0\ddagger} = \tfrac{1}{2}F_s(z_i + \tfrac{1}{2}\Delta z_i) - \tfrac{1}{2}F_s(z_i + \Delta z_i) - \tfrac{1}{2}F_s(z_i) \quad (6.31)$$

By use of an argument similar to that of Appendix (11.I),

$$F_s(z_i) = -\left(\frac{1}{\varepsilon_{op}} - \frac{1}{\varepsilon_s}\right)\frac{(z_i e_0)^2}{2a_i} = Az_i^2 \quad (6.30a)$$

Using Eq. (6.30a) for $F_s(z_i)$ in Eq. (6.31),

$$\Delta F^{0\ddagger} = \frac{(\Delta z)^2}{4}\left(\frac{1}{\varepsilon_{op}} - \frac{1}{\varepsilon_s}\right)\frac{e_0^2}{2a_i}$$

If one adds the image contribution from the activated state ($R =$ distance of the ion from the electrode),

$$\Delta F^{0\ddagger} = \frac{(\Delta z)^2}{4}\left(\frac{1}{\varepsilon_{op}} - \frac{1}{\varepsilon_s}\right)\frac{e_0^2}{2a_1} - \frac{(\Delta z)^2}{4}\left(\frac{1}{\varepsilon_{op}} - \frac{1}{\varepsilon_s}\right)\frac{e_0^2}{R} \tag{6.32}$$

6.9. Comparison between Free Energy of Activation from Continuum Expressions and Experiment

Bockris, Khan, and Matthews[75] pointed out that if one plots the value of $\Delta F^{\ddagger}$ calculated from the solvent fluctuation theory against the corresponding standard free energy ΔF^0 of reaction, one obtains the result of Fig. 6.10. The parameter E_s used for this calculation corresponds to the value found from calculation of this parameter from the equations given by Marcus.[53] The values theoretically predicted in this model for both isotopic ($\Delta F^0 = 0$) and nonisotopic reaction ($\Delta F^0 \neq 0$) are below those observed experimentally. Further, the trends (Fig. 6.10 and Table 6.1) of the experimental results are not followed by the theory either. The lack of

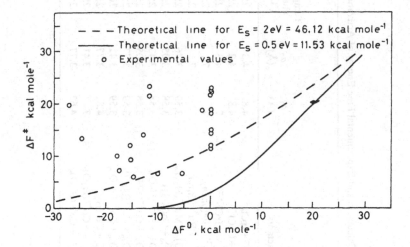

Fig. 6.10. Plot of free energy of activation $\Delta F\ddagger$ against standard free energy ΔF^0 for electron transfer reaction in solution from the theoretical relation of solvent fluctuation model.

Table 6.1

Calculated and Experimental Free Energy of Activation for the Nonisotopic Homogeneous Electron Transfer Reaction in Solution

Reactants	a_1 (Å)	a_2 (Å)	Temperature (°C)	Standard free energy of reaction ΔF^0 (kcal mole^{-1})	E_s (kcal mole^{-1})	$\Delta F^\ddagger$ (continuum) (kcal mole^{-1})	$\Delta F^\ddagger$ (expt.) (kcal mole^{-1})
$Fe(CN)_6^{4-} + IrCl_6^{2-}$	4.8	4.3	25	−15.2	20.3	0.3	9.2
$Fe(CN)^- + OsCl^-$	4.8	4.2	25	−1.6	20.5	4.4	18.5
$Os(dipy)_3^{2+} + IrCl_3^{2-}$	7	4.3	25	−5.4	18.2	2.3	6.5
$Os(dipy)_3^{2+} + Fe(phen)_3^{3+}$	7	7	25	−5.4	13.1	1.2	6.5
$Os(dipy)^{2+} + Ru^{3+}(dipy)$	7	7	25	−10.1	13.1	0.2	6.5
$Fe^{2+}(H_2O)_6 + Mu^{3+}(H_2O)_6$	3.59	3.46	25	−12.9	26.0	1.7	13.9
$V^{2+}(H_2O)_6 + Co(NH_3)_6^{3+}$	3.58	3.05	25	−11.8	28.0	2.9	23.1
$Cr^{2+}(H_2O)_6 + Fe^{3+}(H_2O)_6$	3.58	3.43	25	−27.2	26.2	0.0	19.5
$Cr^{2+}(H_2O)_6 + Co^{3+}(NH_3)_6$	3.58	3.05	25	−11.8	28.0	2.9	21.3
$Fe^{2+}(H_2O)_6 + Ce^{4+}(H_2O)_6$	3.59	3.77	25	−15.5	25.0	0.9	11.8
$Fe^{2+}(H_2O)_6 + Co^{3+}(H_2O)_6$	3.59	3.41	20	−24.7	26.2	0.02	13.2
$Fe^{4-}(CN)_6 + Ce^{4+}(H_2O)_6$	4.81	3.77	25	−17.5	22.0	0.24	7.1
$Fe^{2+}(phen)_3 + Co^{3+}(H_2O)_6$	7	3.41	25	−18.0	22.4	0.22	10.0
$Mo(CN)_8^{4-} + Ce^{4+}(H_2O)_6$	4.62	3.77	25	−14.8	22.3	0.65	5.9

agreement of $\Delta F^{\ddagger}$(continuum) and $\Delta F^{\ddagger}$(experiment) is marked for isotopic reactions (Figs. 6.11 and Table 6.2) for which the standard free energy of reaction is zero.

Bockris, Khan, and Matthews[75] showed that agreement between theoretical and experimental free energy of activation is, however, at once obtainable if the stretching contribution [Eq. (6.9) or (6.10)] of the inner coordinated bond is added to the continuum electrostatic contribution (Fig. 6.12 and Table 6.3). A fair correlation is observed between theory and experiment for the homogeneous and electrochemical cases if the Born contribution to the heat of activation is entirely neglected and only the stretching contribution considered (Figs. 6.13 and 6.14 and Tables 6.3 and 6.4). For the electrochemical redox reactions, better agreement is observed when both stretching and continuum electrostatic contributions are used (Fig. 6.15, Table 6.4), but an estimation based solely on the continuum theory gives no agreement with experiment (Fig. 6.16, Table 6.4). Thus, from these results (see Tables 6.3 and 6.4), the continuum solvent reorganization contribution to the free energy of activation is about one-third of the total value for the heat of activation in redox processes in aqueous solution.

It may be thought that if a quantum mechanical model were used for calculating the polaronic interaction with the ion, instead of the model of ion–dipole interaction used by Bockris and Sen,[71,72] the contribution from the fluctuational situation would be more significant. Indeed, there is some hint of this, because when one calculates (cf. Bockris, Khan and Matthews[75]) the free energy of activation for various redox reactions and compares it with the experimental values, about one-third of the heat of activation does come out to be something other than bond stretching. It

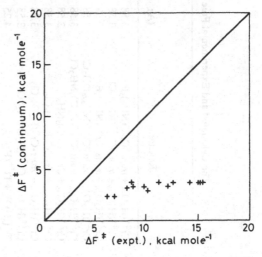

Fig. 6.11. Plot of $\Delta F^{\ddagger}$ (continuum) against $\Delta F^{\ddagger}$ (experimental) for homogeneous electron transfer reactions in solutions involving reactants with water and ammonia molecules as ligands. (Correlation coefficient = 0.41.)

Table 6.2

The Calculated and Experimental Free Energy of Activation for the Homogeneous Electron Transfer Reaction in Solution

Reactants	a_1 (Å)	a_2 (Å)	Temperature (°C)	E_s (kcal mole^{-1})	$\Delta F^{\ddagger}$ (continuum) (kcal mole^{-1})	$\Delta F^{\ddagger}$ (expt.) (kcal mole^{-1})
$Cr^{2+}(H_2O)_6 + trans\text{-}Cr^{2+}(NH_3)_5F$	3.58	3.05	25	28.0	7.0	20.6
$Cr^{2+}(H_2O)_6 + trans\text{-}Cr^{2+}(NH_3)_5Cl$	3.58	3.05	25	28.0	7.0	17.2
$Cr^{2+}(H_2O)_6 + trans\text{-}Cr^{2+}(NH_3)_5Br$	3.58	3.05	25	28.0	7.0	16.4
$Cr^{2+}(H_2O)_6 + trans\text{-}Cr^{2+}(NH_3)_4OH_2Cl$	3.58	3.05	25	28.0	7.0	16.9
$Cr^{2+}(H_2O)_6 + trans\text{-}Cr^{2+}(H_2O)_4NH_3Cl$	3.58	3.4	25	26.3	6.6	14.0
$Cr^{2+}(H_2O)_6 + Cr^{3+}(H_2O)_5NH_3$	3.58	3.4	25	26.3	6.6	22.1
$Cr^{2+}(H_2O)_6 + cis\text{-}Cr^+(H_2O)_4Cl_2$	3.58	3.4	25	26.3	6.6	12.4
$Cr^{2+}(H_2O)_6 + trans\text{-}Cr^+(H_2O)_4Cl_2$	3.58	3.4	25	26.3	6.6	12.2
$Cr^{2+}(H_2O)_6 + Cr^+(H_2O)_5H_2PO_2$	3.58	3.4	25	26.3	6.6	20.2
$Tl^+(H_2O)_6 + Tl^{3+}(H_2O)_6$	4.25	3.81	25	22.9	5.7	21.4
$Cr^{2+}(H_2O)_6 + Cr^+(H_2O)_5CN$	3.56	3.4	25	26.4	6.6	17.4
$Ru^{2+}(NH_3)_6 + Ru^{3+}(NH_3)_6$	3.57	3.47	25	26.0	6.5	11.7
$Co^{2+}(NH_3)_6 + Co^{3+}(NH_3)_6$	3.59	3.05	64.5	28.6	7.1	22.7
$Co^{2+}(H_2O)_6 + Co^{3+}(H_2O)_6$	3.56	3.41	25	26.3	6.6	14.9
$Cr^{2+}(H_2O)_6 + Cr^{2+}(H_2O)_5F$	3.56	3.41	25	26.3	6.6	18.0
$V^{2+}(H_2O)_6 + V^{3+}(H_2O)_6$	3.58	3.385	24.8	26.4	6.6	15.4
$Fe^{2+}(H_2O)_6 + Fe^{3+}(H_2O)_6$	3.59	3.43	0	26.1	6.5	14.3
$Cr^{2+}(H_2O)_6 + Cr^{3+}(H_2O)_6$	3.58	3.4	25	26.3	6.6	22.2
$Mn^{2+}(H_2O)_6 + Mn^{3+}(H_2O)_6$	3.66	3.46	25	25.8	6.4	14.9
$V^{2+}(H_2O)_6 + V^{3+}(H_2O)_6$	3.58	3.385	25	26.4	6.6	18.5

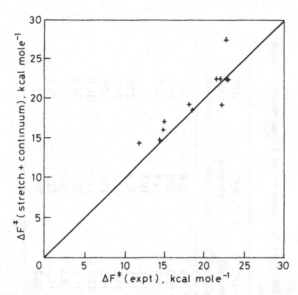

Fig. 6.12. Plot of $\Delta F\ddagger$ (continuum + stretch) against $\Delta F\ddagger$ (experimental) for homogeneous electron transfer reactions in solution. (Correlation coefficient = 0.93.)

seems, however, unlikely that a modification of the classical Bockris and Sen[71,72] calculation of the ion–dipole interaction to the hypothetical quantum mechanical polaronic interaction will increase the average energy of the central ion by an order of magnitude.

6.10. Activation Due to Phonon–Vibron Coupling (PVC)

In the phonon–vibron coupling (PVC) model suggested by Khan and Bockris, an ion can receive energy in a condensed medium due to the transfer of energy from the phonons produced by surrounding solvent oscillators to the ion–solvent bond (inner-sphere) vibrators (vibron). When the ion–solvent bond oscillators receive energy from phonons, it results in an increase in the amplitude of vibration of the ion–solvent bonds. At higher amplitudes, the displacement of the inner-sphere oscillations increases and they move further away from the central ion. The central ion loses negative solvation energy, i.e., gains energy and becomes activated.

The quantum mechanical description of this model can be outlined in the following way. Energy transfer occurs from the surrounding phonon waves by a coupling interaction with the inner-sphere ion–solvent bond oscillators. Thus, phonon waves strain the ion–solvent bond, thereby transferring energy to the inner-sphere ion–solvent oscillators, which

Table 6.3

Calculated Values for Both Continuum and Stretching Contribution to Free Energy of Activation for the Homogeneous Electron Transfer Reaction in Solution

Reactants	Temperature (°C)	$\Delta F^{\ddagger}$ (continuum) (kcal mole^{-1})	$\Delta F^{\ddagger}$ (stretch) (kcal mole^{-1})	$\Delta F^{\ddagger}$ (continuum) (kcal mole^{-1})	$\Delta F^{\ddagger}$ (expt.) (kcal mole^{-1})
$Co^{2-}(H_2O)_6 + Co^{3+}(H_2O)_6$	25	6.6	9.4	16.0	14.8
$V^{2+}(H_2O)_6 + V^{3+}(H_2O)_6$	0	6.6	12.0	18.6	18.5
$Fe^{2+}(H_2O)_6 + Fe^{3+}(H_2O)_6$	25	6.5	8.3	14.8	14.3
$Cr^{2+}(H_2O)_6 + Cr^{3+}(H_2O)_6$	25	6.6	12.7	19.3	22.2
$Co^{2+}(en)_3 + Co^{3+}(en)_3$	50	7.4	15.1	22.5	22.8
$Co^{2+}(en)_3 + Co^{2+}(en)_3OH$	50	7.4	15.1	22.5	21.5
$Mn^{2+}(H_2O)_6 + Mn^{3+}(H_2O)_6$	25	6.4	10.6	17.0	14.9
$Cr^{2+}(H_2O)_6 + Cr^{2+}(H_2O)_6F$	25	6.6	12.7	19.3	18.0
$Co^{2+}(en)_3 + Co^{2+}(en)_3Cl$	50	7.4	15.1	22.5	22.0
$Fe^{3-}(CN)_6 + Fe^{4-}(CN)_6$	25	4.9	9.5	14.4	11.7
$Co^{2+}(en)_3 + Co^{+}(en)_3SO_4$	50	7.4	15.1	22.5	22.8
$Co^{2+}(NH_3)_6 + Co^{3+}(NH_3)_6$	25	7.2	20.2	27.4	22.7

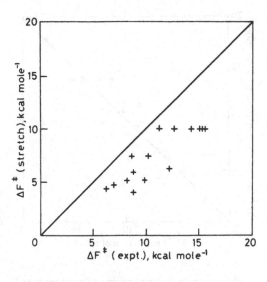

Fig. 6.13. Plot of $\Delta F^{\ddagger}$ (stretch) against $\Delta F^{\ddagger}$ (experimental) for homogeneous electron transfer reactions in solutions. (Correlation coefficient = 0.7.)

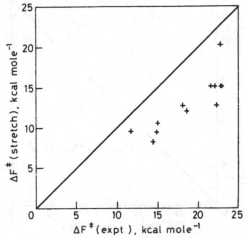

Fig. 6.14. Plot of $\Delta F^{\ddagger}$ (stretch) against $\Delta F^{\ddagger}$ (experimental) for the *electrochemical* electron transfer reactions. (Correlation coefficient = 0.7.)

would undergo transition from the ground state to a higher quantum state (i.e., higher displaced state). The moment the inner-sphere oscillators undergo transition to a sufficiently high quantum state (a displaced state) the central ion loses its stabilization energy (the ion–solvent dipole interaction) and becomes activated. When the ion becomes sufficiently activated, radiationless electron transfer from an electrode can occur (in electrochemical reduction). For a homogeneous exchange reaction, the electron transfer occurs between two activated ions of lower [e.g., $Fe^{z+}(H_2O)$ ion] and higher [e.g., $Fe^{3+}(H_2O)_6$ ion] oxidation states.

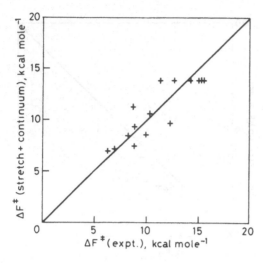

Fig. 6.15. Plot of $\Delta F^{\ddagger}$ (continuum + stretch) against $\Delta F^{\ddagger}$ (experimental) for *electrochemical* electron transfer reactions. (Correlation coefficient = 0.9.)

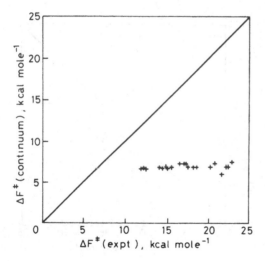

Fig. 6.16. Plot of $\Delta F^{\ddagger}$ (continuum) against $\Delta F^{\ddagger}$ (experimental) for *electrochemical* electron transfer reactions. (Correlation coefficient = 0.33.)

It is thus possible to see how, qualitatively, phonon oscillations (the quantum mechanical analog to the transfer of heat energy by translators in gas-phase systems) can supply energy to an ion. However, the Levich type of difficulty of the *amount* that can be supplied comes in: Is it possible that the phonon oscillation can supply the central ion with the requisite amount of 1 eV? In principle, it is possible because of the frequency of the phonon, which is in the neighborhood of 10^{14} Hz.

Table 6.4

Calculated Values for both Continuum and Stretching Contribution to Experimental Free Energy of Activation for the Electrochemical Redox Reaction

Systems	a_1 (Å)	E_s (kcal mole^{-1})	$\Delta F^\ddagger$ (continuum) (kcal mole^{-1})	$\Delta F^\ddagger$ (stretch) (kcal mole^{-1})	$\Delta F^\ddagger$ (continuum + stretch) (kcal mole^{-1})	$\Delta F^\ddagger$ (expt.) (kcal mole^{-1})
Ce^{4+}(H$_2$O)$_6$	3.77	12.2	3.1	7.6	10.7	10.2
Cr^{3+}(H$_2$O)$_6$	3.4	13.5	3.4	6.3	9.7	12.1
Co^{3+}(NH$_3$)$_6$	3.05	15.0	3.8	10.1	13.9	12.1
[Co(NH$_3$)$_5$NO$_2$]$^{2+}$	3.05	15.0	3.8	10.1	13.9	15.2
V^{3+}(H$_2$O)$_6$	3.385	13.5	3.4	6.0	9.4	8.8
Ti^{4+}(H$_2$O)$_6$	3.52	13.0	3.3	5.2	8.5	8.2
cis-[Co(NH$_3$)$_4$(NO$_2$)$_2$]$^+$	3.05	15.0	3.8	10.1	13.9	14.2
trans-[Co(NH$_3$)$_4$(NO$_2$)$_2$]$^+$	3.05	15.0	3.8	10.1	13.9	15.5
Cr(CN)$_6$$^{3-}$	4.63	9.9	2.5	4.5	7.0	6.2
[Co(NH$_3$)$_5$F]$^{2+}$	3.05	15.0	3.8	10.1	13.9	11.2
Co(en)$_3$$^{3+}$	3.04	15.1	3.8	7.5	11.3	8.6
[CO-(NH$_3$)$_5$ONO]$^{2+}$	3.05	15.0	3.8	10.1	13.9	15.0
Fe^{3+}(H$_2$O)$_6$	3.43	13.3	3.3	4.1	7.4	8.7
Mn^{3+}(H$_2$O)$_6$	3.46	13.2	3.3	5.3	8.6	9.8
Fe(CN)$_6$$^{3-}$	4.65	9.7	2.4	4.8	7.2	6.9

6.11. Formulation of the Probability of Activation from the PVC Model

The probability of the transition of the oscillator (ion–solvent bond) from the initial ground state i to a higher quantum state f by phonon interactions can be calculated. The probability of this transition can be termed the activation probability P_a and can be expressed according to time-dependent perturbation theory as[76]

$$P_a = \frac{2\pi\tau}{\hbar}|\langle\psi_f(x)|V(x)|\psi_i(x)\rangle|^2\rho(E_f)$$

$$= \frac{2\pi\tau}{\hbar}|M|^2\rho(E_f) \tag{6.33}$$

where τ is the time of perturbation of the oscillator by the phonon waves; ψ_i and ψ_f are, respectively, the initial (ground) state and the final (higher quantum) state wave functions of the oscillator; and $\rho(E_f)$ is the density of this activated state, which corresponds to sufficient displacement of the bond to correspond to the final state.

Considering an ion–solvent bond as a one-dimensional harmonic oscillator, the initial normalized wave function can be written (corresponding to the ground state of the harmonic oscillator) as[77]

$$\psi_i = \left(\frac{2m\nu}{\hbar}\right)^{1/4}\exp\left(-\frac{\pi m\nu}{\hbar}x^2\right) \tag{6.34}$$

and the final-state normalized wave function corresponding to the nth quantum state as

$$\psi_f = 2^{-n/2}\left(\frac{1}{n!}\right)^{1/2}\left(\frac{2m\nu}{\hbar}\right)^{1/4}\exp\left(-\frac{\pi m\nu}{\hbar}x^2\right)H_n\left[\left(\frac{2\pi m\nu}{\hbar}\right)^{1/2}x\right] \tag{6.35}$$

where n is the vibrational quantum number, m is the reduced mass of the ion and a solvent particle, ν is the stretching frequency of the oscillator, and $H_n[(2\pi m\nu/\hbar)^{1/2}x]$ is the Hermite polynomial.

Using a similar type of formalism to that used by Goodman, Lowson, and Schiff,[78] the perturbation $V(x)$ of an ion–solvent oscillator due to the interaction with phonon (boson) waves can be expressed (for the one-dimensional case) as

$$V(x) = R_p(x)\frac{d[v_0(x)]}{dx} \tag{6.36}$$

where

$$V_0(x) = \tfrac{1}{2}k_f x^2 \tag{6.37}$$

and

$$R_p = A_p \exp(i\kappa x) \tag{6.38}$$

where k_f is the force constant of the bond, κ is the wave vector of the phonon waves, and A_p is their amplitude.

The phonon waves transfer momentum to bonds of the inner solvent layer and cause transition of the ion–solvent bond to a higher quantum state. The total energy transfer involves a sum over the propagation vectors of the phonon waves. However, energy must be conserved between the phonon and the oscillator[78] so that

$$h\nu_p = E_f - E_i = \Delta E_{fi} \tag{6.39}$$

where ν_p is the frequency of the phonon waves and E_i and E_f are the energies corresponding to the initial and final activated states of the oscillator, respectively.

The amplitude of the phonon waves of Eq. (6.38) is given by the relation,[78] i.e.,

$$A_p = \left\{ \frac{h\nu_p}{2\pi^2 m\nu_p^2 [\exp(h\nu_p/kT)-1]} \right\}^{1/2} \tag{6.40}$$

where m is the mass of a solvent molecule.

Using Eqs. (6.39) and (6.40), one can write A_p as

$$A_p = \left\{ \frac{h\nu_p}{2\pi^2 m\nu_p^2 [\exp(|\Delta E_{fi}|/kT)-1]} \right\}^{1/2} \tag{6.41}$$

Substituting the values of $V_0(x)$ and R_p from Eqs. (6.37) and (6.38) in Eq. (6.36), one obtains the potential energy of perturbation as

$$V(x) = k_f A_p x \exp(i\kappa x) \tag{6.42}$$

Hence, using Eq. (6.42), the matrix element of perturbation M of Eq. (6.33) can be expressed as

$$M = \langle \psi_f(x) | V(x) | \psi_i(x) \rangle$$

$$= \int_{-\infty}^{\infty} \psi_f(x) k_f A_p x \exp(i\kappa x) \psi_i(x) \, dx \tag{6.43}$$

Now, substituting the values of the wave functions $\psi_i(x)$ and $\psi_f(x)$ from Eqs. (6.34) and (6.35) into Eq. (6.43), one obtains

$$M = B \int_{-\infty}^{\infty} \exp\left(-\frac{2\pi m\nu}{\hbar} x^2\right) H_n\left[\left(\frac{2\pi m\nu}{\hbar}\right)^{1/2} x\right] x \exp(i\kappa x) \, dx \tag{6.44}$$

where

$$B = \left(\frac{1}{2^n n!}\right)^{1/2}\left(\frac{2mv}{\hbar}\right)^{1/2} k_f A_p \tag{6.45}$$

Substituting the value of A_p from Eq. (6.41), one obtains

$$B = C\left\{\frac{hv_p}{2\pi^2 m v_p[\exp(|\Delta E_{fi}|kT)-1]}\right\}^{1/2} \tag{6.46}$$

where

$$C = \left(\frac{1}{2^n n!}\right)^{1/2}\left(\frac{2mv}{\hbar}\right)^{1/2} k_f \tag{6.47}$$

Hence, using Eq. (6.44) in Eq. (6.33), the probability of activation becomes

$$P_a = \frac{2\pi B^2 \tau}{\hbar}\left|\int_{-\infty}^{\infty}\exp\left(-\frac{2\pi m v_p}{\hbar}x^2\right)H_n\left[\left(\frac{2\pi m v}{\hbar}\right)^{1/2}x\right]x\,\exp(i\kappa x)\,dx\right|^2 \rho(E_f) \tag{6.48}$$

Substituting Eq. (6.46) in Eq. (6.48), one obtains

$$P_a = \frac{2\pi C^2 \tau}{\hbar}\left|\int_{-\infty}^{\infty}\exp(-\alpha_p x^2)H_n(\alpha^{1/2}x)x\,\exp(i\kappa x)\,dx\right|^2$$

$$\times\left[\frac{hv_p}{\beta[\exp(|\Delta E_{fi}|/kT)-1]}\right]\rho(E_f) \tag{6.49}$$

where

$$\alpha = \frac{2\pi m v}{\hbar}, \qquad \alpha_p = \frac{2\pi m v_p}{\hbar} \tag{6.50}$$

and

$$\beta = 2\pi^2 m v_p^2 \tag{6.51}$$

This probability of activation P_a of Eq. (6.49) can be used in the rate expression (4.79) in Chapter 4 to find the rate of an electron transfer reaction to or from an ion in solution and also from an electrode to an ion in solution and vice versa.

References

1. S. Glasstone, K. J. Laidler, and H. Eyring, *The Theory of Rate Processes*, McGraw–Hill, New York (1941).
2. K. J. Laidler, *Theories of Chemical Reaction Rates*, McGraw–Hill, New York (1969).
3. V. N. Kondriatev (ed.), *Chemical Kinetics of Gas Reactions*, Pergamon Press, Oxford, England (1964).

4. J. B. Hasted, *Physics of Atomic Collision*, Butterworths Publications, London (1964).
5. R. E. Weston, J. R., and H. A. Schwarz, *Chemical Kinetics*, Prentice–Hall, Englewood Cliffs, New Jersey (1972).
6. L. Landau and E. Teller, *Phys. Z. Sowjet Union* **10**, 34 (1936).
7. K. Takayanagz, *Proc. Theoret. Phys. Suppl. (Japan)* **25**, 1 (1963).
8. J. D. Kelley and M. Wolfsburg, *J. Chem. Phys.* **44**, 324 (1966).
9. J. D. Kelley and M. Wolfsburg, *J. Chem. Phys.* **50**, 1894 (1969).
10. D. Rapp, *J. Chem. Phys.* **40**, 3812 (1964).
11. D. Rapp and T. Kassal, *Chem. Rev.* **69**, 61 (1969).
12. C. Zener, *Phys. Rev.* **37**, 556 (1931); *Proc. Cambridge Phil. Soc.* **29**, 136 (1933).
13. C. F. Hansen and W. E. Pearson, *J. Chem. Phys.* **53**, 3557 (1970).
14. W. H. Miller, *J. Chem. Phys.* **53**, 1949, 3578 (1970).
15. W. H. Miller, *J. Chem. Phys.* **54**, 3965, 5386 (1971).
16. T. F. George and W. H. Miller, *J. Chem. Phys.* **56**, 5722 (1972); **57**, 2458 (1972).
17. J. D. Doll, T. F. George, and W. H. Miller, *J. Chem. Phys.* **58**, 1343 (1973).
18. J. M. Bowman and A. Kuppermann, *Chem. Phys. Lett.* **19**, 166 (1973).
19. D. E. Fitz and R. A. Marcus, *J. Chem. Phys.* **59**, 287 (1973).
20. W. Eastes and J. D. Doll, *J. Chem. Phys.* **60**, 297 (1974).
21. R. L. McKenzie, *J. Chem. Phys.* **63**, 1655 (1975).
22. D. Secrest, *J. Chem. Phys.* **51**, 421 (1969).
23. A. E. Roberts, *J. Chem. Phys.* **55**, 100 (1971).
24. L. M. Kopple and J. Lin, *J. Chem. Phys.* **58**, 1869 (1973).
25. J. B. Calvert and R. C. Amme, *J. Chem. Phys.* **50**, 4710 (1966).
26. H. K. Shin, *J. Chem. Phys.* **49**, 3964 (1968).
27. H. K. Shin, *J. Phys. Chem.* **76**, 2006 (1972).
28. M. A. Wartell and R. J. Cross, *J. Chem. Phys.* **55**, 610 (1971).
29. J. D. Doll and W. H. Miller, *J. Chem. Phys.* **57**, 5019 (1972).
30. D. Secrest and B. R. Johnson, *J. Chem. Phys.* **45**, 4556 (1966).
31. W. Eastes and J. D. Doll, *J. Chem. Phys.* **60**, 297 (1974).
32. D. Secrest and W. Eastes, *J. Chem. Phys.* **56**, 2502 (1972).
33. W. S. Liu, *J. Chem. Phys.* **61**, 168 (1974).
34. X. Chapuisat, G. Bergeron, and J. M. Launay, *Chem. Phys.* **20**, 285 (1977).
35. U. Halavee and R. D. Levine, *Chem. Phys. Lett.* **46**, 35 (1977).
36. M. Baer, *J. Chem. Phys.* **60**, 1057 (1974).
37. J. N. Bass, *J. Chem. Phys.* **60**, 2913 (1974).
38. W. A. Lester, Jr. and J. Schaefer, *J. Chem. Phys.* **60**, 1672 (1974).
39. D. J. Kouri and C. A. Wells, *J. Chem. Phys.* **60**, 2296 (1974).
40. S. Green, *J. Chem. Phys.* **62**, 2271 (1975).
41. B. H. Choi and K. T. Tang, *J. Chem. Phys.* **63**, 1783 (1975).
42. S. Chu and A. Dolgarno, *J. Chem. Phys.* **63**, 2115 (1975).
43. J. R. Stallcope, *J. Chem. Phys.* **61**, 5085 (1974).
44. M. D. Pattergill, *J. Chem. Phys.* **62**, 3137 (1975).
45. E. E. Nikitin, in *Physical Chemistry: An Advanced Treatise* (W. Jost, ed.), Volume 6A, Chapter 4, Academic Press, New York (1974).
46. R. E. Amme, in *Advances in Chemical Physics* (J. W. McGowran, ed.), Volume 28, Chapter 3, p. 251, Interscience, New York (1975).
47. E. A. Moelwyn-Hughes, *Chemical Statistics and Kinetics of Solutions*, Academic Press, New York (1971).
48. G. Prigogine and M. Mabicu, *Physica* **61**, 51 (1959).
49. J. Frenkel, *Kinetic Theory of Liquids*, Oxford University Press, Oxford, England (1946).
50. I. D. Clark and R. F. Wayne, in *Comprehension Chemical Kinetics* (C. F. Bunford and C. F. H. Tipper, eds.), Elsevier, New York (1969).

51. A. M. North, *The Collision Theory of Chemical Reaction in Liquids,* Barnes and Noble, New York (1964).
52. P. George and J. S. Griffith, in *The Enzymes* (P. D. Boyer, H. Lardy, and K. Myrback, eds.), Volume 1, Chapter 8, p. 347, Academic Press, New York (1959).
53. R. A. Marcus, *J. Chem. Phys.* **24**, 966 (1956).
54a. D. P. Stevenson, *J. Phys. Chem.* **69**, 2145 (1965).
54b. R. P. Marchi and H. Eyring, *J. Phys. Chem.* **68**, 221 (1964).
55. J. Jortner, *Rad. Res. Supl.* **4**, 24 (1964).
56. J. Jortner and S. Mukamel, *Physical Chemistry* (A. D. Buckingham and C. A. Coulson, eds.), Volume 1, Butterworths Publications, London (1975).
57. R. J. Laidler, *Chemical Kinetics,* McGraw–Hill, New York (1950).
58. R. A. Horne, *J. Inorg. Nucl. Chem.* **25**, 1139 (1963).
59. D. L. Baulch, F. S. Dainton, D. A. Ledward, and H. Suiger, *Trans. Faraday Soc.* **62**, 2200 (1966).
60. E. Nitzan and A. C. Wahl, *J. Inorg. Nucl. Chem.* **28**, 3069 (1966).
61. A. J. Arvia and H. A. Videla, *Electrochim. Acta* **9**, 1199 (1969).
62. M. G. Sustersic, W. E. Triaca, and A. J. Arvia, *Electrochim. Acta* **19**, 1, 19 (1974).
63. E. T. Moisescu and A. Rahmel, *Electrochim. Acta* **20**, 479 (1975).
64. V. N. Chebotin, M. V. Glumov, A. D. Meumin, and S. F. Falgner, *Electrokhim.* **7**, 62 (1971).
65. R. D. Armstrong, T. Dickinson, and K. Taylor, *J. Electroanal. Chem.* **57**, 157 (1974).
66. V. G. Levich, *Recent Advances Electrochemistry,* Volume 4, Chapter 5, Academic Press, New York (1966).
67. V. G. Levich, *Physical Chemistry: An Advanced Treatise in Physical Chemistry* (H. Eyring, D. Henderson, and W. Jost, eds.), Volume 9B, Chapter 12, Academic Press, New York (1970).
68. R. R. Dogonadze, *Reactions of Molecules at Electrodes* (N. S. Hush, ed.), Wiley (Interscience), New York (1971).
69. R. R. Dogonadze and A. M. Kuznetsov, *J. Electroanal. Chim.* **65**, 545 (1975).
70. R. A. M. O'Ferrall, G. W. Koepple, and A. J. Kresge, *J. Amer. Chem. Soc.* **93**, 1 (1971).
71. J. O'M. Bockris and R. R. Sen, *Molec. Phys.* **29**, 357 (1975).
72. J. O'M. Bockris and R. R. Sen, *J. Res. Inst. Catal. Hokkaido Univ.* **20**, 153 (1972).
73. R. D. Mattuck, *A Guide to Feynman Diagrams,* McGraw–Hill, New York (1967).
74. A. J. Appleby, J. O'M. Bockris, R. K. Sen, and B. E. Conway, *M.T.P. International Review of Science,* Volume 6, Electrochemistry (J. O'M. Bockris, ed.), Butterworths Publications, London (1973).
75. J. O'M. Bockris, S. U. M. Khan, and D. B. Matthews, *J. Res. Inst. Catal. Hokkaido Univ.* **22**, 1 (1974).
76. E. E. Anderson, *Modern Physics and Quantum Mechanics,* p. 359, Saunders, Philadelphia (1971).
77. E. Merzbacher, *Quantum Mechanics,* John Wiley and Sons, New York (1970).
78. B. Goodman, A. W. Lawson, and L. I. Schiff, *Phys. Rev.* **71**, 195 (1946).
79. J. Weiss, *Proc. Roy. Soc. A* **222**, 128 (1954).
80. N. S. Hush, *J. Chem. Phys.* **28**, 962 (1958).
81. N. S. Hush, *Trans. Faraday Soc.* **57**, 557 (1961).
82. W. Schmickler, private communication.

7

The Continuum Theory

7.1. Introduction

In the early versions of the continuum theoretical treatment of the rate of a redox process in solution and at electrodes,[1,3-9] there were qualitative introductory remarks about quantum mechanical principles. In particular, the Franck–Condon principle relating to the separation of the electron and solvent components of the Hamiltonian for the electron–solvent system was usually mentioned. However, in the simple pre-1965 formulations, no actual quantum mechanical calculations, nor even formulations of the Hamiltonians, were attempted. In particular, in the sophisticated paper of Marcus[2] of 1965, the actual considerations are mainly in terms of statistical mechanics and electrostatics.

The Russian school, at first under Levich and then Dogonadze, also stressed the quantum mechanical nature of the arguments used in their publications[3-9] and wrote general equations for Hamiltonians of the system concerned. This approach is well exemplified in the comprehensive review by Levich appearing in the *Treatise of Physical Chemistry*, edited by Eyring, Henderson, and Jost.[9] Here, quantum mechanics is often stressed in qualitative terms, and equations for quantum mechanical formulations are made. The equations in final form are classical limiting cases, together with some unevaluated matrix elements (concerning the electron transition). In this chapter, the main contents will be as follows:

(a) The model of the polar solvent used in the continuum approach is discussed, because it is in terms of this model that the Levich–Dogonadze theory of the quantum mechanical aspects of the kinetics of charge transfer in solution is formulated.

(b) The quantum mechanical formulations given by these authors will be presented, and it will be shown at which point a quantum mechanical model has been displaced by a classical solution.

(c) In particular, the transition probability from the continuum treatment will be evaluated for the non-bond-breaking (redox) reaction, and a qualitative discussion for the bond breaking reaction (e.g., proton transfer) is given (cf. also Chapter 13).

(d) Last, some basic aspects associated with continuum models of kinetics at interfaces will be brought out.

7.2. Model of a Polar Solvent

The continuum theory developed by Levich, Dogonadze, and Kuznetsov (LDK)[3-10] depends on their concept of a polar solvent. This arises since they stress that the solvent plays the main role when reaction with charge transfer occurs. Physically, the charge transfer process (chemical and electrochemical) is pictured as involving strong *reorganization* of the polar solvent medium near the ions whose charge undergoes change. Hence, presentation of the model of a polar solvent used by the authors is necessary in the understanding of the continuum theory of charge transfer reactions.

Levich, Dogonadze, and Kuznetsov[3-10] considered that the theoretical formulations available to describe liquids like water were too complex to act as a vehicle for their thoughts concerning the changes that occur in liquids in the vicinity of ions when the electric charge on these change in times of the order of 10^{-15} sec. Because of the difficulty of complexity of the modelistic theories of liquids, and particularly because their requirements were so specialized, they decided to leave out any attempt to use a realistic model for a liquid and assumed, effectively, that their liquids were like ionic crystals.

Correspondingly, within this LDK assumption, there goes the assumption that the motions undergone by the particles are of small displacements and harmonic in nature. Movements concerned with diffusional jumps have been neglected because the timing for these is of the order of 10^{-9} sec, whereas the reorganization time of solvent molecules is in the neighborhood of 10^{-12} sec, so that the translational movements in the liquid can be neglected when one is considering the electron transfer act. LDK refer in discussions of their model of liquids to the two types of vibrations discussed in solids, namely, acoustic and optical[11] (see Chapter 3).

7.2.1. Polarization of the Liquid and Its Fluctuations

Much of the model of the LDK[3-10] approach to electron transfer at interfaces is concerned with the continuum concept of polarization. The polarization can be understood in the following way. It is the dipole moment per unit volume. This does not necessarily refer to any permanent dipole moment in the medium. The induced dipole moment, μ_d, in a *molecule* (as opposed to a medium) is given by $\mu_d = \alpha X$, where α is the polarizability of the molecule and X is the internal field.†

The difference between this induced dipole moment and the polarization is that the polarization is a term referring to a medium as a whole, not to a molecule. Thus the polarization can be written as§

$$P = n\alpha X \qquad (7.1)$$

where n is the number of particles per unit volume. One might say that "polarization" is in a bulk continuum what induced dipole moment is in a molecule.

One of the relations used to relate total polarization to dielectric constant ε is

$$P = \frac{\varepsilon_s - 1}{4\pi\varepsilon_s} D \qquad (7.2)$$

where D is electric induction and ε_s is the static dielectric constant of the medium.

One of the important considerations of the Levich–Dogoṇadze view is that the polarization *fluctuates* on a local scale. Physically, this occurs because particles in the solution are in thermal motion and the distance between the charges in the solution is constantly fluctuating upon a scale that corresponds either to the time range in which orientation changes occur ($\sim 10^{-11}$ sec) or to the region in which vibrational changes occur ($\sim 10^{-14}$ sec).

LDK took the view that there would be only part of this local variation of the polarization in the solvent medium in which they would be interested. They reasoned that it would be the path arising from the orientation, and the reasoning was as follows. Considering the solvent in its average, it has thermal energy of kT, and this corresponds to a frequency in the range of kT/h, or about $10^{12.5}$ at room temperature. Now, LDK pointed out that this was much nearer the range for orientational librational movements (10^{11} sec^{-1} for water) than the vibrational changes (10^{14} sec^{-1}). They argued, therefore, that the only variations in polarization that interested them were those that occurred most frequently, namely, the librational

---

† The distinction between X, the internal field, and the Maxwell (or average) field is important.[61]

§ The relation of the pertinent electrostatic quantities is:
   (1) Dipole moment $\mu = \alpha X$, where α is the polarizability and X is the external field.
   (2) Polarization $P = n\alpha X = (N/V)\alpha X =$ dipole moment per unit volume, and N is the number of particles in the system and V is its volume.
   (3) Electric induction D is εX, where ε is the dielectric constant of the medium.

changes, for these would occur in every water molecule.† Correspondingly, with $\omega_{vib} > kT/h$ at room temperatures, the occupation of higher states will be small. (Note the reversal of reasoning to that in gas kinetics. There, $\omega_{vib} > kT/h$ in the same measure. But it is just these few activated bonds that cause reaction.)

One now considers the effect of local electric fields upon the polarization of the solvent. The local electric fields to which reference is made are those due to charges on ions. The LDK view puts weight upon the electrostatic energy changes due to the ion in the surrounding solvent and no weight upon the bond-breaking properties (as are the principal consideration with most reaction considerations in the gas phase). They are concerned largely with variations in the polarization of the surrounding solvent§ considered in terms of a continuum dielectric. Hence, one needs to find what would happen if one switches on or switches off very quickly a charge, and thus a local electric field, within such a dielectric continuum.

During such quick processes, it is only the optical polarization that is changing. This is given by Eq. (7.2) as above, but we substitute for the static dielectric constant ε_s the optical dielectric constant ε_{op}, which is valid in the optical range of frequencies.

7.2.2. Energy Associated with a Local Fluctuation of Polarization in a Solvent System

In the LDK model of charge transfer reactions in solution, the important aspects of activation processes are associated with local fluctuations of electrostatic energy that occur within the solvent. Thus, it has been pointed out above why there are local fluctuations of electrostatic energy (hence, also of polarization); it is due to the thermal movement of the dipoles in the solvent, a consequence of both vibration (which Levich and Dogonadze tend to discount because of the small number of filled levels) and, particularly, to the orientational changes of one molecule with respect to another.

---

† The nature of this argument is unclear because obviously all the water bonds have at least their ground-state vibrations.

§ One of the characteristics of the LDK view is its preoccupation with average quantities. If a system possesses some quantity on an average to a certain degree and a very large fraction of the molecule possesses this quantity, then it is the only one considered by LDK, and the properties that are only possessed to a small degree by a small number of molecules are neglected by them. This may be an uncertain procedure to use in the calculation of rates. Thus, in the gas phase, it is the exceptional situation, one that happens to a fraction of, say, 10^{-17} (E_a/kT for $E_a > 1$ eV is considered), which is the active part of the mechanism, although LDK would neglect it, for such occurrences are "very infrequent."

In continuum theory, one calculates the polarization energy U_p in the following way:

$$U_p = \int \mathbf{DP} \, dV \tag{7.3}$$

where $\mathbf{D}$ is the induction, $\mathbf{P}$ is the polarization, and dV is the volume element.

Using Eq. (7.2) in Eq. (7.3) one gets†

$$U_p = \frac{4\pi\varepsilon_s}{\varepsilon_s - 1} \int \mathbf{P}^2 \, dV \tag{7.4}$$

Hence, for the change in induction of a solvent system during polarization from zero to $\mathbf{D}$ (as a slow process when a local field is slowly introduced in the solvent system) followed by change in induction from $\mathbf{D}$ to zero (as a fast process when the local field is removed quickly), one can write the energy [using Eq. (7.3)] as§

$$U_p = \int \left(\int_0^{\mathbf{D}} \mathbf{D} \, d\mathbf{P} + \int_{\mathbf{D}}^0 \mathbf{D} \, d\mathbf{P} \right) dV \tag{7.5}$$

$$\text{slow} \qquad \text{fast}$$

Expressing $d\mathbf{P}$ in Eq. (7.5) using Eq. (7.2) and integrating, one gets¶

$$U_p = \frac{1}{8\pi} \left(\frac{1}{\varepsilon_{op}} - \frac{1}{\varepsilon_s} \right) \int \mathbf{D}^2 \, dV \tag{7.6}$$

---

† The integral of the right side of Eq. (7.4) has the dimensions of energy as can be shown:

$$\int \mathbf{P}^2 \, dV = \int \frac{(N\alpha\mathbf{X})^2}{V^2} \, dV$$

$$\equiv -\frac{N(\alpha\mathbf{X})^2}{V}$$

$$\equiv -\frac{N\mu_d^2}{V} = \frac{N e_0^2 r^2}{r^3}$$

$$\equiv -N\frac{e_0^2}{r} = \text{energy}$$

since $\mu_d = e_0 r$ and volume $V = r^3$.

§ The process described by Eq. (7.5) is that in which the solvent undergoes a fluctuation in energy due to the local electric field and therefore in the induction. This first change is regarded as "slow" on the time scale of relaxation movements in liquids, and it is followed by the energy change occurring when the charge on the inducing ion is switched off in neutralization. The net energy is therefore the distortion (or reorganization) energy of the fluctuation.

¶ Thus, the reorganization energy here calculated is for a hypothetical, ideal dielectric fluid, where the effect of changes in field upon the energy are not considered. For example, were this applied to an ionic situation, it would neglect the change of energy of the dipole layer oriented around the ion (when the field due to the ion changes), although this energy is about half the energy of the ion's interaction with the solvent.

Since the total polarization, $\mathbf{P}_{total}$, is the sum of the reorganization polarization, $\mathbf{P}_{reorg}$, and the optical polarization, $\mathbf{P}_{op}$, one can express $\mathbf{P}_{reorg}$, using Eq. (7.2), as

$$\mathbf{P}_{reorg} = \mathbf{P}_{total} - \mathbf{P}_{op}$$

$$= \frac{(\varepsilon_s - 1)}{4\pi\varepsilon_s}\mathbf{D} - \frac{(\varepsilon_{op} - 1)}{4\pi\varepsilon_{op}}\mathbf{D}$$

$$= \frac{1}{4\pi}\left[\frac{1}{\varepsilon_{op}} - \frac{1}{\varepsilon_s}\right]\mathbf{D} \tag{7.7}$$

Using Eq. (7.7) in Eq. (7.6), one gets

$$U_p = \frac{2\pi}{1/\varepsilon_{op} - 1/\varepsilon_s}\int \mathbf{P}_{reorg}^2 \, dV \tag{7.8}$$

Thus, Eq. (7.8) represents the continuum electrostatic version of an energy due to a typical reorganizational polarization fluctuation and would apply to an ionic system in respect to happenings in the outer sphere on electron transfer.

To obtain the energy associated with a local fluctuation, we must calculate the kinetic energy in addition to the potential energy calculated above. We begin by using an equation from mechanics for the kinetic energy as†

$$E_K = \frac{1}{\omega_0}\frac{dU_p}{dt} \tag{7.9}$$

Thus, using U_p from Eq. (7.8) in Eq. (7.9), one gets

$$E_K = \frac{2\pi\omega_0^{-1}}{(1/\varepsilon_{op} - 1/\varepsilon_s)}\int \dot{\mathbf{P}}_{reorg}^2 \, dV \tag{7.10}$$

† A method of obtaining an equation near to Eq. (7.9) begins in general mechanics:

$$U = kx^2 = k(vt)^2$$

Therefore,

$$\frac{dU}{dt} = 2kv^2t = 2k\frac{mv^2}{m}t = 4\frac{k}{m}E_K t$$

But

$$\omega = (k/m)^{1/2}$$

Therefore,

$$\frac{dU}{dt} = 4\omega^2 E_K t = 4\omega^2 E_K \frac{2\pi}{\omega}$$

and

$$\frac{1}{8\pi\omega}\frac{dU}{dt} = E_K$$

It is important to note the physical significance of ω_0, a frequency. It is intended to be an individual frequency that would be associated with what LDK call the "polar model," by which they mean the frequency of oscillation (i.e., the libration) of a dipole in solution. The symbol $\dot{P}_{reorg}$ is the time derivative of the orientational component of the polarization energy.

Thus, adding the Eqs. (7.8) and (7.10), one obtains

$$E_{total} = U_p + E_K$$

$$= \frac{2\pi}{C} \int (P^2_{reorg} + \omega_0^{-1} \dot{P}^2_{reorg}) \, dV \qquad (7.11)$$

where

$$C = \left(\frac{1}{\varepsilon_{op}} - \frac{1}{\varepsilon_s}\right) \qquad (7.12)$$

Expression (7.11), then, is the total energy of a local polarization fluctuation in the solution.†

7.2.3. Hamiltonian for the Pure Solvent

So far our discussion of the LDK[3-10] model has been purely classical, and in order to introduce a quantum mechanical aspect, the most important thing, as usual, is the calculation of the Hamiltonian. Levich and Dogonadze take the attitude that the Hamiltonian of their solvent system (which they insist would be like that of a "polar crystal") would then be the same as the Hamiltonian derived by Pekar[13] and also by Fröhlich.[14a] This Hamiltonian (which is evaluated in Appendix 7.I) is

$$H_s = \sum_k \frac{\hbar \omega_0}{2} \left(q_k^2 + \frac{\partial^2}{\partial q_k^2}\right) \qquad (7.13)$$

where the term q_k is the so-called dimensionless "solvent coordinate," a coordinate by means of which one may measure the variation of a collective mode representing solvent movement during vibrations and librations. ω_0 is the frequency of the solvent libration.

---

† The assumption that the energy of a local polarization fluctuation in solution may be written electrostatically without taking into account the bond-bending energy, for example, that of hydrogen bonding, is difficult to accept. Thus, in 1933, Bernal and Fowler[12] wrote the energy of a solvent in terms of its electrostatic dipole interactions. However, such expressions of a total energy of a liquid are now regarded as inadequate, and the modern expression of distortion or displacement energies within liquids will certainly involve these in addition to simple dipole interactions. The contribution of the bond bending energy is liable to exceed that of the quadrupole and dipole interactions. Such contributions are not considered in the continuum dielectric treatment.

7.2.4. Hamiltonian for the Total Reacting System in a Polar Solution (Ions, Electrons, and a Quasi-Continuum Solvent)

In the previous section we stated the Hamiltonian due to Pekar[13] and Fröhlich[14a] for *crystals* and utilized the assumption due to Levich, Dogonadze, and Kutznezov[3-10] that the crystal will be a model for the solvent in their work.

However, in expressing this Hamiltonian, we have not considered the presence of ionic charges, which obviously must be present in electrolytic solutions, and we have neglected any Hamiltonian for the electron transfer process. In this section we will consider the Hamiltonian for the total system in the continuum model[7,9,10] in a polar solution that consists of ions $A^{z+(1)}$ and $B^{z+(2)}$, the electron, and the surrounding solvent. Let us write the total Hamiltonian of the system consisting of the solvent, two ions, and the electron (which undergoes transfer) as

$$H_{\text{total}} = H_e + H_{\text{ion-solv}} + U_{\text{elec-solv}} \tag{7.14}$$

where H_e is the Hamiltonian of the transferring electron in the field of two ions, $H_{\text{ion-solv}}$ is the Hamiltonian of the solvent in the presence of ions and $U_{\text{elec-solv}}$ describes the interaction of the electron with the solvent. We shall henceforth abbreviate these Hamiltonians to H_e, $H_{i,s}$, and $H_{e,s}$.

Considering the relatively heavy ions to be stationary during electron transfer one can write the Hamiltonian for the electron H_e in terms of the kinetic energy of the electron and its interaction with the ions as

$$H_e = -\frac{\hbar^2}{2m_e}\frac{\partial^2}{\partial r^2} + \frac{(z_1 e_0)^2}{r} + \frac{(z_2 e_0)^2}{|\mathbf{r} - \mathbf{R}|} \tag{7.15}$$

Here, the value of z_1 corresponds to the valency of ion A and z_2 that of ion B; the electronic coordinate $\mathbf{r}$ represents the position of the electron with respect to the ion and $\mathbf{R}$ represents the distance apart of the two ions (Fig. 7.1).

The Hamiltonian of the solvent in the presence of ions can be found as follows. In a polar solvent, as before, the polarization, during reorganization, whereby the induction field is raised from 0 to $\mathbf{D}$, is taken to be switched on slowly, because its corresponds to a dipole movement in solution. In other words, the process of becoming the nonequilibrium state of solvent polarization from the equilibrium state is slow. The reverse

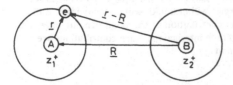

Fig. 7.1. The coordinate representation and separation of two reactant ions, $\mathbf{R}$, and the position vector of electron, $\mathbf{r}$, and the distance between an electron in one ion to the center of another ion, $|\mathbf{r} - \mathbf{R}|$.

process of depolarization (i.e., the returning of the solvent state from a nonequilibrium state to equilibrium state as a result of charge transfer) is taken to occur slowly. In the present case, the solvent system is left in a *polarized* state associated with the induction D_i (due to the presence of charge on the ion). The potential energy of interaction for these processes can be expressed as[†]

$$U = \int \left(\int_0^{\mathbf{D}} \mathbf{D} \, d\mathbf{P} + \int_{\mathbf{D}}^0 \mathbf{D} \, d\mathbf{P} + \int_0^{\mathbf{D}_i} \mathbf{D} \, d\mathbf{P} \right) dV$$

$$\text{slow} \qquad \text{fast} \qquad \text{fast}$$

$$= U_1 + U_2 + U_3 \tag{7.16}$$

Hence, the Hamiltonian $H_{\text{ion–solv}}$ can be expressed as

$$H_{\text{ion–solv}} = E_K + U_1 + U_2 + U_3 \tag{7.17}$$

or

$$H_{\text{ion–solv}} = H_s + U_3 \tag{7.18}$$

where the kinetic energy term E_K together with U_1 and U_2 give rise to Hamiltonian for the pure solvent H_s.

U_3 in Eqs. (7.16) and (7.17) represents an extra interaction for the introduction of new charge into solution[§] and can be expressed using Eq. (7.2) as

$$U_3 = \int \left(\int_0^{\mathbf{D}_i} \mathbf{D} \, d\mathbf{P} \right) dV = \frac{\varepsilon_{\text{op}} - 1}{4\pi\varepsilon_{\text{op}}} \int \left(\int_0^{\mathbf{D}_i} \mathbf{D} \, d\mathbf{D} \right) dV$$

$$= \frac{\varepsilon_{\text{op}} - 1}{8\pi\varepsilon_{\text{op}}} \int \mathbf{D}_i^2 \, dV \tag{7.19}$$

Now, using the Pekar–Fröhlich Hamiltonian for the crystal lattice (taken by Levich and Dogonadze as the value of a solvent Hamiltonian) and Eqs. (7.13) and (7.19) in Eq. (7.16), one can express the Hamiltonian of the solvent in *the presence of ions* as

$$H_{\text{ion–solv}} = \sum_k \frac{\hbar\omega_0}{2} \left(\mathbf{q}_k^2 - \frac{\partial^2}{\partial\mathbf{q}_k^2} \right) + \frac{\varepsilon_{\text{op}} - 1}{8\pi\varepsilon_{\text{op}}} \int \mathbf{D}_i^2 \, dV \tag{7.20}$$

To find $U_{\text{elec–solv}}$ of Eq. (7.14) one uses the general expression for the interaction energy between a charge and solvent with associated induced

---

[†] In reality, the sequence of steps is $\mathbf{D} = 0$ to $\mathbf{D} = \mathbf{D}$ (slow) and $\mathbf{D} = \mathbf{D}$ to $\mathbf{D}_i$ (fast). Expression (7.16) merely divides the integral for convenience in evaluation.

[§] By expression (7.18) is meant the energy of a charge formed "suddenly," i.e., by neutralization (or formation) of one charge by an electron transfer.

field $\mathbf{D}$ and polarization $\mathbf{P}$. Thus [see Eq. (7.3)],

$$U_{\text{elec-solv}} = \int \mathbf{D}_e \mathbf{P} \, dV \tag{7.21}$$

where $\mathbf{D}_e$ is the transient (10^{-15} sec) induced field in the solvent due to the electron in its act of transfer.†

Now, using Eqs. (7.15), (7.20), and (7.21) in Eq. (7.14), one gets the total Hamiltonian for the reacting system in a polar medium as (see Fig. 7.1)

$$H_{\text{total}} = -\frac{\hbar^2}{2m_e} \nabla^2 + \frac{(z_1 e_0)^2}{r} + \frac{(z_2 e_0)^2}{|\mathbf{R} - \mathbf{r}|} + \frac{\hbar \omega_0}{2} \sum \left(\mathbf{q}_k^2 - \frac{\partial^2}{\partial \mathbf{q}_k^2} \right)$$

$$+ \frac{\varepsilon_{\text{op}} - 1}{8\pi\varepsilon_{\text{op}}} \int \mathbf{D}_i^2 \, dV + \int \mathbf{D}_e \mathbf{P} \, dV \tag{7.22}$$

The wave function describing the behavior of the total system (solvent + ion $\mathbf{A}^{z+(1)}$ + ion $\mathbf{B}^{z+(2)}$ + electron) can be found from the solution of the Schrödinger equation

$$H_{\text{total}} \psi(\mathbf{r}, \mathbf{R}, \mathbf{q}_k) = E(\mathbf{r}, \mathbf{R}, \mathbf{q}_k) \tag{7.23}$$

where $\psi(\mathbf{r}, \mathbf{R}, \mathbf{q}_k)$ is the wave function of the system depending on the solvent coordinate $\mathbf{q}_k$, the electron coordinate $\mathbf{r}$, and the interionic distance $\mathbf{R}$ (Fig. 7.1).

7.3. Transition Probability of a Quantum Particle from Ion to Ion in the Original Levich–Dogonadze Treatment[3,4,9,10]

In developing the theory of electron transfer reactions both in homogeneous solutions and at electrodes, Levich and Dogonadze focussed their attention in determining the transition probability of an electron from a donor to an acceptor. In the homogeneous case, the donor of an electron is an ion of lower oxidation number [e.g., $Fe^{2+}(H_2O)_6$ ion] and an acceptor is an ion of higher oxidation number [e.g., $Fe^{3+}(H_2O)_6$ ion]. But in the electrodic case, the donor of the electron is the electrode (in the case of a cathodic reaction) and the acceptor is a positive ion [e.g., H_3O^+ or $Fe^{3+}(H_2O)_6$] in the outer Helmholtz plane (OHP) of the double layer.

In finding the transition probability of the electron per unit time, Levich and Dogonadze[3,4,9,10] used the general expression of time-dependent perturbation theory (i.e., Fermi's golden rule). The general expression of probability of transition of a system from its initial state i to the final

---

† Again, the interaction of the electron with the permanent dipoles in water seems to have been neglected.

state f in unit time is, then,

$$P_{if} = \frac{2\pi}{\hbar} \text{Av} \sum_f \left| \int \psi_f U \psi_i \, d\tau \right|^2 \delta(E_f - E_i) \qquad (7.24)$$

where ψ_i and ψ_f are the wave functions that characterize the initial and the final states of the system, respectively. U is the perturbation operator responsible for making the system undergo a transition from the initial state to the final state. The Dirac δ function of the argument $(E_i - E_f)$ expresses the energy conservation law and corresponds to radiationless transfer. The Dirac δ function in general can be expressed as

$$\delta(E_i - E_f) = \frac{1}{2\pi h} \int_{-\infty}^{\infty} \exp[i(E_i - E_f)t/h] \, dt \qquad (7.25)$$

In Eq. (7.24), the summation is carried over all final states of the system. The property of the δ function is such that when the conservation of energy is not obeyed, i.e., when $E_f \neq E_i$, the δ function becomes zero and thus the transition probability P_{if} becomes zero. In other words, there is no electron transition between states having nonequal energy in accordance with the principle of radiationless transition of an electron between two states.[15] The symbol Av in Eq. (7.24) stands for a statistical averaging (i.e., averaging over the thermal motion) over initial states and this is carried out with the expression

$$Z = \frac{\exp(-E_i/kT)}{\sum_i \exp(-E_i/kT)} \qquad (7.26)$$

where Z represents the normalized probability that the system in statistical equilibrium occupies a state with an energy E_i. In fact, Eq. (7.24) is valid[10] only for a small time interval $t \leqslant \tau$, where t is the mean time interval between transitions.

LDK[3-10] considered the system corresponding to the electron transfer reaction as consisting of two subsystems—one is a rapid subsystem (electron) and the other a slow subsystem (solvent and nuclei of the reacting ions). In quantum mechanics, when considering the behavior of a system (in terms of its wave function) that may be divided into rapid and slow subsystems, one can make use of the Born–Oppenheimer approximation. According to this approximation, one can represent the full wave function $\psi(\mathbf{r}, \mathbf{R})$ of the system as a product of wave function of the slow subsystem $\chi(\mathbf{R})$, depending on the corresponding coordinate $\mathbf{R}$ and the wave function of the rapid subsystem $\phi(\mathbf{r}, \mathbf{R})$ corresponding to coordinates $\mathbf{r}$ and $\mathbf{R}$. Thus, one gets the wave function of the system as

$$\psi(\mathbf{r}, \mathbf{R}) = \chi(\mathbf{R})\phi(\mathbf{r}, \mathbf{R}) \qquad (7.27)$$

Expressing ψ_i and ψ_f of Eq. (7.24) in the form of Eq. (7.27), one gets the transition probability from Eq. (7.24) in the form

$$P_{if} = \frac{2\pi}{\hbar} \text{Av} \sum_f \left| \int \phi_f^*(\mathbf{r}, \mathbf{R}) \chi_f^*(\mathbf{R}) U(\mathbf{r}, \mathbf{R}) \chi_i(\mathbf{R}) \phi_i(\mathbf{r}, \mathbf{R}) \, d\mathbf{r} \, d\mathbf{R} \, \delta(E_f - E_i) \right.$$

(7.28)

where the integration is carried over all the coordinates $\mathbf{r}$ and $\mathbf{R}$ of the *light* and *heavy* particles. One can simplify considerably the expression (7.28) with the assumption that the Franck–Condon principle is applied for this system (consisting of slow and fast particles). Namely, one can consider the wave function of the heavy particle, e.g., a nucleus $\chi(\mathbf{R})$ is localized in space, whereas the wave function $\phi(\mathbf{r}, \mathbf{R})$ and the interaction energy $U(\mathbf{r}, \mathbf{R})$ are distributed over the whole volume of the system. Thus, one may take out the slowly varying function corresponding to $\chi(\mathbf{R})$ from the integrand and assign an average value of this function at some average point.

According to this transformation, one may represent the transition probability in the following form

$$P_{if} = \frac{2\pi}{\hbar} \text{Av} \sum_f \left| \int \phi_f^* U \phi_i \, d\tau \right|^2 \left| \int \chi_f^* \chi_i \, d\mathbf{R} \right|^2 \delta(E_f - E_i) \qquad (7.29)$$

The result of this simplification is that the wave functions for the rapid and the slow subsystem are well separated in the expression (7.29).

Before doing the thermal averaging as represented by Av in Eq. (7.29), one needs to consider that if the rapid subsystem is represented by an electron the energy levels of which are electronic, then spacing between its levels comprise several electron volts (i.e., which corresponds to the order of a few thousand degrees). Therefore, it can be considered that the rapid subsystem is always in the ground electronic state at room temperature. This means that in the averaging process one needs to retain only one term in the sum over the initial state of the rapid subsystem; this term refers to the ground state. Thus, one may write (7.29) as

$$P_{if} = \frac{2\pi}{\hbar} \left| \int \phi_{f0}^* V \phi_{i0} \, d\mathbf{r} \right|^2 \text{Av} \sum_f |\chi_f^* \chi_i \, d\mathbf{R}|^2 \delta(E_f - E_i) \qquad (7.30)$$

where the index zero refers the ground state of the electron. The expression (7.30) containing the averaging only over possible states of slow subsystem (e.g., solvent) for a fixed electronic state constitute the basis of the theory of electron transfer reaction in the continuum treatment. The dependence of transition probability on temperature is only contained in the second factor of Eq. (7.30) and is expressed as

$$\text{Av} \left| \int \chi_f^* \chi_i \, d\mathbf{R} \right|^2 = \frac{\exp(-E_i/kT) |\int \chi_f^* \chi_i \, d\mathbf{R}^2}{\sum \exp(-E_i/kT)} \qquad (7.31)$$

Defining

$$L = \int \phi_f^* U \phi_i \, d\mathbf{r}$$

one can express the transition probability from Eq. (7.30) as†

$$P_{if} = \frac{2\pi}{h} |L|^2 \text{ Av} \sum_f \left| \int \chi_f \chi_i \, d\mathbf{R} \right|^2 \delta(E_{f.} - E_i) \qquad (7.32)$$

This is the general expression given by Levich and Dogonadze from which one could find the transition probability of quantum particles (electrons or protons) in a polar medium using the relevant wave functions of the reacting systems involved. It is worth mentioning that the transition being considered by Eq. (7.32) is not only the probability of transition corresponding to the transfer of an electron from one ion to another ion, where the electronic states have already been arranged so that radiationless transfer occurs. The *probability* of whether the electron in the ion can transfer to the correct state in another ion so that radiationless transfer occurs is taken into account by the averaging represented by Av. This refers to Eq. (7.26), where averaging of the thermal motion over the initial state is carried out with a partition function.

It must be made clear at this point what the expression P_{if} means and what its relation is to an actual electrochemical current. An actual electrochemical current, that which is observed experimentally, will be given by an expression of the type

$$i = e_0 \int^{\infty} \rho(E) f(E) P_t(E) G_i(E) \, dE \qquad (7.33)$$

where $\rho(E)$ is the density of the electron state and $f(E)$ is the Fermi distribution of electrons in the electrode. $P_t(E)$ is the transition probability of an electron from the electrode to an ion in solution or vice versa, and $G_i(E)$ is the probability of having the acceptor or donor in solution having the same electronic energy as that of an electron in the electrode.

What has been formulated in Eq. (7.32) (but not calculated) here by LDK is the latter two terms of this expression, namely, the part that involves $P_t(E)$, and also, by implication, though without evaluation, the expression for $G_i(E)$.

---

† Thus, the quantum mechanical part of the analysis of the electron's movement has been reduced to $|L|^2$, which is not further evaluated (see reference 16).

Transition Probability of an Electron

It may be assumed that the ions $A^{z+(1)}$ and $B^{z+(2)}$ are atomic and that the electron transfer process between them does not involve any bond breaking or bond formation. The states of their internal electronic levels are considered as unchangeable, and the valence electron undergoes transition from the ground electronic state of one into the ground electronic state of another without electronic excitation as noted earlier. The total system for the electron transfer process involves a fast subsystem (electron) and a slow subsystem (ions and solvent). According to the above concept, there is no change in the states of the ion cores, and hence, one may consider the solvent as only the slow subsystem. Thus, one can write Eq. (7.32) as

$$P_{if} = \frac{2\pi}{\hbar}|L|^2 \, Av \sum \left| \int \chi_f(\mathbf{q})\chi_i(\mathbf{q}) \, d\tau \right|^2 \delta(E_f - E_i) \tag{7.34}$$

where $\chi(\mathbf{q})$ is the wave function of solvent oscillators and $\mathbf{q}$ represents the solvent coordinate. The perturbation energy U involved in the expression L of Eq. (7.34) represents the interaction energy of an electron in the reduced ion with the oxidized ion, i.e.,

$$U = \frac{(z_{ox}e_0)^2}{|\mathbf{r} - \mathbf{R}|} \tag{7.35}$$

where z_{ox} is the charge on the oxidized ion.

Now, using Eq. (7.25) in Eq. (7.34), one obtains

$$P_{if} = \frac{1}{\hbar^2}|L|^2 \, Av \sum_f |\langle \chi_f(\mathbf{q})\chi_i(\mathbf{q})\rangle|^2 \int_{-\infty}^{\infty} \exp[i(E_f - E_i)t/h] \, dt$$

$$= \frac{1}{\hbar^2}|L|^2 \, Av \int_{-\infty}^{\infty} \sum |\langle \chi_i|\chi_f\rangle\langle\chi_f|\chi_i\rangle| \exp[i(E_f - E_i)t/h] \, dt$$

$$= \frac{1}{\hbar^2}|L|^2 \, Av \int_{-\infty}^{\infty} \sum_f \left\langle \chi_i \left| \exp\left(\frac{it}{h}E_f\right) \right| \chi_f \right\rangle \left\langle \chi_f \left| \exp\left(-\frac{itE_i}{h}\right) \right| \chi_i \right\rangle \, dt \tag{7.36}$$

To express the Eq. (7.36) in terms of an operator, one uses the relation

$$\exp(\alpha E_n)\phi_n = \exp(\alpha H)\phi_n$$

Now, using this operator relation in (7.36), one obtains

$$P_{if} = \frac{1}{\hbar^2}|L|^2 \, Av \int_{-\infty}^{\infty} \sum_f \left\langle \chi_i \left| \exp\left(\frac{it}{h}H_f\right) \right| \chi_f \right\rangle \left\langle \chi_f \left| \exp\left(-\frac{it}{h}H_i\right) \right| \chi_i \right\rangle \, dt \tag{7.37}$$

With the use of the relation

$$\sum |\chi_f\rangle\langle\chi_f| = 1 \tag{7.38}$$

one gets

$$P_{if} = \frac{1}{\hbar^2}|L|^2 \, \text{Av} \int_{-\infty}^{\infty} \left\langle \chi_i \left| \exp\left(\frac{1}{\hbar}(H_f - H_i)t\right) \right| \chi_i \right\rangle dt \tag{7.39}$$

where the Hamiltonian operators† are

$$H_i = \frac{\hbar\omega_0}{2}\left(\mathbf{q}_i^2 + \frac{\partial^2}{\partial\mathbf{q}_i^2}\right) + Q_i^0 \tag{7.40}$$

and

$$H_f = \frac{\hbar\omega_0}{2}\left(\mathbf{q}_f^2 + \frac{\partial^2}{\partial\mathbf{q}_f^2}\right) + Q_f^0 \tag{7.41}$$

where $\mathbf{q}_i$ and $\mathbf{q}_f$ are the relative (with respect to equilibrium) coordinates of the solvent systems in the initial and in the final states, respectively. Q_i^0 and Q_f^0 are ground-state energies of the solvated ion in its initial and final states, respectively.

The harmonic oscillator wave function[9]

$$\chi_i = A_n H_n(\mathbf{q}_{ki} - \mathbf{q}_{ki}^0) \exp[-\tfrac{1}{2}(\mathbf{q}_{ki} - \mathbf{q}_{ki}^0)^2] \tag{7.42}$$

for the solvent was used be Levich and Dogonadze.[3,4,9,10] In Eq. (7.42), A_n is the normalization constant of the harmonic oscillator wave function, and H_n is the Hermite polynomial; $\mathbf{q}_{ki}^0$ and $\mathbf{q}_{ki}$ are the initial state equilibrium and nonequilibrium solvent coordinates, respectively.

Using the Hamiltonian operation of Eqs. (7.40) and (7.41) and the wave function from Eq. (7.42), LDK[6-10] have quoted an explicit expression that they say they have obtained for the transition probability of the electron. Unfortunately, however, they have not published the derivation of this solution. The equation, which lacks published derivation, is (for the endothermic case)

$$P_{if} = \left(\frac{2\pi}{\hbar^2\omega_0}\right)|L|^2 I_m(Z) \exp\left[-\frac{\hbar\omega_0 m}{2kT} - z \coth\left(\frac{\hbar\omega_0}{2kT}\right)\right] \tag{7.43}$$

where $I_m(Z)$ is the Bessel function of the order m and argument Z, where

$$m = \frac{Q_f - Q_i}{\hbar\omega_0} = \frac{Q}{\hbar\omega_0} \tag{7.44}$$

† The addition of Q_i and Q_f to the Hamiltonian derived in Appendix 7.I arises because the harmonic oscillator for the initial and final system have different energies at the minimum of the potential energy curves.

$$Z = \tfrac{1}{2} \operatorname{csch}\left(\frac{\hbar\omega_0}{2kT}\right) \sum_k (\mathbf{q}_{ki}^0 - \mathbf{q}_{kf}^0)^2$$

$$= \frac{E_s}{\hbar\omega_0} \operatorname{csch}\left(\frac{\hbar\omega_0}{2kT}\right) \tag{7.45}$$

since $E_s = \tfrac{1}{2}\hbar\omega_0 \sum_k (\mathbf{q}_{ki}^0 - \mathbf{q}_{kf}^0)^2$ is the solvent repolarization energy, that is, the energy change when the solvent coordinate changes from its equilibrium value $\mathbf{q}_{ki}^0$ in the initial state to that of final state $\mathbf{q}_{kf}^0$.

The introduction of the expression for the so-called "reorganization energy" in the LDK theory brings conceptual difficulties. The authors refer to a picture described by Levich.[9] The solvent librators, each of energy 0.001 eV, from time to time change their orientation and position (i.e., the solvent coordinate $\mathbf{q}$) in such a way that, in a certain fluctuation, sufficient energy reaches the ion to allow a radiationless transfer of the electron.

The Russian authors are intent to stress the *librational* nature of the solvent movement that they are considering. The reason for this is carefully spelled out in the Russian authors' presentations.[6-10] It is that the average frequency of a librator makes its energy much less than the thermal energy kT (librational frequency is of the order of magnitude 10^{11} Hz), and therefore the librational modes are always "activated," i.e., virtually all the librations are active at room temperature.

However, in the use of the equation

$$E_s = \tfrac{1}{2}\hbar\omega_0 \sum_k (\mathbf{q}_{ki}^0 - \mathbf{q}_{kf}^0)^2 \tag{7.46}$$

where the $\mathbf{q}$ terms represent the dimensionless parameter "solvent coordinate," there appears to be a conceptual difficulty. Thus, Eq. (7.46) implies a simple harmonic vibrator, the potential energy of which is $\tfrac{1}{2}kx^2$ $[\equiv \tfrac{1}{2}\hbar\omega_0(\mathbf{q}_{ki}^0 - \mathbf{q}_{kf})^2]$. A librator is, in fact, a restricted rotator, and the equations for its potential energy are not dependent upon an expression of the form used in Eq. (7.46). For example, it is easy to show that a restricted rotator, operating in a potential energy of interaction with its surroundings of B, has an energy[45]

$$U = kT^2\left(\frac{1}{T} - \frac{B}{kT^2}\right) \tag{7.46a}$$

and if one uses B as $-ze_0\mu/\varepsilon r^2$ for an ion–dipole interaction (for example), the energy that follows is indeed of the order of magnitude 10^{-3} eV (an individual vibrator would have an energy of about 10^{-1} eV). However, it is not dependent on a solvent coordinate variation of the sum of squared terms in the sense of Eq. (7.46) [see Eq. (7.46a)].

Table 7.1
Values of the Energy of Activation

Sign of Q	Condition	Activation energy	Remark
Positive	$E_s \ll |Q|$	$E_a \to \infty$	Impossible
	$E_s \gg |Q|$	$E_a \to E_s/4$	Possible
	$E_s = |Q|$	$E_a = Q$	Improbable
Negative	$E_s \ll |Q|$	$E_a \to \infty$	Impossible
	$E_s \gg |Q|$	$E_a \to E_s/4$	Possible
	$E_s = |Q|$	$E_a = 0$	Possible

The origin of the expression for the so-called reorganization energy that the Russian authors use comes from their assumption of Pekar's Hamiltonian,[13] which, of course, is for a *crystal*. In a solid crystal, the energy is clearly that of *vibrators* and is correctly represented by $(n + \frac{1}{2})h\omega_0$. Is it consistent with the concept of a *librating* solvent molecule, which is, in fact, a restricted rotator?

The expression (7.43) is the general expression of transition probability of an electron applicable for any range of temperature. One may simplify the expression (7.43) by considering the high-temperature limits $kT \gg h\omega_0$, which applies to $T > 10°K$.

In the high-temperature limit, the argument of the Bessel function, $I_m(Z)$ of Eq. (7.43) becomes far greater than unity (i.e., $Z \gg 1$), and hence one may use the asymptotic expansion of the Bessel function. Thus, using this asymptotic expansion of the Bessel function and the high-temperature condition that $kT \gg h\omega_0$, and $kT > h\omega_0 Q/E_s$, one gets from Eq. (7.43):

$$P_{if} = \frac{|L|^2}{h}\left(\frac{\pi}{kTE_s}\right)^{1/2} \exp\left\{-\frac{(E_s + Q)^2}{4E_s kT}\right\} \qquad (7.47)$$

$$P_{if} = \frac{|L|^2}{h}\left(\frac{\pi}{kTE_s}\right)^{1/2} \exp\left\{-\frac{E_a}{kT}\right\} \qquad (7.48)$$

where

$$E_a = \frac{(E_s + Q)^2}{4E_s} \qquad (7.49)$$

is the energy of activation† (Table 7.1).

---

† The expression that has been used by Levich–Dogonadze for energy of activation [Eq. (7.49)] obtained in this complete manner can be obtained in a very simple classical manner as given by Appleby et al.[40] (see also Section 6.6).

Thus, the assumption of a classical (and harmonic) nature of the motion of the solvent dipoles ($\hbar\omega_0 \ll kT$) leads to an Arrhenius-type equation for the transition probability [Eq. (7.48)].

In quantum mechanics one assumes the existence, first of all, of two states of equal energy for a radiationless probable transfer, and then the transition probability is simply the probability that the states will pass from one to the other when they are already at equal energy. In the expression called transition probability (7.43) and (7.47) in the presentation given above, the extra expression for the probability that the two states are equal has been taken into account.

Such an expression differs, for example, from that produced by Khan, Wright, and Bockris,[16] in which time-dependent perturbation theory has been used to calculate the transition probability between two states of equal energy where the states have been assumed to be of equal energy. It is important to note, however, that the LDK workers *have not solved the electronic quantum mechanical part* of their equations [Eqs. (7.43) and (7.47)]. This is displayed in the electronic matrix element $|L|$, which contains the wave function of the electron in the initial and final state and the perturbing operator which causes the transfer.

7.4. Transition Probability in the Bond-Breaking Reactions

In the previous sections, a partial derivation of the transition probability for electron transfer reaction from a redox ion to another is given where no bond breaking is involved. In this section we will discuss basic concepts suggested by Dogonadze, Kuznetsov, and Levich,[8-10] to determine the transition probability of a charge (e.g., the proton) that is accompanied by breaking of a bond (e.g., an H^+—H_2O bond) followed by the forming of a bond (i.e., M—H, a metal–hydrogen bond), but details of the formulation of the transition probability for proton transfer are given in Chapter 10. This is more complicated than the simple electron transfer reaction in a redox system, which does not involve bond breaking. Simple electron transfer reactions are limited, and in most cases the occurrence of a reaction is accompanied by the breaking or forming of a new chemical bond. As an illustration, Dogonadze, Kuznetsov, and Levich (DKL) assumed that in the hydrogen evolution reaction, the slowest process is *proton transition* from the hydroxonium ion (i.e., the H_3O^+ ion) to the electrode.

For the discharge step, these authors make a hypothesis different from those of all previous workers (Gurney, Christov, Bockris), who had supposed that the particles that react are those in higher vibrational energy states. DKL, however, opined that, because the number of bonds in an

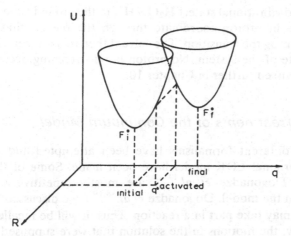

Fig. 7.2. Potential energy surfaces for proton discharge according to the LDK model; q is the solvent coordinate, R is the proton coordinate, F_i^0 and F_f^0 are the free energies in the initial and final states, respectively.

activated state is small, it would not take part in the reaction. In this way, DKL formed a new concept in kinetics, contravening the course of the theory of kinetics from the time of Arrhenius, in which the particles that react had always been the few highly activated ones. Thus, DKL thought that stretching of the H_2O-H^+ and $H-M$ bonds was sufficiently improbable, since for these bonds the $h\omega_0 \gg kT$. They treated the proton as well as the electron as part of a fast subsystem and the solvent as the slow subsystem. The *activation* of the proton system (*not its stretching*) *can arise only due to nonequilibrium solvent polarization fluctuation* in a similar manner as in the electron transfer case. By using the Born–Oppenheimer approximation twice, this approach separates the wave function from the solvent wave functions and treats the electron and proton as a quantum subsystem and the solvent as a classical subsystem.

The potential energy surface assumed for the discharge reaction is schematically shown in Figure 7.2, where q is the solvent coordinate that indicates the state of polarization of the solvent, R is the proton coordinate, and U is the energy of the system. As the value of the solvent coordinate q changes due to the solvent polarization fluctuation, the system moves along the energy surface $U(q, R)$, maintaining the constant value $R = R_0$, while vibrations along the bond H_2O-H^+ remain unexcited.† At $q = q^*$ (activated), the energy of the initial state equals that of the final state and the system undergoes a quantum transition involving proton tunneling

† Later, Dogonadze[10] mentioned that the equilibrium coordinate of the proton changes significantly only in the direction normal to the electrode surface.

from the ground vibrational state,† H_2O-H^+, to the ground state of $M-H$ without change in proton coordinate through the barrier that changes continuously during the transition. The solvent then relaxes, corresponding to the final state of the system. No proton bond stretching occurs. These models are discussed further in Chapter 10.

7.5. Recent Treatments of the Continuum Model

Recently, different formalisms have been attempted and improvements in the original LDK model have been made. Some of the recent treatments of Dogonadze et al.[17-22] are mostly repetitive with little improvement in the model. Dogonadze et al.[18,19] have discussed the types of motion that may take part in a reaction. Thus, it will be recalled that in the earlier view, the motions in the solution that were supposed to effect the electrode reaction were those for which $\hbar\omega_0 \ll kT$. This originated from the fact that it was the authors' wish to see a large number of these motions in action, and their conviction that other theories of reaction kinetics in which it is mainly the activated particles that react is not applicable in electron transfer reaction in solution.

However, more recently, Dogonadze et al.[18,19] have suggested that there is a contribution also from those other types of motion, for example, *vibrations* in addition to librations, may also play a part. In this, they come a little nearer to the current viewpoint of other physical chemists that the main part of the reaction is contributed to by the bonds and the variation of distance during activation.

Nevertheless, in their recent consideration, Dogonadze and Vorotyn-sev say that they believe that bond stretching and $O-H^+$ activation effects could only come into the preexponential factor, though the physical meaning of this statement is unclear. It does seem, however, that stretching frequencies connected with the inner-sphere level in redox processes and the bond-breaking processes in the proton transfer are not supposed to take any part in the Russian authors' model.§

Dogonadze and Kuznetsov[22] have recently started to consider the effects of the inner-sphere degrees of freedom as do authors who do not

---

† According to the Gurney[15] model of the hydrogen evolution reaction, the neutralization of a proton in its ground vibrational state and the transition by an electron from the electrode at the Fermi level is improbable, since when the proton is in the ground vibrational state, its energy in solution is much greater than the electron energy in metal electrode, and thus such a transfer violates the condition of radiationless transition. This condition is only fulfilled if the proton coordinate changes. Only activated protons react (see Chapter 8).

§ According to Bockris, Khan, and Matthews,[42] no parallelism of calculated and experimental $\Delta F^{0\ddagger}$ is observed for redox reactions *unless* bond stretching in the George–Griffith model is taken into account.

take the continuum approach.[24-26] In addition to considering the effect of bonds between the ion and the solvation sheath—previously neglected because of the preoccupation with the *average value* rather than with the contributions of particularly energetic species—the Russian workers[22] added a refinement in that they considered the distribution of frequencies in solvent vibrations, rather than just considering a fixed frequency ω_0.

Dogonadze and Kuznetsov[22] determined the transition probability of the electron when there is a change in the inner coordination sphere for two limiting cases. The two cases are:

(a) A high-temperature approximation with respect to the frequencies of solvent motions (for water these might be the librational movements of the dipoles). When $\hbar\omega_0 < kT$ ("high temperature"), the libration is influenced by thermal effects, and a low-temperature approximation with respect to the intramolecular frequencies of the solvent vibrations' ($\hbar\omega > kT$) is not applicable since the intramolecular vibrations are very weakly excited.

(b) A high-temperature approximation for all frequencies (i.e., for both the librational and inner-sphere vibrations).

Considering the first limiting case ($\hbar\omega_0 > kT$)† Dogonadze and Kuznetsov[23] quote without proof the following expression for the transition probability of electron (both for homogeneous and electrochemical):

$$P_{if} = \omega_p \exp\left[-\sum_{ki} (q_{ki\alpha}^0 - q_{ki\beta}^0)^2\right] \qquad (7.50)$$

where $q_{ki\alpha}^0$ and $q_{ki\beta}^0$ are the equilibrium coordinates of the inner-sphere oscillators in the initial state α and final state β, respectively, and ω_p is equivalent to the probability of electron transfer in the case when there are no changes in the first coordination sphere, i.e., when $q_{ki\alpha}^0 = q_{ki\beta}^0$ and can be expressed as[23]

$$\omega_p = \left(\frac{\pi}{kTh^2E_s}\right)^{1/2} |L|^2 \exp\left[-\frac{(Q+E_s)^2}{4E_skT}\right] \qquad (7.51)$$

where $|L|$ is the electronic matrix element, Q is the heat of reaction, and E_s is the reorganization energy of the continuum solvent. Here also, the quantum mechanical part of the problem, namely the matrix element, in which the heart of the quantum mechanics resides, has remained unevaluated.

As can be seen from Eqs. (7.50) and (7.51), a consideration of the changes in the first coordination sphere leads to the appearance of an

† However, with $\omega \approx 10^{11}$, $T \approx 5°K$. Hence, Dogonadze and Kuznetsov clearly mean "at all temperatures."

exponential factor in the probability of transfer, while the activation energy remains unchanged. To take into account *some* stretching contribution, the Russian workers[23] multiply the normal expression (without stretching) by an expression that involves an exponential in which the stretching of the solvent coordinate is involved, that is, they multiply by a factor

$$\exp\left[-\sum_{ki} (q_{ki\alpha} - q_{ki\beta})^2\right]$$

Some justification for the use of this expression is given by Schmickler and Vielstich.[35] The net probability expression that the Russians also use is thus the normal expression that they use for all their situations with the quantum mechanical matrix element, and they multiply by the unproved exponential term in which stretching of the inner solvent coordinate is given.

Outside the Russian group there are some other workers, namely, Schmickler and Vielstich,[24,25] Kestner, Logan, Ulstrup, and Jortner,[26,27] and Schmidt,[28a,b-31] who have examined the theory of electron transfer reaction using basically the LDK continuum model. They have tried to add to it the accounting for the inner sphere.

It might be thought that the more important advance that is needed is to examine the *physical basis* of the continuum model, with the various difficulties of a fundamental kind with which it is associated[40] and which apply to all the rest of the treatments that have followed it. *Numerical tests*[16,42] are what is primarily required.

7.5.1. Schmickler–Vielstich Treatment

Schmickler and Vielstich[24] used the same model and an analogous approach to that of LDK[4,9,10] to find the transition probability of an electron in redox reactions in solution. They found expressions for the transition probability that are identical with the Levich and Dogonadze[3,4] expression for the high-temperature limit but differed for the low-temperature approximation. Schmickler and Vielstich[24] thought that in the original LD treatment,[4] an approximation was made during the calculation (which LDK have unfortunately never published) that is not valid for the low-temperature case.†

Later Dogonadze and Kuznetsov[32] pointed out that the discrepancy between Schmickler and Vielstich treatment[24] arises from the fact that the German authors made an error in neglecting the noncommutativity of the solvent Hamiltonians for the initial and final states. In other words, they

---

† However, in view of the fact that "low temperature" here means ~10°K, the relevance of the calculation is less clear.

put

$$\exp\!\left(\frac{it}{\hbar}H^f\right)\exp\!\left(-\frac{it}{\hbar}H^i\right) = \exp\!\left[\frac{it}{\hbar}(H^f - H^i)\right] \qquad (7.52)$$

which is inadmissible, since the operators H^i and H^f do not commute. The commutator (H^i, H^f) can be ignored *only* in a high-temperature approximation.[33,34] † For this reason the *exponential factor* in the expressions of transition probability in both the Levich and Dogonadze and the Schmickler treatments coincides in the high-temperature limit. Schmidt[34] supported the view of Dogonadze and Kuznetsov[32] regarding the neglect of noncommutivity of the solvent Hamiltonian in the work of Schmickler and Vielstich.[24]

Moreover, Dogonadze and Kuznetsov[32] mentioned that the Schmickler–Vielstich[24] formula does not satisfy the detailed balancing condition, i.e.,

$$W_{if} = W_{fi}\exp(-Q/RT) \qquad (7.53)$$

since the Schmickler–Vielstich expression leads to the incorrect relation,

$$W_{if} = W_{fi}\exp(-Q/\tfrac{1}{2}\hbar\omega_0) \qquad (7.54)$$

In response to remarks of Dogonadze and Kuznetsov,[32] Schmickler and Vielstich[35] pointed out that the neglect of the commutator (H^f, H^i) is not an error but an approximation known as the *semiclassical Franck–Condon principle*,[33] and this approximation is valid if

$$E_s \gg \hbar\omega_0 \tanh\frac{\hbar\omega_0}{kT} \qquad (7.55)$$

This inequality is satisfied in the case of most redox reactions, since $\omega_0 \sim 10^{11}$ sec and energy of reorganization E_s is of the order of 0.5 eV. Of course Schmickler and Vielstich[35] agree that their result does not satisfy the detailed balancing conditions [Eq. (7.53)].

Schmickler[25] attempted to advance the original LDK model by incorporating the inner coordination sphere contributions with that of continuum outer sphere in a unified manner. To find the transition probability of an electron for an electrochemical redox reaction, he used the general expression for the transition probability[24,25]

$$W_{if} = \frac{|L|^2}{Z\hbar^2}\int dt \sum_\alpha \langle \chi_\alpha^i|\exp[it(H^f - H^i)]|\chi_\alpha^i\rangle \exp\!\left(\frac{-E_\alpha^i}{kT}\right) \qquad (7.56)$$

where H^i and H^f are the Hamiltonian of the heavy particle (solvent and ion system) in the initial and final states, respectively. χ_α^i and E_α^i are,

---

† The situation with ω_0 as a librator frequency of 10^{11} Hz implies $kT > \hbar\omega_0$ at all practical temperatures, i.e., the high temperature approximation is always applicable.

respectively, the wave function and energy of the heavy particle in the initial state. Z is the partition function for the initial state, and α is the quantum number.

In this unified treatment, the Hamiltonian for the heavy particle system has been expressed as a sum of two Hamiltonians, i.e.,

$$H = H_{\text{out}} + H_{\text{in}} \tag{7.57}$$

where H_{out} is the Hamiltonian without any inner-sphere effect and is identical with that used by LD[1] [Eq. (7.13)]. H_{in} is the Hamiltonian for the *inner sphere* and in the harmonic approximation has been expressed for the initial state i and final state f, respectively, as

$$H_{\text{in}} = \sum_{\mu=1}^{n} \left(\frac{\hbar^2}{2m_\mu} \frac{\partial^2}{\partial S_\mu^2} + \frac{m_\mu}{2} (\omega_\mu^i)^2 S_\mu^2 \right) \tag{7.58}$$

$$H_{\text{in}}^f = \sum_{\mu=1}^{n} \frac{h^2}{2m_\mu} \frac{\partial^2}{\partial S_\mu^2} + \frac{m_\mu}{2} (\omega_\mu^f)^2 (S_\mu - \varepsilon_\mu)^2 \tag{7.59}$$

where the sum is taken over all normal coordinates S_μ, m_μ is the reduced mass, ω_μ^i and ω_μ^f are the frequencies of the inner-sphere oscillation, and ε_μ is the dielectric constant of the medium in the initial and final states. Using Eqs. (7.57), (7.58), and (7.59) in Eq. (7.56), Schmickler[25] gave the following expression for the transition probability

$$W_\mu = \frac{|L|^2}{\hbar^2} \int \exp\left(it \frac{Q_0^f - Q_0^i + E_s}{h} \right) \exp\left(-i^2 \frac{\hbar\omega_0}{2} \coth \frac{\hbar\omega_0}{2kT} \right)$$

$$\times \prod_{\mu=1}^{n} \left\{ \left[\frac{(ith/2)(\omega_\mu^{f2} - \omega_\mu^{i2})}{m_\mu \omega_\mu^i \tanh(h\omega_\mu^i/2kT)} \right]^{-1/2} \exp\left(it \frac{m_\mu \omega_\mu^{f2} \varepsilon_\mu^2}{2\hbar} \right) \right.$$

$$\left. \times \exp\left(-t^2 \frac{m_\mu (\omega_\mu^f)^4 (\varepsilon_\mu^2/4\hbar)}{\omega_\mu^i \tanh(\hbar\omega_\mu^i/2kT) - (it/2)[(\omega_\mu^f)^2 - (\omega_\mu^i)^2]} \right) \right\} \tag{7.60}$$

where E_s is the free energy of the reorganization and has been used as that given by Marcus.[2] Q^i and Q^f are the minima of the potential energy of the initial and final state, respectively, and they depend linearly on the over-potential.

Schmickler[25] calculated the dependence of the rate constant on temperature and overpotential using Eq. (7.60) for aquo and cyano complexes of ferric ion in water and ethanol, respectively, taking into account that the inner sphere lowers the current by *a factor of two to three*. It was also predicted that if the ligands of an oxidized species are more firmly bound than in a reduced species, the transfer coefficient α is less than 0.5. However, this will be the situation in most redox systems, since a smaller bond length and a higher bond energy are observed in oxidized species than

Table 7.2
Influence of the Inner Sphere

Outer hydration sphere (Levich–Dogonadze)	Inner and outer sphere
$\alpha = 0.5$	$\alpha = 0.4\text{–}0.6$
$\alpha \neq \alpha(T)$	$\alpha = \alpha(T)$
$\alpha = \alpha(\eta)$, Tafel region for small overpotentials	$\alpha = \alpha(\eta)$, greater Tafel region
E_s decreases slightly with temperature	E_s increases with temperature

in reduced species, although the transfer coefficient is usually about 0.5. Table 7.2 summarizes the effects caused by the inclusion of inner-sphere contribution. The important difference is that the inner sphere causes quantum effects for electron transfer that manifest themselves in the temperature dependence of the transfer coefficient and in the energy of activation. It is important to méntion that though Schmickler took into account the inner-sphere contribution along with the continuum contribution of the LD model, he did not succeed in obtaining the linear Tafel line as shown in Fig. 7.3 [calculated from Eq. (7.60)]. Recently, in a similar treatment, Schmickler[36] concluded that the energy of activation for the electrochemical electron transfer reaction of a redox system decreases with temperature due to an increase in tunnel transition of the system, and also that the curvature of Tafel lines is greater at lower temperatures. (However, no curvature is observed experimentally at low temperatures.[59,60])

The reason Schmickler put forward for the decrease of activation energy due to increased tunneling may be less clear than is necessary. This is because activation energy is the barrier for the system to go from the

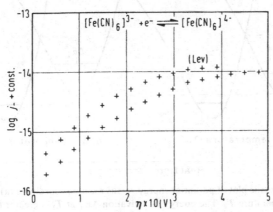

Fig. 7.3. Tafel plot for the reduction current density.

initial state to the final state. It is the property of the system itself. It should not be dependent upon whether the system undergoes tunneling or not. The decrease of activation energy with increase of temperature can be explained. The activation energy is a measure of the energy difference in the energy of the system in the initial state and that in the activated state. Since the system will be initially in a higher energy state at higher temperatures, it is possible that the difference between the energy of the initial state at higher temperatures and that of an activated state will be less, and thus the apparent activation energy will decrease (Fig. 7.4).

Later, Schmickler[37] left the idea of following the continuum treatment of LDK. He included a consideration of activation of the inner layer. Schmickler agrees with the view of the Bockris group[38,40] that use of the polaron theory to describe a polar solvent is doubtful, since the polar solvent differs greatly from an ionic crystal for which the polaron theory was originally developed.[13] In particular, a polar solvent has only a local order of a temporary nature, and in the vicinity of the dissolved ion this local order differs considerably from that in the bulk; it is unlikely, therefore, that long-range, undamped polarization waves can exist in a polar solvent.

Thus, Schmickler[37] reintroduced a consideration of a molecular model in treating the activation of both the inner and the outer layer. It is a collection of polarizable dipoles. Using this approach, he derived an expression for an energy of reorganization in terms of modelistic concepts.

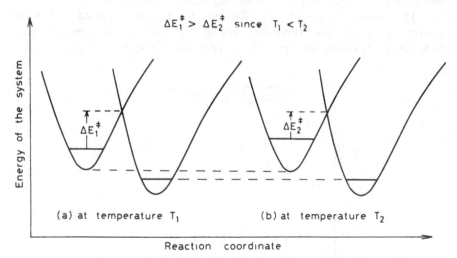

Fig. 7.4. The schematic plot of potential energy versus reaction coordinate (a) at temperature T_1 and (b) at temperature T_2. The energy of activation $\Delta E_1^{\ddagger}$ at T_1 is greater than $\Delta E_2^{\ddagger}$ at T_2 since $T_1 < T_2$.

He did not quantitatively calculate the value of the reorganization energy because the *result* of his calculation is essentially that of Marcus.

Schmickler[37] criticized the expression used by Bockris and Sen[38] to estimate the electrostatic energy of ion–solvent interaction. He claimed that it leads to serious errors in the numerical estimates and invalidates the conclusion that the contribution to the activation process due to fluctuation of the energy of the ion as a result of interaction with the surrounding librating dipoles is too infrequent to be of consequence. He suggested that the correct expression for the free energy of interaction would involve

$$\Delta U = \Delta F - T\frac{\partial \Delta F}{\partial T} = -\frac{e_0^2}{2r_i}\left(1 - \frac{1}{\varepsilon_s} - \frac{T}{\varepsilon_s^2}\frac{\partial \varepsilon}{\partial T}\right) = -46 \text{ kcal mole}^{-1} \quad (7.61)$$

where the free energy

$$\Delta F = -\frac{1}{8\pi}\left(1 - \frac{1}{\varepsilon_s}\right)\int \mathbf{P}^2 \, dV \quad (7.62)$$

where r_i is the radius of the ion with first solvation shell, ε_s is the static dielectric constant, and P is the polarization of the medium. Thus, Schmickler obtained the ion–solvent interaction energy as -46 kcal mole^{-1} compared to -3.3 kcal mole^{-1} in the Bockris and Sen work and remarked that their estimate was too low by an order of magnitude, and the same error perpetuates to their estimate of energy fluctuations. This circumstance invalidates their conclusion that the outer-sphere solvation contributes only negligibly to the activation process.

Schmickler may have misunderstood the situation of activation due to solvent fluctuation, since his espression (7.61) gives the solvation energy due to ion–solvent interaction when an ion is brought *from the vacuum to a solution*. But the energy concerned in considering the probability of a fluctuation is not this energy. It is the extra energy that the ion undergoing electron transfer experiences due to interaction with the surrounding dipoles when the ion is *already* in solution. Thus, in the calculation of Bockris and Sen, the potential energy calculated was that between a typical ion and surrounding dipoles at various distances from the ion, where Schmickler refers to the Bornian transfer process between an ion in a vacuum to an ion in a solvent. Moreover, Schmickler's expression (7.61) does not give the *average* energy $\langle E \rangle$ of the ion–solvent interaction, which one needs to use in the probability of fluctuation expression used by Bockris and Sen.[38]

7.5.2. Kestner, Logan, and Jortner (KLJ) Treatment

The treatment of Kestner, Logan, and Jortner[26] regarding the electron transfer reaction in polar solvents is virtually an extension of the original

Levich–Dogonadze–Kuznetsov[3–10] theory. They derived a quantum mechanical expression for nonadiabatic electron transfer reactions where the role of configurational changes in the first coordination layer is incorporated. In this treatment, both the initial and final states consisting of reacting species (ions) and the solvent were considered to be in the ground electronic and ground vibrational state. The electron transfer process is visualized as the irreversible decay of the initial state into a manifold of final states. Second-order perturbation theory and the generating function method of Kubo[39] were used to find the following thermally averaged transition probability for the homogeneous electron transfer reaction

$$W_{if} = \frac{|L|^2}{h} \left(\frac{2\pi}{h^2 D^2} \right)^{1/2} \exp\left[\frac{(Q - E_s)^2}{2 D^2 h^2} \right] \qquad (7.63)$$

where Q is the difference in the energies of the initial and final states in their equilibrium configurations, and reorganization energy is given as

$$E_s = \frac{1}{2} \sum_j \mu_j \omega_j^2 \left[(r_j^0)_i - (r_j^0)_f \right]^2 \qquad (7.64)$$

where μ_j and ω_j are the reduced mass and the frequency of the jth normal mode of the system, $(r_j^0)_i$ and $(r_j^0)_f$ are the dimensional equilibrium coordinates of the jth normal mode in the initial and final states of the solvent system, and D is a term having the dimension of frequency.

The term L in Eq. (7.63) is the electronic matrix element involving the initial state and the final state wave functions of the electron. KLJ also did not determine the electronic matrix element L. They separated out this matrix element from the vibrational overlap term in the second-order perturbation matrix element using the Condon approximation. The expression for the transition probability in the high-temperature limit reduces to the equation obtained by LDK,[3–10] though KLJ[26] took into account the frequency dispersion (i.e., the distribution of frequencies) in the solvent. To take into account the inner-sphere contribution, KLJ used the generalized nuclear coordinates of various normal modes. But this approach remains valid only when the vibrational quantum number of any normal mode is much greater than unity. This condition is met by the low-frequency solvent mode but not by the high-frequency inner-sphere mode at normal temperature.

Thus, to take into account the inner-sphere contribution, KLJ modified the approach where the total potential energy of the system was written as the sum of inner-sphere and the continuum solvent contribution. The inner-sphere part was estimated by considering only the symmetric breathing mode as that of George and Griffith (formulated at a higher level by Marcus) (see Section 4.3) on the nearest solvent molecule in the harmonic approximation. The continuum solvent contribution was estimated using the LD Hamiltonian. The transition probability was then derived using the above-mentioned method but giving no quantitative estimate that could be used in comparisons with experiment.

KLJ,[26] however, estimate quantitatively the contribution of the energy of activation both from the inner sphere and the continuum for the Fe^{2+}/Fe^{3+} system. To find the inner-sphere contribution they used the harmonic approximation and expressed the energy of the reorganization, inner sphere, E_c as

$$E_c = \tfrac{1}{2}\hbar\omega_c\Delta_c^2 \qquad (7.65)$$

where Δ_c is the difference in the dimensionless normal coordinate of the inner sphere in the oxidized and reduced species. This expression (7.65) is unclear in that it does not take the sum over all the inner coordinated ion–solvent bonds.

To find the continuum contribution to the energy of activation, KLJ modified the original expression of repolarization energy given by Marcus[1] and written in the form for the isotopic exchange reaction and for one electron transfer as

$$E_s = e_0^2\left(\frac{1}{\varepsilon_{op}} - \frac{1}{\varepsilon_s}\right)\left[\frac{z^2}{a_z} - \frac{(z^2-1)}{a_{z-1}} - \frac{1}{R}\right] \qquad (7.66)$$

where z and $(z-1)$ are the charges on the reduced and oxidized ions, respectively.

The terms a_z and a_{z-1} are the radii (along with first coordination shell) of the reduced and oxidized ion, respectively, and R is the distance of closest approach between the two ions, so that the $1/R$ term takes into account the repulsive interaction. The term ε_{op} is the optical dielectric constant of the medium.

It seems that although KLJ[26] used a more realistic expression in Eq. (7.66) for the free energy of reorganization of the continuum solvent system than that of Marcus[2] (Chapter 6), there is still doubt about its validity. Were the simple Born-charging expression applied in the original Libby manner without taking into account the activated state approach of Hush or the irreversible thermodynamics of Marcus, the expression would be:

$$E_s = e_0^2\left(\frac{1}{\varepsilon_{op}} - \frac{1}{\varepsilon_s}\right)\left[\frac{z^2}{a_z} - \frac{(z-1)^2}{a_{z-1}} - \frac{z(z-1)}{R}\right] \qquad (7.67)$$

KLJ gave the expression for activation energy in terms of the inner energy change E_c and the outer energy change E_s as

$$E_a = \frac{(Q - E_s - E_c)^2}{4(E_s + E_c)} \tag{7.68}$$

The numerical values of E_a were calculated from Eq. (7.68), where E_c and E_s were calculated from Eq. (7.65) and (7.66), respectively. The results agree fairly well with the experimental energies of activation for the systems considered in agreement with the finding of Bockris, Khan, and Matthews[42] that *when the stretching term is introduced, agreement can be obtained.*

Ulstrup and Jortner[27] advanced a general quantum mechanical formalism for the role of quantum modes of vibration on the electron transfer reaction where any bond breaking or bond formation is involved. They considered the electron donor–electron acceptor pair together and the entire polar solvent as a "supermolecule" undergoing a nonradiative decay process. The main result in the work of Ulstrup and Jortner[27] is that they found the activation energy *to be asymmetric with a slower decrease energy with increasing heat of reaction.* Their result does not resemble the experimental result for the dependence of free energy of activation on the free energy of reaction as plotted by Bockris *et al.*[40-42] (Fig. 6.10).

7.5.3. Schmidt's Treatment

Schmidt[28,29] examined the LDK model using a new and different treatment called the nonequilibrium statistical mechanical linear response formalism of Kubo,[39] which has been generalized by Yamamato[43] to find the rate constant of any chemical reaction. Schmidt used Yamamato's general expression[43] for the rate constants of a chemical reaction.

Schmidt[29,30] has shown that the linear response formulation yields the same results as obtained by Marcus[2] and by the Dogonadze[9,10] group when the same approximations involved in these treatments are introduced back into the Schmidt treatment. He claims this nonequilibrium statistical mechanical treatment gives rise to more general results compared to the Dogonadze[3,4,9,10] treatment. But one notices that the treatment involves some drastic approximations that appear to be unrealistic. Also, it would be very difficult to compare his final results with those of the experiment, because of the complex form of the final expressions. The author has not published computations from his final results.

However, apart from the mathematical obscurity of Schmidt's work the following approximations appear unrealistic. The first major approximation is that Schmidt[29] considered the harmonic binding of a transferring electron to the donor ion. This is unrealistic. For example, for electron transfer between a transition metal redox couple the transferring electron is mainly the 3d-level electron. The wave function of the 3d electron is quite different[16] from the harmonic oscillator wave function. If an electrode is a donor of an electron, the wave function of the electron in the metal is not harmonic either.[16]

Schmidt also used the dielectric continuum model for the solvent system, as did LDK. Moreover, the total Hamiltonian used by him[28] is inappropriate,[40] as it omits crucial terms that involve the important ion-solvent interaction.[26] It is desirable to write the Hamiltonian in the Schrödinger representation before proceeding to the second quantization formulation. In his formal treatment, Schmidt disregards the reverse reaction which seems inconsistent with the general formalism.

Schmidt[28a,b,29] attempted to introduce the effect of ionic diffusion in electron transfer reactions. However, ionic diffusion should not be of importance specifically for electron transfer reactions where there is no bond breaking or formation, since the electron transfer time is much shorter than the diffusion time of the ions. Of course, ionic diffusion could always become important at sufficiently high rates, but all the reactions to which Schmidt's calculations could be applied if they were amenable to numerical tests would be in a region in which diffusion would not play a role.

Schmidt[30] applied the Kubo response theory[39] to the electron transfer process at electrodes and gave an expression for the electrochemical exchange current i_0. In this treatment, Schmidt included the (1) inner-sphere vibronic interaction, (2) the effect of a reactant center of mass motion and transport to a most favorable electron exchange configuration, and (3) the effect of a strong exchange interaction. This last effect does not allow the use of the quantum mechanical time-dependent perturbation treatment, since the latter is applicable only for weak interactions.

Though the Schmidt treatment[30] involves complex mathematical sophistication and gives a more general expression for the rate constant and the exchange current, his expressions are cumbersome to use in a computation which would compare them with experiment. In the present stage of the continuum theory, where it cannot even represent Tafel's law, sophistication is not yet needed.

In another paper, Schmidt[31] used the response theory to extend the LDK model for outer-sphere electron transfer reactions (reactions without bond breaking or forming bonds) and includes the inner-sphere vibronic degree of freedom. The inner-sphere terms are considered in the harmonic approximation. It is further assumed[31] that there is no coupling between the inner-sphere vibrations and the neighboring solvent. This is a gross approximation. There should be in fact coupling between inner-sphere vibrations and the vibrations of the outside solvent system when they are augmented (see Section 6.10). The total wave function is then expressed as the product of the wave functions of the electronic, inner-sphere vibronic, and solvent subsystems. The rate expression derived in this manner is applicable for nonadiabatic reactions[31] primarily, though Schmidt claimed applicability also to adiabatic reactions.

The nature of the effect due to inclusion of the inner-sphere vibronic terms depends on whether the vibrational potential energy surfaces for the donor and acceptor species undergo severe displacement, distortion, or both, and also on whether the fundamental frequencies of the inner-sphere normal modes are greater or less than $4kT$. Three cases have been distinguished, namely:

(a) In the high-temperature limit ($\hbar\omega_0 \ll 4kT$), the inner-sphere effect enters through the preexponential part of the rate constant expression.

(b) In the low-temperature limit ($\hbar\omega_0 \gg 4kT$), the effect of the inner-sphere vibronic terms is negligible and only the continuum solvent modes contribute in the rate constant expression, in agreement with the claims of the LDK group.[3-10] If $\nu_{vib} = 1.1 \times 10^{14}$, as for $O—H^+$, the low-temperature condition would usually be applicable. On the other hand, for redox systems with, for example, $\nu = 1 \times 10^{13}$, the high-temperature limit would apply.[†]

(c) At both high- and low-temperature limits, if there is a strong distortion in the vibrational potential energy surfaces (as would be expected), the inner-sphere effect appears in both the preexponential and the exponential part of the rate expression.

Schmidt[31] has, in the continuum solvent system on the rate of electron transfer reaction, introduced a damping factor as a parameter in the solvent Hamiltonian. The damping processes result in a decrease in the activation energy of the process. It should be mentioned, however, that the modified LDK solvent Hamiltonian[3,4] takes both frequency dispersion and damping effects into account implicitly. The results obtained using this Hamiltonian should show the same tendencies as observed by Schmidt.[31]

7.6. Difficulties of the Continuum Theory

Methodological Difficulties

Before we see what the specific confrontations are that make acceptance of the continuum approach a difficult one, it is desirable to make

---

[†] Thus, the equation indicates a maximum probability of transition when $q^0_{inner\ \alpha} = q^0_{inner\ \beta}$, i.e., when there is *no* stretching. This of course is diametrically opposed to Bockris, Khan, and Matthews,[42] who found agreement with experiment in redox reactions only *if stretching* occurs.

some general observations concerning the nature of the theory:

(a) First, it is unusual to have a theory of a kinetic process in terms of a continuum model. Thus, for example, continuum theories of phenomena such as elasticity and the strength of materials are of considerable use, but the couching of a theory of a *kinetic* process in continuum terms is not met with in other fields.

(b) Correspondingly, the theory as presented hitherto has been largely nonmodelistic. It is true that after about 1971 some attempt has been made[17-31] to provide some mention of inner-sphere activation, but these explanations usually lack the treatment of molecular models related to physical entities known to be present.

(c) Although the solvent system given in the continuum theory has been treated with a harmonic approximation in a quantum mechanical way, no solution of the matrix element of transition for the electron is provided, and this is the heart of the quantum mechanical aspect of the theory. One of the main reasons for the absence of quantum mechanical solutions is that the authors have complained (cf. Schmickler[25]) that knowledge of the wave functions involved in the situation is insufficient for the purpose intended. Wave function for electrons in various states of ions has been given successfully by Yunta[44] and this can be fairly easily (cf. Khan *et al.*[16] and Chapter 13) modified to take into account perturbation due to the presence of the solvent when the ion is in solution.

(d) The continuum theory results are distinguished by a dearth of comparison with experiment of numerical results of the equations. Thus, even gross discrepancies between theory and experiment, such as the lack of ability to yield even a small section of linearity between log rate and overpotential, is seldom brought out and discussed in the literature.

(e) Conceptual errors inhibit acceptance of the work. Thus, the idea that there is no activation of higher energy levels from the surrounding solvent to the central ion arose because Levich and his colleagues did not comprehend that the energy differences between the vibrational levels in the $O-H$ bonds are not separated by quantum jumps of $\frac{1}{2}h\nu_{0,vib}$, but that the rotational modes produce in solution a certain smoothed-outness of levels, as is readily emphasized by the spectroscopic work on, e.g., H_3O^+ in solution.[46-48]

(f) Last, the concept of a few particularly active or energetic molecules as the reactants ("Arrhenius kinetics") is rejected, and a stress is laid upon a species having an average energy as the energy supplier.

Specific Difficulties

Although the continuum theory is highly developed, little discussion has been devoted to the molecular aspect. Those parts of the discussion that are so devoted sometimes seem to involve concepts which differ from the mainstream of thought in modelistic kinetics. Thus:

(a) There is a concept, mentioned above, that there cannot be any transfer of thermal energy from the solvent to the reacting ion. We show in Table 7.3 the spectra of water, whereby it can be seen that some of the water modes are of the order of thermal motions. Hence, there *is* a "continuous" aspect of the distribution of available thermal states in water,[44-46] i.e., modes of the order of kT. There is no reason to assume that there is not a transfer to the continuumlike water motion of the ion. Hence, thermal activation would be expected to contribute to the energy of inner bonds.

(b) The solvation shell of ions[49,50] has been neglected as a source of energy upon the ion in its activation. It is as though, in the electrostatic theories offered for the reaction between ions in the study of organo-chemical reactions, ion–dipole forces were neglected, the surrounding Born solvation energy only being treated, although the Born energy is less than half the total energy. Moreover, the interaction of the ion with its surrounding dipoles is charge dependent, so that there will be an effect in this energy upon charge transfer, even in a case (if any exist) where there is no ion-bond stretching.

(c) Neglect of *activated* bonds, particularly in bond-breaking reactions, and the rejection of the concept that in reactions only hot molecules react is unfamiliar to most persons engaged in kinetic mechanism investigations.

(d) In the original quantum mechanical theory of LDK,[3-10] the continuum solvent system outside the first solvation shell is represented by the *polaron model*.† The application of this model (which has its origin in solid state concepts) to electrolyte solution for the purpose of explaining long-range energy transfer is doubtful, as in a liquid there is no periodicity. Solvents like water have local order of only a temporary nature, and in the vicinity of dissolved ions this local order differs considerably from that in the bulk.[40,51] Very considerable damping of energy transfer by wet polarons would be expected.

(e) The Born–Landau Hamiltonian arises when the electronic adiabatic approximation is used to treat the problem of an electron trapped in a polar solvent. This approximation can be applied if the binding energy of the trapped electron is much smaller than the binding energy of the medium electrons. It is a satisfactory approximation in a *polar crystal* where the binding energy of the trapped electron has been shown by Markham and Seitz[52] and Fröhlich[14] to be of the order of 0.1 eV. With *polar liquids*, however, the situation is different. The binding energy of the

† The polaron is described in Chapter 3.

Table 7.3
Calculated and Observed Frequencies for Solutions of H₃O⁺ Salts in SO₂ and Crystalline H₉O₄Br⁻ [62]

Frequency	H_3O'/SO_2 Observed Infrared	H_3O'/SO_2 Observed Raman	H_3O'/SO_2 Calculated	D_2O^+/SO_2 Observed infrared	D_2O^+/SO_2 Calculated	$H_9O_4^+/Br^-$ Observed infrared	$H_9O_4^+/Br^-$ Calculated	$D_2O_4^+Br^-$ calculated
ν_3	3405	3415	3410	2580	2425	2060	2640	1880
ν_1	3470	3447	3455	2677	2550	2630	2060	1520
ν_2			785		593	1313	1320	985
ν_4	1670	1700	1690		1230	1845	1840	1320
ν_L						902	870	615
ν_L						738	920	660
ν_r						569	405	400
ν_r						556	570	550
ν_b						78	97	97

medium electrons (4–5 eV) is of the same order of magnitude as the binding energy of the trapped electrons (1 eV);[51] the electronic adiabatic approximation thus becomes less than valid.

7.6.1. Fluctuation in Continuum Theory

We have mentioned that a part of the conceptualization of the continuum theory of electron transfer reaction rests on the probability of a fluctuation in the solvent polarization energy to supply an energy of activation of the order of 1.0 eV. The frequency of this fluctuation must be consistent with experiment. The frequency of the abnormal perturbation of librators has remained undiscussed until recent time. However, Bockris and Sen[38] made an attempt to compute, with classical statistics, the probability of fluctuation of solvent polarization energy to supply free energy of activation of the order of 1.0 eV. Their procedure was as follows:

The continuum theory[3,4,9,10] assumes that an electrostatic fluctuation in the ion–solvent polarization interaction is the origin of activation. Thus, one can consider a situation where an ion of radius r_1 (including the diameter of a water molecule in the inner solvation sheath) and charge e_0 is surrounded by N water dipoles. The interaction energy between the ion and the dipole is

$$E = -\frac{e_0\mu \cos \theta}{\varepsilon_s R^2} \tag{7.69}$$

where μ is the dipole moment, θ is the angle the dipole makes with the field direction, and R is the distance between the ion and the dipole. ε_s is the static dielectric constant of the medium.

The number of dipoles in a shell at $\mathbf{R}$ and $\mathbf{R} + d\mathbf{R}$ from the ion is

$$N = \frac{4\pi \mathbf{R}^2 \, d\mathbf{R}}{\frac{4}{3}\pi r_\omega^3} = \frac{3\mathbf{R}^2 \, d\mathbf{R}}{r_\omega^3} \tag{7.70}$$

where r_ω is the radius of the water dipole.

Let r_∞ be the radius of a circular area around the ion containing the N dipoles. Then

$$N = \int_{r_i}^{r_\infty} \frac{3}{r_\omega^3} \mathbf{R}^2 \, d\mathbf{R} = \frac{1}{r_\omega^3}(r_\alpha^3 - r_i^3) = \frac{r_\infty^3}{r_\omega^3} \tag{7.71}$$

since $r_\infty \gg r_i$.

Hence,

$$r_\infty = N^{1/3} r_\omega \tag{7.72}$$

As a zeroth approximation (neglect of lateral interaction), the number of dipoles $dN(\theta, \mathbf{R})$ between $\mathbf{R}$ and $\mathbf{R} + d\mathbf{R}$ from the ion having an angle between θ and $\theta + d\theta$ is[†]

$$dN(\theta, \mathbf{R}) = \frac{4\pi \mathbf{R}^2 \, d\mathbf{R}}{\frac{4}{3}\pi r_\omega^3} \exp\left(\frac{e_0\mu \cos \theta}{\varepsilon_s \mathbf{R}^2 kT}\right) \frac{2\pi \sin \theta \, d\theta}{4\pi}$$

$$= \frac{3}{2r_\omega^3} \exp\left(\frac{e_0\mu \cos \theta}{\varepsilon_s \mathbf{R}^2 kT}\right) \mathbf{R}^2 \, d\mathbf{R} \sin \theta \, d\theta \tag{7.73}$$

Therefore, the probability that a dipole has an angle between θ and $\theta + d\theta$ with an ion at a distance between $\mathbf{R}$ and $\mathbf{R} + d\mathbf{R}$ from the ion is

$$P_i = \frac{dN(\theta, \mathbf{R})}{N} = \frac{(3/2r_\omega^3) \exp(e_0\mu \cos \theta / \varepsilon_s \mathbf{R}^2 kT) \mathbf{R}^2 \, d\mathbf{R} \sin \theta \, d\theta}{\int_0^\pi \int_{r_i}^{r_\infty} (3/2r_\omega^3) \exp(e_0\mu \cos \theta / \varepsilon_s \mathbf{R}^2 kT) \mathbf{R}^2 \, d\mathbf{R} \sin \theta \, d\theta}$$

$$\tag{7.74}$$

Equation (7.74) is for noninteracting dipoles. However,[54] one represents water as a system of noninteracting dipoles[§] having the effective dipole moment μ_{eff} given as[¶]

$$\mu_{\text{eff}} = 1 + g \overline{\cos \gamma} \tag{7.75}$$

where g is the number of neighboring water molecules around a water molecule, and $\overline{\cos \gamma}$ is the average of the cosines of the angles between dipole moments of the central water molecule and those of its neighbors. In evaluating $g \overline{\cos \gamma}$, Kirkwood[54] considered the nearest neighbors, but Pople[55] extended Kirkwood's[54] treatment and considered the contribution of first and second layers of water molecules, obtaining $\mu_{\text{eff}} = 1.53\mu_\omega$ as the effective dipole moment of water in water.

Equation (7.74) becomes

$$P_i = \frac{(3/2r_\omega^3) \exp(e_0\mu_{\text{eff}} \cos \theta / \varepsilon_s \mathbf{R}^2 kT) \mathbf{R}^2 \, d\mathbf{R} \sin \theta \, d\theta}{\int_0^\pi \int_{r_i}^{r_\infty} (3/2r_\omega^3) \exp(e_0\mu_{\text{eff}} \cos \theta / \varepsilon_s \mathbf{R}^2 kT) \mathbf{R}^2 \, d\mathbf{R} \sin \theta \, d\theta} \tag{7.76}$$

The average value of the interaction energy between the ion and the dipole is thus

$$\langle E \rangle = \frac{\displaystyle\int_0^\pi \int_{r_i}^{r_\infty} (3/2r_\omega^3)(-e_0\mu_{\text{eff}} \cos \theta / \varepsilon_s \mathbf{R}^2) \mathbf{R}^2 \, d\mathbf{R}}{\displaystyle\int_0^\pi \int_{r_i}^{r_\infty} (3/2r_\omega^3) \exp(e_0\mu \cos \theta / \varepsilon_s \mathbf{R}^2 kT) \mathbf{R}^2 \, d\mathbf{R} \sin \theta \, d\theta} \quad \times \exp(e_0\mu_{\text{eff}} \cos \theta / \varepsilon_s \mathbf{R}^2 kT) \sin \theta \, d\theta \tag{7.77}$$

[†] The energy to be considered is the ion–dipole energy. It is not the solvation energy, i.e., the energy needed to transfer from a vacuum, but from infinity in the solution.

[§] The dipoles only interact electrostatically but not by specific short-range molecular forces.

[¶] The value of the dielectric constant to be used in these expressions is that for a time average evaluation, i.e., 80.

The integrals of Eq. (7.77) can be evaluated to give

$$\langle E \rangle = -\frac{1}{N}\frac{e_0^2\mu_{\text{eff}}^2}{\varepsilon_s^2 r_\omega^3 r_i kT} \tag{7.78}$$

The total average interaction energy is then

$$\langle U \rangle = N\langle E \rangle = -\frac{e_0^2\mu_{\text{eff}}^2}{\varepsilon_s^2 r_\omega^3 r_i kT} \tag{7.79}$$

Following a similar procedure, we can evaluate the average square of the energy,

$$\langle U^2 \rangle = \frac{e_0^2\mu_{\text{eff}}^2}{\varepsilon_s^2 r_\omega^3 r_i} \tag{7.80}$$

The mean square deviation, then, is given as

$$\sigma^2 = \langle U^2 \rangle - \langle U \rangle^2 \tag{7.81}$$

One can use the Gaussian distribution to evaluate the probability of solvent polarization fluctuation to give rise to an amount of energy $(U - \langle U \rangle)$ as

$$P_{\text{fluctuation}} = \frac{1}{(2\pi)^{1/2}}\exp\left(-\frac{(U-\langle U \rangle)^2}{2\sigma^2}\right) \tag{7.82}$$

For the proton-discharge reaction, a typical measured activation energy at the reversible potential is 20 kcal mole^{-1}. Thus, we need a fluctuation $(U - \langle U \rangle) = 20$ kcal mole^{-1}. Thus, using Eq. (7.82), $P_{\text{fluctuation}}$ comes out to be 10^{-41}. Thus, the probability of a fluctuation in solvent polarization energy due to the organization together of many solvent librators is too small to supply the energy of activation for the typical electron transfer reaction.

7.6.2. Comparison of the Continuum Theory with Experiment

It has been noted above that the continuum theory has been little compared with experiment, but this has recently been done by Bockris *et al.*[40-42] One finds a nonlinear log current–potential relation[62,63] from the continuum solvent fluctuation model (Fig. 7.5), since a parabolic relation between energy and displacement is assumed in the model. The motions supposed to be the origin of the ion's energy are small perturbations in the libratory oscillations of the solvent molecules far from the ion, and these must be harmonic, so that introduction of anharmonicity to the present continuum model (often quoted as a solution to the problem of a varying β) is not very plausible. Thus, the continuum theory remains saddled with a lack of consistence with the first law of electrode kinetics, i.e., Tafel's law.

Table 7.4
Predictions of the Continuum (C) Model (Based on Developments of Libby's Theory) and the
Molecular (M) Model (Based on Developments of Gurney's Theory)

Test	Comments
1. (a) Current density is exponentially proportional to the overpotential.	M predicts experiment. C shows continuous curvature (Fig. 7.5).
(b) Variation of β with overpotential.	M gives negligible variation over 1.5 V for HER and thus nearly agrees with experimental free energy of activation. C shows continuous variation (Fig. 7.6).
2. Magnitude of separation factor on different metals.	
3. Variation of the separation factor with potential.	No calculation done on C model. M model reproduces experiment for high η.
4. For reaction that involves adsorbed intermediates, the current is a function of the heat of adsorption of one of the intermediates involved in the rate-determining step.	M reproduces $dS/d\eta$ better than C (Fig. 7.7). For HER both models are consistent with experiment (Fig. 7.8).
5. The rate is a function of the solvent dielectric constant.	Predicted well by M model and not by C model (Fig. 7.9).
6. Plot of ΔF^*_{expt} versus ΔF^*_{calc}.	No correlation on C theory (Fig. 7.10).

The results of continuum theory originated by Libby[56] and the molecular model originated by Gurney[15] are compared with experiment[40-42] and given in Table 7.4 and Figs. 7.5–7.10. Some other comparisons of continuum theory with experiment were shown in Figs. 6.10–6.16. These comparisons indicate significant shortcomings in the results of the continuum approach when compared with experiment.

7.7. Summary

The continuum theory began with a simple qualitative suggestion by Libby[56] in terms of the changes in the Born energy that would occur when a charge was annihilated on an ion. Concepts associated with the continuum theory grew via the paper of Weiss[57] and were then developed first by R. A. Marcus,[1,2] then by Levich, and more recently by Dogonadze and Kutznetsov. In inorganic chemistry, it is usual to refer to "the Marcus theory" and to speak of "outer-sphere reactions," although there seems at present a lack of experimental evidence that there are any reactions, the rate of which is dependent only on the electrical properties of the sphere outside the primary solvation sheath.

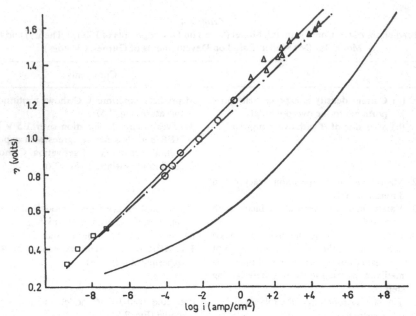

Fig. 7.5. Tafel lines from thermal and electrostatic approaches compared with experiment. From the experiment of Nürnberg[65] ($\triangle$); Bockris and Azzam[62] ($\bigcirc$); Bowden and Grew[64] ($\square$); Despic and Bockris[63] ($-\cdot-$) (calculated with $ax_0 = 10$, $D_1 = 100$ kcal mole^{-1}, $D_2 = 45$ kcal mole^{-1}); from the continuum theory of Levich[9] ($---$) ($E_s = 2$ eV).

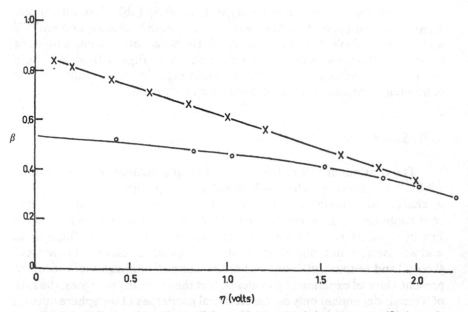

Fig. 7.6. Variation of transfer coefficient with potential; ($\times$) electrostatic, ($\bigcirc$) thermal.

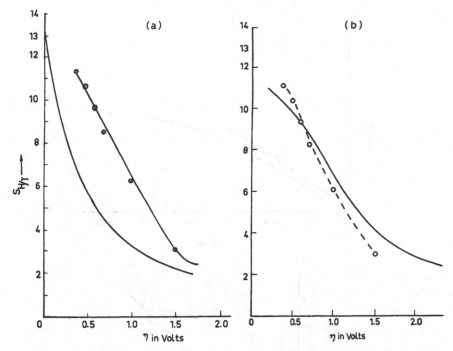

Fig. 7.7. Variation of separation factor with (a) potential–electrostatic approach, potential–thermal approach. The circles give experimental data and the solid curves give the theoretical results, the Levich theory ($E_s = 2$ eV) for (a) and Bockris and Mathews theory for (b).

Correspondingly, in the recent book on elementary electrode kinetics by Vielstich and Schmickler,[58] the continuum theory is presented (briefly) as the only type of quantum mechanical approach to electrode kinetics, although it does not give rise to Tafel's law. It is as though there had been a theory at a high quantum mechanical level of reactions in the gas phase that did not yield consistency with the Arrhenius exponential law between rate and temperature, but which nevertheless was accepted and extended while competing theories that predict experiment were less used.†

It seems reasonable, in the early retrospect being given here, to see the development of this dielectric continuum theory of solution kinetics as an example of overspecialization. The high degree of mathematical sophistication possible has been an attraction to those to whom the demands of

---

† Historically, it is remarkable to note the slow progress of the Gurney quantal theory on the rate of electron transfer at the interface, among the first three or four bases of quantum chemistry, but which was effectively dropped in the 1950s in favor of the Born-continuum-oriented work (this being named "quantum mechanical").

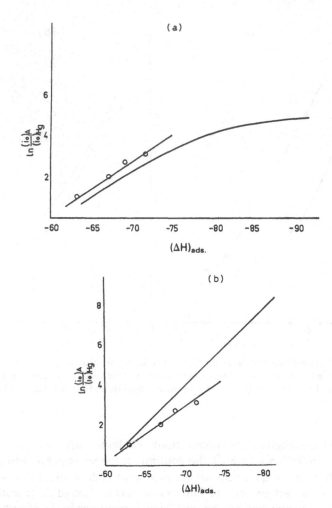

Fig. 7.8. Variation of $\ln[(i_0)_A/(i_0)_{Hg}]$ with the heat of adsorption for the (a) electrostatic approach; (b) thermal approach. The circles give experimental data and the solid curves give the theoretical results.

comparison with experimental trends and parameters have been less important. The alternative molecular models, developed in a less advanced mathematical way, seem less attractive to workers from mathematical physics. Figures 7.5–7.10 bring into doubt the path in theoretical electrode kinetics begun by Libby. One might end this chapter by asking whether that continuum path has already exhausted its usefulness.

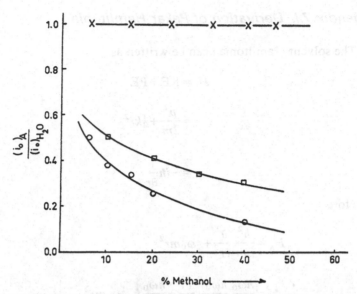

Fig. 7.9. Variation of $\ln[(i_0)_A/(i_0)_{H_2O}]$ with the variation of methanol concentration. □, Thermal approach; ×, electrostatic approach; ○, experimental.

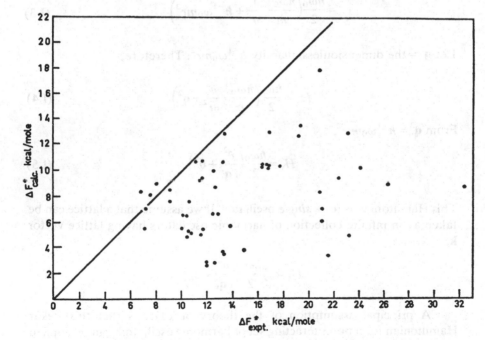

Fig. 7.10. Plot of ΔF_{calc}^* (from electrostatic approach) against ΔF_{expt}^* for electron transfer reactions.

Appendix 7.I. Derivation of Pekar Hamiltonian

The solvent Hamiltonian can be written as

$$H_s = KE + PE$$

$$= \frac{p^2}{2m} + \tfrac{1}{2}kr^2 \tag{I.1}$$

$$p = -i\hbar\frac{\partial}{\partial r} \tag{I.2}$$

Therefore,

$$H_s = \frac{\hbar^2}{2m}\frac{\partial^2}{\partial r^2} + \tfrac{1}{2}\omega_0^2 mr^2$$

$$= \frac{\hbar\omega_0}{2}\frac{\hbar\omega_0^{-1}}{m}\frac{\partial^2}{\partial r^2} + \frac{\hbar\omega_0}{2}\hbar^{-1}\omega_0 mr^2$$

$$= \frac{\hbar\omega_0}{2}\left(\frac{\hbar\omega_0^{-1}}{m}\frac{\partial^2}{\partial r^2} + \hbar^{-1}\omega_0 mr^2\right) \tag{I.3}$$

Let $\mathbf{q}^2 =$ the dimensionless quantity $\hbar^{-1}\omega_0 mr^2$. Therefore,

$$H_s = \frac{\hbar\omega_0}{2}\left(\frac{\hbar\omega_0^{-1}}{m}\frac{\partial^2}{\partial r^2} + \mathbf{q}^2\right) \tag{I.4}$$

From $\mathbf{q}^2 = \hbar^{-1}\omega_0 mr^2$,

$$H_s = \frac{\hbar\omega_0}{2}\left(\frac{\partial^2}{\partial \mathbf{q}^2} + \mathbf{q}^2\right) \tag{I.5}$$

This Hamiltonian is for a *single* oscillator. If we assume that a lattice can be taken as an infinite collection of harmonic oscillators having lattice vector **k**,

$$H_s = \sum_{\mathbf{k}} \frac{\hbar\omega_0}{2}\left(\frac{\partial^2}{\partial \mathbf{q_k}^2} + \mathbf{q_k}^2\right) \tag{I.6}$$

A principal assumption of the theory of LDK is that this Pekar Hamiltonian for a noninteracting set of harmonic oscillators can be applied to liquids. This was the assumption of Mie (1903)[14b] and is not accepted in modern theories of liquids.

References

1. R. A. Marcus, *J. Chem. Phys.* **24**, 966 (1956); *Can. J. Chem.* **37**, 138 (1959); *J. Phys. Chem.* **67**, 853, 2889 (1963).
2. R. A. Marcus, *J. Chem. Phys.* **43**, 679 (1965).
3. V. G. Levich and R. R. Dogonadze, *Dokl. Akad. Nauk. SSSR* **124**, 123 (1959).
4. V. G. Levich and R. R. Dogonadze, *Coll. Czech. Chem. Commun.* **26**, 293 (1961).
5. R. R. Dogonadze and Y. A. Chizmudzher, *Dokl. Akad. Nauk. SSSR* **144**, 1077 (1962).
6. R. R. Dogonadze, A. M. Kuznetsov, and Y. A. Chizmadzher, *Zhur. Fiz. Khim.* **38**, 1195 (1964).
7. V. G. Levich, *Recent Advances in Electrochemistry*, Volume 4, Chapter 5, Interscience, New York (1966).
8. R. R. Dogonadze, A. M. Kuznetsov, and V. G. Levich, *Dokl. Akad. Nauk SSSR* **188**, 383 (1969).
9. V. G. Levich, *Physical Chemistry; An Advanced Treatise* (H. Eyring, D. Henderson, and W. Jost, eds.), Volume 9B, Chapter 12, Academic Press, New York (1970).
10. R. R. Dogonadze, in *Reactions of Molecules at Electrodes* (N. S. Hush, ed.), Chapter 3, Wiley (Interscience), New York (1971).
11. C. Kittel, *Solid State Physics*, Wiley (Interscience) (1971).
12. J. Bernal and R. Fowler, *J. Chem. Phys.* **1**, 515 (1933).
13. S. I. Pekar, *Investigations of Electronic Theories and Crystals*, Fitmatgiz, Moscow (1951).
14a. H. Fröhlich, *Advan. Phys.* **3**, 325 (1954).
14b. A. Mie, *Ann. Phys.* **11**, 657 (1903).
15. R. W. Gurney, *Proc. Roy. Soc. (London)* **A134**, 137 (1931).
16. S. U. M. Khan, P. Wright, and J. O'M. Bockris, *Electrokhim.* **13**(6), 914–923 (1977).
17. R. R. Dogonadze, A. M. Kuznetsov, and M. A. Vorotyntsev, *Phys. Stat. Solidi.* **54**, 125 (1972).
18. M. A. Vorotyntsev, R. R. Dogonadze, and A. M. Kuznetsov, *Electrokhim.* **10**, 687 (1974).
19. M. A. Vorotynsev, R. R. Dogonadze, and A. M. Kuznetsov, *Electrokhim.* **10**, 887 (1974).
20. R. R. Dogonadze and A. M. Kuznetsov, *Progr. Surface Sci.* **6**, 1 (1975).
21. R. R. Dogonadze and A. M. Kuznetsov, *Electrokhim.* **11**, 3 (1975).
22. R. R. Dogonadze and A. M. Kuznetsov, *J. Electroanal. Chem.* **65**, 545 (1975).
23. R. R. Dogonadze and A. M. Kuznetsov, *Electrokhim.* **3**, 324 (1967).
24. W. Schmickler and W. Vielstich, *Electrochim. Acta* **18**, 883 (1973).
25. W. Schmickler, *Ber. Bunsenges Phys. Chem.* **77**, 991 (1973).
26. N. R. Kestner, J. Logan, and J. Jortner, *J. Phys. Chem.* **78**, 2148 (1974).
27. J. Ulstrup and J. Jortner, *J. Chem. Phys.* **63**, 4358 (1975).
28a. P. P. Schmidt, *J. Chem. Phys.* **56**, 2775 (1972).
28b. P. P. Schmidt, *J. Chem. Phys.* **57**, 3749 (1974).
29. P. P. Schmidt, *J. Chem. Phys.* **58**, 8389 (1973).
30. P. P. Schmidt and H. B. Mark, Jr., *J. Chem. Phys.* **58**, 429 (1973).
31. P. P. Schmidt, *J. Chem. Soc., Faraday Trans. II* **69**, 122, 1104 (1973).
32. R. R. Dogonadze and A. M. Kuznetsov, *Electrochim. Acta* **19**, 961 (1974).
33. M. Lax, *J. Chem. Phys.* **20**, 1752 (1952).
34. P. P. Schmidt, *J. Phys. Chem.* **78**, 1684 (1974).
35. W. Schmickler and W. Vielstich, *Electrochim. Acta* **19**, 963 (1974).
36. W. Schmickler, *Electrochim. Acta* **21**, 161 (1976).
37. W. Schmickler, *Ber. Bungen Ges.* **80**, 834 (1976).
38. J. O'M. Bockris and R. R. Sen, *J. Res. Inst. Catal. Hokkaido Univ.* **21**, 55 (1973); *Molec. Phys.* **29**, 357 (1975).
39. R. Kubo, *J. Phys. Soc. Japan* **12**, 570 (1957).

40. A. J. Appleby, J. O'M. Bockris, and R. K. Sen, *MTP International Review of Science* (J. O'M. Bockris, ed.), Volume 6, Butterworths, London (1973).
41. J. O'M. Bockris, R. K. Sen, and K. L. Mittal, *J. Res. Inst. Catal., Hokkaido Univ.* **20**, 153 (1972).
42. J. O'M. Bockris, S. U. M. Khan, and D. M. Matthews, *J. Res. Inst. Catal., Hokkaido Univ.* **22**, 1 (1974).
43. T. Yamamoto, *J. Chem. Phys.* **33**, 281 (1960).
44. J. Yunta, E. R. Mayquez, and C. S. Rio, *Phys. Rev.* **4**, 1483 (1974).
45. E. A. Moelwhyn-Hughes, *Physical Chemistry*, Pergamon Press, Oxford (1964).
46. M. Falk and P. A. Giguerre, *Can. J. Chem.* **35**, 1195 (1957).
47. M. Falk and T. A. Ford, *Can. J. Chem.* **44**, 169 (1966).
48. R. E. Moore, O. Ferral, G. W. Koeppl, and A. J. Kresge, *J. Am. Chem. Soc.* **93**, 1 (1971).
49. J. O'M. Bockris and P. P. S. Saluja, *J. Phys. Chem.* **76**, 2140 (1972).
50. P. P. S. Saluja, Ph.D. Thesis, University of Pennsylvania, Philadelphia, Pa. (1972).
51. W. Schmickler, *Electrochim. Acta* **70**, 12, 163 (1975).
52. J. R. Markham and F. Seitz, *Phys. Rev.* **74**, 1014 (1948).
53. J. Jortner, *Radiat. Res. Suppl.* **4**, 24 (1964).
54. J. G. Kirkwood, *J. Chem. Phys.* **4**, 592 (1936).
55. A. J. Pople, *Proc. Roy. Soc.* **A205**, 163 (1951).
56. W. Libby, *J. Phys. Chem.* **56**, 863 (1952).
57. J. Weiss, *Proc. Roy. Soc.* **A222**, 128 (1959).
58. W. Vielstich and W. Schmickler, *Electrochemie Kinetik elektrochemischer Systeme*, Dr. Dietrich Steinkopff Verlag, Darmstadt (1976).
59. J. O'M. Bockris and D. B. Matthews, *J. Chem. Phys.* **44**, 298 (1966).
60. J. O'M. Bockris, P. Parsons, and H. Rosenberg, *Trans. Faraday Soc.* **47**, 766 (1951).
61. F. Bötcher, *Dielectric Polarization* (2nd ed.), p. 161, Elsevier, Amsterdam (1952).
62. J. O'M. Bockris and A. M. Azzam, *Trans. Faraday Soc.* **48**, 145 (1952).
63. A. R. Despic and J. O'M. Bockris, *J. Chem. Phys.* **32**, 389 (1960).
64. F. P. Bowden and K. W. Grew, *Discuss. Faraday Soc.* **2**, 81, 91 (1947).
65. N. Nürnberg, quoted by V. Levich.[9]

8

Interfacial Electron Tunneling

8.1. Introduction

A complete structure of quantum mechanics materialized in the brief period between 1925 and 1928 with the development of matrix mechanics by Heisenberg in 1925 and wave mechanics by Schrödinger in 1926.[1,2] Gamow[1-3] in 1928 first put forward the theory of radioactive α particle decay on the basis of quantum mechanics. It was made independently by Ronald Gurney and Condon[4] in 1929. The theory of radioactive decay provides one of the more important examples of the application of the idea of tunneling; i.e., it concerns the escape of α particles from a nucleus when the total energy of the particle is less than its potential energy of binding to the nucleus.

Two years after the publication of the theory of the decay of the nucleus, Ronald Gurney[5] in 1931 gave his *pioneering quantum mechanical theory of electrochemical kinetics* concerning the tunneling of an electron from its bound states in the metal through the electrode–solution interface to an ion in solution.

The quantum mechanical theory of electrode processes, pioneered at such an early time in the history of quantum mechanics by Gurney, did not get much attention from electrochemists for about three decades, until the papers of Gerischer in the early 1960s. This was partly due to the confusion brought about by a simplifying assumption in Gurney's theory that there is no interaction of either the ion or the atom with the metal at the interface. This made the calculated activation energy too high (pointed out by Butler[6]), and it seems that this discrepancy with experiment discredited the theory. Electrochemists, with the negligible knowledge of quantum mechanics spread among them in the generation prior to the 1960s, took the discrepancy to mean that the quantum mechanical tunneling concept was not justified in its application to electrodes. "Gurney's theory is wrong" was the overreaction to Butler's correction.[6] This misunderstanding held basic electrode kinetic theory back for a generation. It encouraged the growth of concepts[7,8] often referred to as quantum mechanical, but in which central parts of the quantum aspects were often left unevaluated or reduced to their classical analogs.

In this chapter, we will introduce the general theory of the transfer of particles through a potential barrier when the total energy of a particle is less than that of the potential barrier. Using this theory, we will formulate the tunneling probability expression for different kinds of potential (barrier) shapes on the basis of a particular approximation for the solution of the Schrödinger equation. Finally, we will explain the basics of Gurney's theory of electron tunneling at the interface applied to electrode processes and develop the subsequent use of the tunneling theory in modern times.

8.2. Solution of Schrödinger Equation: Particle at a Rectangular Potential Barrier

In this section we shall solve the Schrödinger equation for a particle in a simplified prototype situation relevant to tunneling, such as a particle passing through a rectangular potential barrier, when the kinetic energy of the particle is less than the potential energy it has within the barrier. We consider the one-dimensional rectangular potential illustrated in Fig. 8.1, which is expressed as [9,10]

$$U(x) = 0, \qquad x < 0 \qquad \text{in region I}$$

$$U(x) = U_0, \quad 0 < x < a \qquad \text{in region II} \qquad (8.1)$$

$$U(x) = 0, \qquad x > a \qquad \text{in region III}$$

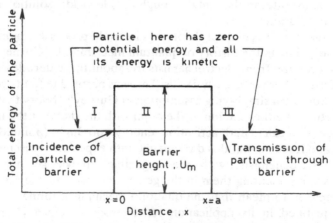

Fig. 8.1. A schematic diagram of a rectangular potential energy barrier for a tunneling particle. In regions I and III, the particle has total energy equal to kinetic energy. In region II, the particle experiences a potential energy barrier of height U_m and width a. Within the barrier, the potential energy is constant.

Classically, a particle of total energy E in region I incident upon the barrier from the left in the direction of increasing x will have unit probability of being reflected back if $E < U_0$. That is, the probability of entering into the potential region II and then coming out in region III will be zero. There will be unit probability of transmission if $E > U_0$; i.e., there will be zero probability of reflection. But neither of these results is obtained in quantum mechanics.

The Schrödinger equation for the particle in the potential energy regions I and III can be written in one dimension as[9,10]

$$-\frac{\hbar^2}{2m}\frac{d^2\psi(x)}{dx^2} = E\psi(x) \tag{8.2}$$

and that in region II of potential energy U_0 is

$$\frac{\hbar^2}{2m}\frac{d^2\psi(x)}{dx^2} = (U_0 - E)\psi(x) \qquad \text{for } E < U_0 \tag{8.3}$$

In regions I and III, the Schrödinger equation (8.2) is that of a free particle of total energy E (which is its kinetic energy, since in regions I and III the particle has zero potential energy). Equation (8.2) has the solutions[9,10]

$$\psi_I(x) = A\,\exp(ik_1x) + B\,\exp(-ik_1x) \qquad \text{in region I} \tag{8.4}$$

$$\psi_{III}(x) = C\,\exp(ik_1x) + D\,\exp(-ik_1x) \qquad \text{in region III} \tag{8.5}$$

where

$$k_1 = (2mE/\hbar^2)^{1/2} \tag{8.6}$$

is the wave number of the particle in terms of its energy E and mass m. The two terms, $\exp(ik_1x)$ and $\exp(-ik_1x)$ of Eqs. (8.4) and (8.5) describe an incident particle that moves from left to right and right to left, respectively.

There is important physical meaning in the two terms $\exp(ik_1x)$ and $\exp(-ik_1x)$ of Eq. (8.4). Thus, the state of the wave that represents the existence of the particle in region I is not that of a simple situation in which a particle proceeds in a vacuum without interaction with its surroundings. In fact, the particle in region I would have experienced, in a quantum mechanical sense, the presence of the barrier. One of the effects of the barrier is to reflect the particles from the barrier back in the direction from *right to left.* The second term in Eq. (8.4), that with $B\,\exp(-ik_1x)$, represents those particles that are traveling not from left to right but from right to left, owing to reflection at the barrier.

Similarly, in Eq. (8.5), where the wave function is given for particles in region III, the first term represents the particles that go from left to right.

However, the second term here from region III only represents the formal possibility that particles can go from right to left. As there is postulated to be no further barrier in region III, there is no actual movement of particles from right to left in region III, and the second term in Eq. (8.5) is thus taken as zero.

In region II, the Schrödinger equation (8.3) for $E < U_0$ has the solution†

$$\psi_{II}(x) = F \exp(+k_2 x) + G \exp(-k_2 x) \qquad (8.7)$$

where

$$k_2 = \left[\frac{2m(U_0 - E)}{\hbar^2} \right]^{1/2} \qquad (8.8)$$

is the wave number of the particle in region II. In region II, the first and the second terms in Eq. (8.7) represent the probability amplitudes of the waves moving from the left to right and from right to left, respectively. Thus, one observes that there is a finite probability amplitude, $\psi_{II}(x)$ and $\psi_{III}(x)$ that the particle will be found in the classically forbidden region II and III, even when the particle total energy $E < U_0$.

To obtain the probability that the particle is transmitted through region II into region III when it strikes the barrier in the boundary of regions I and II, the idea of probability flux is introduced. This is defined as the velocity of the incident particle times the probability of the existence of the particle moving from left to right in region I (for the incident wave). For the transmitted wave, the flux is defined as the velocity of the transmitted particle times the probability of the existence of the particle moving from left to right in region III. Thus, flux means flow, and it seems reasonable to define the probability of transmission P_T as[10]

$$P_T = \frac{\text{(velocity of the particle in region III)} \times \text{(probability of the existence of the transmitted particle moving from left to right in region III)}}{\text{(velocity of the particle in region I)} \times \text{(probability of the existence of the incident particle moving from left to right in the region I)}}$$

$$= \frac{v_{III} |\psi_{III}(x)|^2}{v_I |\psi_I(x)|^2} \qquad (8.9)$$

Since in regions I and III the total energy of the particle is the same and equal to E, and this E is entirely kinetic energy, the velocity in the region I, v_I is equal to the velocity in the region III v_{III}; i.e., $v_I = v_{III}$

† We have not considered the case for $E > U_0$, since we are interested in the tunneling situation when $E < U_0$.

[E = total energy = kinetic energy in the region I and III, since the potential energy in these regions is zero (see Fig. 8.1)]. This is termed *elastic tunneling*.

Note that the velocity of the *tunneled* particle in region III has the same kinetic energy as in region I. Physically, this means that the particle that has passed through the barrier has the same kinetic energy (goes at the same speed) as before it struck the barrier. Neither before or afterward does it have any potential energy. Of course, this strikes one as unreal, for one would expect physically that the particle would do work on passing through the barrier and thus convert some of the kinetic energy it had before it struck the barrier into potential energy, which would have interacted with the atoms that form whatever is the physical version of the barrier in the problem concerned. What the tunneling theory says is that, from the physical point of view, some particles get through and some particles do not. The ones which get through, i.e., the tunneling particles, are assumed to have the same energy on the right in region III as on the left in region I.

The reasoning as to why particles can pass through a barrier and come out on the other side with the same energy, i.e., no work apparently done, comes from the basic assumptions about the wave nature of particles. Considering now a beam of light when it strikes a medium (i.e., a barrier), it is reduced in intensity, but the light that comes through (so long as there has been no absorption, i.e., only elastic scattering) is unchanged in wavelength, i.e., the quanta have the same energy. Likewise, the particles (which have a wave nature!) can also tunnel through a barrier and can *have the same energy before and after tunneling* (zero energy loss on passing through the barrier).

Now, using Eq. (4.31), we get

$$P_T = \frac{|\psi_{III}(x)|^2}{|\psi_I(x)|^2} = \frac{|C|^2}{|A|^2} = \frac{C^*C}{A^*A} \tag{8.10}$$

where C^* and A^* are the complex conjugates of C and A, respectively. The coefficient C can be found in terms of A as follows:

For the continuity (i.e., equality) of the wave functions $\psi_I(x)$ and $\psi_{II}(x)$ at the boundary of the region I and II (Fig. 8.1) and their first derivatives $d\psi_I(x)/dx$ and $d\psi_{II}(x)/dx$ at the same boundary at $x = 0$ (i.e., at the beginning of the barrier) we get from Eqs. (8.4) and (8.7) that

$$A + B = F + G \tag{8.11}$$

$$ik_1(A - B) = k_2(F - G) \tag{8.12}$$

Likewise, for the continuity (i.e., equality) of the wave functions $\psi_{II}(x)$ and $\psi_{III}(x)$ and their first derivatives $d\psi_{II}(x)/dx$ and $d\psi_{III}(x)/dx$ at $x = a$ (namely, at the other boundary of the barrier of Fig. 8.1) we get from Eqs. (8.5) and (8.7) that

$$F \exp(+k_2 a) + G \exp(-k_2 a) = C \exp(ik_1 a) \qquad (8.13)$$

$$k_2[F \exp(+k_2 a) - G \exp(-k_2 a)] = ik_1 C \exp(ik_1 a) \qquad (8.14)$$

When Eq. (8.14) is divided by k_2 on both sides and the result is added to and subtracted from (8.13), we get two equations:

$$F = \frac{1}{2} C \left(1 + \frac{ik_1}{k_2}\right) \exp(ik_1 - k_2)a \qquad (8.15)$$

$$G = \frac{1}{2} C \left(1 - \frac{ik_1}{k_2}\right) \exp(ik_1 + k_2)a \qquad (8.16)$$

Similarly, if Eq. (8.12) is divided by k_2 and the result is added or subtracted from Eq. (8.11), one obtains

$$F = \frac{1}{2} \left[A \left(1 + \frac{ik_1}{k_2}\right) + B \left(1 - \frac{ik_1}{k_2}\right) \right] \qquad (8.17)$$

$$G = \frac{1}{2} \left[A \left(1 - \frac{ik_1}{k_2}\right) + B \left(1 + \frac{ik_1}{k_2}\right) \right] \qquad (8.18)$$

If the left-hand side of Eqs. (8.17) and (8.18) are replaced by the right-hand side of Eqs. (8.15) and (8.16), respectively, we get

$$C \left(1 + \frac{ik_1}{k_2}\right) \exp(ik_1 - k_2)a = \left(1 + \frac{ik_1}{k_2}\right) A + \left(1 - \frac{ik_1}{k_2}\right) B \qquad (8.19)$$

$$C \left(1 - \frac{ik_1}{k_2}\right) \exp(ik_1 + k_2)a = \left(1 - \frac{ik_1}{k_2}\right) A + \left(1 + \frac{ik_1}{k_2}\right) B \qquad (8.20)$$

If we divide Eq. (8.19) by $(1 - ik_1/k_2)$ and divide Eq. (8.20) by $(1 + ik_1/k_2)$ and subtract the result, we find that

$$\left[\frac{1 + ik_1/k_2}{1 - ik_1/k_2} \exp(ik_1 - k_2)a - \frac{1 - ik_1/k_2}{1 + ik_1/k_2} \exp(ik_1 + k_2)a \right] C$$

$$= \left(\frac{1 + ik_1/k_2}{1 - ik_1/k_2} - \frac{1 - ik_1/k_2}{1 + ik_1/k_2} \right) A \qquad (8.21)$$

or

$$\frac{C}{A} = \frac{(\beta - 1/\beta)\exp(-ik_1a)}{\beta\exp(-k_2a) - 1/\beta\,\exp(k_2a)}$$

$$= \frac{(\beta - 1/\beta)\exp(-ik_1a)}{(\beta - 1/\beta)\cosh k_2a + (-\beta - 1/\beta)\sinh k_2a} \tag{8.22}$$

where

$$\beta = \frac{1 + ik_1/k_2}{1 - ik_1/k_2} \tag{8.23}$$

On simplifying, we get

$$C = A\frac{(4ik_1/k_2)\exp(-ik_1a)}{-2(1 - k_1^2/k_2^2)\sinh k_2a + (k_1k_1/k_2)\cosh k_2a} \tag{8.24}$$

Hence, from Eq. (8.10), the transmission coefficient is

$$P_T = \frac{C^*C}{A^*A} = \frac{4k_1^2k_2^2}{(k_1^2 + k_2^2)^2\sinh^2 k_2a + 4k_1^2k_2^2} \tag{8.25}$$

When the product $k_2a = [2m(U_0 - E)/\hbar^2]^{1/2}a$ (i.e., when $U_0 - E = 10\ \text{eV}$ and $a \simeq 10\ \text{Å}$) is larger than unity, i.e., when the barrier is very high and wide, P_T will have a very small value and is given approximately by

$$P_T = \frac{16E}{U_0}\left(1 - \frac{E}{U_0}\right)\exp(-2k_2a) \tag{8.26}$$

Equations (8.25) and (8.26) provide proof of the *remarkable fact* that a material particle[†] of mass m and total energy E, which is incident on a potential barrier having a height U_0 and finite thickness a, actually does have a certain probability P_T of penetrating through the barrier and appearing on the other side, in region III. Such an event would be entirely *impossible* in classical mechanics. This phenomenon is called barrier penetration or "tunneling" through the barrier. Historically, the establishment of tunneling in 1928 was the first of many problems in which quantum mechanics made clear that events impossible according to classical mechanics do have a finite probability of occurring.

The plot of P_T for different values of E/U_0 from Eq. (8.26) (see Ref. 10) shows that for $E/U_0 \ll 1$, P_T is very small, but when E/U_0 is only a little smaller than unity, P_T is appreciable. It is worth mentioning here that one does not in general use Eq. (8.26) in the actual electrochemical case to

---

[†] In a sense typical of quantum mechanics, this statement is not *quite* true. A pure *particle* could not tunnel. Tunneling occurs because each particle is also a wave. Waves do pass through objects with photons that get through unchanged in wavelength.

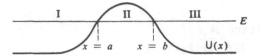

Fig. 8.2. A smoothly rounded potential barrier.

account for the tunneling of electrons across the interface and its contribution to the current density. One generally uses the WKB expressions for the tunneling probability. This is due to the fact that Eq. (8.26) is an exact solution obtained for an ideal rectangular barrier (Fig. 8.1), but in the real situation, e.g., at the interface, the shape of the barrier is not, in fact, rectangular but rather a complicated one. The approximation allows expressions for the tunneling probability through more realistic barriers to be deduced, and these are discussed in the next section.

8.3. WKB Approximation and Tunneling through Barriers

8.3.1. Derivation of the WKB Wave Functions

In the previous section, we solved the Schrödinger equation exactly for a situation [as shown in Eqs. (8.4), (8.5), and (8.7)] in which the potential energy barrier under consideration is taken to be independent of distance in region II (Fig. 8.1). But if the potential energy barrier does not have a simple form (e.g., constant with respect to distance x) the solution of the Schrödinger equation even in one dimension is quite complicated. The WKB method,[11] named after its originators Wentzel, Kramer, and Brillouin, is a treatment for obtaining an approximate solution of the Schrödinger equation when the potential varies only fairly slowly with distance x (Fig. 8.2).

When the potential is not constant in a given region, we can write the Schrödinger equation not in the form of Eq. (8.3) but in a modification of it [for $E^x < U(x)$][11] as

$$\frac{\hbar^2}{2m}\frac{d^2\psi}{dx^2} - [U(x) - E]\psi(x) = 0$$

or

$$\frac{d^2\psi(x)}{dx^2} - \kappa^2(x)\psi(x) = 0 \tag{8.27}$$

where

$$\kappa(x) = \left\{\frac{2m[U(x) - E]}{\hbar^2}\right\}^{1/2} \tag{8.28}$$

The general solution of the differential equation (8.27) can be written as[11]

$$\psi(x) = A(x) e^{iS(x)} \qquad (8.29)$$

where $A(x)$ is a function of x and $S(x)$ is a function $k(x)$ and needs to be determined. It is clear that $S(x)$ must have the magnitude of a wave vector. When we insert Eq. (8.29) into Eq. (8.27), the result becomes

$$A''(x) - A(x)[S'(x)]^2 - \kappa^2(x)A(x) + i[2A'(x)S'(x) + A(x)S''(x)] = 0 \qquad (8.30)$$

where single prime and double primes represent the single and the double differential with respect to x. From Eq. (8.30), we can write

$$A(x)S''(x) + 2A'(x)S'(x) = 0 \qquad (8.31)$$

and

$$A''(x) - A(x)[S'(x)]^2 - \kappa^2(x)A(x) = 0 \qquad (8.32)$$

In general, if $a + ib = 0$, because one cannot add together a real and an imaginary function, *both* of these functions must be 0, so that $a = b = 0$. Now, multiplying Eq. (8.31) by A, we get

$$A^2(x)S''(x) + 2A(x)A'(x)S'(x) = \frac{d}{dx}[A^2(x)S'(x)] = 0 \qquad (8.33)$$

We know that the differential coefficient of any constant is zero, hence it can be written that $A^2(x)S'(x)$ is a constant C_0; i.e.,

$$A^2(x)S'(x) = C_0 \qquad (8.34)$$

Therefore,

$$A(x) = \frac{C_0^{1/2}}{[S'(x)]^{1/2}} \qquad (8.35)$$

Hitherto, in the discussion of this section there has been no approximation. Considering Eq. (8.35), one observes that $A(x)$ is a function of $S'(x)$. From Eqs. (8.28) and (8.29), it is seen that $S'(x)$ is a function of $U(x)$, and if $U(x)$ changes little over a distance of the order of the de Broglie wavelength, it seems reasonable to suppose that $A(x)$, which is a function of $S(x)$ will vary little with respect to x. Hence, the derivative $A'(x)$ is small, and we shall therefore take the second derivative $A''(x)$ to be negligible and equal to zero.† This is the well-known WKB approximation,

---

† In general, $A''(x)$ is not zero. However, in the WKB method it is taken as zero.

and using it, we get from Eq. (8.32)

$$[S'(x)]^2 \simeq -\kappa^2(x) \tag{8.36}$$

or

$$S(x) \simeq \pm i \int \kappa(x)\,dx \tag{8.37}$$

Using Eqs. (8.35) and (8.37) in Eq. (8.29), we get, using $C' = (C_0/i)^{1/2}$

$$\psi(x) \simeq \frac{C'}{[\kappa(x)]^{1/2}} \exp\left[\pm \int \kappa(x)\,dx \right] \tag{8.38}$$

for the region where $E < U(x)$ (i.e., the tunneling case).

Equation (8.38) is the approximate wave function for a system in a region where the potential energy of the particle varies slowly across a de Broglie wavelength of the particle with the added condition (which would correspond to the classical situation) that

$$U(x) > E$$

For the region where the potential energy $U(x)$ is less than the total energy E (i.e., the classical case), the same result is found, except that

$$\kappa(x) = i \left\{ \frac{2m[E - U(x)]}{\hbar^2} \right\}^{1/2} = ik(x) \tag{8.39}$$

and hence the wave function with new constant C'' becomes

$$\psi(x) \simeq \frac{C''}{[k(x)]^{1/2}} \exp\left[\pm i \int k(x)\,dx \right] \tag{8.40}$$

The wave functions given in Eqs. (8.38) and (8.40) were obtained by solving the Schrödinger equation (8.27) using the WKB approximation. Now, let us investigate under what situations the WKB approximation is applicable, i.e., when in Eq. (8.32) $A''(x) \simeq 0$. *Thereafter, we shall use the* approximate wave functions to find the probability of passing through a barrier where the height of it varies with distance.

8.3.2. Nature of the WKB Approximation

The approximate solution of Schrödinger equation (8.27) was obtained taking $A''(x)$ to be negligible. Under what conditions will this occur? We get from Eqs. (8.32) and (8.39)

$$\frac{A''(x)}{A(x)} = [S'(x)]^2 + \kappa^2(x)$$

$$= [S'(x)]^2 - \kappa^2(x) \tag{8.41}$$

Equations (8.38) and (8.40) will be good approximate wave functions if the difference between the terms on the right-hand side of Eq. (8.41) is much smaller than either of the terms on the left-hand side of Eq. (8.42); i.e.,

$$|[S'(x)]^2 - \kappa^2(x)| = \left|\frac{A(x)''}{A(x)}\right| \ll [S'(x)]^2 \tag{8.42}$$

We now replace $A(x)$ and $A''(x)$ by means of Eq. (8.35) in the inequality (8.42), and after some algebra, we obtain the relation

$$|\tfrac{3}{4}[S''(x)]^2 - \tfrac{1}{2}S'(x)S'''(x)| \ll [S'(x)]^4 \tag{8.43}$$

Now, let us see under what conditions the inequality (8.43) is satisfied. We find that if we get the inequalities

$$[S'(x)]^4 \gg \tfrac{3}{4}[S''(x)]^2 \tag{8.44}$$

and

$$|[S'(x)]^3| \gg \tfrac{1}{2}|S'''(x)| \tag{8.45}$$

then the inequality (8.43) is satisfied. Thus, the inequalities (8.44) and (8.45) express two of the conditions under which the inequality (8.43) is satisfied. The inequalities (8.44) and (8.45) are not the only conditions under which the inequality (8.43) is satisfied, but we can satisfy ourselves with the two conditions (8.44) and (8.45). We notice that when condition (8.44) is satisfied, condition (8.45) is automatically satisfied, since the differentiation of (8.44) gives the condition (8.45). Hence, we consider the condition (8.44) and write it (removing the factor $\tfrac{3}{4}$, which is unimportant when inequality sign $\gg$ is used) as

$$[S'(x)]^4 \gg [S''(x)]^2 \tag{8.46}$$

Now, replacing the value of $S'(x)$ and $S''(x)$ in terms of $k(x)$ and $k'(x)$ using Eq. (8.36), we get from Eq. (8.46) that

$$k^4(x) \gg [k'(x)]^2$$

or

$$k^4(x) \gg \left[\frac{dk(x)}{dx}\right]^2$$

or

$$k^2(x) \gg \left|\frac{dk(x)}{dx}\right|$$

or

$$k(x) \gg \frac{1}{k(x)}\left|\frac{dk(x)}{dx}\right| \tag{8.47}$$

Multiplying both sides of (8.47) by $\hbar$ and replacing† $1/k(x)$ by $\lambda(x)/2\pi$, we get

$$\hbar k(x) \gg \frac{\lambda(x)\hbar}{2\pi}\left|\frac{dk(x)}{dx}\right|$$

or

$$p(x) \gg \frac{\lambda(x)}{2\pi}\left|\frac{dp(x)}{dx}\right| \tag{8.48}$$

or

$$p(x) \gg \frac{1}{k(x)}\left|\frac{dp(x)}{dx}\right| \tag{8.48a}$$

Thus, condition (8.48) implies that the WKB approximation is valid for the situation where the change of momentum over one wavelength is small compared with the momentum itself.

Thus, from (8.48), the WKB approximation applies to the electron transfer situation if

$$\frac{p(x)}{\Delta p(x)} \gg \frac{\lambda}{2\pi}\frac{1}{\Delta x}$$

$$\gg \frac{5\times 10^{-8}}{2\pi}\frac{1}{5\times 10^{-8}} = \frac{1}{2\pi}$$

for the electrode situation. But obviously, $p(x)/\Delta p(x) > 1$ and is hence $\gg 1/2\pi$. The condition is hence fulfilled. It would break down if $\Delta x \ll \lambda$, i.e., if the barrier width were less than the particle wavelength.

The condition (8.48a) obviously breaks down if $k(x) = [2m(U_x - E)/h^2]^{1/2} = 0$, i.e., when $U(x) = E$ or at a value of x (Fig. 8.2) such that the particle contacts the barrier [where, of course, $E = U(x)$; then k_x would become zero].

8.3.3. Transmission through a Barrier: WKB Tunneling Expression

The WKB approximate wave function is valid when $A'' = 0$ or $p \gg [\lambda(x)/2\pi]|dp/dx|$ can be applied to calculate the transmission coefficient through a barrier upon which particles are incident from the left with insufficient energy to pass over the top of the barrier classically; i.e., $E < U(x)$ of Fig. 8.2. The problem is similar to that of the rectangular potential barrier of Section 8.2 with the exception that no special assumption is made here concerning the shape of the barrier.

† It is in fact $\hbar k = p = h/\lambda = hk/2\pi$. Therefore, $k = 2\pi/\lambda$.

In region I, of Fig. 8.2, which extends from the left up to $x = a$, the potential energy $U(x)$ is zero or small, and the total energy E is always greater than $U(x)$. We can, therefore, use the WKB wave function, Eq. (8.40), to find the WKB wave function for the particle in region I, and it will be

$$\psi_I(x) = \frac{C''}{[k(x)]^{1/2}} \exp\left[\pm i \int k(x)\, dx\right] \qquad (8.40a)$$

Now the wave represented by Eq. (8.40a) can be regarded as going from left to right (for the positive sign in the exponential), but it can also be regarded being reflected, i.e., going from right to left (for the negative sign in the exponential). The constant C'' ($=A$) of Eq. (8.40a) represents the amplitude (or intensity) of the left-to-right wave and another constant (B) would have to be used for the amplitude of the right-to-left wave, and these two values of C'' correspond to the coefficients of positive and negative signs of $\exp[i \int_a^x k(x)\, dx]$ in Eq. (8.40), respectively. Thus, the total wave function is

$$\psi_I(x) = \frac{A}{[k(x)]^{1/2}} \exp\left[i \int_a^x k(x)\, dx\right] + \frac{B}{[k(x)]^{1/2}} \exp\left[-i \int_a^x k(x)\, dx\right]$$
$$(8.49)$$

The wave function for the particle in region II (Fig. 8.2), where $U(x) > E$, can be obtained using the WKB wave function as given by Eq. (8.38), i.e.,

$$\psi_{II} = \frac{C'}{[\kappa(x)]^{1/2}} \exp\left[\pm \int \kappa(x)\, dx\right] \qquad (8.38a)$$

Expressing the constant C' in this region for the wave traveling from left to right [corresponding to the positive sign in the exponential in Eq. (8.38a)] by C and that for the wave from right to left [corresponding to the negative sign in Eq. (8.38a)] by D, we get

$$\psi_{II} = \frac{C}{[\kappa(x)]^{1/2}} \exp\left[+\int_a^x \kappa(x)\, dx\right] + \frac{D}{[\kappa(x)]^{1/2}} \exp\left[-\int_a^x \kappa(x)\, dx\right]$$
$$(8.50)$$

Similarly, the wave function in the region III where $U(x) < E$ can be written in the same form as in Eq. (8.49) [which is *also* true for $U(x) < E$] but with constants F and G for the wave from left to right and from right to left, respectively. The form of the wave function becomes

$$\psi_{III}(x) = \frac{F}{[k(x)]^{1/2}} \exp\left[i \int_b^x k(x)\,dx\right] + \frac{G}{[k(x)]^{1/2}} \exp\left[-i \int_b^x k(x)\,dx\right]$$

$$(8.51)$$

Now, using Eqs. (8.49)–(8.51) for the regions I, II, and III (Fig. 8.2), we can find some relation between A, B, F, and G via C and D as was done in Section 8.2 for the rectangular barrier. The details are given in Appendix 8.I. The result is[12]

$$A = \frac{1}{2}\left[F\left(2\theta + \frac{1}{2\theta}\right) + iG\left(2\theta - \frac{1}{2\theta}\right)\right] \tag{8.52}$$

where

$$\theta = \exp\left[\int_a^b \kappa(x)\,dx\right] \tag{8.53}$$

Using Eq. (8.25) for $\kappa(x)$ in Eq. (8.53), one gets

$$\theta = \exp\left\{\int_a^b \left[\frac{2m[U(x) - E]}{h^2}\right]^{1/2} dx\right\} \tag{8.54}$$

Assuming that there is no wave incident from the right, i.e., when $G = 0$, we get from Eq. (8.52) that

$$A = \frac{1}{2}F\left(2\theta + \frac{1}{2\theta}\right) \tag{8.55}$$

where the (8.55) relates the coefficients of the incident wave A and that of transmitted wave F.

The transmission coefficient is defined [see Eq. (8.9) and Section 8.2] as

$$P_T = \frac{v_{trans}|\psi_{trans}|^2}{v_{incident}|\psi_{incident}|^2} = \frac{|F|^2}{|A|^2} \tag{8.56}$$

Since the transmitted velocity $v_{trans} = v_{incident}$, putting the value of $|A|^2$ from Eq. (8.55) in Eq. (8.56), we get

$$P_T = \frac{4}{(2\theta + 1/2\theta)^2} \tag{8.57}$$

Equation (8.57) is general within the WKB approximation. However, we can now make a further approximation by assuming the condition of a high and a broad barrier. The mathematical expression of this condition is obtained from Eq. (8.54). It is

$$\theta = \exp\left\{ \int_a^b \left[2m\frac{[U(x)-E]}{\hbar^2} \right]^{1/2} dx \right\} \gg 1 \tag{8.58}$$

from which it follows that the condition is†

$$\frac{1}{2\theta} \ll 2\theta$$

Hence, from Eqs. (8.53), (8.57), and (8.58),

$$P_T = \frac{1}{\theta^2} = \exp\left[-\int_a^b \kappa(x)\, dx \right]^2$$

$$= \exp\left[-2\int_a^b \kappa(x)\, dx \right] \tag{8.59}$$

Thus, utilizing Eq. (8.28) for the value of $\kappa(x)$, one finds

$$P_T = \exp\left(-2\int_a^b \left\{ \frac{8\pi^2 m}{h^2}[U(x)-E] \right\}^{1/2} dx \right) \tag{8.60}$$

This is the expression for the transmission coefficient utilizing the WKB general expression of Eq. (8.57) and the conditions that the barrier height and width are both large. This is the probability of tunneling. It was first obtained by Gamow for the transmission coefficient of α particles from the nucleus and thus is called the Gamow equation.

So far we have left $U(x)$ inexplicit in Eq. (8.60); i.e., we have made the expression (within the realms of the approximations used) general for any shape of barrier, because we have not defined the form of $U(x)$. Let us now take a few likely forms and express the functionality of $U(x)$, whereupon we can derive practical expressions for tunneling through barriers of definite shapes.

---

† If one puts the value of barrier width as 5 Å, barrier height as 10 eV with respect to bottom of the conduction, and take electron energy $E = 5$ eV corresponding to Fermi level energy, the value of θ becomes 9.4 and $1/\theta \approx 0.1$ and hence condition (8.58) is satisfied. This condition is also satisfied for various other values of barrier width and height met in chemical problems.

8.3.3.1. The Parabolic Barrier $[U(x) = U_0 - \frac{1}{2}fx^2]$

Let us consider a parabolic barrier of the type given in Fig. 8.3. From Eq. (8.60), the term to be integrated in the tunneling probability expression is

$$\int_a^b [U(x) - E]^{1/2} \, dx = \int_a^b [U_0 - \tfrac{1}{2}fx^2 - E]^{1/2} \, dx$$

$$= \left(\frac{f}{2}\right)^{1/2} \int_a^b (a^2 - x^2)^{1/2} \, dx \qquad (8.61)$$

where

$$a^2 = 2(U_0 - E)/f \qquad (8.62)$$

The positive and negative values of a (in Fig. 8.3) correspond respectively to a and b. Because the barrier is symmetrical we can write

$$\int_a^b [U(x) - E]^{1/2} \, dx = \left(\frac{f}{2}\right)^{1/2} \int_{-a}^a (a^2 - x^2)^{1/2} \, dx = 2\left(\frac{f}{2}\right)^{1/2} \int_0^a (a^2 - x^2)^{1/2} \, dx$$

$$= \left(\frac{f}{2}\right)^{1/2} [a^2 \sin^{-1}(x/a) + x(a^2 - x^2)^{1/2}]_0^a$$

$$= \left(\frac{f}{2}\right)^{1/2} \left(\frac{\pi a^2}{2}\right) \qquad (8.63)$$

Putting this value into Eq. (8.61) from Eq. (8.63), we get

$$\int_a^b [U(x) - E]^{1/2} \, dx = \frac{\pi}{2} a(U_0 - E)^{1/2} \qquad (8.64)$$

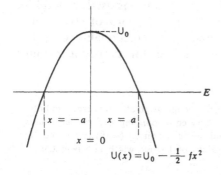

$$U(x) = U_0 - \frac{1}{2}fx^2$$

Fig. 8.3. The parabolic potential barrier.

Putting this value into Eq. (8.60), we get the tunneling probability expression for the parabolic barrier as

$$P_T = \exp\left(-2\int_a^b \left\{\frac{8\pi^2 m}{h^2}[U(x)-E]\right\}^{1/2} dx\right)$$

$$= \exp\left\{-2\left(\frac{8\pi^2 m}{h^2}\right)^{1/2}\int_{-a}^a [U(x)-E]^{1/2}\, dx\right\}$$

$$= \exp\left[-\frac{4\pi}{h}(2m)^{1/2}\frac{\pi}{2}a(U_0-E)^{1/2}\right]$$

$$= \exp\left\{-\frac{4\pi^2}{h}\frac{2a}{4}[2m(U_0-E)]^{1/2}\right\}$$

$$= \exp\left\{-\frac{\pi^2 l}{h}[2m(U_0-E)]^{1/2}\right\} \tag{8.65}$$

where the width of the barrier $l = 2a$ and U_0 represents the barrier maximum (Fig. 8.3).

8.3.3.2. The Rectangular Barrier $[U(x) = U_0]$

We have already given an expression for the tunneling probability of a particle through a rectangular barrier by solving the Schrödinger equation exactly for constant potential in Section 8.2. Here, we derive an analogous expression for the tunneling probability, using the WKB expression (8.60). We assume that $U(x)$ is zero when x is more negative than $x_1 = -a$, and when x is more positive than $x_2 = +a$ (Fig. 8.4). Between these values of x, $U(x)$ is assumed to have the constant value of U_0, the height of the barrier. The tunneling probability for the rectangular barrier is [from Eq. (8.60)]

$$H = \exp\left\{-2\left(\frac{8\pi^2 m}{h^2}\right)^{1/2}\int_{-a}^a [U(x)-E]^{1/2}\, dx\right\}$$

$$= \exp\left\{-\frac{4\pi l}{h}[2m(U_0-E)]^{1/2}\right\} \tag{8.66}$$

where $2a = l$ is the thickness, or the width of the barrier.

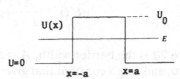

Fig. 8.4. A rectangular potential barrier.

Table 8.1
Tunneling Probability for Various Barrier Height and Width

Types of barrier	Barrier height (w.r.t. Fermi level) (eV)	Barrier width (Å)	Tunneling probability, P_T
Parabolic barrier	5	5	1.2×10^{-4}
	10	10	9.1×10^{-12}
	15	15	5.2×10^{-21}
	20	20	6.0×10^{-32}
Rectangular barrier	5	5	1.0×10^{-5}
	10	10	8.8×10^{-15}
	15	15	1.5×10^{-26}
	20	20	1.7×10^{-40}

The values of tunneling probability P_T for electron passage through different types of barriers and their heights and width are given in Table 8.1 to give an idea how the value of the tunneling probability is sensitive on these barrier parameters.

8.4. Tunneling through an Eckart Potential Barrier

So far in previous sections we have given the method of exact solution of the Schrödinger equation for a constant potential energy (Section 8.2) and the approximate solution of the Schrödinger equation for a potential energy that varies slowly with respect to position using the WKB approximation (Section 8.3). This solution has been used to find the tunneling probability for a particle incident on specific potential barriers. Here, we will give the tunneling probability expression for the particle incident on a general potential [Eq. (8.67)] given by Eckart that is neither constant nor slowly varying with position. The Schrödinger equation with this potential [Eq. (8.67)] is soluble in terms of hypergeometric functions[13] (see Appendix 8.II).

Eckart[14] considered a particle of mass m moving in the potential that he constructed empirically with an eye to the need to present barriers of different shapes. The expression for the potential he chose is,

$$U(x) = \frac{A\, e^{2\pi x/d}}{1 + e^{2\pi x/d}} + \frac{B\, e^{2\pi x/d}}{(1 + e^{2\pi x/d})^2} \qquad (8.67)$$

where $2d$ is the barrier width, A is the difference between U at $-\infty$ and U at $+\infty$, and B is a constant that gives the measure of the barrier height. This

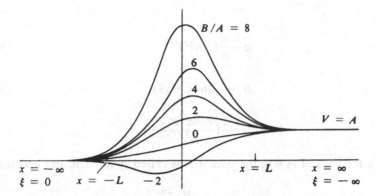

Fig. 8.5. The Eckart potential. $V(x) = -A\xi(1-\xi)^{-1} - B\xi(1-\xi)^{-2}$, where $\xi = e^{2\pi x/L}$ for various values of the ratio B/A. If $A = 0$, the potential is symmetrical about $x = 0$. The Eckart potential varies smoothly from a plateau of $V = 0$ at $x = -\infty$ to a plateau of $V = A$ at $x = +\infty$.

potential is illustrated in Fig. 8.5, which shows that the Eckart potential encompasses a variety of forms of potential depending on the ratio of B/A. The Schrödinger equation with this potential [Eq. (8.67)], is

$$\frac{d^2\psi}{dx^2} + \frac{2m}{h^2}\left[E - \frac{Ae^{2\pi x/d}}{1+e^{2\pi x/d}} - \frac{Be^{2\pi x/d}}{(1+e^{2\pi x/d})^2}\right]\psi = 0 \tag{8.68}$$

where E is the energy and m is the mass of the particle.

Eckart[14] has solved this Schrödinger equation to find the wave functions in the region $x \rightarrow -\infty$ (from where the particle is incident) and also in the region $x \rightarrow +\infty$ (to where the particle is transmitted). He defined the coefficients of the incident wave as M and the reflected wave as N. The reflection coefficient R is

$$R = \frac{|M|^2}{|N|^2} \tag{8.69}$$

Eckart[14] found the transmission coefficient [for details of solution of Eq. (8.68) and derivation of P_T see Appendix 8.II] to be

$$P_T = 1 - R$$

$$= \frac{\cosh[2\pi(\alpha+\beta)] - \cosh[2\pi(\alpha-\beta)]}{\cosh[2\pi(\alpha+\beta)] + \cosh 2\pi\gamma} \tag{8.70}$$

where

$$\alpha = \frac{d}{h}(2mE)^{1/2} \tag{8.71}$$

$$\beta = \frac{d}{h}2m(E-A)^{1/2} \tag{8.72}$$

$$\gamma = \frac{1}{2}\left(\frac{8md^2B}{h^2}-1\right)^{1/2} \tag{8.73}$$

In the case of $A = 0$, the potential is called symmetric and becomes

$$U(x)=\frac{Be^{2\pi x/d}}{(1+e^{2\pi x/d})^2} \tag{8.74}$$

and for this symmetrical Eckart barrier [Eq. (8.74)], the expression for tunneling probability [obtained putting $A = 0$ in Eq. (8.72) so that β becomes equal to α] becomes

$$P_T = \frac{\cosh 4\pi\alpha - 1}{\cosh 4\pi\alpha + \cosh 2\pi\gamma} \tag{8.75}$$

It is important to note that in an Eckart barrier when it is used for an activation barrier (Fig. 8.5) the value of A corresponds to the heat of reaction, and hence, $A = 0$ corresponds to isotropic reactions, where the heat of reaction is zero.

The WKB approximate expression for the tunneling probability can be only found for symmetric and very simple type of barrier potentials, such as the symmetrical, parabolic, triangular, or rectangular. But for asymmetric potential barriers, as shown in Fig. 8.5, the WKB solution would be difficult, because it would be necessary to substitute $U(x)$ into Eq. (8.61) in an inordinately complex way and then integrate. An advantage of the Eckart potential and its solution is that the corresponding potential function on which it is based has two variable parameters, A and B, respectively. Adjusting their values, one may get a realistic shape for a potential barrier (Fig. 8.5) and not only for the symmetrical shape to which the WKB methods have been applied. The barrier shapes in Fig. 8.5 are those seen in chemical and electrochemical kinetics. Such barriers can be used for the treatment of the proton and electron tunneling situation. Eckart barriers can be also used, depending on the shape of the barrier that exists at the interface, which is a function of the reaction, electrolyte, and electrode.

This barrier model has been used for the determination of the tunneling probability in electrochemical hydrogen, deuterium, and tritium evolution reactions by several authors.[15–21] Christov[17–19] and also Bockris and Srinivasan[21] and Bockris and Matthews[16,20] made theoretical evaluations of separation factors for hydrogen and deuterium, and also hydrogen and tritium, using the Eckart barrier (Chapter 12).

8.5. Gurney's Application of Gamow's Tunneling Theory to Electron Transfer at Interfaces

Gurney's pioneer work[5] on tunneling at electrodes was stimulated by that of Bowden and Rideal,[22] who published in 1928 a widespread series of data showing the exponential dependence of current upon overpotential. Gurney uses as his starting point the discussion by Oliphant and Moon[23] of the neutralization of gaseous ions at an electrode. These workers considered the tunneling of an electron from an electrode to an ion. To allow for the lack of radiation detected during an electrode process (unlike that of α decay), Gurney assumed that the energy of the electron on the left and right of the barrier at the electrode–solution interface is the same, i.e., that the energy of the electron in the donor state [e.g., an electron in the electrode (metal) state] is equal to that of the electron in the acceptor (e.g., a hydrogen atom).

8.5.1. Neutralization of a Gaseous Ion at an Electrode

The situation of neutralization of a gaseous ion[5,24] by electron tunneling is illustrated in Fig. 8.6, in which the potential energy of an electron along a line perpendicular to the surface of the metal is sketched. The work function of the metal is denoted by ϕ and the ionization potential of the ion by I.

Consider the radiationless transfer of an electron to an atom. It is necessary to find, with respect to a common reference state, the energy of the electron in the hydrogen atom. Consider first, the energy of the electron in the metal with respect to the reference state, which by common agreement in this work we take as the energy of the electron in a vacuum, when the potential energy is zero. One of the well-known properties of a metal is its work function. ϕ is a positive quantity, the energy needed to knock an electron out of the metal's Fermi level to a vacuum. Hence, the energy of the electron at the metal's Fermi level with respect to that of an electron *in vacuo* is $-\phi$ (Fig. 8.6).

Having defined, thus, the energy of an electron in a metal, let us turn to deal with the energy of an electron in the hydrogen atom. We can put an electron into a vacuum from a hydrogen atom by ionizing it, and the work to do this will be $+I$, where I is the ionization energy and is a positive

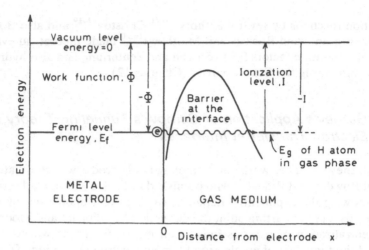

Fig. 8.6. The schematic diagram of the potential energy barrier at the interface of a metal electrode and a gas medium of H^+ ions. The diagram shows also the ground-state hydrogen atom levels, E_g, in the gas phase.

quantity. Hence, if we want to consider an electron and find out how much the electronic energy of it is in a hydrogen atom with respect to the energy of the electron in a vacuum, the energy is simply the negative value of the ionization energy. Hence, the energy needed to take an electron from a vacuum and put it onto a hydrogen ion, thus forming a hydrogen atom, is $-I$. The energy of an electron in the hydrogen atom with respect to that of an electron in vacuum is therefore $-I$ (Fig. 8.6).

Gurney's radiationless tunneling condition for the neutralization of a hydrogen ion in vacuum by an electrode at an electrode is†

$$\phi \leq I \qquad (8.76)$$

This condition (8.76) means that the magnitude of electron energy in the

† It may at first be thought that the condition "less than" in Eq. (8.76) (rather than equal to) is a breakdown of the condition of radiationless transfer. However, this is not the case. The electron emits from a region near to the Fermi level in the metal and must find a state that is equal to it in energy on the other side of the barrier (to obey the condition of radiationless transfer). Thus, a condition in which the value of the ionization energy I is smaller than the electron energy is not acceptable, for the electron then has no place to go, no state that can accept it at equal energy. Now, however, apart from the condition in which $I = \phi$, there is also an acceptable radiationless condition $\phi < I$. Thus, in such a case the electron can transfer across the barrier, at first to a state of equal energy, but then not finding a satisfactory *stable* state there, can go to a *lower* energy level than it had in the electrode. The rest of the energy is absorbed in the solution in the form of heat. (This mechanism is not applicable to the ideal vacuum case where there is no heat sink. Thus, the electron does not tunnel in the vacuum case but rather forms a space charge.)

metal electrode (with respect to vacuum) must be less than or equal to the magnitude of the electron energy (with respect to vacuum) in the hydrogen atom in the gas medium. The Fermi level of the electron in a metal electrode and a single electron level in a hydrogen atom in the gas medium are shown in Fig. 8.6.

The parabolic line between the metal and the hydrogen atom in this figure represents the energy barrier, and it arises, physically, because when the electron is emitted from the metal it interacts with the metal in an image-force manner. The image energy decreases the *negative* potential energy of the electron as the distance between the metal and electron increases; i.e., as the electron moves to the right in Fig. 8.6. Hence, the *positive* potential energy of the electron increases as it moves from the metal to the ion. The ion is attracted by the positive (H^+) ion, so that the potential energy of the electron decreases as it moves toward the ion. The net result is a parabolic shape of the potential energy barrier (Fig. 8.6).

8.5.2. Neutralization of an Ion in Solution at the Electrode

When the ion is no longer in the gas phase, but in solution, the situation is more complicated. Thus, the energy of the electron with respect to the potential energy of the electron in the metal at the Fermi level (with respect to the energy in the vacuum) remains unchanged. However, the energy of the electron in a hydrogen atom produced by addition of an electron to H^+ (that is, the state that has to be equal to the energy of the electron in the metal for the radiationless condition to apply) is different when the hydrogen ion finally exists in the solution than when it exists in the gas phase.

We can rationalize this difference in the following way. We have previously seen that, when the hydrogen atom is in the gas phase, the energy difference between an electron in the hydrogen atom and in the vacuum is $-I$. Let us follow up what will happen if we find out the energy corresponding to $-I$ in the solution. Consider first a hydrogen atom in solution. Let us ionize it. The work done is *at first* $+I$, but now the hydrogen ion is being created in solution and one must immediately add to the energy of the atom the energy of the hydrogen ion in solution, so that the changed energy is $I + L$, where L is the solvation energy. The solvation energy, which is the energy needed to take an ion from the gas phase and put it in the solution, is itself a negative quantity. Thus, the energy needed to form the hydrogen ion in solution is $I + L$, so that the energy required to take an electron from the gas phase and neutralize a H^+ ion *in solution* is $-(I + L)$.

The condition for the neutralization of the hydrogen ion in solution, with the present degree of discussion, and corresponding to the Gurney

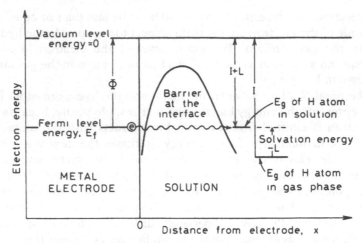

Fig. 8.7. The schematic diagram of the potential energy barrier at the interface of a metal electrode and solution of H^+. The diagram shows also the effect of solvation energy, L, on the ground-state energy level, E_g, of hydrogen atoms in solution.

radiationless condition, is therefore

$$-\phi \geqslant -(I+L) \quad \text{or} \quad \phi \leqslant (I+L) \tag{8.77}$$

This is illustrated in Fig. 8.7.

8.5.2.1. Effect of Electrode Potential

It is easy to understand, in the context of the basic condition of Gurney for radiationless tunneling, what happens when we apply an external potential to the electrode (for, in discussion hitherto, we have implicitly assumed that there is zero potential difference between the electrode and the solution).

If we flow in electrons (cathodic polarization) or extract electrons from the electrode (anodic polarization) by an outside source of potential, the energy of the outcoming surface electrons changes by $e_0 V$ and the work function becomes $\phi + e_0 V$.† This is because the work function ϕ is

---

† Since in the definition of work function the Fermi level is involved, a change in work function by $e_0 V$ looks as if the Fermi level has been changed by $e_0 V$ at the electrode potential. We will retain this convention for simplicity since the change in Fermi level is mathematically equivalent to change in work function and can be easily pictured (see Fig. 8.8). But, in fact, the Fermi level does not change with potential; rather, the work function changes. The origin of the change of the work function is in the change in electron energy when it comes out of the electrode surface through the field fall or rise (depending on cathodic or anodic polarization) across the interface.

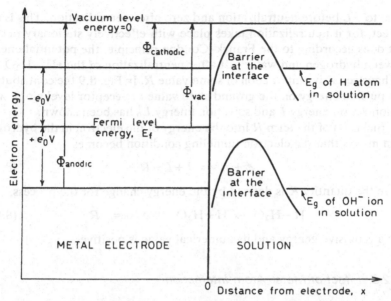

Fig. 8.8. The schematic diagram of the potential energy barrier at the interface of a metal electrode and the solutions of H^+ and OH^-, respectively. The diagram shows the change in the magnitude of work functions in the presence of cathodic and anodic polarizations compared with that of vacuum work function. Also shown are the ground-state energy levels, E_g, of hydrogen atom and OH^- ion solutions.

the energy required to knock out an electron (having its energy corresponding to the Fermi level) from an electrode surface across the interface. When an outside potential is applied, it becomes easier (in the case of cathodic polarization) or more difficult (in the case of anodic polarization) to knock out an electron from the electrode surface across the interface. The amount of energy involved corresponding to easiness or difficulty is $e_0 V$ at the electrode potential V. Hence, the effective work function at the electrode potential becomes $(\phi + e_0 V)$, where the sign of V must be considered.

Thus, reformulating the Gurney condition for radiationless transfer with electrode potential V, we have:

$$-(\phi + e_0 V) \geq -(I + L) \quad \text{or} \quad (\phi + e_0 V) \leq (I + L) \qquad (8.78)$$

where e_0 is the electronic charge and the sign of the potential V should be taken into account.

8.5.2.2. Effect of the Hydrogen Atom–Water Interaction

Up to this point, Gurney was working with the model that the potential energy of interaction of the ion and the adjacent water molecules is equal to $-L$ before neutralization and zero after neutralization. This is not correct, for if neutralization takes place with effectively stationary nuclei, as it does according to the Franck–Condon principle, the potential energy between hydrogen and water just after neutralization of the H^+—H_2O ion will have repulsive interaction of some value R. In Fig. 8.9 the contribution of repulsion energy on the ground state value of acceptor level along with the ionization energy I and solvation energy L_0 has been shown.

Inclusion of the term R into the energy of the electron in the hydrogen atom means that the electron tunneling condition becomes

$$\phi + e_0 V \leq I + L - R \tag{8.79}$$

where the quantity R is defined by the energy change for the process,

$$H—H_2O \rightarrow H + H_2O \qquad \Delta H_0 = -R \tag{8.80}$$

R is a repulsive energy and its numerical value is positive.

8.5.2.3. Effect of Metal–Atom Interaction

Butler[6] reexamined the theory of Gurney and besides the repulsion between hydrogen and water, he considered attraction A between the hydrogen atom and the metal surface, which Gurney had neglected.

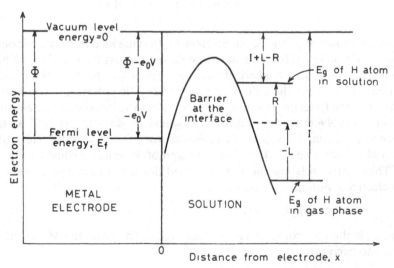

Fig. 8.9. The schematic diagram of the potential energy barrier at the interface of a metal electrode and solution of H^+. The diagram shows the effect of cathodic potential on the work function and also the effect of both solvation energy L and repulsion of energy between hydrogen atom and water molecule on the ground-state energy level of hydrogen atom in solution.

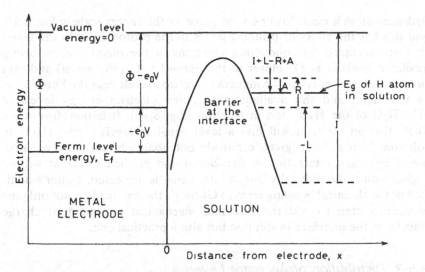

Fig. 8.10. The schematic diagram of the potential energy barrier at the interface of a metal electrode and solution of H^+. The diagram shows the effect of the cathodic potential on the work function and also the effect of solvation energy L repulsion energy R, and the attraction energy between the metal and hydrogen atom for the adsorption A on the ground-state energy level E_g of hydrogen atom in solution.

Inclusion of this metal–atom interaction energy† A changes the work done in bringing an electron from vacuum to a hydrogen ion in solution (Fig. 8.10).§ Hence, the electron tunneling condition becomes

$$\phi + e_0 V \leqslant I + L - R + A \qquad (8.81)$$

8.5.2.4. Results of Making the Energy in the Hydrogen Atom More Negative by Considering Metal–Hydrogen Interaction

It is easy to realize the effect of this process on the likelihood that an electron will transfer from the metal to a proton in solution. In Fig. 8.10 there are shown two cases, one in which the hydrogen atom is supposed to have the zero interaction with the metal, and one in which the interaction energy A is supposed to be high; i.e., a large negative heat of adsorption [in the region of some 50 kcal (or 2.2 eV) per gram atom] is introduced. It will be seen that the effective level of the ground state of the electron in the

---

† A is defined by energy change for the process

$$MH \rightarrow M + H \qquad \Delta H_0 = A \qquad (8.82)$$

A is attraction energy and is thus negative in numerical value.
§ A more detailed argument is in Bockris and Reddy.[47]

hydrogen atom is made lower on the curve for the energy scale in Fig. 8.10, and this has the effect of bringing levels in the proton close to the levels that are occupied by conduction electrons in the metal (i.e., it brings available levels in H_3O^+ closer to the Fermi level in the metal) and thus there is a greater overlap of conduction-electron levels near the Fermi level in the metal and the available distributed electron energy levels of $H^+—H_2O$ of the H_3O^+ ion in solution (Fig. 8.10). It is therefore more likely that an electron will find a level equal to levels in the H_3O^+ in solution. Hence, for a given electrode potential, when a large negative metal–hydrogen interaction is considered, the predicted current will be higher than if the metal–hydrogen interaction is neglected. Butler's addition of the chemical bonding term to Gurney's theory[5] made it not only the pioneering step toward the quantum mechanical treatment of charge transfer at the interface in solution but also a practical one.

8.5.3. Distribution of Acceptor Levels

In the discussion given hitherto of the Gurney radiationless condition for the neutralization of ion in solution at electrodes, we have modified the basic condition set up by Oliphant and Moon[23] to allow for the effect of solvation energy and also for the effect of electrode potential, adsorption energy, and repulsion energy.

However, in introducing solvation energy, we have made too rough and ready a statement by implying that the solvation only has a single value.† The *measured* solvation energy predominantly refers to the situation in which protons exist in the ground vibrational–rotational states and interact with the surrounding solvent. In fact, there is a series of vibrational states set up when the solvated proton interacts with the surrounding solvent molecules.

In solution, however, although on a short time scale of less than 10^{-12} sec the energy of the proton with respect to a water molecule is still quantized over longer times, the quantization is made more diffuse. This is a phenomenon seen when one goes from the gas phase to the solution phase and is an example of a particular kind of line broadening. The bonds that give rise to quantized energy levels are subject to a number of collisions with the surrounding molecules, and this smooths out the difference between the energy levels of the vibrational–rotational states.

---

† It might be thought that there would be a question of the appropriate dielectric constant to use in the Born term. However, the state formed by donation of an electrode electron to H_3O^+ has no charge and hence the question does not arise. The solvation energy referred to here is the time-average solvation energy of H_3O^+ contributing to the energy of the various vibrational–rotational states in H_3O^+.

The appropriate model for representing this situation (pointed out by Gurney[5] in 1931) is to express the energy levels of the proton–solvent, H^+—H_2O, bond in solution in terms of a classical Boltzmann distribution, i.e., to recognize the fact that the separation of vibration–rotation states has been smoothed out.

When we come to write down a condition corresponding to Eq. (8.78) we should not write a single value for L but a series of values L_n, which correspond to various acceptor levels, where $n = 1, 2, \ldots$ corresponds to the vibrational levels in the H_3^+O ion (the difference between which has been made so small by the line broadening).

We are calculating the energy of an electron in a hydrogen *atom* and yet talking about the vibration–rotational levels of the interaction between protons and the surrounding water molecules. The fact that they play an important part in the calculated energy of the electron of the hydrogen atom in solution arises because we calculate the energy of the electron when it has formed a hydrogen atom, starting from a proton in solution, and relate the zero state of potential energy to the energy of an electron *in vacuo*. The deduction of the final state energy has been outlined above. Mentally, we start with a hydrogen atom in solution and ionize it, forming an electron in the gas phase and a hydrogen ion, i.e., a proton, in solution. It is at this point that the energy levels for the proton ion–water interaction enter. There is, therefore, a series of values that we must allow for, and thus the energy to take the electron from the gas phase to the hydrogen ion in solution also contains a series of electron acceptor levels corresponding to smeared-out vibrational–rotational levels of H^+—H_2O bond in solution. This situation is shown in Fig. 8.11. It facilitates the situation for the electron and its donation to a proton in solution. Thus, were the situation such that only the ground state levels were available (Fig. 8.10), there would be only one energy level at which the electron could be produced from a metal electrode and thus undergo a *radiationless* transition to a single hydrogen ion level in solution.

However, there is a continuous series of acceptor energy levels in solution (Fig. 8.11) and as the electrode potential changes, and hence the energy level of the surface electrons changes, electrons from the levels around the "Fermi level" can undergo radiationless transfer to the corresponding acceptor levels in solution. Of course, for each level (and hence for each new electrode potential) there is a different intensity of transfer (or current density), for it is not true that there is an equal availability of these various vibrational–rotational levels. At the higher states in energy in the solvation proton, compared with the ground state energy, they become more remote, and less in occupancy by the protons; i.e., the number of protons having higher vibrational–rotational energy become less. Thus, the number of protons available to conduction-band electrons decreases as the energy separation from the ground state increases. This is the key point about the dependence of the rate of an electrode reaction upon potential. Indeed, it is the basis of Tafel's law, which expresses the exponential dependence of rate upon the potential.

According to the Boltzmann distribution function, the electron energy levels in H_3O^+ corresponding to its different vibrational–rotational energy U levels can be expressed as:

$$N(U) = N(U_0) \exp[-(U - U_0)/kT] \qquad (8.83)$$

We need, thus, to express Eq. (8.83) in terms of the energy of the electron E.

From Fig. 8.12 we see that $(GC - FB)$ represents $(E - E_0)$, where E_0 is the electron energy when $H^+—H_2O$ bond is in the ground vibrational–rotational state and E_n is the electron energy when $H^+—H_2O$ bond is in any other higher nth vibrational–rotational state other than the ground state. U_0 is the vibrational–rotational energy of the $H^+—H_2O$ bond when it is in ground state, and U_n is that when it is in the nth vibrational–rotational state (Fig. 8.12). Since $(E_n - E_0)$ is greater than $(U_n - U_0)$ (Fig. 8.12), Gurney equated them by the equation

$$U_n - U_0 = \beta(E_n - E_0) \qquad (8.84)$$

where β is a factor greater than zero and less than unity.

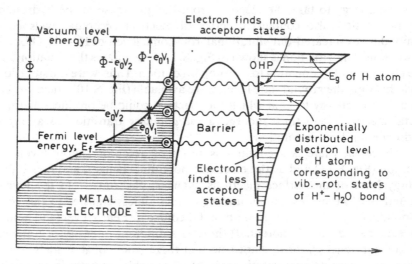

Fig. 8.11. The schematic diagram of the potential energy barrier at the interface. The electron energy levels and their distribution in the electrode and in the solution are shown as are the ground-state electron energy level of the hydrogen atom and the exponentially distributed electron level of hydrogen atom corresponding to vibrational–rotational states of $H^+—H_2O$ bond.

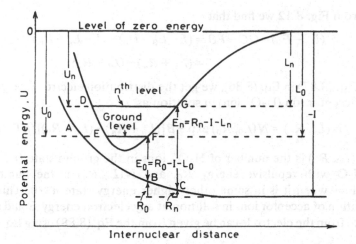

Fig. 8.12. Variation of potential energy of H^+-H_2O and e^- and of $H-H_2O$ with internuclear separation; R_0 and R_n are the $H-H_2O$ interaction energies produced by adiabatic electron transfer to H^+-H_2O with solvation energies L_0 and L_n, respectively.

Equation (8.84) can be rearranged to give

$$\beta = \frac{U_n - U_0}{E_n - E_0} = \frac{U_n - U_0}{(R_n - L_n) - (R_0 - L_0)} \tag{8.85}$$

These magnitudes depend on the relative *slopes* of the curves *FG* and *BC*. For example, if *BC* is sloping downward at the same rate that *FG* is sloping upward, β will have the value of $\frac{1}{2}$, but if their rate of sloping downward and upward are not the same, one will observe that the value of β deviates from $\frac{1}{2}$ and ranges from 0 to 1. This factor β is called the *symmetry factor*, because its value depends on the symmetry of the two potential energy surfaces, the energies of that are being represented by *FG* and *BC* (Fig. 8.12).

In Eq. (8.85), the value of β depends on the applied potential, since the reactant potential-energy curves shift vertically as a consequence of the change of applied potential. So, we may say in a physical sense that β gives a measure of the fraction of electrical energy used to give rise to *available* energy states of H_3O^+, so that electron tunneling from the cathode becomes possible to neutralize H_3O^+ ions to form hydrogen atoms. Thus, β gives a measure of the *efficiency* with which the electrical energy can be applied to accelerate the rate of electrochemical reactions.

Now, using $\beta(E_n - E_0)$ in place of $(U_n - U_0)$ we get from Eq. (8.84) that

$$N(E) = N(E_0) \exp[-\beta(E - E_0)/kT] \tag{8.86}$$

since from Fig. 8.12 we find that

$$(E_n - E_0) = GC - FB = (I - L_0 - R_0) - (I - L_n - R_n)$$
$$= (L_n + R_n) - (L_0 + R_0) \qquad (8.87)$$

Using Eq. (8.87) in Eq. (8.86), we get the distribution electron energy level in the acceptor ion (H_3O^+ ion) in solution as

$$N(L_n, R_n) = N(L_0, R_0) \exp - \beta[(L_n + R_n) - (L_0 + R_0)]/kT \qquad (8.88)$$

where (L_0, R_0) is the number of H_3O^+ ions in the ground state ($U_n = U_0$) $H^+ - H_2O$ with repulsive energy R_0, and $N(L_n, R_n)$ is the number of H_3O^+ ions when it is in some other higher energy state $n > 0$. Thus, the distribution of acceptor ions in solution at any electron energy E and at any distance from the electrode can be given from the Eq. (8.88) using Eq. (8.87) as

$$N(E) = N(E_0) \exp[-\beta(E - E_0)/kT] \qquad (8.89)$$

8.5.4. Velocity of Interfacial Electron Transfer in the Original Gurney Quantum Mechanical Model

What has been presented up to now is preliminary, to make clear the physical basis of the quantum mechanical model for the rate of transfer of electrons from metal to acceptor levels in the solution given by Gurney.[5] The aim hitherto has been to show the various influences upon the basic condition for radiationless transfer. The condition for radiationless transition that takes into account all the influences is given by Eq. (8.81). The distribution and the availability of acceptor level is obtained from Eq. (8.89).

An expression for the actual rate of transfer of electrons from an electrode (metal) to an acceptor in solution will contain several elements. First, we shall have to use Gamow's expression for the probability of tunneling through the barrier (Fig. 8.11). Second, we need to use the expression in Eq. (8.89) for the probability of existence on the solution side of the barrier of acceptor levels of energies equal to those energies that exist in the electrode. There will be other factors that affect the result, such as the number of electrons in the Fermi level and the number of acceptors in the ground state for acceptors in solution—almost equal to the number of (e.g., H_3O^+) ions in solution.

Gurney gave an expression for the current density i as proportional to

$$\iint n(E, V)N(E, x)P_T(E, x)\, dE\, dx \qquad (8.90)$$

where $n(E, V)$ is the number of electrons in the metal having the energy E at the electrode potential† V. It is given as the product of the density of states $\rho(E, V)$, of electrons in the electrode and the Fermi distribution $F(E, V)$

$$n(E, V) = \rho(E)F(E, V) = B(E + \rho_0 V)^{1/2}\{1 + \exp[(E + e_0 V - E_f)/kT]\}^{-1}$$
(8.91)

where E_f is the value of E at the Fermi level of the metal and is a constant at a given temperature and $N(E, x)$ is the number of acceptor ions having electron energy E at a distance x from the electrode. It is given by Eq. (8.84). $P_T(E, x)$ in Eq. (8.90) is the probability of tunneling of electron at energy E at a distance x from the electrode. It is given by the WKB expression for the rectangular barrier and height U_m and width l and given by Eq. (8.66).

Using the number of acceptor ions at a fixed distance at OHP and the expression (8.91), the current density can be written as proportional to

$$\int_0^{E_0} A(E + e_0 V)^{1/2}\{1 + \exp[(E + e_0 V - E_f)/kT]\}^{-1}N(E_0)$$
$$\times \exp[-\beta(E - E_0)/kT]\exp\left\{-\frac{2\pi l}{h}[2m(U_m - E)]^{1/2}\right\} dE \quad (8.92)$$

where U_m is the electron energy corresponding to barrier maximum (Fig. 8.11) and A is a constant.

A more detailed expression for the current density using the Gurney model is given in Chapter 10. The field effect on the barrier is neglected above. The utilization of the basic expression for current density [Eq. (8.92)] in its various improved forms has been discussed in Chapters 10, 12, and 13.

8.6. Tunneling through the Adsorbed Layer in Free Space

In this section, we will discuss the tunneling of electrons from the electrode surface covered with an adsorbed layer of atoms to a free space (vacuum) under the influence of the applied field. The discussion regarding the vacuum case will help in understanding the tunneling through the

† It is cogent to discuss what value of V, the electrode potential, is relevant here. It is, of course, in no way the conventional (or relative) electrode potentials, which are the potentials of certain cells used in relative measurements against hydrogen electrodes. V should be the absolute metal solution potential difference. It is difficult to obtain this quantity. If one takes V as the difference between the actual electrode potential and the electrode potential when the charge on the electrode surface is zero (p.z.c.), namely the rational potential, an error is made: the value neglects the sum of the surface potentials due to dipoles at the p.z.c. and that due to the electron overlap potential at p.z.c.[25]

adsorbed layer in solution. The pioneering theory in this field is by Stern, Gossling, and Fowler.[26] The interpretations of experimental measurement of field emission current through adsorbed layers of atoms have been based almost entirely on the Fowler–Nordheim analysis[27] and its extensions to include image forces and atomic polarizabilities. All these analyses are based on a model that is explicitly constructed to describe clean surfaces. The deviations of the calculated field emission current from the experimental results for clean surfaces are usually attributed to changes in the effective value of the work function used in the Fowler–Nordheim equation. In work by Duke and Alferieff,[28] it was pointed out that the assumption that adsorption alters the emission only via a work function modification is inadequate both because the adsorbate changes the qualitative shape of the electrostatic potential seen by the tunneling electron and *because the adsorbate can act as an energy momentum source* (in some cases as a sink) *for those electrons.*

This energy momentum change, which Duke and Alferieff[28] suggest is caused by adsorbed layers, has been observed in experiments such as those in which very thin films of metals are used as a means of determining the adsorption of materials from solution. In the work of Gileadi, and Cahan, Stoner, and Bockris,[29,30] the resistance of a thin platinum film was examined as a function of its potential. In contact with a solution of phosphoric acid, the conduction changed markedly with the potential. The adsorption of the various entities associated with phosphoric acid will, of course, change with the potential, and it is likely that the change in conductivity that was observed by these workers[29,30] arises from an energy momentum effect of the adsorbed phosphoric acid, which causes injection or subtraction of electrons from the lattice. In a formalistic sense, the electrochemist would say that "the electron overlap potential[25] changes with the adsorption of the material."

To examine the influences of an adsorbate on electron field emission apart from the effect of the adsorbate on the work function, Duke and Alferieff[28] constructed a one-dimensional schematic model of emission tunneling in which the adsorbate potential is replaced by a potential well of width W and depth $(E_f + \Phi - e_0 V - V_R)$ centered at a distance d outside the metal surface, as shown in Fig. 8.13. This model thus treats the adsorbed atom as an additional potential outside the metal surface composed of a finite-range attractive interaction and a repulsive core associated with the point–ion model.

The one-dimensional potential seen by the electron due to the presence of an adsorbate is

$$V(x) = \theta(x)\left[(E_f + \Phi) - e_0 X x + V_a(x - d) - \frac{e_0^2}{16\pi\varepsilon_0 x} + V_p(x)\right] \quad (8.93)$$

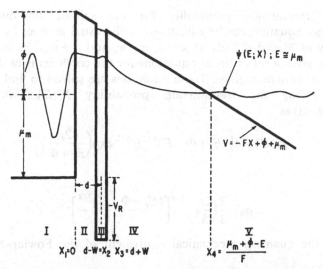

Fig. 8.13. Schematic illustration of the one-dimensional, one-electron pseudopotential used to describe field emission from a metal in the presence of an adsorbed atom.

where E_f is the Fermi energy of the metal, Φ is the work function, X is the electric field strength, e_0 is the electronic charge, and d is the distance of the center of the adsorbate potential from the mirror plane of the metal at $x = 0$. The adsorbate potential is V_a. A polarization potential results from the polarization of an adsorbed atom by the electric field and is defined as V_P, and $\theta(x)$ is the step function, i.e.,

$$\theta(x) = \begin{cases} 1 & \text{for } x > 0 \\ 0 & \text{for } x < 0 \end{cases} \tag{8.94}$$

In their mathematical development, Duke and Alferieff[28] neglect image potential and polarization potential V_P. To find the transmission probability, they solved the Schrödinger equation with the simplified potential (where V_a is not included) and obtained the outgoing wave in the region $\underline{V}$ (Fig. 8.13) as

$$\psi_{\underline{V}}(E, x) = C[\text{Bi}(\eta) + i\,\text{Ai}(\eta)] \tag{8.95}$$

where

$$\eta = k_F\{[(E_f + \Phi - E)/X] - x\} \tag{8.96}$$

and

$$k_F = (2mX/h^2)^{1/3} \tag{8.97}$$

The functions $\text{Ai}(\eta)$ and $\text{Bi}(\eta)$ are the Airy functions.[13]

The transmission probability for the soluble one-dimensional Schrödinger equation can be calculated in the usual manner by imposing continuity of Ψ and $d\Psi/dx$ at $x = 0$, $d - w$, and $d + w$ (Fig. 8.10). This procedure gives a set of linear equations for the coefficients of the eigenfunctions in various regions. These equations are solved to find C of Eq. (8.95) yielding the transmission probability $P_0(E)$ (after much simplification) as

$$P_0(E) = 4\left[(E_f + \Phi - E)^{1/2} E^{1/2} \exp\left(\frac{-2\rho_0}{E_f + \Phi}\right) \right] \qquad (8.98)$$

where

$$\rho_0 = \frac{2}{3}\left(\frac{2m}{\hbar^2}\right)^{1/2}\left[\frac{(E_f + \Phi - E)^{3/2}}{X} \right] \qquad (8.99)$$

which is the quantum mechanical equivalent of the Fowler–Nordheim equation.

Duke and Alferieff[28] used the potential due to an adsorbed layer V_a in two ways. One was to have a potential with a square well of finite width and another with an attractive delta function potential having its bound state at the energy $E = -B_e$. In the second case, V_a is expressed as

$$V_a(x) = (2\hbar^2 B_e/m)^{1/2}\delta(x - d) \qquad (8.100)$$

and obtain the transmission probability as:

$$P(E) = \frac{4\nu_1 \exp(-2\rho_0)}{[1 - \nu_2(1 - J^2)]^2 + \nu_1^2[1 - \nu_2(1 + J^2)]^2} \qquad (8.101)$$

where

$$\nu_1 = [(E_f + \Phi - E)/E]^{1/2} \qquad (8.102)$$

$$\nu_2 = [B_e/(E_f + \Phi - E - e_0 Xd)]^{1/2} \qquad (8.103)$$

$$J = \exp(-\alpha\rho_0) \qquad (8.104)$$

$$\alpha = \tfrac{3}{2}[e_0 Xd/(E_f + \Phi - E)] \qquad (8.105)$$

Although the transmission probability [Eq. (8.101)] is obtained with a simplified delta function potential, it suffices to predict the effect of the adsorbate and give a major qualitative modification of the Fowler–Nordheim[27] transmission probability. If the substance that adsorbs is a metal, the primary new effect is the *resonance transmission*. This occurs when the energy of an electron in the Fermi level is equal to an energy state of an electron in the adsorbed atom. If the energy of the transmitting particle drops below the resonance energy, the transmission coefficient becomes zero. For neutral adsorbates,[30,31] the transmission probability can be either enhanced or diminished by the presence of the adsorbate, depending upon the effect of the potential set up by the adsorbate, which can either raise or lower the barrier.† Duke and Alferieff[28] have also considered the situation where the barrier has a finite width, and is not a δ function.

The resonance tunneling through adsorbed layers was treated by Gadzuk[33] and by Gadzuk and Plummer[34] by considering the transition of an electron from a metallic to the free state induced by the perturbation of an applied field. The kinetics were treated by means of time-dependent perturbation theory in which the transition probability (in the absence of the adsorbed atom) was characterized by a matrix element of transition:

$$T_0 = \langle \Psi_f | V_F | \Psi_m \rangle \tag{8.106}$$

where the perturbing potential $V_F = e_0 X x$ is the potential of the field X between the electrode and the collector. Ψ_m is the wave function of the metal electron and Ψ_f is that of the free electron.

In the presence of an adsorbed atom, an additional term can be included in the Hamiltonian for field emission. The atom may effect the potential changing V_m to V'_m, where $V_m = 0$ for $x < 0$ and equals $E_f + \Phi$ for $x > 0$. The new Hamiltonian will then appear as

$$H = E_K + V'_m + V_F + V_a \tag{8.107}$$

Gadzuk[33,34] considers the possibilities of resonance tunneling to an intermediate state, e.g., the adsorbed atom core, and expresses the tunneling matrix as

$$T = \langle \Psi_f | V_F | \Psi_m \rangle + \frac{\langle \Psi_f | V_F | \Psi_a \rangle \langle \Psi_a | V_F | \Psi_m \rangle}{E - E_Q - i\Gamma} \tag{8.108}$$

where Ψ_a is the wave function of the transferring electron in the adsorbed atom. The energy of this transferring electron in the atom is

$$E_Q = \varepsilon_a^0 + \Delta E \tag{8.109}$$

$\varepsilon_a^0 = $ unperturbed eigenvalue

$\Delta E = $ energy shift of the atom near the metal surface $\tag{8.110}$

$= \langle \Psi_a | V'_m | \Psi_a \rangle$

† According to this prediction for the effect of additives on the emission rate into a vacuum (i.e., both positive and negative effects), the emission rate into an electrolytic solution is found to be either increased or decreased by the adsorption of materials onto the surface.[31,32]

Γ is the broadening of the atomic level due to presence of the metal and is given by

$$\Gamma = 2\pi \sum_m \delta(E - E_m)|\langle \Psi_m | V'_m | \Psi_a \rangle|^2 \qquad (8.111)$$

where the summation is over all metallic states m and E_m is the energy of the metallic states.

Equation (8.108) is a matrix element describing the transition matrix involving the resonance process. The transition probability is obtained using the Fermi expression

$$P_T = \frac{2\pi}{\hbar}|T|^2\rho(E_f) \qquad (8.112)$$

where $\rho(E_f)$ is the density of final free state of the electron.

8.7. Electron Transfer through Adsorbed Layers in Solution

To take into account the influence of tunneling probability on anodic oxygen evolution and other redox reactions at oxide-covered platinum electrodes, Schultze and Vetter[35-38] used the simplified rectangular WKB expression for tunneling probability,

$$P_T(d, E) \simeq \exp\left[-\frac{4\pi d}{h}(2m_e \, \Delta E_t)^{1/2}\right] \qquad (8.113)$$

where d is the thickness of the oxide layer (i.e., of the barrier) and ΔE_t is the mean height of the potential barrier for the electron transfer.

It was found by Vetter and Schultze[36,37] that at all oxide-covered electrodes the current density decreases with increasing d up to 10–15 Å. In the case of thicker layers (e.g., AuO or Fe_2O_3), however, the current density is nearly independent of d. This is because the thicker oxide layers behave as semiconductor electrodes. The thickness of the potential barrier becomes independent of d and also becomes smaller than the earlier important thickness of the oxide, because tunneling of electrons proceeds to the conduction band of the oxide from donors in solution. The rate of electron transfer thus becomes dependent on the semiconductor properties of the oxide layer.

Recently, Schultze and Stimming[39] investigated the electron transfer across the passive oxide layers on iron electrodes for the reaction system $[Fe(CN)_6]^{4-}$–$[Fe(CN)_6]^{3-}$. The dependence of current densities on the thickness of passive layer d were determined. At constant d, a Tafel relation was observed.

Schultze and Stimming[39] mentioned that electron transfer in such situations can be explained only by rate-determining tunnel processes with participation of the conduction band of an oxide layer. A comparison of the current–potential curves of the redox system at passive iron electrodes with those at other free-metal and semiconductor electrodes indicates that the oxide-covered electrode represents a transition between metal and semiconductor electrodes. The tunnel calculations were all carried out using the WKB expression (8.113).

Recently, Schmickler and Ulstrup[40a] gave a quantum mechanical theory of electron transfer reactions at film-covered metal electrodes. The theory is presented for elastic (i.e., no electron–electron or electron–phonon interaction is considered) tunneling through the film (see Memming and Mollers[40b]). The calculations are based on linear response theory and transfer Hamiltonian formalism.

Schmickler and Ulstrup[40a] considered the total reacting system as consisting of a film (oxide)-covered metal electrode in contact with a redox electrolyte (Fig. 8.14). According to transfer (or perturbation) Hamiltonian formalism, the three subsystems, i.e., the metal, the film, and the depolarizer (redox ions), are considered to be separable, and the total Hamiltonian for the system has been written as

$$H = H_L + H_R + H_T \qquad (8.114)$$

where H_L is the Hamiltonian of the isolated electrode, H_R that of the redox system, and H_T is an operator that transfers the electron from the depolarizer (redox ions) to the electrode or vice versa.

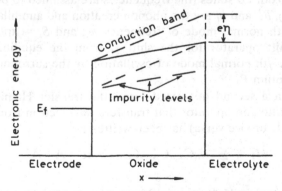

Fig. 8.14. Model for an oxide-covered electrode where the oxide contains impurities. The diagram represents the barrier for the electrons in the oxide layer only. No consideration of the barrier at the interface is made.

The Hamiltonian H_L in the second quantized form (see Landau and Lifschitz,[45] p. 221), has been written as

$$H_L = \sum_{k,\sigma} E_k C^+_{k,\sigma} C_{k,\sigma} \tag{8.115}$$

where k denotes the momentum of the electron in the metal electrode; σ is the spin of an electron in a given electronic state; E_k is the corresponding energy, which is independent of σ in the effective mass approximation; and $C^+_{k,\sigma}$ and $C_{k,\sigma}$ are the creation and annihilation operators, respectively (see Landau and Lifschitz[45]). The Hamiltonian for the redox system H_R is expressed in the form

$$H_R = \sum_{\gamma} N_{\gamma} h_{R,\gamma} \tag{8.116}$$

where γ describes the position of the depolarizer (redox ions) in the outer Helmholtz plane, N_{γ} ($= 0$ or 1) is the number of particles at site γ, and $h_{R,\gamma}$ is given by

$$h_{R,\gamma} = P_{\mathrm{red}} h_{\mathrm{red}} + P_{\mathrm{ox}} h_{\mathrm{ox}} \tag{8.117}$$

where, for the reduced and oxidized ions,

$$h_{\mathrm{red}} = E^0_{\mathrm{red}} + \sum_j \hbar \omega_j b^+_j b_j \tag{8.118}$$

and

$$h_{\mathrm{ox}} = E^0_{\mathrm{ox}} + \sum_j \hbar \omega_j S_j b^+_j b_j S^+ \tag{8.119}$$

P_{red} and P_{ox} are the so-called projection operators[†] onto the electronic states of reduced and oxidized ion, respectively, and E^0_{red} and E^0_{ox} are the corresponding ground-state electronic energies, the ω_j values are the frequencies of the jth normal mode of ion solvent–vibration in both reduced and oxidized states (the frequencies are assumed to be the same[§] in both states), b^+_j and b_j are the phonon creation and annihilation operators for the jth normal mode of frequency ω_j, and $S_j = \exp(i\xi_j P_j)$ is the coordinate shift operator for the shift ξ_j from the equilibrium values induced in the jth normal mode of oscillation by the surrounding solvent having momentum P_j.

Finally, in a second quantized form, the transfer Hamiltonian, H_T (i.e., the Hamiltonian operator that transfers the electron from the redox ion to electrode or vice versa) has been written as[40a]

$$H_T = \sum_{k,\sigma,\gamma} \Lambda_{fi} C^+_{k,\sigma} C^+_{\mathrm{ox},\gamma} C_{\mathrm{red},\gamma} + hc = A + A^+ \tag{8.120}$$

† Projection operator—see *Mathematics for Quantum Chemistry* by J. M. Anderson.[46]
§ For simplicity, the frequencies were considered the same. It is worth mentioning that the spectroscopic result for such frequencies is different in reduced and oxidized states.[41,42]

where Λ_{fi} is the electronic matrix element of transition. $C^+_{ox,\gamma}$ and $C_{red,\gamma}$ denote electronic creation and annihilation operators for a depolarizer (redox system) molecule at the site γ. The first term of Eq. (8.120) describes an electron transfer from the redox ion to the electrode, and the second term the reverse process. The function Λ_{fi} is determined inside the effective barrier† (oxide film layer) region [which is considered trapezoidal in shape (Fig. 8.14)].

Using the linear response formalism the expression for the current density for the anodic process, given by Schmickler and Ulstrup,[40a] follows from their Hamiltonians as

$$\vec{j} = \frac{4e_0 m_B}{\hbar^3} n_0 \left(\frac{\pi}{kTE_s}\right)^{1/2} \int_0^\infty dE \int_0^E dE_\| |\Lambda_{fi}(E, E_\|, e_0\eta)|^2$$

$$\times \left\{ 1 - n(E) \exp\left[\frac{(-E_s + E - \mu_L - e_0\eta)^2}{4E_s kT}\right]\right\} \tag{8.121}$$

where m_B is the effective mass of the electron in the barrier region, E_s is the reorganization energy of the nuclei of ions and dipoles, η is the overpotential, and $E_\|$ is the kinetic energy of the electron corresponding to the momentum parallel to the boundary of the barrier. The expression for the cathodic current density is given[40a] analogously.

The current was computed for different barrier parameters, and the Tafel plots for both anodic and cathodic current density for different film thickness and barrier height of 1.5 eV above the Fermi level are given in Fig. 8.15.§ The reorganization energy E_S was considered to be 1.0 eV for the process.[39]

Recently, Schmickler[43,44] applied the theory of resonance tunneling (developed by Duke and Alferieff[28] and also by Gadzuk[33,34]) for tunneling in a vacuum through an adsorbed layer (see Section 8.6) to explain a number of interesting features of Tafel plots in electrochemical situations when the electrode is covered by an oxide layer. These features are the following as observed by Schultze and Stimming[39] on a passive iron electrode with films of 10–30 Å thick.

(a) The Tafel plot shows a downward bending of the anodic curves at low overpotentials.

---

† It seems that the barrier at the film layer–electrolyte interfaces has been neglected in this treatment.

§ The authors[40a] do not obtain a Tafel equation. Although there is no guarantee that they would obtain a Tafel equation of transfer through metal oxide, the difficulty is that they do not obtain the Tafel equation even for a thickness of 4 Å, which is the typical thickness of the barrier at the electrode solution interface (see Fig. 8.15).

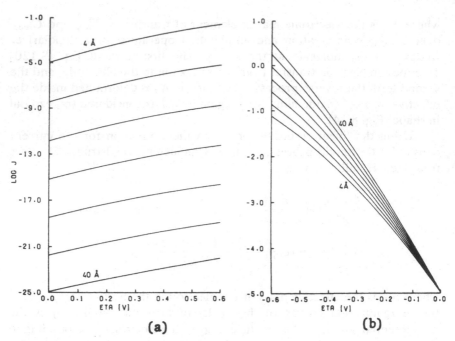

Fig. 8.15. Tafel plots of the (a) anodic and (b) cathodic current densities for film thickness varying from 4 to 40 Å in steps of 6 Å. The cathodic curves have been normalized for $\eta = 0$. Barrier height = 1.5 eV. VL = VR = 1.5; harmonic.

(b) Upward bending of both anodic and cathodic curves at high over-potentials.

(c) Small transfer coefficient in the anodic and cathodic Tafel regions (e.g., $\alpha_c + \alpha_a = 0.6$ for oxide thickness of 17.1 Å[39]).

Schmickler[43,44] considered a metal electrode covered by an oxide film of uniform thickness a in contact with a redox electrolyte to find the resonance tunneling current through the adsorbed layer. The film is assumed to be uniformly *doped* with impurities, the energy levels of which lie at a fixed distance below the bottom of the conduction band V_B. V_B varies linearly with distance x from the metal–oxide layer interface, and $V_R = V_B$ at $x = a$ at the oxide layer–electrolyte interface.

Using the above model for the metal-adsorbed-layer electrolyte system, Schmickler gave an expression based on the Franck–Condon approximation and second-order perturbation theory with respect to the electronic matrix element for the anodic current density as

$$\vec{j}_r(\eta) = 2\pi C[1 - n(E_r)]D(E_r, \eta)\rho(E_r)|\langle e|V_1|r\rangle|^2|\langle r|V_2|m\rangle|^2\Gamma^{-2}(E_r)$$

$$(8.122)$$

where C is a constant, $n(E_r)$ is the Fermi function corresponding to the energy at the impurity state E_r. $D(E_r, \eta)$ is the probability of finding a level E_r for an electron donor at an overpotential η in the electrolyte. $\rho(E_r)$ is the density of electronic states in the metal. $|e\rangle$, $|r\rangle$, and $|m\rangle$ represent wave functions for the electron in the ion in solution, in the localized impurity in the oxide layer, and in the metal, respectively. V_1 is the perturbation operator coupling $|e\rangle$ and $|r\rangle$, and V_2 couples $|r\rangle$ and $|m\rangle$. Γ represents the lifetime broadening of the impurity state in the oxide.

The lifetime broadening due to the transition from metal, the impurity in the oxide, and oxide to a solution to a first-order perturbation is given by[43]

$$\Gamma = (2\pi/\hbar)[\rho(E_r)|\langle m|V_2|r\rangle|^2 + |\langle r|V_1|e\rangle|^2] \qquad (8.123)$$

To give a workable expression for the resonance transition, Schmickler[43] expressed the two matrix elements in Eq. (8.122) in terms of the WKB approximation[45] (for the impurity positioned at $x = l$ and with a barrier having an effective width in respect to its potential energy well of $2w$) as

$$|\langle m|V_2|r\rangle|^2 \simeq \frac{1}{\rho(E_r)} C_1 \exp\left(-2\int_0^{l-w} k(x)\,dx\right) \qquad (8.124a)$$

and

$$|\langle r|V_1|e\rangle|^2 \simeq C_2 \exp\left(-2\int_{l+w}^a k(x)\,dx\right) \qquad (8.124b)$$

where

$$k(x) = \left(\frac{2m_{\text{eff}}[V_B(x) - E_r(l)]}{\hbar^2}\right)^{1/2} \qquad (8.125)$$

is the wave number of the electron in the barrier and m_{eff} is the effective mass of the electron in the oxide. C_1 and C_2 are slowly varying functions of energy.

To express $D(E_r, \eta)$ for the electrolyte, the usual harmonic approximation has not been used but has been expressed by an empirical form (termed "a linear approximation"), i.e.,

$$D(E_r, \eta) = \exp\{-[E_a + (E_r - e_0\eta)/2]/kT\} \qquad (8.126)$$

corresponding to a constant symmetry factor of 0.5. This step appears to be inconsistent with the basic model used, which does not allow for a constant transfer coefficient, because it involves an expression for the dependence of rate on potential that in fact includes a squared term in potential. The neglect of this term to obtain experiment-consistent results does not appear justified (see Table 7.1).

The final expression for the contribution of an impurity situated at $x = l$ to the anodic current is

$$\Delta \vec{j}(l) = C_1 2\pi C [1 - n(E_r)] \exp\{-[2E_a^0 + (E_r - e\eta)]/2kT\}[R_1(l)R_2(l)$$
$$+ R_2(l)] \tag{8.127}$$

where

$$R_1(l) = \exp\{-\tfrac{4}{3}(2m_{\text{eff}})^{1/2}/\hbar g[V_2 - E_r(l) + g(l - w)^{3/2} - (V_L - E_r)^{3/2}]\}$$
$$\tag{8.128a}$$

$$R_2(l) = \exp\{-\tfrac{4}{3}(2m_{\text{eff}})^{1/2}/\hbar g(V_R - E_r)^{3/2} - [V_L + g(l + w - E_r)]^{3/2}\}$$
$$\tag{8.128b}$$

and $g = (V_R - V_L)/a$ is the slope of the barrier and

$$E_r(l) = E_r^0 + gl \tag{8.129}$$

The average contribution of a single impurity to the anodic resonance current is

$$\vec{j}_r = \frac{1}{a} \int_0^a \Delta \vec{j}(l) \, dl \tag{8.130}$$

The corresponding expression for the cathodic direction is given as[43]

$$\overleftarrow{j}_r = \frac{1}{a} 2\pi C C_1 \int_0^a n(E_r) \exp\left\{-\frac{[E_a^0 - (E_r - e_0\eta)/2]}{kT}\right\} \frac{R_1(l)R_2(l)}{R_1(l) + R_2(l)} \, dl$$
$$\tag{8.131}$$

Schmickler[43] computed both the anodic and cathodic current densities from Eqs. (8.130) and (8.131), respectively. Figure 8.15 shows the Tafel plots of $\vec{j}_r$ and $\overleftarrow{j}$. Here the barrier is considered flat at the equilibrium and relatively low ($V_B = 0.2$ eV at $\eta = 0$); E_r^0 (the energy level in the adsorbed layer) is taken to vary over the range -0.05 to -0.2 eV. The result is in quantitative agreement with experiment.[39] The value of the transfer coefficient obtained by Schmickler is in the range of 0.23 to 0.31 for α_a and 0.35 to 0.43 for α_c for the oxide thickness $a = 20$ Å at $\eta = 400$ mV for the anodic, and -200 mV for the cathodic case, having $V_L = V_R = 0.2$ eV for different energy of E_r^0 varying from -0.05 to -0.2 eV.

Appendix 8.I. Derivation of Eq. (8.52) for the WKB Transmission Coefficient from WKB Wave Functions

We have the WKB wave functions for various regions [see Fig. 8.2 and Eqs. (8.49), (8.50), and (8.51)]:

$$\psi_I(x) = \frac{A}{[k(x)]^{1/2}} \exp\left[i\int_a^x k(x)\,dx\right] + \frac{B}{[k(x)]^{1/2}} \exp\left[-i\int_a^x k(x)\,dx\right], \quad x < a$$

$$\text{(I.1)}$$

$$\psi_{II}(x) = \frac{C}{[\kappa(x)]^{1/2}} \exp\left[\int_a^x \kappa(x)\,dx\right] + \frac{D}{[\kappa(x)]^{1/2}} \exp\left[-\int_a^x \kappa(x)\,dx\right],$$

$$a > x < b \quad \text{(I.2)}$$

$$\psi_{III}(x) = \frac{F}{[k(x)]^{1/2}} \exp\left[i\int_b^x k(x)\,dx\right] + \frac{G}{[k(x)]^{1/2}} \exp\left[-i\int_b^x k(x)\,dx\right],$$

$$x > b \quad \text{(I.3)}$$

To find the transmission coefficient P_T from Eq. (8.56), we need to express the coefficient A of incident wave in terms of coefficient F of transmitted wave. To solve this we utilize the following connection formulas.[†] First,

$$\psi(x) = \frac{1}{[k(x)]^{1/2}} \exp\left[\pm i\int_b^x k(x)\,dx\right] \qquad \text{for } x > b \text{ in region III} \quad \text{(I.4)}$$

then $\psi(x)$ of region III can be expressed in terms of wave vector $\kappa(x)$ of region II, i.e.,

$$\psi(x) = \frac{1}{[\kappa(x)]^{1/2}}\left\{\exp\left[\int_x^b \kappa(x)\,dx \pm \frac{i\pi}{4}\right]\right.$$

$$\left. + \frac{1}{2}\exp\left[-\int_x^b \kappa(x)\,dx \pm \frac{i\pi}{4}\right]\right\} \qquad \text{for } a < x < b \text{ in region II} \quad \text{(I.5)}$$

The second connection formula needed is

$$\psi(x) = \frac{1}{[\kappa(x)]^{1/2}} \exp\left[\pm\int_a^x \kappa(x)\,dx\right] \qquad \text{for } a < x < b \text{ in region II} \quad \text{(I.6)}$$

Then $\psi(x)$ of region II can be expressed in terms of wave vector $k(x)$ of region I, i.e.,

$$\psi(x) = \frac{1}{[k(x)]^{1/2}} \exp\left[\int_x^a k(x)\,dx + \frac{\pi}{4}\right] \qquad \text{for } x < a \text{ in region I} \quad \text{(I.7)}$$

[†] The connection formulas have been taken from F. Constantinescu and P. Magyari.[48]

For simplification, we express

$$\int_x^b \kappa(x)\,dx = \int_a^b \kappa(x)\,dx - \int_a^x \kappa(x)\,dx$$

$$= \alpha - \int_a^x \kappa(x)\,dx \tag{I.8}$$

where $\alpha = \int_a^b \kappa(x)\,dx$.

Now, using the first connection formula (I.4) and (I.5) and also (I.8), one can express the first term of (I.3) as

$$\psi(x) = \frac{F}{[k(x)]^{1/2}} \exp\left[i\int_b^x k(x)\,dx\right] \qquad \text{for } x > b$$

$$= \frac{F}{[\kappa(x)]^{1/2}} \left\{ \exp\left[\int_x^b \kappa(x)\,dx - \frac{i\pi}{4}\right] + \frac{1}{2}\exp\left[-\int_x^b \kappa(x)\,dx + \frac{i\pi}{4}\right] \right\}$$

$$= \frac{F}{2[\kappa(x)]^{1/2}} \left\{ e^{i\pi/4}\, e^{-\alpha} \exp\left[\int_a^x \kappa(x)\,dx\right] - 2\, e^{-i\pi/2}\, e^{\alpha} \exp\left[-\int_a^x \kappa(x)\,dx\right] \right\}$$

$$= \frac{F}{2[\kappa(x)]^{1/2}} \left\{ e^{i\pi/4}\, e^{-\alpha} \exp\left[\int_a^x \kappa(x)\,dx\right] \right.$$

$$\left. -2i\, e^{\alpha} \exp\left[-\int_x^a \kappa(x)\,dx\right] \right\} \qquad \text{for } a < x < b \tag{I.9}$$

since $e^{-i\pi/2} = i$.

Using (I.6) and (I.7), one can express (I.9) as

$$\psi = \frac{F}{2[k(x)]^{1/2}}\, e^{i\pi/4} \left\{ e^{-\alpha} \cos\left[\int_x^a k(x)\,dx + \frac{\pi}{4}\right] - 4i\, e^{\alpha} \sin\left[\int_x^a k(x)\,dx + \frac{\pi}{4}\right] \right\}$$

$$= \frac{F}{2[k(x)]^{1/2}}\, e^{i\pi/4} \left\{ e^{-\alpha}\frac{1}{2} \exp\left[i\int_x^a k(x)\,dx + \frac{i\pi}{4}\right] \right.$$

$$+ e^{\alpha}\frac{1}{2} \exp\left[-i\int_x^a k(x)\,dx - \frac{i\pi}{4}\right]$$

$$- 2\, e^{\alpha} \exp\left[i\int_x^a k(x)\,dx + \frac{i\pi}{4}\right] + 2\, e^{\alpha} \exp\left[-i\int_x^a k(x)\,dx - \frac{i\pi}{4}\right] \right\}$$

$$= \frac{F}{2[k(x)]^{1/2}}\, e^{i\pi/4}\, e^{-i\pi/4} \left\{ \frac{i}{2}\, e^{-\alpha} \exp\left[i\int_x^a k(x)\,dx\right] \right.$$

$$+ \frac{1}{2}\, e^{-\alpha} \exp\left[-i\int_x^a k(x)\,dx\right]$$

$$-2i\,e^{\alpha}\,\exp\left[i\int_{x}^{a}k(x)\,dx\right]+2\,e^{\alpha}\,\exp\left[-i\int_{x}^{a}k(x)\,dx\right]\right\}$$

$$=\frac{F}{2[k(x)]^{1/2}}\left\{\left(2\,e^{\alpha}+\frac{e^{-\alpha}}{2}\right)\exp\left[i\int_{a}^{x}k(x)\,dx\right]\right.$$

$$\left.-i\left(2\,e^{\alpha}-\frac{e^{-\alpha}}{2}\right)\exp\left[-i\int_{a}^{x}k(x)\,dx\right]\right\} \tag{I.10}$$

Similarly, the second term of (I.3) can be expressed as

$$\psi=\frac{G}{[k(x)]^{1/2}}\exp\left[-i\int_{b}^{x}k(x)\,dx\right]\qquad\text{for region }x>b$$

$$=\frac{G}{2[k(x)]^{1/2}}\left\{\left[2\,e^{\alpha}+\frac{e^{-\alpha}}{2}\right]\exp\left[-i\int_{x}^{a}k(x)\,dx\right]\right.$$

$$\left.+i\left(2\,e^{\alpha}-\frac{e^{-\alpha}}{2}\right)\exp\left[i\int_{a}^{x}k(x)\,dx\right]\right\}\qquad\text{for region }x<a \tag{I.11}$$

We notice that expression (I.10) and (I.11) represent the wave function of region I for $x<a$ in terms of coefficients F and G, respectively. Hence, we can equate the coefficients of the terms

$$[k(x)]^{-1/2}\exp\left[i\int_{a}^{x}k(x)\,dx\right]\quad\text{and}\quad[k(x)]^{-1/2}\exp\left[-i\int_{a}^{x}k(x)\,dx\right]$$

of (I.10) and (I.11) with those of (I.1) [since this expression (I.1) also represents the wave function of region I for $x<a$]. We get

$$A=\frac{1}{2}F\left(2\,e^{\alpha}+\frac{e^{-\alpha}}{2}\right)+iG\left(2\,e^{\alpha}-\frac{e^{-\alpha}}{2}\right)$$

$$B=-\frac{i}{2}F\left(2\,e^{\alpha}-\frac{e^{-\alpha}}{2}\right)+G\left(2\,e^{\theta}+\frac{e^{-\alpha}}{2}\right) \tag{I.12}$$

Writing $\theta=e^{\alpha}$, the result becomes

$$\begin{pmatrix}A\\B\end{pmatrix}=\frac{1}{2}\begin{bmatrix}2\theta+\dfrac{1}{2\theta} & i\left(2\theta-\dfrac{1}{2\theta}\right)\\[2ex]-i\left(2\theta-\dfrac{1}{2\theta}\right) & 2\theta+\dfrac{1}{2\theta}\end{bmatrix}\begin{pmatrix}F\\G\end{pmatrix} \tag{I.13}$$

Hence,

$$A=\frac{1}{2}\left[F\left(2\theta+\frac{1}{2\theta}\right)+iG\left(2\theta-\frac{1}{2\theta}\right)\right] \tag{I.14}$$

Appendix 8.II.   Solution of Eq. (8.68) and Derivation of Transmission Coefficient P_T for Eckart Barrier

Expressing $\xi = -\exp 2\pi x/d$ in Eq. (8.68), one gets

$$\frac{d^2\psi}{dx^2} + \frac{2m}{\hbar^2}\left[\frac{A\xi}{1-\xi} + \frac{B\xi}{(1-\xi)^2} + E\right]\psi = 0 \qquad (II.1)$$

we change the variable in Eq. (II.1) from x to ξ, and to do this we note that

$$\frac{d^2\psi}{dx^2} = \frac{d^2\psi}{d\xi^2}\left(\frac{d\xi}{dx}\right)^2 + \frac{d\psi}{d\xi}\frac{d^2\xi}{dx^2} \qquad (II.2)$$

and since $d\xi/dx = 2\pi\xi/d$, $d^2\xi/dx^2 = 4\pi^2\xi/d^2$, Eq. (II.1) becomes

$$\xi^2\frac{d^2\psi}{d\xi^2} + \xi\frac{d\psi}{d\xi} + \frac{2md^2}{\hbar^2}\left[E - \frac{A\xi}{1-\xi} + \frac{B\xi}{(1-\xi)^2}\right]\psi = 0 \qquad (II.3)$$

We now define a new variable $y = (1-\xi)^{-1}$. After multiplying through by $y(1-y)^{-1}$, Eq. (II.3) becomes

$$y(1-y)\frac{d^2\psi}{dy^2} + (1-2y)\frac{d\psi}{dy} + \frac{2md^2}{\hbar^2}\left(\frac{E}{y(1-y)} - \frac{A}{y} - B\right)\psi = 0 \qquad (II.4)$$

To solve the differential equation (8.68) we first examine its asymptotic condition to give the solution for the wave function when $x \to -\infty$ and when $x \to +\infty$, since we consider the incident particle is from the left in the region $-\infty$ and is transmitted in the region $+\infty$. As $x \to -\infty$ or $+\infty$, $v(x) \to 0$, so Eq. (8.68) becomes

$$\frac{d^2\psi}{dx^2} + \frac{2m}{\hbar^2}E\psi = 0$$

the solution of this equation is

$$\psi_{x\to-\infty} = a_1\exp(ik_1x) + a_2\exp(-ik_1x) \qquad (II.5)$$

where

$$k_1 = (2mE/\hbar^2)^{1/2} \qquad (II.6)$$

The term $a_1\exp(ik_1x)$ and $a_2\exp(-ik_1x)$ represent the incident and reflected waves, respectively, and

$$\psi_{x\to\infty} = b_1\exp(ik_2x) + b_2\exp(-ik_2x) \qquad (II.7)$$

where

$$k_2 = \left[\frac{2m(E-A)}{\hbar^2}\right]^{1/2} \quad \text{for } E > A \tag{II.8}$$

We deal with the case where particles approach the barrier from $x = -\infty$ so there can be no wave moving to the left at $x = +\infty$ (see Section 8.2). The term $b_1 \exp(ik_2 x)$ represents the transmitted wave. We can put $b_1 = 1$, since one is at liberty to choose the absolute value of any of the parameters a_1, a_2, b_1, b_2, as one can solve for the ratios of the coefficients a_1, a_2, and b_1. The asymptotic form of Eq. (II.5) can be expressed as a function ξ as

$$\exp(ik_2 x) = [\exp(2\pi x/d)]^{ik_2 d/2\pi} = (-\xi)^{i\beta} \tag{II.9}$$

where $\beta = k_2 d/2\pi$. Thus $\psi \rightarrow (-\xi)^{i\beta}$ as $\xi \rightarrow -\infty$ (i.e., $x \rightarrow \infty$).
   Similarly, from Eq. (II.7),

$$\psi \rightarrow a_1(-\xi)^{i\alpha} + a_2(-\xi)^{-i\alpha} \tag{II.10}$$

as $\xi \rightarrow 0$ (i.e., $x \rightarrow -\infty$), where $\alpha = k_1 d/2\pi$.
   So far we have found out asymptotic forms of the solution of Eqs. (II.1) and (II.4). This is because we need the behavior of the wave function in the incident region ($x = -\infty$) and the transmitted region ($x = +\infty$). Now let us consider the solution of Eq. (II.3) as a product of two functions, i.e.,

$$\psi = S(\xi)U(\xi) \tag{II.11}$$

where $S(\xi)$ is the asymptotic function and represents the asymptotic behavior of the wave function ψ. Hence, $S(\xi)$ should be a function that satisfies the above two asymptotic conditions given in Eqs. (II.9) and (II.10). The functions $S(\xi)$ can be written as

$$S(\xi) = (1-\xi)^{i\beta}\left[\frac{\xi}{(\xi-1)}\right]^{i\alpha} \tag{II.12}$$

since the right-hand side of Eq. (II.12) goes to $(-\xi)^{i\beta}$ as $\xi \rightarrow \infty$ and goes to $(-\xi)^{\alpha}$ as $\xi \rightarrow 0$. Thus, we can write the solution of Eq. (II.4) from Eqs. (II.11) and (II.12) as

$$\psi = (1-\xi)^{i\beta}\left(\frac{\xi}{1-\xi}\right)^{i\alpha} U(\xi) \tag{II.13}$$

Now, we need still to determine the $U(\xi)$. To do this, we first transform Eq. (II.13) in terms of y [since $y = 1/(1-\xi)$] as

$$\psi(y) = (y)^{-i\beta}(y-1)^{i\alpha}U(y) \tag{II.14}$$

and put Eq. (II.14) in Eq. (II.4). After detailed algebra and multiplying through by $y^{i\beta}(y-1)^{i\alpha}$, we obtain

$$y(1-y)\frac{d^2U}{dy^2}+[1-2i\beta-(2+2i\alpha-2i\beta)y]\frac{dU}{dy}$$

$$+\left[-2\alpha\beta-B'+(1-y)^{-1}(-\alpha^2y+i\alpha y-i\alpha)\right.$$

$$\left.+y^{-1}(-A'+i\beta y+\beta^2y-\beta^2)+\frac{E'}{y(1-y)}\right]U=0 \qquad (II.15)$$

where $A'=2md^2A/\hbar^2$, $B'=2md^2B/\hbar^2$ and $E'=2md^2E/\hbar^2$. We can simplify the coefficient of U in Eq. (II.15) as

$$-2\alpha\beta-B'+I(\beta-\alpha)+\frac{A'-E'+\beta^2+(\alpha^2+\beta^2)y(y-1)+(\alpha^2-\beta^2-A')y}{y(y-1)}$$

$$(II.16)$$

But from the definition of α and β, we get $\alpha^2=E'$, $\beta^2=E'-A'$, and $\alpha^2-\beta^2=A'$. Hence, using these values in Eq. (II.16), one gets the coefficient of U as

$$(\alpha-\beta)^2-B'+i(\beta-\alpha) \qquad (II.17)$$

and putting (II.17) in place of coefficient U in Eq. (II.15), we get

$$y(1-y)\frac{d^2U}{dy^2}+[1-2i\beta-(2+2i\alpha-2i\beta)y]\frac{du}{dy}$$

$$+[(\alpha-\beta)^2-B'+i(\beta-\alpha)]U=0 \qquad (II.18)$$

This expression (II.18) can be identified with the standard form of the hypergeometric equation of the type

$$y(1-y)\frac{d^2U}{dy^2}+[c-(a+b+1)]\frac{du}{dy}-abU=0 \qquad (II.19)$$

for

$$c=1-2i\beta$$

$$a+b+1=2+2i\alpha-2i\beta \qquad (II.20)$$

$$ab=B'-(\alpha-\beta)^2-i(\alpha-\beta)$$

if we choose

$$a=\tfrac{1}{2}+i(\alpha-\beta+\gamma)$$

$$b=\tfrac{1}{2}+i(\alpha-\beta-\gamma) \qquad (II.21)$$

with $\gamma=\tfrac{1}{2}(4B'-1)^{1/2}$, it is found that the second and third equations of (II.20) are satisfied.

The general solution of Eq. (II.19) is

$$U(y) = C_1 F(a, b, c, y) + C_2 y^{1-c} F(1 + a - c + 1 + b - c + 2 - c - y) \quad \text{(II.22)}$$

where C_1 and C_2 are arbitrary constants and F is the hypergeometric function. To evaluate C_1 and C_2 we must fit Eq. (II.22) to the boundary conditions of the problem by determining the behavior of $\psi(y)$ as $y \to 0$ (i.e., $x \to +\infty$) and $y \to 1$ (i.e., $x \to -\infty$). It can be shown from the polynomial expression of the hypergeometric function that

$$F(a, b, c, 0) = 1 \quad \text{(II.23)}$$

Hence, from Eq. (II.22), we get

$$U(y) = C_1 + C_2 y^{1-c} \quad \text{when } y \to 0 \quad \text{(II.24)}$$

From Eq. (II.24) for $y \to 0$, we get

$$U(y) = [C_1(1 - \xi)^{i\beta} + C_2(1 - \xi)^{-i\beta}]\left(\frac{\xi}{\xi - 1}\right)^{i\alpha}$$

$$= C_1(-\xi)^{i\beta} + C_2(-\xi)^{i\beta} = C_1 \exp(ik_2 x) + C_2 \exp(-ik_2 x) \quad \text{(II.25)}$$

since when $y \to 0$, $\xi \to -\infty$ and $\xi/(\xi - 1) \to 1$.

Equations (II.25) and (II.7) are identical. Hence, according to Eq. (II.7), we can choose $C_1 = 1$ and $C_2 = 0$. Thus, in general, (II.22) can be written as

$$U(y) = F(a, b, c, y) \quad \text{(II.26)}$$

and from Eqs. (II.14) and (II.26), we get

$$\psi = y^{-i\beta}(1 - y)^{i\alpha} F(a, b, c, y) \quad \text{(II.27)}$$

The function $F(a, b, c, y)$ is conversant for $6 < y < 2$ and can be expressed as the linear combination of two hypergeometric functions such that[49]

$$F(a, b, c, y) = MF(a, b, a + b - c + 1, 1 - y)$$
$$+ (1 - y)^{c - a - b} NF(c - a, c - b, c - a - b + 1, 1 - y) \quad \text{(II.28)}$$

where

$$M = \frac{\Gamma(c)\Gamma(c - a - b)}{\Gamma(c - a)\Gamma(c - b)} \quad \text{(II.29)}$$

and

$$N = \frac{\Gamma(c)\Gamma(a + b - c)}{\Gamma(a)\Gamma(b)} \quad \text{(II.30)}$$

where Γ represents the gamma function. We obtained the hypergeometric function F equal to unity for $x \to +\infty$ (i.e., when $y \to 0$); we need to find out the hypergeometric function when $x \to -\infty$ (i.e., when $y \to 1$). We see when $y \to 1$, $(1-y) \to 0$, so both F functions on the right-hand side of Eq. (II.28) go to unity. Thus,

$$\lim_{y \to 1} F(a, b, c, y) = M + (1-y)^{-2i\alpha} N \qquad (\text{II.31})$$

since $c - a - b = -2i\alpha$ (from Eq. II.20 and II.21).

Therefore, as $y \to 1$, one gets from Eq. (II.26).

$$U(y) = M + (1-y)^{-2i\alpha} b_2 \qquad (\text{II.32})$$

and for Eq. (II.27),

$$\psi(y) = M y^{-i\beta}(1-y)^{i\alpha} + N y^{-i\beta}(1-y)^{-i\alpha} \qquad (\text{II.33})$$

and in the limit $\xi \to 0$ (i.e., $y \to 1$), Eq. (II.33) can be written in the form

$$\psi(\xi) = M(-\xi)^{i\alpha} + N(-\xi)^{-i\alpha} \qquad (\text{II.34})$$

Computing Eq. (II.34) with (II.10) one finds that $a_1 = M$ and $a_2 = N$.

The reflection coefficient R of the particle can be obtained in terms of coefficients a_1 and a_2 as

$$R = \frac{|a_2|^2}{|a_1|^2} = \frac{|N|^2}{|M|^2} \qquad (\text{II.35})$$

Now, putting the value of N and M from Eqs. (II.29) and (II.30) in (II.35), one gets

$$R = \left| \frac{\Gamma(a+b-c)}{\Gamma(a)\Gamma(b)} \frac{\Gamma(c-a)\Gamma(c-b)}{\Gamma(c-a-b)} \right|^2 \qquad (\text{II.36})$$

From Eqs. (II.20) and (II.21), one finds that $a + b - c = 2i\alpha$ and $c - a - b = -2i\alpha$. Hence, one gets

$$\left| \frac{\Gamma(a+b-c)}{\Gamma(c-a-b)} \right|^2 = \left| \frac{\Gamma(0+i2\alpha)}{\Gamma(0-i2\alpha)} \right|^2 = 1 \qquad (\text{II.37})$$

Since in general when any quantity, say, u and v, are real

$$\frac{\Gamma(u+iv)}{\Gamma(u-iv)} = 1 \qquad (\text{II.38})$$

Therefore, Eq. (II.36) becomes

$$R = \left| \frac{\Gamma(c-a)\Gamma(c-b)}{\Gamma(a)\Gamma(b)} \right|^2 \qquad (\text{II.39})$$

It can be shown that in general

$$\Gamma(\tfrac{1}{2}+iv) = \tfrac{1}{2}(\cosh \pi v)^{-1/2} \qquad \text{for real } v \qquad (II.40)$$

We get from Eqs. (II.20) and (II.21) that

$$c - a = \tfrac{1}{2} + i(-\alpha - \beta - \gamma)$$
$$c - b = \tfrac{1}{2} + i(-\alpha - \beta + \gamma)$$
$$a = \tfrac{1}{2} + i(\alpha - \beta + \gamma)$$
$$b = \tfrac{1}{2} + i(\alpha - \beta - \gamma) \qquad (II.41)$$

α and β are certainly real, and for γ real one gets

$$R = \frac{\cosh \pi(\alpha - \beta + \gamma) \cosh \pi(\alpha - \beta - \gamma)}{\cosh \pi(\alpha + \beta + \gamma) \cosh \pi(\alpha + \beta - \gamma)} \qquad (II.42)$$

Since $\cosh(x+y)\cosh(x-y) = \tfrac{1}{2}(\cosh 2x + \cosh 2y)$, Eq. (II.42) can be reduced to

$$R = \frac{\cosh 2\pi(\alpha - \beta) + \cosh 2\pi\gamma}{\cosh 2\pi(\alpha + \beta) + \cosh 2\pi\gamma} \qquad (II.43)$$

The transmission coefficient P_T is given by

$$P_T = 1 - R$$
$$= \frac{\cosh 2\pi(\alpha + \beta) - \cosh 2\pi(\alpha - \beta)}{\cosh 2\pi(\alpha + \beta) + \cosh 2\pi\gamma} \qquad (II.44)$$

This above expression for P_T is applicable when γ is real. γ is real only when $B' > \tfrac{1}{4}$ and $B' > (2d)^2/4\chi^2$, and thus B' is proportional to the ratio of the width $2d$ of the potential (barrier) and the wavelength λ of the incident particle. When $B' > \tfrac{1}{4}$, λ is small compared to $2d$ and γ is real, and this is true for many physical cases. But for the case where γ is imaginary, the Γ function of Eq. (II.39) may be evaluated by the method given by Nielsen.[50]

References

1. A. Beiser (ed.), *The World of Physics*, McGraw–Hill, New York (1960).
2. W. H. Cropper, *The Quantum Physicists, and an Introduction to Their Physics*, Oxford University Press, London (1970).
3. G. Gamow, *Z. Phys.* **51**, 204 (1928).
4. R. W. Gurney and E. U. Condon, *Phys. Rev.* **33**, 127 (1929).

5. R. W. Gurney, *Proc. Roy. Soc.* **A134**, 137 (1931).
6. J. A. V. Butler, *Proc. Roy. Soc.* **A125**, 423 (1936).
7. R. A. Marcus, *J. Chem. Phys.* **24**, 966 (1956).
8. R. A. Marcus, *J. Chem. Phys.* **43**, 679 (1965).
9. S. Gasiorowicz, *Quantum Physics*, John Wiley and Sons, New York (1974).
10. R. M. Eisberg, *Fundamentals of Modern Physics*, John Wiley and Sons, New York (1967).
11. D. Park, *Introduction to Quantum Theory*, McGraw–Hill, New York (1974).
12. E. Merzbacher, *Quantum Mechanics*, John Wiley and Sons, New York (1970).
13. M. Abramowitz and I. A. Stegun (eds.), *Handbook of Mathematical Functions*, Dover Publications, New York (1964).
14. C. Eckart, *Phys. Rev.* **35**, 1303 (1930).
15. B. E. Conway, *Can. J. Chem.* **37**, 178 (1959).
16. J. O'M. Bockris and D. B. Matthews, *J. Chem. Phys.* **44**, 298 (1966); *Proc. Roy. Soc. (London)* **A297**, 479 (1966).
17. St. G. Christov, *Ann. Phys.* **15**, 87 (1965).
18. St. G. Christov, *J. Res. Inst. Catal. Hokkaido Univ.* **16**, 169 (1968).
19. St. G. Christov, *J. Res. Inst. Catal. Hokkaido Univ.* **24**, 27 (1976).
20. D. B. Matthews and J. O'M. Bockris, *Modern Aspects of Electrochemistry* (B. E. Conway and J. O'M. Bockris, eds.), Volume 6, Plenum Press, New York (1971).
21. J. O'M. Bockris and S. Srinivasan, *J. Electrochem. Soc.* **111**, 844, 853, 858 (1964).
22. P. M. Bowden and E. K. Rideal, *Proc. Roy. Soc. (London)* **A120**, 59 (1928).
23. M. Oliphant and P. B. Moon, *Proc. Roy. Soc. (London)* **A127**, 388 (1930).
24. D. B. Matthews and S. U. M. Khan, *Aust. J. Chem.* **28**, 253 (1975).
25. J. O'M. Bockris and M. A. Habib, *J. Electroanal. Chem.* **68**, 367 (1976).
26. T. E. Stern, B. S. Gossling, and R. H. Fowler, *Proc. Roy. Soc. (London)* **A124**, 699 (1929); R. H. Fowler, *Proc. Roy. Soc. (London)* **A122**, 36 (1929).
27. R. H. Fowler and L. Nordheim, *Proc. Roy. Soc. (London)* **A119**, 173 (1928); L. Nordheim, *Proc. Roy. Soc. (London)* **A121**, 626 (1928).
28. C. B. Duke and M. E. Alferieff, *J. Chem. Phys.* **46**, 923 (1967).
29. E. Gileadi, G. Stoner, and J. O'M. Bockris, *J. Electrochem. Soc.* **113**, 585 (1965).
30. J. O'M. Bockris, B. D. Cahan, and G. Stoner, *Chemical Inst.* **1**, 273 (1969).
31. J. O'M. Bockris and B. E. Conway, *Trans. Faraday Soc.* **45**, 989 (1949).
32. J. O'M. Bockris, J. McBreen, and L. Nanis, *J. Electrochem. Soc.* **113**, 637 (1966).
33. J. W. Gadzuk, *Phys. Rev. B (Solid State)* **1**, 2110 (1970).
34. J. W. Gadzuk and E. W. Plummer, *Revs. Mod. Phys.* **45**, 487 (1973).
35. J. W. Schultze and V. K. J. Vetter, *Electrochem. Acta* **18**, 889 (1973).
36. V. K. J. Vetter and J. W. Schultze, *Ber. Bunsenges Phys. Chem.* **77**, 945 (1973).
37. V. K. J. Vetter and J. W. Schultze, *Ber. Bunsenges Phys. Chem.* **77**, 945 (1973).
38. P. Kohl and J. W. Schultze, *Ber. Bunsenges Phys. Chem.* **77**, 953 (1973).
39. J. W. Schultze and U. Stimming, *Z. Phys. Chem. N.F.* **98**, 285 (1975).
40*a.* W. Schmickler and J. Ulstrup, *Chem. Phys.* **19**, 217 (1977).
40*b.* W. Memming and H. Möllers, *Ber. Bunsen-gesell.* **76**, 475 (1972).
41. K. Nakamato, *Infrared Spectra of Inorganic Coordination Compounds*, Wiley (Interscience), New York (1963).
42. J. Nakaguwa and T. Shimanonchi, *Spectrochim. Acta* **20**, 429 (1969).
43. W. Schmickler, *J. Electroanal. Chem.* **82**, 65 (1977).
44. W. Schmickler, *J. Electroanal. Chem.* **83**, 387 (1977).
45. L. D. Landau and E. M. Lifshitz, *Quantum Mechanics*, p. 24, Pergamon Press, Oxford (1965).
46. T. M. Anderson, *Mathematics for Quantum Chemistry*, p. 9, W. A. Benjamin, New York (1966).

47. J. O'M. Bockris and A. K. Reddy, *Modern Aspects of Electrochemistry*, Volume 2, pp. 961–972, Plenum Press, New York (1973).
48. F. Constantinescu and P. Magyari, *Problems in Quantum Mechanics*, Chapter IV, Pergamon Press, Oxford (1971).
49. D. Rapp, *Quantum Mechanics*, Holt, Rinehart and Winston, New York (1971).
50. N. Neilson, *Handbuch der Theorie der Gammafunction*, Teubner, Leipzig (1906).

9

Proton Transfer in Solution

9.1. Introduction

The theory of the mobility of protons has been discussed for many years, because protons have an abnormal mobility, being under a unit electric field some six times faster in their motion than other ions of the same size as a H_3O^+ ion. This effect has been discussed non-quantum-mechanically, but since 1933 it has been discussed in a quantum mechanical way. Ideas and calculations of Bernal and Fowler[1] and Conway, Bockris, and Linton,[2] and the development of these ideas by Eigen and de Maeyer[3] are well known.

The field is one that relates to a wider situation. It is not only concerned with the jumping of protons from one position in the solution to another, but also to transfers connected with acid–base reactions and reactions involving organic molecules. Here, the early contributions of R. P. Bell are outstanding.[4] It may be that proton transfer reactions are important in biology.

9.2. First Theory of Quantum Mechanical Transfer of Protons[1]

Bernal and Fowler[1] treated the transfer of protons in solution in a quantum mechanical way not long after the beginning of the applications of quantum mechanics to chemistry. Thus, in 1931 the first applications of quantum mechanics was made to electrodics in the paper by R. W. Gurney,[5] and in 1933 the first application to ionics by Bernal and Fowler[1] was made in a famous paper in the first edition of the *Journal of Chemical Physics*.

Bernal and Fowler[1] treated two adjacent water molecules separated by a distance R as being alternative sites for a proton. The wave equation for the proton in the two sites was written, and it was supposed that a field energy $e_0\mathbf{X}r$ perturbs the potential energy of the proton but is small compared with the bond energy of the proton to the oxygen. The wave functions ψ_1 and ψ_2 represent the proton in its two locations. The transfer distance δ is the radius of the water molecule, and the potential energy $U(r)$ is a function of the position of the proton between the two oxygen atoms. Bernal and Fowler found the mobility in terms of a distance Δ over which the particle transfers in N exchange events per unit time. Thus,

$$\Delta = \frac{NX\delta e_0 R}{4} \cdot \frac{\int \psi_1 \psi_2 \, d\tau}{\left(\int \psi_1 \psi_2 \, d\tau \right) \int U(r) \psi^2 \, d\tau - \int \psi_1 \psi_2 U(r) \, d\tau} \tag{9.1}$$

where the integral with respect to $d\tau$ denotes the integration over the configuration space for the proton, and Δ is the distance covered by the proton down the field after N events.

The final state is identical to the initial one, so that one may write

$$\int \tfrac{1}{2} R \psi_1^2 \, d\tau = - \int \tfrac{1}{2} R \psi_2^2 \, d\tau \tag{9.2}$$

Bernal and Fowler[1] took the field applied to the proton X as the local Lorenz field and estimated the interaction integral given in Eq. (9.1) by assuming a rectangular well through which the protons would have to pass. The value of the distance between two adjacent water molecules R taken by Bernal and Fowler was too small, being only 0.16 Å. An estimate of the value of the potential well led to 15 kcal mole^{-1}, and the numerical result becomes consistent with the observed mobility if the number of exchange events N_1 per second is 2.5×10^{12}.

An unclear conclusion arose from these calculations. They were consistent with rate of transfer well above that observed, and the authors say that "the speed of migration is *so high* that it is controlled by the transmission time." The indication seems to be that because the actual transition due to tunneling of the proton is so rapid, then it must be that the overall observed velocity is dependent on the presentation of suitable configurations that allow fast jumping to take place.

One of the contradictions of Bernal and Fowler's[1] treatment was that the paper contained the first model in which water was presented as *bound* in the structure, but at the same time, in the proton transfer model, water molecules were allowed to rotate freely. On the other hand, in a model given in another part of the same paper, they are restricted in translation by hydrogen bonding. One of the less successful calculations that came out of the paper was that of the ratio of the mobilities of hydrogen and deuterium, which was 6. Just after the publication appeared, the mobility ratio was found for the first time experimentally to be about 1.4.†

9.3. Conway, Bockris, and Linton (CBL) Model of 1956[2]

Based partly upon measurements of the degree of extinction of the anomalous proton mobility in mixtures of organic liquids with water, Conway, Bockris, and Linton suggested a detailed model in 1956, and this§ remains the principal model for proton transfer in water at this time. The model chosen by CBL was that of a water molecule in the primary hydration sheath of a hydroxonium ion[7] oriented so that one of its 2p orbitals is in line with an O—H bond of a neighboring H_3O^+ ion. The model taken is that for the minimum distance of movement of a proton in transfer between two water molecules. The proton transfer model is shown in Fig. 9.1.

It is necessary to construct potential energy curves relating the proton motion to distance and a barrier for the proton to tunnel through, and this was done by the use of Morse-type curves, resulting in the barrier shown in Fig. 9.2. To obtain the barrier height, resonance of the activated complex was required, and the height of the barrier was found to be 3.85 kcal mole^{-1} for the proton and 5.1 kcal mole^{-1} for the deuteron.

† Bernal and Fowler's paper was received by the *Journal of Chemical Physics* on April 29, 1933. Lewis and Doody sent their results on the isotope mobility ratio to the *Journal of the American Chemical Society* on July 24, 1933, and it was published on August 5 of that year.

§ The CBL model is often erroneously attributed to Eigen and de Maeyer,[3] who in 1958 discussed and criticized the method of averaging used by Conway and Bockris in 1956.

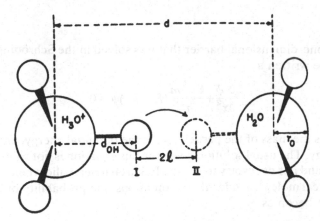

Fig. 9.1. Model for proton transfer between an H_3O^+ ion and a favorably oriented H_2O molecule in its solvation sheath.

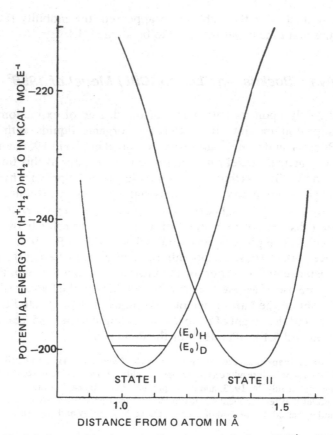

Fig. 9.2. Potential energy curves for proton transfer from H_3O^+ to H_2O.

The one-dimensional barrier that was solved in the Schrödinger equation can be written as

$$\frac{\partial^2 \psi}{\partial x^2} + \frac{8\pi^2 m}{h^2}(E - V_x)\psi = 0 \tag{9.3}$$

where m is the mass of the particle, V_x is the potential energy, and E is the total energy. The usual solution given for this equation is not valid for small barriers,[4] and it is necessary to use the Eckart barrier, valid when $(V_x - E)$ is some 10 kcal mole^{-1}. Under these conditions, the probability of tunneling per impact becomes

$$P_T = \frac{\cosh 2\pi(\alpha + \beta) - \cosh 2\pi(\alpha - \beta)}{\cosh 2\pi(\alpha + \beta) + \cosh 2\pi\delta} \tag{9.4}$$

where

$$\alpha = \frac{l}{h}(2mE)^{1/2}, \qquad \beta = \frac{l}{h}[2m(E-A)]^{1/2}$$

$$\delta = \frac{1}{2}\left(\frac{B-C}{C}\right)^{1/2}, \qquad C = \frac{h^2}{8ml^2}$$

(9.5)

$2l$ is the barrier width ($0.35 \ 10^{-8}$ cm), $B (= 4E^{\ddagger}$, where $E^{\ddagger} = 3.85$ kcal mole^{-1}) is the barrier height, and A is the difference in energy of the initial and final states.

The applied field must obviously be Lorenzian, so that if X_{int} is the internal field acting at a particular molecule due to applied field, then (Böttcher[27])

$$X_{int} = \frac{3kT + \mu R^*}{3kT} \frac{1}{1 - g\alpha} \frac{3\varepsilon}{2\varepsilon + 1} X_{ext}$$

(9.6)

where R^* is the reaction field of the dipole moment μ, i.e.,

$$R^* = \frac{\mu F}{1 - g\alpha}$$

(9.7)

where g is the number of dipoles in a cluster and α is the polarizability.

For water, $\varepsilon_s = 78$ in Eq. (9.6) and $X_{int} = 11.2 X_{ext}$. However, we shall take 6 as the saturation value of ε_s near the ion, and it follows that $X_{int} = 6.9 X_{ext}$. The height of the barrier is diminished for the forward transfer by the fraction $le_0 X$. Thus, the energy of the particle is a particle of initial energy W_x through a barrier of height $E_x^{\ddagger}$ is $E_x + \beta le_0 X$, where $0 < E_x < \infty$ and β is the fraction $0 < \beta < 1$.

When $\beta le_0 X = E_x$ and $1/h(2m)^{1/2} = K$, then we have $E_x \ll W_x$; there will be two values of P_T, one for the forward reaction and one for the reverse reaction. These two equations are then given (for a barrier) as

$$P_T(\text{forward}) = \frac{\cosh 2K(W_x)^{1/2} - \cosh(E_x/W_x)K(W_x)^{1/2}}{\cosh 2K(W_x)^{1/2} - \cosh \pi[4(E_x^{\ddagger} - E_x - C)/C]^{1/2}}$$

(9.8)

and

$$P_T(\text{reverse}) = \frac{\cosh 2K(W_x)^{1/2} - \cosh(E_x/W_x)K(W_x)^{1/2}}{\cosh 2K(W_x)^{1/2} - \cosh \pi\{[4(E_x^{\ddagger} + E_x) - C]/C\}}$$

(9.9)

However, *in the absence of a field*, the total rate of transfer across a barrier is clearly given by

$$N = \frac{N_0}{kT}\int_0^\infty P_T(\text{forward}) \exp(-W_x/kT) \, dW_x$$

(9.10)

where N_0 is the number of particles that strike a given cross section per unit time and pass the barrier.

When *the field is applied*, P_T is changed according to Eq. (9.8) and Eq. (9.9) and the distribution of particles with respect to their energies is changed by $\exp(\pm E_x/kT)$. The net rate of transfer across the barrier in the foward direction in the presence of field is given by

$$N(\text{forward}) - N(\text{reverse}) = \frac{N_0}{kT}\left[\int_0^\infty P_T(\text{forward}) \exp\left(\frac{-W_x}{kT}\right)\left(1 + \frac{E_x}{kT}\right) dW_x\right.$$

$$\left. - \int_0^\infty P_T(\text{reverse}) \exp\left(\frac{-W_x}{kT}\right)\left(1 - \frac{E_x}{kT}\right) dW_x\right]$$

$$(9.11)$$

under the condition that $E_x \ll kT$.

The value of the quantity E_x as a function of distance from the initial state is measured from the equilibrium position of the transferring proton. Values of this E_x, which correspond not only to distance but also to various values of W_x, were evaluated using a model such as that in Fig. 9.2. For $x = 0$, $W_x = 0$, and $\beta = 1$ (when $l = 0.35A/2$), E_x at the activated state is found to be 1.93×10^{-20} erg molecule^{-1}. For $W_x > 10^{-13}$ erg molecule^{-1}, $E_x = 1.93 \ 10^{-20}$ erg molecule^{-1}, since electrical energy is gained by the proton in going *down* the gradient from $\beta = 0$ to $\beta = 1$. Of course, in the use of Eq. (9.11), each factor $P_T \exp(-W_x/kT)$ must be multiplied by the value of $(1 \pm E_x/kT)$ to obtain the forward and reverse rate. The integrals were evaluated graphically. The net rate of proton transfer is proportional to the field strength at low fields, and one obtains

$$N(\text{forward}) - N(\text{reverse}) = \frac{N_0}{kT} \frac{2E_x}{kT} \int_0^\infty P_T(\text{forward}) \exp\left(\frac{-W_x}{kT}\right) dW_x$$

$$(9.12)$$

The term

$$\int_0^\infty P_T(\text{forward}) \exp(-W_x/kT) \, dW_x \qquad (9.13)$$

is evaluated graphically and is shown in Fig. 9.3.

From this it is possible to calculate the mobility, *recalling that a proton can be transferred at each H_3O^+ ion*, so that the net rate of transfer u is

$$u = \frac{v\delta}{kT} \frac{3E_x}{kT} \int_0^\infty P_T(\text{forward}) \exp\left(\frac{-W_x}{kT}\right) dW_x \quad \text{cm sec}^{-1} \qquad (9.14)$$

Equation (9.14) gives the net rate of proton passage, i.e., the mobility

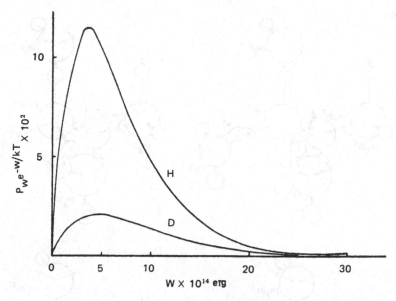

Fig. 9.3. $P_w e^{-w/kT}$ as a function of W for H˙ and D˙ quantum mechanical tunneling.

per volt, as 7.58×10^{-1} cm sec^{-1} for H$^+$ and 1.2×10^{-1} cm sec^{-1} for D$^+$, and in conventional terms this gives $U_{H^+} = 73,000$ and $U_{D^+} = 12,000$, whence $U_H/U_D = 6$. The ratio is wrong and the values of the mobilities are far too high. Two conclusions arise. First, Bernal and Fowler[1] are confirmed in the conclusion that very rapid proton transfer occurs; second, that the proton transfers according to quantum mechanical tunneling alone are *far too high*. In the treatment given so far, each H$_3$O$^+$ could transfer a proton. In reality, each one *matched to an appropriately orientated* bond can. Thus, although the quantum mechanical tunneling indeed occurs in proton transfer in solution and is the vital process that accounts for the high velocity, it is not the rate-determining step.

CBL[2] were then able to show that it is the rotation of a water molecule into the appropriate position (Fig. 9.4) that allows proton tunneling to take place. This is the relevant process that controls the rate of proton transfer. Thus, the quantum mechanical transfer, as described, occurs, but the rotation of a water molecule into the correct position must occur before it can take place. This rotation occurs at a rate consistent with the observed mobility only under the influence of a field due to H$_3$O$^+$. The rotation necessary to allow the H$_3$O$^+$ to be in a position such that an orbital is available for the quantum mechanical transfer of a proton to occur is shown in Fig. 9.4.

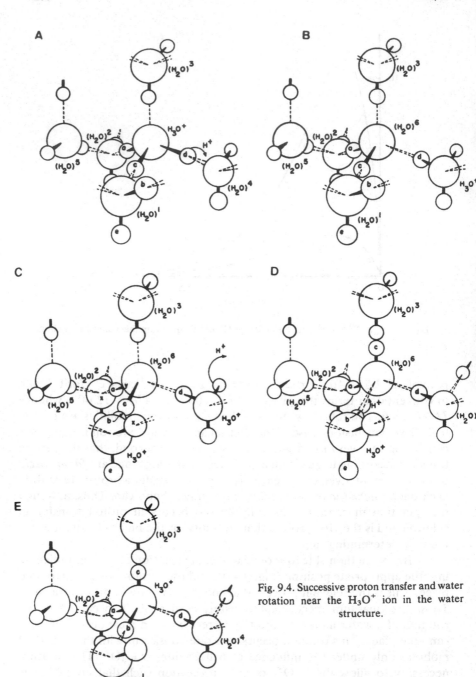

Fig. 9.4. Successive proton transfer and water rotation near the H_3O^+ ion in the water structure.

The CBL[2] calculation showed that it was *the effect of the field of the* H_3O^+ *upon the water molecule rotation*, not the effect of the field upon the tunneling, that controlled the mobility. The calculated value of the conventional mobility (the velocity of the proton under $1\ V\ cm^{-1}\ sec^{-1} \times$ the Faraday) was 270, in good agreement with the experimental value of 300. In addition to this reasonably good numerical result, *obtained without iteration or any kind of scaling*, the rate-determining water rotation theory predicts the ratio between the mobility of a proton and a deuteron as 1.4 at room temperature, in agreement with experiment. The mobility of the OH^- ion is satisfactorily explained, and the mobility of the OH^- ion comes out to be about $\sqrt{\frac{1}{4}} = 0.5$ times that of the H^+ ions, in reasonable agreement with experiment. Temperature and pressure effects, the absence of an anomalous electrical conductivity of ammonium ions, and the conductance of HCl in mixed alcohols, are well explained, mostly quantitatively.

One of the better tests of the CBL water rotation model is the comparison of the experimental relaxation times from dielectric measurements with those calculated in the theory concerned. The slow water rotation process, under the influence of the nearby proton, comes out to be reasonably consistent with the observed results.

Proton Mobility in Ice

The mobility of protons in undoped ice would be expected to be connected with the local structure of ice. *The proton concentration in ice is less than that in water.*[8] If the proton concentration in the H_2O network is sufficiently small, the average water molecule will have rotated back into a suitable position to receive a new proton after the passage by the last one. The necessity for a rate-determining rotation of water molecules, then, is removed and the velocity tends to rise toward that much greater velocity calculated at the beginning of the section for proton transfer *not* dependent upon H_2O rotation.

The calculations indicate that the changeover from the rate-determining water mechanism (effected by the field of the advancing proton) and the quantum mechanical tunneling mechanism should occur when the proton concentration is below about $5 \times 10^{-11}\ M$.

Thus, the original 1956 calculation of CBL[2] gives the quantum mechanical transition rate of protons as $8 \times 10^{-1}\ cm^2\ V^{-1}\ sec^{-1}$, and mobility in ice is experimentally $3 \times 10^{-1}\ cm^2\ V^{-1}\ sec^{-1}$, a not unreasonable match. A similar mechanism to that of CBL was published by Eigen and de Maeyer[3] two years after the 1956 mechanism publication of Conway and Bockris.[2] However, the Conway and Bockris results were quantitative, and those of Eigen and de Maeyer descriptive.

9.4. Eigen and De Maeyer's Model

Eigen and de Maeyer published several papers[3] on the mechanism of proton mobility. Essentially, the qualitative Eigen and de Maeyer model,[3] published two years after publication of the calculation of Conway, Bockris, and Lynton, appears to be similar to that earlier published, though couched in different terms. However, Eigen and de Maeyer[3] objected to various details of the calculation earlier published by CBL. From this time onward, most of the references in the literature to the proton transfer mechanism in water in which the rate-determining step is understood to be field-induced rotation of water have been to the later work of Eigen and de Maeyer.[3]

The criticisms made by Eigen and de Maeyer[3] to the calculations of CBL[2] are as follows: In the CBL work,[2] a continuous range of energy states was assumed. Thus, Eq. (9.10) of CBL[2] could be replaced by an equation of the type (as treated by Conway[10]):

$$N = N_0 \left[\sum_{n=0}^{\infty} P_T(\text{forward}) g \exp\left(\frac{-(n+\frac{1}{2})h\nu_0}{kT} \right) \right] \left[\sum_{n=0}^{\infty} g \exp\left(\frac{-(n+\frac{1}{2})h\nu_0}{kT} \right) \right]$$

(9.15)

Here, g is the quantum weight of each state in the vibrational quantum number, and ν_0 is the fundamental frequency. The summation involves the vibrational states, the energies of which are not exactly "integral plus $\frac{1}{2}$" multiples of ν_0 because of the anharmonisity effect. For the present problem, $h\nu_0 \gg kT$, so that the summation need not be extended beyond the second term. The considerable line broadening in solution will reduce greatly the distinguishability of the vibrational states and diminish the predicted results of quantization.†

Most of the tunneling occurs from the zero-point energy state, and the details of the distribution then become rather unimportant. The final mobility expression, correctly quantized by Conway,[10] becomes

$$u = \frac{\nu_0 \delta \phi_x}{kT} \frac{\sum_{n=0}^{\infty} P_T(\text{forward}) \exp[-(n+\frac{1}{2})h\nu_0/kT]}{\sum_{n=0}^{\infty} g \exp[-(n+\frac{1}{2})h\nu_0/kT]}$$

(9.16)

Slight differences from the original CBL calculations are wrought, but no

---

†This Eigen and de Maeyer point is similar to one made later by Levich.[28] The latter maintained that it would not be possible to explain a smooth relation of current to potential if electrons had to leave a metal and find H^+ ions in given vibration–rotation states in H_3O^+. As these are ~ 0.5 eV ($\approx \frac{1}{2}h\nu_0$) apart, the current–potential relation would consist of a series of steps. Levich had also neglected the effect of collision in liquids on smoothing out $(\Delta E)_{\text{vib}}$.

differences in mechanistic conclusions. Correspondingly, the apparent activation energy will be temperature dependent[11] because of the change of structure of the surrounding water.

Another part of Eigen and de Maeyer's[3] criticisms of CBL,[2] made on the way to agreeing with its conclusions, is the assignment of the proton-jumping distance as 0.35 Å, based upon an O–O distance of 2.45Å. Thus, it was argued[3] that for the O–O distance of 2.45 Å, there would be no barrier for proton transfer if the calculation of Huggins[12] were accepted, for these indicated a symmetry of the hydrogen bonds for O–O distances of less than 2.65 Å. However, this work is based upon old calculations that have been superseded by the experimental fact that KH_2PO_4, where the O–O bond distance for hydrogen is 2.55 Å (i.e., less than Huggins' critical distance), exhibits residual entropy associated with the degeneracy of the hydrogen positions, thus indicating a symmetry of positions in the hydrogen bond.

A more subtle question concerns which of two rotational rates should be used in considering dipole reorientation in water. CBL[2] *chose to use the dipole reorientation rate for ice*, and as this was too low, used without field-induced dipole reorientation to explain the facts of mobility development in their model with field-induced orientation of water. Eigen and de Maeyer[3] suggest that the normal rate of reorientation value in liquid water, i.e., 10^{-11} sec^{-1}, should be used so that, then, the field-induced orientation would not be necessary.

However, this choice would not allow for an explanation of the facts of the higher proton movement within ice. Further, the heat of activation of dipole orientation in ice is 13.3 kcal mole^{-1}; for water, 3.9 kcal mole^{-1}. It seems that this must be due to the higher proton concentration in water than in ice. In pure ice, the dielectric relaxation behavior can be interpreted[13] as being caused by libration–rotation transitions in the absence of the proton fields, while in water it is determined mainly by rotations induced by fields on the mobile H_3O^+. Hence, the heat of activation for relaxation in acid solutions is lower, because the fields of the migrating protons assist the rotation of the water molecules. Thus, the rate of relaxation of 10^{-11} sec^{-1} in water can be regarded as associated with a field-induced rotation *and therefore should not be used as though it were associated with water rotation alone.*

Some statements made by Eigen and de Maeyer[3] have referred to "structural diffusion" as a mechanism for proton transfer, but this appears to be only an *omnis coverendum* for mechanisms similar to those originated and calculated by CBL[2]. Erdey-Gruz and Lengyal[14] have recently given a convincing description of the priority of the CBL[2] model over the usually quoted Eigen and de Maeyer work.[3]

9.5. Polarization of the Hydrogen Bond and the Proton Transfer Mechanism

Weidermann and Zundel[15] have compared two processes: (a) the process they attribute to Eigen and de Maeyer[3] (the CBL model[2]), in which a proton transfer occurs between two water molecules by means of quantum mechanical tunneling, thereby arranging the bridges for the transfer ("structural migration"), and (b) a field-dependent mechanism of their own, which they describe as "the polarization of H_3O^+," for a doubly degenerate state with respect to its protons across H_3O^+ when one of the bridges is linearized. This degeneracy is removed by the external field, since the status preferred for H_3O^+ for excess charge lies at the hydrogen nucleus in bridge b (Fig. 9.5). The migration after structure breaking of the excess proton in the opposite field direction is prevented. The point of this mechanism is that the field removes the degeneracy that has developed, and thus the proton moves.

It appears that this view involves a misunderstanding of the CBL theory.[2] What Weidermann and Zundel[15] argue is that the effect of the field

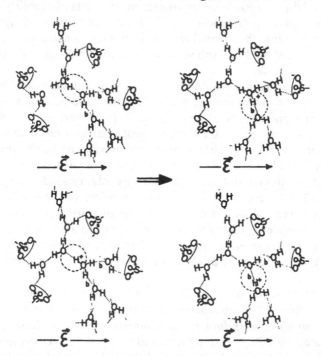

Fig. 9.5. (Left) Two proton boundary structures of the excess proton in the $H_5O_2^+$ group in polyestyrenesulfonic acid. (Right) Two proton boundary structures of the excess proton after one step of structure diffusion.

upon the tunneling distances in both directions is very small, and because of this, they go to another mechanism to explain the field dependence. This, of course, neglects the fact that the field dependence studied by CBL[2] is not that of the tunnel transfer, which is, indeed, almost field independent. CBL[2] point out that the field dependence comes in the seventh place in the energy term, *a conclusion similar to that of Weidermann and Zundel,*[15] although pointed out many years earlier. What CBL[2] did was to take into account the field dependence of the *rotation* of the water molecules under the local field of the hydrogen ions.

Correspondingly, the further calculations that Weidermann and Zundel[15] give, which deal with the transition dipole moment affected by the large polarizability in the hydrogen bond, seem to some extent irrelevant. Thus, the conclusion they make that the protons tunnel more in one direction than the other when a field is applied is qualitatively obvious. The Weidermann and Zundel work[15] does not appear to effect the likelihood that the breaking of a water molecule out of its structure under the influence of the local protons is the rate-determining step in proton transfer in aqueous solution. Field-induced rotation of water is not dealt with in their theory.

Weidermann and Zundel's theory[15] involves the assumption that the proton movement itself occurs independently of the local proton field. They then try to make the known potential dependence of mobility depend upon the field. In fact, CBL[2] show that it is the structural movement of the water in the field in the H_3O^+ ion that is potential dependent in a rate-determining way. In the CBL theory,[2] it is *the time of rotation of the dipole* that is field dependent. The net force exerted on a rotating dipole is dependent on the energy needed to break the hydrogen-bond binding the rotating water molecule; it *then* has to accelerate and move to its final position under the influence of the field of the neighboring H_3O^+ ion, but also that of the applied field, apart from that of hydrogen bonds themselves.

Thus, in the absence of an applied field, the displacement time in rotation is 2.4×10^{-4} sec for a water molecule bonded in the structure of water. When an external field is applied, rotation will be either slightly hindered or slightly accelerated, depending on its direction compared with the vector of the applied field. The average couple on the component μ of the HOH dipole in the direction of the O—H bond being rotated due to the integrated average applied force is:

$$f' = \mu' E_i/3 \tag{9.17}$$

where $\mu' = \mu_{H_2O} \cos(105/2)$, 105° being the HOH angle in the water molecules. The average tangential force rotating the dipole of length $\sim l$ is $f = f'/l$. $X_i = 6.9X$, X is taken in electrostatic units for 1 V cm^{-1}. In Eq. (9.17), $\mu' = 1.11 \times 10^{-18}$ esu and $2l = \mu'/e_0 = 0.23$ Å. Hence, $f = 0.74 \times$

10^{-11} dyn molecule^{-1}. The final result is that the net rate of rotation *down* the field is field dependent, and thus, so is proton transfer. This field dependence has not been treated by Weidermann and Zundel.[15]

9.6. Lifetime of H_3O^+

The lifetime of the ground state varies between 1×10^{-12} and 4×10^{-12} sec, indicating that the proton spends about 99% of its time within its well.[2] The excited state lifetime is in the region of $1 - 7 \times 10^{-14}$ sec.[2] Hence, only about 2% of the time of a proton is spent jumping and the H_3O^+ form is present for about 98% of the time.

We investigate a system in which a proton transition between two adjacent wells is impossible. We consider that H_0 is a sum of two commuting operators H_a and H_b, the eigenfunctions of which have nonvanishing amplitudes over only one of the regions, (a) or (b), respectively. If the solutions $|\psi_m\rangle$ of the eigenvalue equation

$$(H_a + H_b)|\psi_m\rangle = H_0|\psi_m\rangle = E_m|\psi_m\rangle \tag{9.18}$$

are known, a proton that is, for example, in well (a) (Fig. 9.6) with certainty at time $t = t_0$ can be described by the statistical operator

$$p^{(a)}(t_0) = \sum_m^a |\psi_m\rangle p_m \langle\psi_m| \tag{9.19}$$

The quantities p_m are the probabilities of finding the particle in a state represented by $|\psi_m\rangle$; the index a or b on the summation indicates that it is over the eigenstates of H_a or H_b, respectively. The time dependence of the density operator $p^{(a)}$ is determined by the unitary time transformation

$$U(t, 0) = \exp[-(i/h)Ht] \tag{9.20}$$

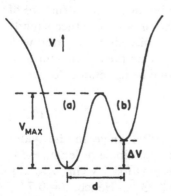

Fig. 9.6. Double-minimum potential V and characteristic potential parameters V_{max}, ΔV, and d (schematically).

The probability $W^{(b)}(t)$ of finding the particle represented by $p^{(a)}(t)$ at a particular time t in the well (b) results as the expectation value of the projection operator.

If t' is the time after which the probability of finding the particle becomes equal to the average value, $\bar{W}^{(b)}$ for the first time, the mean lingering time $\tau^{(a)}$ of a proton in well (a) can be evaluated approximately[17] as

$$\tau^{(a)} = t'/\bar{W}^{(b)} \qquad (9.21)$$

Analogous formulas apply for the proton transition from (b) to (a).

The lifetimes determined[17] thus are of the same order of magnitude as those that result from the quasiclassical methods used by other authors.[2,18] Thus, τ is there found from the classical frequency ν of oscillation of the proton in its well and from the penetration coefficient P_T of the potential barrier in the form

$$\tau = 1/(\nu P_T) \qquad (9.22)$$

where P_T is determined using Gamow's tunneling expression.

9.7. Recent Quantum Mechanical Work on Proton Transfer in Solution

Tunneling in the proton transfer between two water molecules has been treated by Flanigan and De la Vega[19] using a SCF CNDO (complete neglect of differential overlap) method. The potential function that describes the proton between two water molecules is represented by a harmonic potential with a hindering barrier as

$$V = A_2 x^2 + V_0 \exp(\alpha x^2) \qquad (9.23)$$

where A_2, V_0, and α are the three parameters needed to describe such a potential. The coefficients of the quadratic term are found from a least squares analysis of that portion of the curve that approaches a parabolic behavior and where the Gaussian part is negligible. The authors use a method similar to that of Swalen and Ibers.[20]

The authors[19] imply that from a knowledge of the energy barrier $E^{\ddagger}$ and the distance from the center at which the minimum occurs, the parameters V_0 and α can be obtained. Considering the parameters V_0 and α as known, they write the Hamiltonian as

$$H = -\frac{1}{2m}\frac{d^2}{dx^2} + A_2 x^2 + V_0 \exp(-\alpha x^2) \qquad (9.24)$$

with the quantities given in atomic units.

The matrix element of this Hamiltonian was evaluated using the eigenfunction of a harmonic oscillator as a basis set. The use of the transformation $\xi^2 = (2A_2m^2)^{1/2}x^2$ reduces to diagonal form the Hamiltonian found by including the quadratic term only. The calculation of the contribution of the Gaussian involves standard integrals. They use[19] an eleven-function basis set in the expansion. The matrix obtained was diagonalized using the Jacobi method from which the eigenvalues and eigenvectors were obtained. The eigenvalues obtained are referred to the minimum of the parabolic potential: but the subtraction of $V_0 - E^{\ddagger}$ from the obtained eigenvalues gives the energy referred to as the minimum of the potential energy curve. In addition, the squares of the eigenfunctions for different positions of the protons were obtained.

The authors show that as the energy barrier increases, the eigenvalues increase, but the eigenvalues of the even functions increase faster than those of the odd functions and form a symmetric–antisymmetric doublet. The presence of this doublet indicates tunneling, and the proton oscillates between the two wells with the frequency $(E_1 - E_2)/\hbar$, where E_1 and E_2 are the energies of the proton when it is in the well 1 and 2, respectively, of the double-well potential.

The results indicate that tunneling occurs optimally when the proton is oscillating between the two wells at 8.7×10^{-13} sec^{-1}. In Fig. 9.7, the values of ψ^2 for the proton are given as a function of the distance from the proton to the three interoxygen distances. The proton density is a maximum at the center. This type of work, therefore, indicates more refined methods of representing the hydrogen bond, though it has not contributed as yet directly to the problem of proton mobility.

In the treatment of Carbonell and Kostin,[21] the rate of proton transfer is obtained by following the work of Pshenichnov and Sokolov,[22] who studied proton transfer in water. They picture the transition of the proton as due to perturbation of the steric situation in terms of dipole–dipole forces. The dipole moments of the surrounding water molecules play a major role in stimulating the perturbed positions of the protons, a model that is clearly a generalization of the more specific model expressed by CBL.[2]

As a nearby dipole changes its orientation, it produces a fluctuating potential on the proton in the hydrogen bond. The correlation coefficients in the corresponding spectral density distributions can then be calculated. The results are averaged over all molecules of the surroundings, and the following expression is obtained for the spectral density of the fluctuating potential:

$$j(\omega) = \frac{8\pi N \mu_d^2}{9d_c^3} \frac{\tau}{1 + \omega^2\tau^2} \exp\left(-\frac{\hbar\omega}{2kT}\right) \tag{9.25}$$

where N is the number of molecules per unit volume, d_c is the distance of closest approach between the proton in the hydrogen bond and a molecule

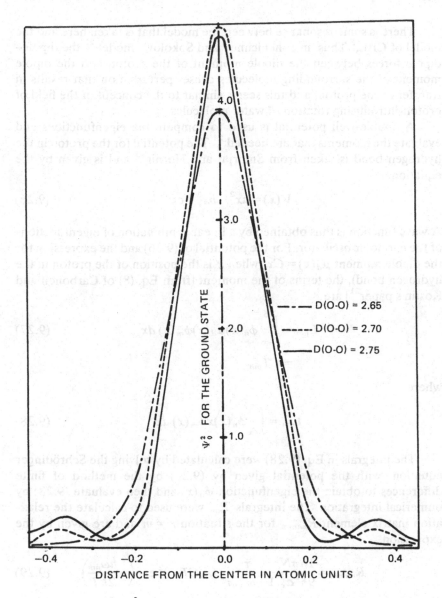

Fig. 9.7. The value of ψ_0^2 as a function of the position of proton for three interoxygen distances.

of the surroundings, τ is the correlation time of the fluctuating potential, μ_d is the dipole moment of a molecule of the surroundings, and ω is the frequency of the proton corresponding to its energy difference in two potential wells of the double-well potential considered.

There is some resonance between the model that is taken here and the model of CBL.[2] Thus, in Pshenichnov and Sokolov's model,[22] the dipole–dipole forces between the dipole moment of the proton and the dipole moment of the surrounding molecules causes perturbation that results in transfer of the proton, and this seems similar to the concept of the field of proton-introducing rotation of water molecules.

A double-well potential is used to compute the eigenfunctions and evaluate the moments that are needed.[21] The potential for the proton in the hydrogen bond is taken from Smorjai and Hornig[23] and is given by the equation

$$V(x) = ax^2 + bx^3 + cx^4 \tag{9.26}$$

A wave function is thus obtained by a linear combination of eigenfunctions of the harmonic oscillator. For the potential of (9.26) and the expression for the dipole moment $\mu_d(x) = Cx$ (when x is the position of the proton in the hydrogen bond), the terms of the moment [from Eq. (8) of Carbonell and Kostin's paper[21]] are

$$L_{nm} = \int_{-\infty}^{\infty} \phi_n(x) \mu_d(x) \phi_m(x)\, dx \tag{9.27}$$

$$= CT_{mn}$$

where

$$T_{nm} = \int_{-\infty}^{\infty} \phi_n(x) x \phi_m(x)\, dx \tag{9.28}$$

The integrals in Eq. (9.28) were calculated by solving the Schrödinger equation, with the potential given by (9.26) by the method of finite differences to obtain the eigenfunction $\phi_n(x)$ and then evaluate (9.28) by numerical integration. The integrals T_{nm} were used to calculate the relaxation matrix element R_{nnmm} for the situation $n \neq m$ and are given by the expression

$$R_{nnmm} = \frac{16\pi N}{9h^2 d_c^3} \frac{\tau}{1 + \omega_{nm}^2 \tau^2} (\varepsilon_d CT_{nm})^2 \exp\left(-\frac{\hbar \omega_{nm}}{2kT}\right) \tag{9.29}$$

$$\omega_{nm} = (E_n - E_m)/\hbar \tag{9.30}$$

The application to which this expression is put is that of an exothermic reaction of protons between two water molecules, and the barrier maximum is calculated to be 1.09 eV above the minimum of the reactant potential well, and the minimum product well is 0.31 eV below the minimum of the reactant well, which corresponds to the heat of reaction. The distance between the minimum of the reactant well and the minimum of the product well is taken to be 1 Å. The wave function was obtained by solving the Schrödinger equation numerically and was found to be concentrated in the product region and zero in the reactant region for $n = 1, 3$, and 5, but for $n = 2, 4$, and 6, they were highly concentrated in the reactant region.

For each temperature, the eigenvalues and eigenvectors were determined numerically, and finally, the probability that the particle is in the product region was obtained from the following expression

$$P_{\text{product}}(t) = \sum_n I_{nn} \cdot \sigma_{nn'}(t) + 2 \sum_{m>n} \sum_n I_{nn} |\sigma_{nm}(t)| \cos \theta_{nm}(t) \qquad (9.31)$$

where

$$I_{nm} = \int_{\text{product}} \phi_n(\mathbf{r}) \phi_m(\mathbf{r}) \, d\mathbf{r} \qquad (9.32)$$

and $\theta_{nm}(t)$ is the phase of the density matrix $\sigma_{nm}(t)$. The probability that a proton transfers from one well to another as a function of time at different temperatures is shown in Fig. 9.8.

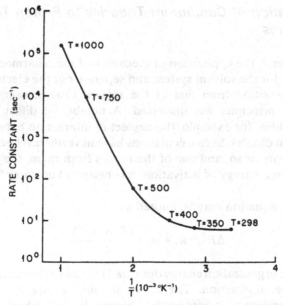

Fig. 9.8. Rate constant $k_0(T)$ for a proton in a double-well potential versus reciprocal temperature.

Carbonell and Kestin[21] defined the initial value of the reaction rate constant k_0 at $t = 0$ as

$$k_0 = -\left(\frac{1}{P_{\text{product}}(0)}\right)\left(\frac{d}{dt}P_{\text{product}}(0)\right) \tag{9.33}$$

The result of the final calculations shows that at temperatures called "high" (namely, above 500°K), the Arrhenius law should be observed in proton transfer, indicating that classical behavior obtains and little tunneling occurs. On the other hand, below about 500°, the Arrhenius law flattens out to give little temperature variance at room temperature (see Fig. 9.8).

One difficulty in this work[21] is the neglect of specific structural considerations, for example, whether rotation of an actual water molecule in a solvent occurs. Similarly, it neglects the change of the structure in the solvent medium as the temperature changes; this change is likely to be the main reason for temperature dependence in the lower region of temperature. It is left to the reader to find out what happens if one does realize that 500°K brings one into the vapor phase. The fact that rates of proton transfer do depend upon temperature in the room temperature region is entirely neglected. This lack of an established connection between the calculations and reality does not avoid the point that they give techniques that may be applicable to practical situations.

9.8. Application of Continuum Theories to Proton Transfer Reactions

In Chapter 7, the application of theories that use a harmonic oscillator wave function for the solvent system and separate out the electronic matrix element of transition from that of the solvent system, using the Born–Oppenheimer principle, was discussed. A number of difficulties inhabit these calculations, for example, the neglect of interaction between the ion and permanent dipoles. Such calculations have nevertheless been popular in the past 15 years or so, and one of the results from them, the relationship between the free energy of activation, has become known as "the Marcus equation."

Thus, the equation may be quoted as

$$\Delta F^{\ddagger} = w_r + w_p + \frac{(E_s + \Delta F^0)^2}{4E_s} \tag{9.34}$$

where E_s is the organization energy discussed in Chapter 7 and ΔF^0 is the free energy change in reaction. The terms w_r and w_p are the energies of interaction between reactants as they approach each other and between products as they retreat from each other, respectively. It is a frequent (if seldom justified) approximation to take these two terms as equal and opposite, or small, and to take, therefore,

$$\Delta F^{\ddagger} = \frac{(E_s + \Delta F^0)^2}{4E_s}$$

$$= \frac{E_s}{4}\left(1 + \frac{\Delta F^0}{E_s}\right)^2 \tag{9.35}$$

There are several difficulties with this equation, e.g., those which have been pointed out when it was applied to electrode reactions (Chapter 7).

The situation here is worse, because the basic assumption of no overlap between the ion and the inner solvent is clearly inapplicable to proton–water interactions, there being strong bonding between the proton and one solvent molecule. The neglect of interactions of the H_3O^+ ion with the surrounding solvent, except the Born one, during the transfer of protons appears not to be worthy of the term rough approximation but to be simply wrong. In addition, in many of the discussions of rate, the term $(\Delta F^0/E_s)^2$ in Eq. (9.35) is dropped on the assumption that E_s is much larger than ΔF^0. This is not usually the case (Table 9.1).

Several workers[25,26] have tried to apply Eq. (9.35). This was done, for example, by Koeppl and Kresge.[25] They assumed that AH and BH bonds have the same force constant and constant horizontal displacement and obtain expectedly the same type of expression as Eq. (9.35). The assumption of equal force constants is obviously not correct and the degree to which it will cause discrepancies will have to be worked out case by case.

They found that values of E_s have a discrepancy of a factor of about 2, if they are calculated from the observed value of the Bronsted exponent $\alpha = (\partial \Delta F^{\ddagger}/\partial F^0)_{E_s, w_r, w_p}$ and from the theory. Correspondingly, Kreevoy and Oh[26] attempted to apply Eq. (9.35) to the hydrolysis of diazoacetate ion. They attempted to obtain a relationship between the rate constant and the free energy change of the reaction. They found it necessary to use three adjustable parameters before they could find a fit.

It was maintained in Chapter 7 that Eq. (9.35) should not be applied to any reaction and that a bond-stretching equation of the type first introduced by George and Griffith[29] (cf. Despic and Bockris[30]) must always be used.†
However, the continuum equation, with its insistence upon the Born solvation energy and complete neglect of bond activation, would be particularly inapplicable to a *proton* transfer reaction, where the bonding properties are predominant in determining the O—H motion and hence reaction. There

---

† Detailed results illustrating this conclusion for redox reactions are given by Bockris, Khan, and Matthews.[31]

Table 9.1

Degree of Approximation in the Marcus Equation for Activation Energy in Redox Reactions Arising from Neglect of Free Energy of Reaction

Reaction	ΔF^0 (expt.) (eV)	E_s (calc.) (eV)	Value from Marcus equation (eV)	Value neglecting ΔF^0 (eV)	Percent error
$Fe(CN)_6^{4-} + IrCl_6^{2-}$	-0.65	0.87	0.06	0.22	266
$Fe^{2+}(H_2O)_6 + M_n^{3+}(H_2O)_6$	-0.56	1.1	0.07	0.28	300
$V^{2+}(H_2O)_6 + Co^{3+}(WH_3)_6$	-0.51	1.2	0.1	0.3	200
$Cr^{2+}(H_2O)_6 + Fe^{3+}(H_2O)_6$	-1.17	1.1	0.006	0.28	>1000
$Fe^{2+}(H_2O)_6 + Ce^{4}(H_2O)_6$	-0.67	1.0	0.03	0.25	800
$Fe^{2+}(phen)_3 + Co^{3+}(H_2O)_6$	-0.78	0.96	0.01	0.24	>1000

have been too many attempts to obtain quantitative relationships between rate constants and equilibrium constants, as remotely connected as is the difference in the depths of two valleys with the height of the mountain between.

References

1. J. D. Bernal and R. H. Fowler, *J. Chem. Phys.* 1, 515 (1933).
2. B. E. Conway, J. O'M. Bockris, and H. Linton, *J. Chem. Phys.* 24, 834 (1956).
3. M. Eigen and L. de Maeyer, *Proc. Roy. Soc. (London)* A247, 505 (1958); *The Structure of Electrolytic Solutions* (J. Hamer, ed.), Chapter 5, John Wiley and Sons, New York (1959); cf. *Symposium on Hydrogen Bonding, Ljubjana, July 29–August 3, 1957* (Hadzi Dusan, ed., with the cooperation of H. W. Thomson), Pergamon Press, Oxford (1959).
4. E. Caldin and V. Gold (eds.), *Proton-Transfer Reactions*, Chapman and Hill, London (1975).
5. R. W. Gurney, *Proc. Roy. Soc. (London)* A134, 137 (1931).
6. G. N. Lewis and F. G. Doody, *J. Electrochem. Soc.* 55, 3504 (1933).
7. J. O'M. Bockris, *Quart. Rev. Chem. Soc. London*, 111, 173 (1949).
8. N. Bjerrum, *Science* 115, 385 (1952); cf. *K. Danske Vidensk. Selsk.*, 27, No. 1 (1951).
9. B. E. Conway and J. O'M Bockris, *J. Chem. Phys.* 28, 354 (1958).
10. B. E. Conway, *Can. J. Chem.* 37, 178 (1959); *Symposium on Charge Transfer Processes*, Preprint No. 7, Chemical Institute of Canada, Ottawa (1958).
11. R. P. Bell, J. A. Fendley, and J. R. Hulett, *Proc. Roy. Soc. (London)* A235, 453 (1956).
12. M. L. Huggins, *J. Phys. Chem.* 40, 723 (1936).
13. B. E. Conway, *Can. J. Chem.* 37, 613 (1959).
14. E. Erdey-Gruz and P. Lengyal, *Modern Aspects of Electrochemistry* (B. E. Conway and J. O'M. Bockris, eds.), Volume 12, Plenum Press, New York (1977).
15. E. G. Weidermann and G. Zundel, *Z. Naturforsch* 25a, 627 (1970).
17. J. Brickmann and H. Zimmermann, *Z. Naturforsch* 23a, 11 (1968).
18. P. O. Löwdin, *Advances in Quantum Chemistry*, Volume 3, p. 216, Academic Press, New York (1965).
19. M. C. Flanigan and J. R. De la Vega, *Chem. Phys. Lett.* 21, 521 (1973).
20. J. D. Swalen and J. A. Ibers, *J. Chem. Phys.* 36, 1914 (1962).
21. R. G. Carbonell and M. D. Kostin, *J. Chem. Phys.* 60, 2047 (1974).
22. E. A. Pshenichnov and N. D. Sokolov, *Int. J. Quant. Chem.* 1, 855 (1967).
23. R. L. Somarjai and D. F. Hornig, *J. Chem. Phys.* 36, 1980 (1962).
24. A. J. Kresge, *Proton Transfer Reaction* (E. Caldin and V. Gold, eds.), Chapter 7, Chapman and Hill, London (1975).
25. G. W. Koeppl and A. J. Kresge, *J. Chem. Soc., Chem. Commun.*, p. 371 (1973).
26. M. M. Kreevoy and S. W. Oh, *J. Amer. Chem. Soc.* 95, 4805 (1973).
27. L. Böttcher, *Theory of Electric Polarization*, p. 181, Elsevier, New York (1952).
28. B. Levich, Theory of electrode processes, in *Treatise on Physical Chemistry* (H. Eyring, L. Henderson, and M. Justi, eds.), Volume 9, Academic Press, New York (1970).
29. P. George and J. Griffith, in *The Enzymes*, 1959 (P. D. Boyer, H. Lardy and K. Myrberg, eds.), Volume 1, Chapter 8, p. 347, Academic Press, New York (1959).
30. A. Despic and J. O'M. Bockris, *J. Chem. Phys.* 32, 389 (1960).
31. J. O'M. Bockris, S. U. M. Khan, and D. B. Matthews, *J. Res. Inst. Catal. Hokkaido Univ.* 22, 1 (1974).

10

Proton Transfer at Interfaces

10.1. Introduction

During the first half of the century, the majority of the experimental work on electrode kinetics was in fact on the hydrogen evolution reaction. From 1960 the work became oriented toward redox reactions, which were thought to be easier to understand, and to some extent this diminished the integration of the larger picture in electrode kinetics, for in hydrogen evolution many kinds of combinations and possibilities are present, not only electron transfer. In this chapter, by discussing specifically proton discharge and its relation to hydrogen evolution, a balance toward the larger picture will be given.

10.2. Historical Perspective

The first quantitative assay into electrode kinetics was carried out by Tafel.[1] In modern terminology, Tafel[1] found that the relationship between rate and the departure of the potential from the equilibrium one would be represented by the following law:

$$i = A \exp(-\alpha e_0 \eta / kT) \tag{10.1}$$

However, Tafel's assumption was that the combination of hydrogen atoms on the surface was the slow state of such hydrogen evolution reactions, and his neglect of consideration of the electron transfer stage as rate determining led to a phenomenological calculation of the incorrect value of α.

Indeed, until about 1930, electrode kinetics was hindered by the avoidance of what was called at that time "slow discharge" as a rate-determining step. The origin of this was clear; electrochemists could see that in the case of hydrogen and oxygen evolution, where departures from equilibrium are marked, the possibilities of chemical combination on the surface as a slow state existed. In the case of the redox reactions, where only

315

slow discharge was a possible rate-determining stage, the early measurements did not detect overpotential, the values thereof being small and the applicability of the Tafel relation less clear. Hence, in the 1930 picture, a fundamental cause for a departure from equilibrium in redox reactions (i.e., very high i_0 assumed) was rejected; the small departures from ideality were attributed to the influence of rate control by mass transport. As there was no surface recombination reaction possible in redox reactions, it was clear, to the 1930 view, that overpotential phenomena were associated with chemical combination on surfaces, where (for hydrogen and oxygen evolution) large overvoltage was observed.

It was Smits[2] in 1922 who first suggested that the slow discharge could be a rate-determining step in hydrogen discharge reactions.† The celebrated Gurney theory,[3] the first quantum mechanical theory to enter kinetics, appeared in 1931 and, as explained in earlier chapters, failed to meet with resonance among electrochemical theorists of the time.

The next quantum mechanically oriented contribution in respect to proton transfer was the work of Bawn and Ogden,[5b] who measured separation factors of deuterium and hydrogen. It is obvious that the separation factor is of particular importance for any contemplation of the quantum mechanical aspects of charge transfer kinetics involving protons, because the relative rate of the passage of two particles with mass difference as great as one to two through a barrier would immediately involve large differences. Bawn and Ogden,[5b] then, immediately invoked the possibility of quantum mechanical transfer to explain the large separation factors that they had measured.

In 1936, Butler[6] implicitly supported the quantum mechanical theory of Gurney[3] by taking it and providing it with an essential improving point, namely the point that the hydrogen atom in H_3O^+ discharge that resulted from the quantum mechanical transfer of an electron to a proton in solution was bound to the electrode surface, i.e., had undergone chemibonding. The quantum mechanically connected work of Butler was preceded in this respect by the phenomenological paper of Horiuti and Polanyi,[7] who had pointed out that considerations of the potential energy–distance variation in proton discharge with neglect of the quantum mechanical barrier penetrating aspects would give rise to connections between the heat of adsorption of the hydrogen atom on the surface and the rate of reaction.

During the 1950s, the Butler–Gurney-oriented view of proton transfer was developed by Parsons and Bockris[34] and by Bockris and Conway.[35] In

---

† A theory by Butler[4] tried in a general way to explain the relationship between current and potential, in a somewhat charge-transfer-oriented way, without the participation of the factor β, so that the rate of variation of current with potential still could not be quantitatively interpreted.

these papers the stress was, however, the chemical bonding situation upon the energy states of the proton in solution and the particle that would be formed by it after proton transfer, rather than the details of the quantum mechanical penetration of the barrier. The assumption was implicitly made that quantum mechanical penetration occurs easily, i.e., that the transition probability is very high and that the process is adiabatic.

During the 1960s, there intervened the erudite and quantum mechanical contributions made by St. G. Christov.[8-10] These works began with the assumption that the rate-determining step in the proton discharge reaction was in fact proton transfer and then consisted of construction of detailed models with the application of quantum mechanical calculations involving the Gamow application of the WKB approximation to the velocity of transfer through the barrier as a function of the barrier width and height. The intensely quantum mechanical work and prior work of Christov, with its numerical evaluations in comparison with experiment, particularly those of the continuum school, remained unquoted and neglected by many later workers.[11,12] Work related to that of Christov was carried out by Bockris, Srinivasan, and Matthews.[16,17] Bockris and Srinivasan,[13-15] relating their work to that of Keii and Kodera[19] and thus to Horiuti's school of a catalytic approach to electrochemical kinetics, carried out calculations of the statistical thermodynamics of the situation for the separation factor that would arise, including the difficulties of proton penetration of the barrier. Thus, Bockris and Srinivasan[13-16] utilized a model in which quantum mechanics was the implicit background within the formulation of the theory of absolute rates of reactions to calculate the numerical values associated with various rate-determining steps and paths of hydrogen evolution mechanisms at the surface. The principal result of this work was to show that the numerical value of the separation factor, according to this theoretical approach, was highly mechanism dependent and that the differences in the values that would be theoretically expected with various mechanisms differed more than any discrepancies in experimental values, so that the differences in mechanism could be examined merely by a determination of the separation factor.

The work of Bockris and Matthews[17,18] then came to add the tunneling component to the work of Bockris and Srinivasan,[13-15] and combined the quantum orientation of the Christov work,[8,9] to the statistical thermodynamic calculations that Bockris and Srinivasan[13-15] had carried out. Thus, Bockris and Matthews[17,18] calculated the quantum contribution to the separation factor, as Bell and co-workers[20,21] had done earlier in chemical kinetics.

However, the work of Bockris and Matthews[17,18] did more than this, because it led to the rediscovery of a fact that had been presented in earlier literature[22-25] but remained uninterpreted, namely the interesting (and potentially practically important) fact that the separation factor is greater at smaller overpotentials than at higher ones. This was used by Bockris and Matthews[17,18] for a detailed theory of quantum mechanical effects of penetration, because with the extra experimental knowledge of how the separation factor varied with potential (measured in an experimental system that made the theoretical work easier to carry out, namely, in the mercury–solution system), it was possible to attain parameters that helped in the elucidation of the theory. Thus, the Bockris and Matthews[17,18] treatment of the penetration in the barrier is focussed in respect to its comparison with experiment upon the separation factor and its variation with potential, but at the same time keeping the Tafel slope constant over the requisite variation of potential range (about 1.0 V), and this is difficult to do when the quantum effects are taken into account.†

The proton-oriented view of the quantum theory of electrode processes was stressed by Sen and Bockris[26] in their examination of the applicability of the WKB theory, which had earlier been applied without reservation to processes at electrodes. The criticism by Levich[11] in 1970 of the processes used by Bockris and Matthews[27,28] led to the Sen and Bockris[26] investigation, which defined the range and applicability of the WKB approximation (Section 10.14).

While these quantum mechanical developments were occurring in proton-oriented electrode process theory, there had been a development of a totally different kind of approach to the quantum mechanics of electrode processes. This different view was taken, apparently in ignorance of the previous work, by Libby[29] in 1952. The Libby model was quantum mechanically oriented in that it stressed the effect of the Born–Oppenheimer theory of the separation of the operators for the nuclear and electron motions. The Franck–Condon principle was invoked in the quantum mechanical concept of an electron transfer taking place so quickly that the movement of the nucleii during this change would not be significant.

Thus, Libby[29] pointed out that there would be an energy change during electron transfer that would correspond to a change in the Born equation, only of charge, and not of the value of the solvated radius, i.e., the momentary configuration of the product ion would be that of the reactant, in accordance with the Franck–Condon principle.

A parallel paper to that of Libby—in which the contributions of Libby[29] are not quoted—was that of Weiss,[31] which discussed "the solvation of electrons." The concept of the polarization changing in an optical manner in

---

† Thus an early theory by Conway[25] for quantum mechanical penetration of protons through barriers had a Tafel line with a considerable curvature.

redox processes was brought in for the first time by Weiss.[31] The theory of Weiss[31] is not generally quoted in the literature concerned.† What is usually quoted—and what, indeed, became famous, as a part of the "quantum mechanical" theory of electrode kinetics—is a theory by R. A. Marcus,[32] which involves in a principal way the equation of Born using the optical and static dielectric constants (cf. Weiss[31]).

The early continuum model[32] became well known. (See Bockris, Khan, and Matthews.[33]) It was used by inorganic chemists to discuss redox reactions. In fact, according to the first theory of Marcus[32a,b] there is no effect of the stretching of bonds in the activation situation—no effect of any parameters of the system except the dielectric of the solvent and the radius of the ion.

The theories of Marcus,[32] and the earlier theories of Levich[11] and Dogonadze,[12] were called "quantum mechanical," although the quantum mechanical aspect (as, indeed, was so with the 1950 work of Parsons and Bockris[34] and Conway and Bockris[35]) was restricted to the assumption of the Franck–Condon principle; no quantum mechanical equations were formulated. There was, for example, in the Marcus[32] 1965 paper (the culmination of the contributions by Marcus) no explicit consideration of barrier penetration or probability for available acceptor levels in solution: Everything was dependent upon a so-called reorganization energy, written in terms of the continuum Born equation.

Thus, by the mid-1960s a quite different (and much more rudimentary) mode of consideration of the quantum mechanics of electrode processes than the earlier (but molecular-modeled) quantum mechanical considerations of the Gurney[3] processes had developed. This theory can be said to (a) destress any quantum mechanical evaluation in terms of equations; (b) destress the molecular picture, being continuum and dielectric theory-oriented; and (c) avoid consideration of bonding and electrocatalysis. One of the less strong aspects of this continuum treatment of electrode processes was the lack of ability to reproduce the first law of electrode kinetics. The repeated appearance in the relevant publications of the equation

$$i = A \exp\left[-\frac{(E_s + \Delta F_0)^2}{4E_s kT}\right] \tag{10.2}$$

where E_s is the continuum solvent reorganization energy and ΔF^0 is the free energy of reaction, in contrast to the Tafel equation, made it difficult for

---

† A similar misquotation is made in respect to the accounting for the bond-stretching contribution in redox reactions. This is also often attributed to Marcus.[32] However, bond stretching in electron transfer in radii systems was first taken into account by George and Griffith.[99]

those in touch with the experimental data, e.g., that in Vetter's book, to accept such theories. Numerical comparison with experiment tended to be sparse until the mid-1970s.[33]

From the late 1960s and particularly from about 1970, the continuum work was oriented by Russian workers, particularly toward the proton transfer situation. To a small extent, overlap between the two approaches in electrode kinetics (the proton-oriented quantum mechanical molecular path beginning with Gurney and the redox-system-oriented continuum path commenced by Libby) began.

This beginning was inhibited in 1970 by a review published by Levich.[11] In this, the author treats liberally the proton transfer case but neglects states above the ground state of the proton because of his error in considering the vibrational states in H_3O^+ *in solution* to be distinct and separate as in the gas phase, so that no smooth rate–potential relation would be possible if these states are to be activated. (Levich apparently did not understand line broadening and the smoothing out of the vibration–rotation spectrum in water.) Correspondingly, his stress upon the Hamiltonian for a solid crystal to be used in liquids, a stress upon the so-called "quantum mechanical transport of the whole system" (metal ions, too), and the continued violation of the first law of electrode kinetics, made some of the material in this review difficult to accept.

After the breakup of the Levich, Dogonadze, and Kuznetsov team in 1972–1973, because of the politically based expulsion of Levich from cooperation with other Russian scientists, the taking into account of proton transfer from levels above the ground state began to be more stressed, although electrocatalytic effects were still not treated. Thus, in the presentation made by Dogonadze[12] in Hush's collection of continuum papers, there is an attempt to analyze the effects of higher vibrational levels on proton transfer.†

10.3. The Proton-Associated Aspects of the Gurney Quantum Mechanical Model

This model has been presented in Chapter 8, but we shall bring out points that serve to contrast the molecularly oriented Gurney[3] view with that of Libby[29] and followers.[11,12,32]

---

† Continuum theorists have not been unaware of the negative impression made by the lack of ability of their basic equations to reproduce the first law of electrode kinetics. They suggest that this arises because they assume harmonic vibrations. The effect of these anharmonic vibrations, however, had been taken into account as early as 1960 by Despic and Bockris,[37] who had examined the anharmonicity of the metal ion–water bond interaction on the applicability of Tafel's law and had calculated its effect.

(a) Gurney[3] connected the rate of H^+ neutralization to bonding of the proton to O and the result of the solvation energy of H_3O^+. It was not associated only with the solvation energy of water layers outside H_3O^+. There is a clear difference between this model and the continuum attempt developed later, which dealt (perhaps rather surprisingly) only with the energy effect of electron transfer on the external (Born) energy.

(b) The Gurney model used classically distributed states in solution and uses the so-called β factor, which may be defined[37] in terms of the conversion of electrical to chemical energy. There is no difficulty with the relative constancy of β,[37] which for low i_0 situations should exist for a change of overpotential of more than 1 V.

(c) Gurney asserted the lack of distinction in vibrational levels in 1931. Contributions made by Ackerman[39] and O'Ferrall *et al.*,[40] as by Rudolph and Zimmerman,[41] which indicate the quasicontinuum vibrational levels in H_3O^+ from spectral evidence, were still to come. The Gurney paper was a seminal one and is to quantum electrode kinetics what the Heitler and London paper of 1927 is to valency theory.

10.4. Butler's Modification of Gurney's Model: Electrocatalysis

Butler[46] realized in 1936 that Gurney's model did not give reasonable values of the heat of activation and espied the reason, namely, that Gurney had not bonded hydrogen to the electrode after neutralization, i.e., had not taken into account chemisorption. In the Gurney model, hydrogen atoms were left neutralized as free hydrogen atoms a few angstroms from the electrode. The addition of the bond to the electrode pulled down the energy of activation into a reasonable range (see Section 8.5.2.3).

However, Butler's modification[43] of Gurney's quantum mechanical theory[3] gave rise to the underlay of electrocatalysis, because, for charge-transfer reactions rate-determining in proton transfer, the major point about the electrocatalytic aspect of an electrode is its bonding to hydrogen. Such views (the forerunner of which, however, was in the theory of Horiuti and Polanyi[7]) were basic to concepts of electrocatalysis published in the 1960s.[44]

10.5. Rate-Determining Step and Path in Hydrogen Evolution

The kinetics of proton discharge are connected with the kinetics of the overall hydrogen evolution reaction. It is not possible to discuss the former without discussing the latter. It is assumed by many workers[11,12] (and always by contributors who regard the Born energy as the only relevant energy) that proton transfer to the electrode is the rate-determining step in hydrogen evolution, whereas this is often not the case.[45-47]

A distinction was made between the two types of mechanism simultaneously and independently by Conway and Bockris[48-50] and by Gerischer.[51-53] The dependence upon the heat of adsorption was the key determining role. If the electron transfer to the proton was rate determining, the Gurney-Butler mechanism was the rate-determining step, then the exchange current density at the equilibrium potential should increase with increase of the chemibonding of the hydrogen to the metal.

On the other hand, if the electrochemical desorption of an adsorbed hydrogen atom, with proton transfer to the site and electron transfer to it at the same time, were a rate-determining step (at high hydrogen coverage), then the exchange current density at the equilibrium potential would become smaller with increase of the bonding. A similar course would occur if chemical combination on the surface were rate determining. Bockris showed in 1946 and 1951[49a] that two distinct classes could be seen and the rationalization was given by Conway and Bockris in 1957. The two different trends of i_0 as a function of bonding are shown in Fig. 10.1.

Thus, simple proton discharge in acid solutions is only rate determining at mercury, probably also at lead, tin, cadmium, and other soft metals in acid solution. However, on the transition metals in acid solution, the rate-determining step is the electrochemical desorption step

$$H_3O^+ + H_{ads} + e_0 \rightarrow e_0 \rightarrow H_2 + H_2O \qquad (10.3)$$

The diagnostic isotopic separation factors associated with some mechanisms for hydrogen evolution are shown in Table 10.1.

Thus, it is important in discussions of the quantum character of proton transfer to make sure in relating the hydrogen evolution results to the proton transfer results what mechanism is rate determining, for *qualitatively* different dependence upon parameters such as bond strength for the metal–hydrogen interaction may be expected in either case.

10.6. The Basic Role of the Calculation of Separation Factors

Separation-factor calculations represent a direct method of knowing paths and mechanisms of the hydrogen evolution reaction[13-15] (hence, proton transfer is taken under the given circumstances). The separation factor varies with potential, and this is the most direct key to the quantum properties of protons in their transfer to electrodes.

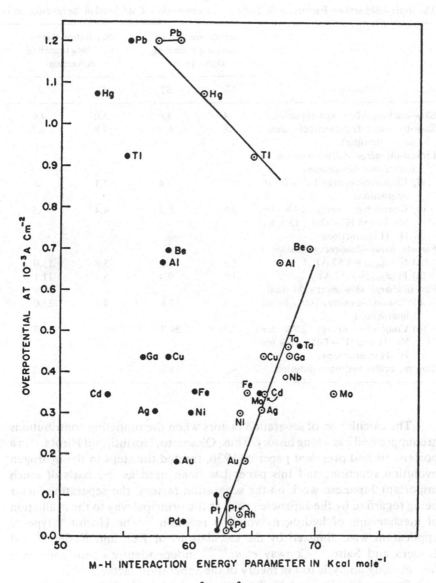

Fig. 10.1. Hydrogen overpotential at 10^{-3} A cm^{-2} plotted as a function of the metal–hydrogen interaction energy parameter. (●) Rüetschi and Delahay's values of $\frac{1}{2}D_{MH}$ + 51.7 kcal mole^{-1}[= true D_{MH} − 23.06 × $(X_M − X_H)^2$]. (◎) True values of D_{MH} calculated from Pauling's equation using known electronegativity values. Values given are calculated from Eqs. (10.8) or (10.9) or directly from Gordy's values of X_M. Values of η at 10^{-3} A cm^{-2} are calculated from the experimental Tafel parameters recorded by J. O'M Bockris [*J. Res. Natl. Bur. Standards* **524**, 243 (1954)].

Table 10.1
Theoretical Separation Factors for the Different Mechanisms as Calculated in the Investigation

Mechanism	Separation factors excluding tunneling corrections		Separation factors including tunneling corrections	
	S_D^*	S_r^*	S_D	S_r
Slow-discharge–fast-recombination	2.4	3.4	3.0	4.6
Slow-discharge–fast-electrochemical (in equilibrium)	3.8	6.2	3.8	6.2
Linked-discharge–electrochemical (either rate-determining)				
(i) Coulombic energy 100% for all interactions	3.4	5.4	4.1	7.0
(ii) Coulombic energy 20% for $M-H$ and H^+-OH_2; 15% for $H-H$ interactions	3.6	5.7	4.4	7.5
Fast-discharge–slow-recombination				
(i) Ni ($d_{Ni-Ni} = 3.52$ Å)	5.5	11.3	5.8	13.0
(ii) Pt ($d_{Pt-Pt} = 2.77$ Å)	4.9	9.4	5.5	11.1
Fast-discharge–slow-electrochemical				
(i) Coulombic energy 100% for all interactions	8.3	19.8	9.1	23.0
(ii) Coulombic energy 20% for $Ni-H$ and H^+-OH_2; 15% for $H-H$ interactions	9.7	24.7	10.7	28.7
Slow molecular hydrogen diffusion	4.5	8.2	4.5	8.2

The calculation of separation factors when the tunneling contributions are suppressed has a long history. Thus, Okamoto, Horiuti, and Hirota,[22] in a portentous and prescient paper of 1936, treated the steps in the hydrogen evolution reaction, and this paper has been used as the basis of much important Japanese work on the separation factors, the separation factor being regarded by the Japanese group as the principal way to the evaluation of mechanisms of hydrogen evolution reaction.[54] The Horiuti[23] type of application was climaxed by the calculations of Keii and Kodera[19] and Kodera and Saito.[24] Conway et al.[55–58] independently contributed in a far-reaching analysis to the theory of the separation factor.

That the precise model is determinative in the calculation and thus that the calculated values can give information on the mechanism is illustrated by the difference between the calculations carried out by Keii and Kodera[19] and those of Bockris and Srinivasan,[13–15] where the difference of the stretching frequency in the activated state caused the value to change in a direction that could make proton discharge (rather than the combination of H atoms) acceptable as a rate-determining step in some examples of the hydrogen evolution reaction.

Elements of the methods used in such calculations are important to understand, because their strength is the basis of mechanism assignment and hence the quantum mechanical calculation of rates. Thus, for example, the hydrogen (H) and deuterium (D) separation factor (S_D) can be expressed as

$$S_D = \left(\frac{C_H}{C_D}\right)_g \bigg/ \left(\frac{C_H}{C_D}\right)_s \tag{10.4}$$

where $(C_H/C_D)_g$ and $(C_H/C_D)_s$ are the ratios of the atomic concentrations of hydrogen to deuterium in the gas phase and solution, respectively.

On a quasi-phenomenological basis, the hydrogen–deuterium separation factor S_D can be expressed as

$$S_D = \frac{1}{2} \frac{P_{T,H}}{P_{T,D}} \frac{f_H^{\ddagger}}{f_D^{\ddagger}} \frac{f_{HDO(g)}}{f_{H_2O(g)}} K_D \tag{10.5}$$

where $P_{T,H}/P_{T,D}$ is the ratio of the tunneling factor for hydrogen and deuterium, $f_H^{\ddagger}/f_D^{\ddagger}$ is the partition function ratio of isotopic activated complexes, $f_{HDO(g)}/f_{H_2O(g)}$ is the partition function ratio of isotopic water molecules, and K_D is the equilibrium constant for the reaction

$$H_2O(l) + HDO(g) \rightleftharpoons HDO(l) + H_2O(g) \tag{10.6}$$

Similarly the hydrogen (H) and tritium (T) separation factor (S_T) is given by

$$S_T = \frac{1}{2} \frac{P_{T,H}}{P_{T,T}} \frac{f_H^{\ddagger}}{f_T^{\ddagger}} \frac{f_{HTO(g)}}{f_{H_2O(g)}} K_T \tag{10.7}$$

where K_T is the equilibrium constant for the reaction

$$H_2O(l) + HTO(g) \rightleftharpoons HTO(l) + H_2O(g) \tag{10.8}$$

10.6.1. Calculation of the Separation Factor

To determine S_D and S_T, one needs to determine the terms involved in Eqs. (10.5) and (10.7).

10.6.1.1. The Tunneling Ratios ($P_{T,H}/P_{T,D}$ and $P_{T,H}/P_{T,T}$)

The tunneling ratios depend on potential. The tunneling factor was determined[13] at the barrier height corresponding to current densities of 10^{-2} A cm^{-2} and at the corresponding overpotential $\eta = 1$ V. The barrier height has been taken in the activation energy E_a, which is obtained using the relation

$$E_a = E_a^0 + \beta e_0 \eta \tag{10.9}$$

where E_a^0 is the activation energy at the reversible potential and β is the symmetry factor. $E_a = 10$ kcal mole^{-1} was taken for $E_a^0 = 21.5$ kcal and $\beta = \frac{1}{2}$ at $\eta = 1$ V.

The tunneling factors were obtained both for the symmetrical Eckart and parabolic barriers for both smaller proton transfer distance (1.15 Å for nonisotopic ions) and for higher proton transfer distance (2.86 Å), which follows from the double-layer model,[59] where a layer of water molecules is assumed between the electrode surface and the outer Helmholtz plane.

For the smaller proton transfer distance, the tunneling factor ratios $P_{T,H}/P_{T,D}$ and $P_{T,H}/P_{T,T}$ for a symmetrical Eckart barrier are 4.8 and 8.4, respectively. The corresponding values for parabolic barrier are 14 and 33, and in the case of a high degree of tunneling, the parabolic barrier overestimates the tunneling factor ratios.

For the higher proton transfer distance, the tunneling factor rates $P_{T,H}/P_{T,D}$ and $P_{T,H}/P_{T,T}$ are 1.24 and 1.35, respectively, for a symmetrical Eckart barrier, and these are in agreement with the corresponding ratios of 1.25 and 1.34 obtained for a parabolic barrier.

10.6.1.2. Partition Function Ratio of Isotopic Activated Complexes (f_H/f_D and f_H/f_T)

Total Partition Function Ratios. The activated complex $H_2O\cdots H\cdots M$ (or its isotopes) is analogous to a linear triatomic molecule.[34] The partition function ratio of the isotopic activated complexes is given by

$$\frac{f_{H^\ddagger}}{f_{D^\ddagger}} = \frac{f_{t,H^\ddagger}}{f_{t,D^\ddagger}} \frac{f_{r,H^\ddagger}}{f_{r,D^\ddagger}} \frac{f_{v,H^\ddagger}}{f_{v,D^\ddagger}} \tag{10.10}$$

where $f_t^\ddagger$, $f_r^\ddagger$, and $f_v^\ddagger$ represent the translational, rotational, and vibrational contributions, respectively, of the indicated isotopes.

Translational Partition Function Ratio. The activated complex may be regarded as immobile. Under these conditions, the translational partition function ratio is unity.

Rotational Partition Function Ratio. The partition function ratio due to this restricted rotation in activated complex is expressed as

$$\frac{f_{r,H^\ddagger}}{f_{r,D^\ddagger}} = \frac{\sinh^2 h\nu_D/2kT}{\sinh^2 h\nu_H/2kT} \tag{10.11}$$

Since the observed rotational frequencies (e.g., for water 600 cm^{-1}) are small and are inversely proportional to the square roots of the corresponding moments of inertia, one gets

$$\frac{f_{r,\text{H}^\ddagger}}{f_{r,\text{D}^\ddagger}} = \frac{I_{\text{H}^\ddagger}}{I_{\text{D}^\ddagger}} \tag{10.12}$$

where $I_{\text{H}^\ddagger}$ and $I_{\text{D}^\ddagger}$ are the moments of inertia of the hydrogen and deuterium activated complexes, respectively, about an axis perpendicular to the axis of the molecule and through their respective centers of gravities. The calculated rotational partition function ratios are

$$\frac{f_{r,\text{H}^\ddagger}}{f_{r,\text{D}^\ddagger}} = 0.983 \tag{10.13}$$

Similarly,

$$\frac{f_{r,\text{H}^\ddagger}}{f_{r,\text{T}^\ddagger}} = 0.962 \tag{10.14}$$

Vibrational Partition Function Ratios. The vibrational partition function ratio of the isotopic activated complexes[60] is given by

$$\frac{f_{v,\text{H}^\ddagger}}{f_{v,\text{D}^\ddagger}} = \frac{\sinh(h\nu_\text{D}/2kT)_{st}}{\sinh(h\nu_\text{H}/2kT)_{st}} \frac{\sinh^2(h\nu_\text{D}/2kT)_b}{\sinh^2(h\nu_\text{H}/2kT)_b} \tag{10.15}$$

The subscripts st and b stand for stretching and bending frequencies, respectively.

The bending frequency (ν) of a linear triatomic molecule XYZ is given by[61]

$$\lambda = 4\pi^2\nu^2$$
$$= \frac{1}{l_1^2 l_2^2}\left[\frac{l_1^2}{m_\text{Z}} + \frac{l_2^2}{m_\text{X}} + \frac{(l_1+l_2)^2}{m_\text{Y}}\right]k_b \tag{10.16}$$

where l_1 and l_2 are the distances of X and Z from central atom Y; m_X, m_Y, and m_Z are the masses of atoms X, Y, and Z, respectively, and k_b is the bending force constant. In the case under consideration, $\text{X} = \text{M}$; $\text{Y} = \text{H, D,}$ or T; and $\text{Z} = \text{H}_2\text{O}$.

Generally, bending frequencies are small ($<600 \text{ cm}^{-1}$). We may hence assume that

$$\frac{\sinh^2(h\nu_\text{D}/2kT)_b}{\sinh^2(h\nu_\text{H}/2kT)_b} = \frac{\nu_\text{D}^2}{\nu_\text{H}^2} = \frac{m_\text{H}}{m_\text{D}} = \frac{1}{2} \tag{10.17}$$

$$\frac{\sinh^2(h\nu_\text{T}/2kT)_b}{\sinh^2(h\nu_\text{H}/2kT)_b} = \frac{\nu_\text{T}^2}{\nu_\text{H}^2} = \frac{m_\text{H}}{m_\text{T}} = \frac{1}{3} \tag{10.18}$$

as was done originally by Melander.[62] The potential energy of the system can then be expressed as a function of displacement from the saddle point of the activated complex by the equation (Fig. 10.2)

$$V = V_0 + \tfrac{1}{2}k_{11}(r_1 - r_1^{\ddagger})^2 + \tfrac{1}{2}k_{22}(r_2 - r_2^{\ddagger})^2 + k_{12}(r_1 - r_1^{\ddagger})(r_2 - r_2^{\ddagger}) \qquad (10.19)$$

where k_{11} is the force constant for the stretching of the bond between the water molecule and H^+ ion, k_{22} that for the stretching of the bond between the hydrogen and metal atoms, and k_{12} is a coupling constant. The kinetic energy of the system is given by

$$T = \tfrac{1}{2}m_1 r_1^2 + \tfrac{1}{2}m_2 r_2^2 + \tfrac{1}{2}m_3 r_3^2 \qquad (10.20)$$

where m_1, m_2, and m_3 are the masses of the water molecule, hydrogen, and metal atoms, respectively.

Using Lagrange's equation of motion, it can then be shown that the stretching frequencies are given by the equation

$$\lambda^2 - \lambda\left[\left(\frac{1}{m_1} + \frac{1}{m_2}\right)k_{11} + \left(\frac{1}{m_2} + \frac{1}{m_3}\right)k_{22} - \frac{2}{m_2}k_{12}\right]$$

$$+ \frac{m_1 + m_2 + m_3}{m_1 m_2 m_3}(k_{11}k_{22} - k_{12}^2) = 0 \qquad (10.21)$$

where

$$\lambda = 4\pi^2 \nu^2 \qquad (10.22)$$

One of these frequencies is imaginary for the activated complex. Two of the approximate methods used are the following:

Method 1. *Case of low real stretching frequencies.*[62] If the imaginary frequency be taken as zero (one vibration becomes a translation along the reaction path), the last term in Eq. (10.21) can be set equal to zero. Therefore,

$$k_{11}k_{22} = k_{12}^2 \qquad (10.23)$$

Under these conditions, the real frequency of the activated complex is given by

$$\lambda = 4\pi^2 \nu^2 = \frac{k_{11}}{m_1} + \frac{k_{22}}{m_3} + \frac{k_{11} + k_{22} - 2k_{12}}{m_2} \qquad (10.24)$$

Using Eq. (10.23), Eq. (10.24) becomes

$$\lambda = 4\pi^2 \nu^2 = \frac{k_{11}}{m_1} + \frac{k_{22}}{m_3} + \frac{k_{11} + k_{22} - (k_{11}k_{22})^{1/2}}{m_2} \qquad (10.25)$$

If one now makes the assumption that $m_1 \gg m_2$, $m_3 \gg m_2$, and $k_{11} \gg k_{22}$, Eq. (10.25) reduces to

$$\lambda = 4\pi^2\nu^2 = \frac{k_{11} + k_{22} - 2(k_{11}k_{22})^{1/2}}{m_2} \qquad (10.26)$$

If one assumes that these frequencies are small, the vibrational partition function ratio of the hydrogen to deuterium (or hydrogen to tritium) activated complexes is equal to the ratio of the corresponding frequencies. Thus,

$$\frac{f_{v,\mathrm{H}^\ddagger}}{f_{v,\mathrm{D}^\ddagger}} = \frac{m_\mathrm{H}^{1/2}}{m_\mathrm{D}^{1/2}} = \frac{1}{2^{1/2}} \qquad (10.27)$$

$$\frac{f_{v,\mathrm{H}^\ddagger}}{f_{v,\mathrm{T}^\ddagger}} = \frac{m_\mathrm{H}^{1/2}}{m_\mathrm{T}^{1/2}} = \frac{1}{3^{1/2}} \qquad (10.28)$$

Using Eqs. (10.13), (10.14), (10.17), and (10.18), the complete partition function ratio is

$$\frac{f_{\mathrm{H}^\ddagger}}{f_{\mathrm{D}^\ddagger}} = 0.3479 \qquad (10.29)$$

$$\frac{f_{\mathrm{H}^\ddagger}}{f_{\mathrm{T}^\ddagger}} = 0.1851 \qquad (10.30)$$

The use of this method for the calculation of the vibrational partition function ratio gives an upper limit for the separation factors.

Method 2. *Case of moderate or high frequencies.*[63] In this method, the potential energy of the linear three atom system (Fig. 10.2) is given by the Heitler–London expression, viz.,

$$V = K_{12} + K_{23} + K_{31} - \left\{ \tfrac{1}{2}[(J_{12} - J_{23})^2 + (J_{23} - J_{31})^2 + (J_{31} - J_{12})^2] \right\}^{1/2} \qquad (10.31)$$

where the K values are the Coulombic and J values the exchange contributions to the total energies for interactions between H_2O and H^+, hydrogen and metal and metal and H_2O.[24]

The potential energy V_{12} of the initial state $H_2O-H^+ + e_{0,\mathrm{M}} + M$, expressed as a function of the distance between H^+ ion and H_2O molecule (pseudo atom) can then be obtained by considering the following:

$$
\begin{array}{ccc}
H + H_2O + M & \xrightarrow{\;V_{12}\;} & H_3O^+ + e_M + M \\
{\scriptstyle \uparrow -I} & & {\scriptstyle \downarrow -D_{12}} \\
H^+ + H_2O + e + M & \xleftarrow{\;\Phi\;} & H^+ + H_2O + e_M + M
\end{array}
\qquad (10.32)
$$

Therefore,

$$V_{12} = I + D_{12} - \Phi = K_{12} + J_{12} \tag{10.33}$$

where I is the ionization energy of the hydrogen atom, Φ is the electronic work function of the metal, D_{12} is the energy of the H^+–OH_2 interaction as a function of the internuclear distance and is given by

$$D_{12} = D_{12}^0 \{1 - \exp[-a(r_1 - r_1^0)]\}^2 - L \tag{10.34}$$

where L is the heat of solvation of a proton and D_{12}^0 is the proton affinity of water.

Similarly, the potential energy of the final state $MH + H_2O$, expressed as a function of the distance between the metal and hydrogen atoms, is obtained by considering the reaction

$$H + M + H_2O \rightarrow MH + H_2O \tag{10.35}$$

where

$$D_{23} \simeq D_{23}^0 \{\exp[-2a_2(r_2 - r_2^0)] - 2\exp[-a(r_2 - r_2^0)]\}$$
$$= V_{23} = K_{23} + J_{23} \tag{10.36}$$

Since the metal atom and water molecule are far apart for all values of r_1 and r_2, and because the interaction between metal and H_2O is much less than between hydrogen and metal or H^+ and H_2O, the M–H_2O interaction was taken as zero for all distances of separation, i.e.,

$$V_{31} = K_{31} + J_{31} = 0 \tag{10.37}$$

As in the method of Eyring *et al.*,[63] a percentage of each of the energies V_{12}, V_{23}, and V_{31} is taken to coulombic and the balance as exchange for all distances of separation. Thus, the energy of the system $H_2O\cdots H\cdots M$ was calculated, using Eq. (10.31), as a function of r_1 and r_2 (Fig. 10.2). All calculations were carried out with $M = Hg$. The constants used in the calculation of V_{12} and V_{23} with the aid of Eqs. (10.33), (10.34), and (10.36) are given in Table 10.2.

Three calculations were carried out varying the percentage coulombic

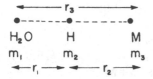

Fig. 10.2. Activated complex for slow-discharge mechanism.

Table 10.2

Force Constant (k's), Stretching Vibrational Frequencies (ω's), Partition Function Ratios ($f_{H^\ddagger}/f_{D^\ddagger}/f_{T^\ddagger}$) of Isotopic Activated Complexes for the Slow-Discharge Mechanism

Parameter	Calculation number		
	1	2	3
p_1 (%)	20	5	39
p_2 (%)	20	5	3
$r_1^\ddagger$ (Å)	1.05	1.05	1.05
$r_2^\ddagger$ (Å)	3.40	2.92	3.30
$E^\ddagger$ (kcal mole^{-1})	2	6	4
k_{11} (kcal mole^{-1} Å^{-2})	688.6	635.1	654.8
k_{22} (kcal mole^{-1} Å^{-2})	−3.9	−12.7	−10.6
k_{12} (kcal mole^{-1} Å^{-2})	0	0	0
ω_H (cm^{-1})	2906	2773	2820
ω_D (cm^{-1})	2110	2015	2048
ω_T (cm^{-1})	1763	1686	1714
$\omega_{H^\ddagger}$ (cm^{-1})	52i	90i	84i
$\omega_{D^\ddagger}$ (cm^{-1})	50i	88i	83i
$\omega_{T^\ddagger}$ (cm^{-1})	50i	89i	81i
$\dfrac{\sinh(h\nu_D/2kT)_s}{\sinh(h\nu_H/2kT)_s}$	0.1465	0.1605	0.1552
$\dfrac{\sinh(h\nu_T/2kT)_s}{\sinh(h\nu_H/2kT)_s}$	0.0633	0.0725	0.0672
$(f_{H^\ddagger}/f_{D^\ddagger})\,10^2$	7.207	7.897	7.636
$(f_{H^\ddagger}/f_{T^\ddagger})\,10^2$	2.030	2.325	2.155

energies p_1 and p_2 in V_{12} and V_{23}, respectively. In the first calculation, p_1 and p_2 were taken as 20%. The energy V given by Eq. (10.31) was calculated as a function of the variable distances r_1 and r_2. From the table of values of V as a function of r_1 and r_2, the reaction path and saddle point were determined. Thus, $r_1^\ddagger = 1.05$ Å and $r_2^\ddagger = 3.40$ Å. This calculation showed that the activation energy excluding zero point energies is 2 kcal mole^{-1}. The force constants k_{11}, k_{22}, and k_{12} appearing in Eq. (10.19) were then calculated by evaluating the second derivatives of V (Eq. 10.31) at the saddle point. The coupling constant k_{12} is zero for the present model, since the interaction between M and H_2O was ignored. The stretching vibrational frequencies, calculated using Eq. (10.21), for the isotopic activated complexes (hydrogen, deuterium, and tritium) in this case are given in Table 10.2.

In the second calculation, the percentage coulombic energy used for each interaction was reduced to 5%, since the first calculation gave too small an activation energy. The activation energy increased to 6 kcal mole^{-1} in this case. In the third calculation, the percentage coulombic energy for the oxygen–H$^+$ bond was taken as 39% and 3% for the mercury–hydrogen bond. The activation energy became 4 kcal mole^{-1}.

The results of all three calculations, including partition function ratios of the isotopic activated complexes, are given in Table 10.2. The quantities p_1 and p_2 are the percentage coulombic energies of the oxygen–H$^+$ and hydrogen–mercury bonds, respectively, $r_1^{\ddagger}$ and $r_2^{\ddagger}$ are the oxygen–H$^+$ and mercury–hydrogen internuclear distances, respectively, in the activated complex, $E^{\ddagger}$ is the activation energy, and $\sinh(h\nu_D/2kT)_s/\sinh(h\nu_H/2kT)_s$ and $\sinh(h\nu_T/2kT)_s/\sinh(h\nu_H/2kT)_s$ are the partition function ratios of the isotopic activated complexes due to the real stretching frequencies.

Table 10.2 reveals that the variation in percentage coulombic energy has little influence on the real vibrational frequencies (ω_H, ω_D, ω_T) of the activated complexes. Further, the calculated activation energies are much less than the experimentally observed values. This discrepancy is due to the neglect of the H$_2$O–mercury interaction in the calculation of V.

10.6.1.3. The Equilibrium Constant (K_D or K_T)

The equilibrium constant K_D is known over a wide range of temperatures.[66] K_T is known at temperatures ranging from 0° to 90°C. Its value at 25°C is 1.093.

10.6.1.4. Partition Function Ratio of Isotopic Water Molecules in Gas Phase ($f_{HDO(g)}/f_{H_2O(g)}$ and $f_{HTO(g)}/f_{H_2O(g)}$)

The partition function ratio ($f_{HDO(g)}/f_{H_2O(g)}$) is given by

$$\frac{f_{HDO(g)}}{f_{H_2O(g)}} = \frac{[H_2O]}{[HDO]} \frac{m_{HDO}^{3/2}}{m_{H_2O}^{3/2}} \frac{(I_A I_B I_C)_{HDO}^{1/2}}{(I_A I_B I_C)_{H_2O}^{1/2}} \frac{\prod\limits^{3} \sinh(h\nu_i/2kT)_{H_2O}}{\prod\limits^{3} \sinh(h\nu_i/2kT)_{HDO}} \tag{10.38}$$

An expression may be written for the ratio ($f_{HTO(g)}/f_{H_2O(g)}$). Using the spectroscopic data $f_{HDO(g)}/f_{H_2O(g)}$ is 59.16[64] at 25°C and $f_{HTO(g)}/f_{H_2O(g)}$ is 289.54.[65] Together with K_D and K_T given in the previous section, one gets

$$\frac{1}{2} K_D \frac{f_{HDO(g)}}{f_{H_2O(g)}} = 31.68 \tag{10.39}$$

$$\frac{1}{2} K_T \frac{f_{HTO(g)}}{f_{H_2O(g)}} = 158.23 \tag{10.40}$$

10.6.2. Results of Separation-Factor Calculations

The results of separation factor calculations for both the cases (one with exclusion of tunneling correction and the other with inclusion of tunneling correction) are given in Table 10.3. The mechanism considered here is the slow discharge of H_3O^+ and fast electrochemical desorption of hydrogen atom from the electrode, respectively. The results of separation factor calculations for the slow discharge of H_3O^+ and fast recombination hydrogen atom mechanism are in Table 10.4.

It may be seen from Table 10.4 that the separation factor for a slow-discharge mechanism depends on the desorption step as well. For a desorption step by the combination of adsorbed hydrogen atoms, the separation factor is determined by the ratio of rate constants for the discharge step. For a desorption by the electrochemical mechanism, the separation factors depend on whether this step is treated as in equilibrium or not in equilibrium. In the former case, the separation factors (S_D and S_T) become respective equilibrium constants of the H_2/HDO_1 and H_2/HTO_1 interchange reactions. In the latter case, the separation factors are the same as for a coupled discharge-electrochemical desorption mechanism. The H/D separation factors are not sufficiently separated to be of value in determining the desorption step. The H/T separation factors may be useful for this purpose.

A slow-discharge–fast-electrochemical desorption mechanism cannot be distinguished from a fast-discharge–slow-electrochemical desorption mechanism at high overpotentials when the reverse currents of the discharge and electrochemical desorption steps may be neglected in comparison to their forward currents. The only way of distinguishing between these two mechanisms is by a determination of degree of coverage (θ) of atomic hydrogen on the electrode, since $\theta \to 0$ for a slow-discharge mechanism and $\theta \to 1$ for a slow-electrochemical desorption mechanism.

The tunneling corrections are probably overestimated. In this case, the H/T separation factors for the three cases are nearly equal. There are two ways of examining the tunnel effect: (a) influence of potential on the separation factors,[68] and (b) influence of temperature on the separation factor at constant potential. Assuming no change of mechanism, with variation of potential, there should be no change of separation factor with current density if the desorption step is the electrochemical desorption mechanism and is treated as in equilibrium. The other mechanisms will be affected, since the tunneling factors change with potential. A temperature effect on S will distinguish the slow-discharge–fast-electrochemical desorption mechanism (with the latter treated as in equilibrium) from the others. The importance of the tunneling contribution is given in the above calculations, but more important aspects of the separation factor, such as its dependence upon potential, are given in Section 10.8.

Table 10.3

Separation Factors for the Slow-Discharge-Fast-Electrochemical Desorption Mechanism at High Overpotentials

Calculation number for S_D, x_1 or S_T, x_1	Percentage Coulombic energy (p)			Separation factors excluding tunneling corrections			Separation factors including tunneling corrections		
	M—H	H⁺—OH₂	H—H	$S_{D,1}^*$	S_{D,E_1}^*	S_D^*	$S_{D,1}$	S_{D,E_1}	S_D
3	100	100	100	2.4	11.8	3.4	3.0	13.0	4.1
4	20	20	15	2.4	14.9	3.6	3.0	16.4	4.4
				$S_{T,1}^*$	S_{T,E_1}^*	S_T^*	$S_{T,1}$	S_{T,E_1}	S_T
3	100	100	100	3.4	27.9	5.4	4.6	29.5	7.0
4	20	20	15	3.4	34.1	5.7	4.6	39.5	7.5

Table 10.4

Theoretical Separation Factors for the Slow-Discharge–Fast-Recombination Mechanism

Mechanism	Separation factors excluding tunneling corrections		Separation factors including tunneling corrections	
	$S_D^{\ddagger}$	$S_T^{\ddagger}$	S_D	S_T
Slow-discharge–fast-recombination	2.4	3.4	3.0	4.6
Slow-discharge–fast-electrochemical	3.8	6.2	3.8	6.2
Linked-discharge–electrochemical				
(i) Coulombic energy: 100% for all interactions	3.4	5.4	4.1	7.0
(ii) Coulombic energy: 20% for M—H and H⁺—OH₂; 15% for H—H interactions	3.6	5.7	4.4	7.5

10.7. Quantum Character of Proton Transfer: Contributions of Christov

The contributions of St. G. Christov[8–10,69,70] to electrode kinetics are valuable because the author is a quantum chemist and because his considerations were from the point of view of chemical kinetics and structure, rather than from the point of view of continuum dielectrics. In some respects, the stress made upon the quantum theory by Christov extends (but corrects) the early work of Bawn and Ogden,[5b] who interpreted the separation factor in values in terms of the quantum theory.

In considering the velocity of protons in passage and through barriers, Christov[8–10,69,70] calculated the separation factor as a function of the energy differences of the two isotopic bonds to oxygen (this had already been done by the Japanese workers[22–24] and also by Bockris[46]), but Christov stressed the quantum mechanical contribution to this for the first time after the calculations of Bawn and Ogden[5b] in 1934.

Christov's work[8–10,69,70] fine tuned the calculations to find the shape and height of the barrier that corresponded to the facts. He based his work upon two facts—the separation factor for mercury and the *b* value of the Tafel equation that does not change with potential over a certain range. With these facts in mind, Christov calculated the barrier shape consistent with the effects and distinguished between the classical activation energy, the heat of reaction, the parabolic barrier approximation, and the Eckart barrier, which he used in his calculations (Fig. 10.3).

Christov's stress upon the quantum mechanical nature of proton transport was historically of help to electrochemical theory and represented a

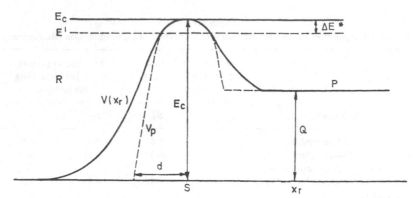

Fig. 10.3. $V(x_r) =$ energy profile along the reaction coordinate x_r; $E_c =$ classical activation energy; $Q =$ heat of reaction at $T = 0$; $E' = E_c - \Delta E^*$; $V_p =$ parabolic barrier approximation in the energy range $E > E'$; $d =$ half-width of the parabolic barrier V_p; $R =$ reactants region; $P =$ products region.

continuation of the Gurney[3] approach. His picking out of separation facts as determinative in mechanism greatly furthered the work of Bawn and Ogden.[5b] However, the most important evidence of quantum mechanical tunneling was still to come—that of the dependence of separation factor upon potential (Section 10.8).

The work of Christov[8-10] is noteworthy for its quantum mechanical and structural character, thereby differing importantly from the parallel continuum work. The difficulties it faced were in relating itself to experiment. It used the value of the separation factor (a satisfactory criterion) but also the value of the Tafel slope b. This latter quantity is not a very satisfactory criterion, because it had to be explained how the quantum mechanical character fitted into the fact that the slope b did not change with potential over the experimental range. In addition, the difference between the actual numerical value of the quantum calculations of the separation factor and those that arose only from taking into account the zero point energy differences was not sufficiently great to be determinative of the quantum character (in contrast to the potential dependence of the separation factors).

10.8. Quantum Character of Proton Transfer: Bockris and Matthews

10.8.1. Variation of the Separation Factor with Potential

There is no doubt that the work of Christov gave rise to the work of Bockris and Matthews.[18] Christov[9] had suggested that an important effect upon separation factors would be obtained by working at low temperatures, and Bockris and Matthews had measured to $-71°C$ by using concentrated HCl solutions. However, during this work, they, essentially accidentally, ran into an effect that was more important and diagnostic than the effects expected at low temperatures, namely, the considerable variation of the separation factor with potential. The results are exemplified in Fig. 10.4.

Although the significance of the potential dependence of the separation factor is clear to workers with quantum mechanical orientations, it is necessary to discuss other possible dependences of the separation factor upon potential before drawing the conclusion that they are quantum mechanically caused. Alternate possibilities were examined by Bockris and Matthews,[18] for example,

(a) The symmetric stretching frequency of the activated state had been shown earlier by Bockris and Srinivasan[13-15] to be important in determining the numerical value of the separation factor. Bockris and Matthews[18] examined to what extent the symmetric stretching frequency depended upon electrode charge and therefore on potential, and they obtained an expression for the potential dependence of the separation factor. The expression is given by

$$S_{H.T} = \text{const} \exp[-(h/4\pi kT)(0.42)(k_s^*/\mu_H)^{1/2}] \qquad (10.41)$$

Bockris and Matthews[18] used the variation of force constant of the symmetric stretching of the bonds of activated complex k_s^* with potential to give a possible effect. The theory indicates that as the potential becomes

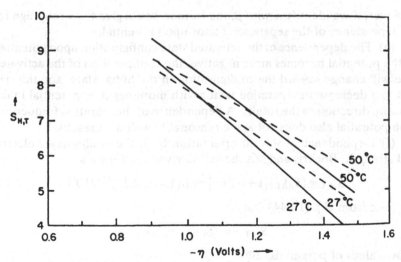

Fig. 10.4. Comparison of calculated (dashed lines) and observed (solid lines) dependence of $S_{H.T}$ on potential. $\varepsilon_0^* = 1.5 \times 10^{-12}$ erg, $2d = 4.0$ Å, $A_0 = -0.6 \times 10^{-12}$ erg.

more negative, S should increase as the force constant decreases. However, experiment shows that S decreases with more negative potential.

(b) Alternatively, the restricted rotation of water molecules at the metal–solution interface might be affected by the electric field. However, since the solvent is the same for H_2TO^+ and for H_3O^+ ions, any effect that would thereby arise on the energy for solvent libration would affect the solvation energy of the isotopically different ions by the same amount. Such a theory, therefore, would not lead to any isotopic effect at all.

(c) Dependence of the type of metal site for hydrogen adsorption potential. Surfaces of metals give single or multiple sites for adsorption. When the adsorption site is only a single one, one metal atom interacts with an adsorbed species, but in the multiple-site adsorption, two or more metal atoms interact with an adsorbed species. If one uses a hard-sphere model for the H_3O^+ ion discharge, it will take place on that site that offers the strongest metal–hydrogen bond. Thus, electrostriction would cause the H_3O^+ to move closer to the electrode, but the site for easiest deposition of hydrogen would be unchanged, so that this cannot explain the potential effect.

(d) Dependence of the symmetry factor on the isotope. Post and Hisky[71] found that

$$(\alpha_H - \alpha_D) = -0.014 \tag{10.42}$$

and this looks similar to the formula of the Bockris and Matthews work, which was

$$(\beta_H - \beta_T) = 0.061 \tag{10.43}$$

However, if we identify α with β, the former would give the wrong sign for the dependence of the separation factor upon potential.

(e) The dependence of the activated state configuration upon potential. As the potential becomes more negative, the configuration of the activated state will change toward the configuration in the initial state, and this will lead to a decrease in separation factor with more negative potential in the opposite direction to the results. A dependence of the vibration frequencies upon potential also does not agree reasonably well with results.

(f) Dependence of solvent orientation from the double-layer electric field strength. Bockris and Matthews[18] derived the formula

$$S_{H,T} = (S_0)_{H,T}\{1 + 0.266[\tanh(1.402\,\Delta\Phi)]^{1/2}/kT\} \tag{10.44}$$

It is seen from Eq. (10.44) that

$$S_{H,T} \rightarrow (S_0)_{H,T} = 3.4$$

at low values of potentials, $\Delta\Phi$.

$$S_{H,T} \rightarrow 0.73$$

at high potentials. The plot of S against potential is nonlinear according to this equation, and gives rise to experimentally inconsistent results at low potentials. The effect of specific adsorption on chloride ions was also found to be inconsistent in its predicted effect upon the potential dependence of the isotope separation factor with experiment.

A summary of the theories for the dependence of separation factor on potential is shown in Table 10.5. It appears that the effects observed by Bockris and Matthews[18] are indeed due to the different degrees of quantum mechanical tunneling undergone by the hydrogen and the tritium particles in the interfacial transfer through a barrier.†

10.9. Quantum Mechanical Interpretation of the Evolution of the Separation Factor with Potential

10.9.1. Quantum Mechanical Correction to Separation Factor

The quantum mechanical correction of separation factor is defined by Bockris and Matthews[17] as

$$\Gamma = Sq/S_c \tag{10.45}$$

where S_q and S_c are the quantum mechanical and classical separation factors, respectively. S_q for protium–tritium can be expressed as

$$(S_q)_{H,T} = \frac{\tau_H}{\tau_T}(S_c)_{H,T} \tag{10.46}$$

where τ_H and τ_T are the quantum mechanical transmission coefficient for protium and tritium transfer, respectively.

τ can be defined as

$$\tau = i_q/i_c \tag{10.47}$$

where i_q and i_c are quantum mechanical and classical rate of electrochemical electron transfer for the discharge step

$$H_3O^+ + M(e) \rightarrow M_H + H_2O \tag{10.48}$$

One may write i_q as

$$i_q = k_1 C_{H^+} \int_{E_0}^{\infty} P_T(E) \exp[-(E - E_0)/kT] \, dE \tag{10.49}$$

---

† Some of the results of the separation factor measurements obtained at low temperatures by Bockris and Matthews[18] are certainly affected by chloride ion adsorption.

Table 10.5
Theories for the Dependence of S on Potential[18]

Theory	Assumptions	Predictions	
		$dS/d\Delta\phi$	Other
Dependence of the symmetric stretching frequency of the activated state on the electrode charge.	Low bending frequency and high symmetric stretching frequency of the activated state.	−	$S_{H,T}$ tends to 3.4 as $\Delta\phi$ tends to zero.
Dependence of librational behavior of the solvent on the double layer electric field.	Low librational frequency.	0	
Dependence of the type of metal site for H adsorption on potential.	Change from single- to multiple-site adsorption due to electrostriction.	0 or −	
Effect of electrostriction on the interaction between solvent and the activated state.	Old double-layer model.	−	
Specific adsorption.	Adsorbed anions interact with the transition state causing an isotope effect.	+	$S_{H,T}$ tends to 3.4 as $\Delta\phi$ tends to large cathodic values. S depends on anion and solution concentration.
Dependence of activated state configuration on potential.	Small bending and large symmetric stretching frequencies of the activated state.	+	$S_{H,T}$ tends to unity as $\Delta\phi$, tends to large cathodic values. Nonlinear dependence of S on $\Delta\phi$.
Symmetry factor.	$\beta_H \neq \beta_D \neq \beta_T$.	+	$\beta_T - \beta_H = 0.069$. $S_{H,T}$ tends to 3.4 as $\Delta\phi$ tends to zero.
Dependence of the solvent orientation on the double-layer electric field strength.	Old double-layer model. Large bending frequency in the activated state.	+	$S_{H,T}$ tends to 3.4 as $\Delta\phi$ tends to zero. Nonlinear dependence of S on $\Delta\phi$.
Proton tunneling.	BDM model of double layer.	+	$S_{H,T}$ tends to 3.4 as $\Delta\phi$ tends to large cathodic values. Linear dependence of S on potential.

where k_1 is a frequency factor, $P_T(E)$ is the probability of proton tunneling at energy level E, E_0 is the zero-point energy level for the H^+-OH_2 bond stretching.

Equation (10.49) reduces to the classical current i_c in the limit of low and wide barriers. Thus,

$$i_c = k_1 C_{H^+} \int_{E^\ddagger}^{\infty} \exp[-(E-E_0)/kT]\, dE$$

$$= k_1 C_{H^+} kT \exp[-(E^\ddagger - E_0)/kT] \tag{10.50}$$

where $E^\ddagger$ is the energy of the barrier maximum.

Using Eqs. (10.49) and (10.50) in Eq. (10.47), one gets

$$\tau = (kT)^{-1} \exp\left[\frac{(E^\ddagger - E_0)}{kT}\right] \int_{E_0}^{\infty} \exp[-(E-E_0)/kT] P_T(E)\, dE \tag{10.51}$$

or

$$\tau = (kT)^{-1} \exp[(E^\ddagger - E_0)/kT] J \tag{10.52}$$

where

$$J = \int_{E_0}^{\infty} \exp[-(E-E_0)/kT] P_T(E)\, dE \tag{10.53}$$

Using Eqs. (10.45) and (10.46), one gets

$$\Gamma_{H,T} = \tau_H / \tau_T \tag{10.54}$$

From Eqs. (10.52) and (10.54),

$$\Gamma_{H,T} = \exp[(E_{0,T} - E_{0,H}) kT] J_H / J_T \tag{10.55}$$

To find Γ, one needs to evaluate J_H and J_T, and these terms involve the tunneling probability factor $P_T(E)$ [see Eq. (10.53)] for the unsymmetrical Eckart barrier. The tunneling probability $P_T(E)$ is given by [see Eqs. (8.67) and (8.70)]

$$P_T(E) = \frac{\cosh[2\pi(\lambda+\mu)] - \cosh[2\pi(\lambda-\mu)]}{\cosh[2\pi(\lambda+\mu)] + \cosh 2\pi\sigma} \tag{10.56}$$

where

$$\lambda = (d/h)(2mE)^{1/2} \tag{10.57}$$

$$\mu = (d/h)[2m(E-A)]^{1/2} \tag{10.58}$$

$$\sigma = \tfrac{1}{2}[(8md^2B/h^2) - 1]^{1/2} \tag{10.59}$$

where $2d$ is the width of the barrier.

$A = A_0 + e_0\eta$, $B = 4(E_0^\ddagger + \beta e_0\eta)$, and η is the overpotential. B has its minimum value for $A = 0$ when $B = 4E^\ddagger$. For $\eta = 0$, $E^\ddagger = 0.8 \times 10^{-12}$ erg and $2d = 3.0$ Å; the value of σ is found to be 0.87. Since $2d \geqslant 3.0$ Å and

$E^{\ddagger} \geqslant 0.8 \times 10^{-12}$ erg, the above figure represents the minimum value of σ. Thus, for the values of σ found, one may write

$$2 \cosh 2\pi\sigma \simeq \exp(2\pi\sigma) \tag{10.60}$$

The minimum value of λ corresponds to $\varepsilon = \varepsilon_0$. For $2d = 1.5$ Å and $\varepsilon = \varepsilon_0$, then $\lambda = 2.242$. Thus, the approximation

$$2 \cosh 2\pi\lambda = \exp(2\pi\lambda) \tag{10.61}$$

is valid. Since μ is positive, then $\lambda + \mu > \lambda$ and the approximation

$$2 \cosh 2\pi(\lambda + \mu) = \exp[2\pi(\lambda + \mu)] \tag{10.62}$$

is valid. Substituting Eqs. (10.60), (10.61), and (10.62) into (10.56) gives

$$P_T(E) = \frac{\exp[2\pi(\lambda + \mu - \sigma)] - \exp[2\pi(\lambda - \mu - \sigma)] - \exp[-2\pi(\lambda - \mu + \sigma)]}{\exp[2\pi(\lambda + \mu - \sigma)] + 1} \tag{10.63}$$

10.9.2. Results of Tunneling

Bockris and Matthews[17] used an iterative approach to determine Γ, due to uncertainties in the value of the proton transfer distance $2d$ and the barrier height. Values of J for protium, deuterium, and tritium were calculated numerically on a UNIVAC Solid-State 80 digital computer. The calculations were carried out for temperatures of 323°, 298°, 273°, 240°, and 198°K; for barrier heights of 1.6, 1.8, and 2.0×10^{-12} erg; for barrier widths 3.0, 4.0, 5.0, and 6.0 Å; for A_0 values of 0.8, 0.4, 0.0, -0.4, and -0.8×10^{-12} erg; and for overpotentials of 0.0, 0.4, 0.8, 1.2, 1.4, 1.6, 1.8, 2.0, and 2.2×10^{-12} erg.

From the computed values of J for protium, deuterium, and tritium, the value of Γ was calculated according to Eq. (10.55). The experimental dependence of Γ on potential is given in Fig. 10.5. A comparison between the computed and experimental dependence of separation factor $S_{H,T}$ on potential is given in Fig. 10.4.

10.9.3. Conclusion

It is important to mention that in the Bockris and Matthews'[17] calculation it was necessary to agree with the following parameters:

(a) The Tafel slope is constant with potential over ~ 1 V.
(b) The numerical value of the separation factor.
(c) The potential dependence of the separation factor.

Consistence with these parameters gave rise to the following probable barrier [see Eqs. (10.56)–(10.59)] parameters: $E^{\ddagger} = 1.5 \times 10^{-12}$ erg, $2d =$

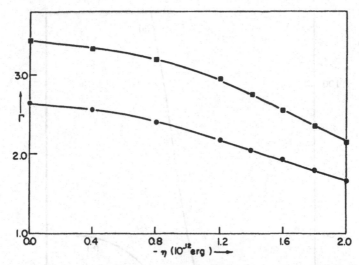

Fig. 10.5. Dependence of Γ on potential. $A_0 = 0$, $T = 298°K$, $2d = 4.0$ Å, $\varepsilon_0^* = 1.6 \times 10^{-12}$ erg; $\bullet = \Gamma_{H,D}$, $\blacksquare = \Gamma_{H,T}$.

4.0 Å, and $A_0 = -0.6 \times 10^{-12}$ erg. The barrier parameters deduced by Christov[9] were $E^{\ddagger} = 1.6 \times 10^{-12}$ erg and $2d = 4.7$ Å. In the work of Christov,[9] the value of A_0 was not considered as a parameter and the zero-point energy of the initial state was not included in the analysis. For comparison to the present work, one should then give Christov's barrier height as 1.9×10^{-12} erg. This work of Bockris and Matthews[17] therefore shows (a) the importance of the potential effect of the separation factor upon the quantal character of the proton transfer, and (b) the helpful effect of such studies upon the calculation of barrier height and shape.

10.10. Variation of the Symmetry Factor with Potential

It is difficult to obtain clear information upon the experimental variation of β with potential because of the influence of possible other effects, (for example, the effect of specific adsorption of ions) near the potential of zero charge, which Bockris, Ammar, and Huq[72] first showed may give rise to a dependence of β upon the variation of the electrokinetic potential with the electrode potential.

It is necessary to work at relatively high negative potentials, where the effect of potential upon the shape of the barrier would become important, to establish the effect of β upon potential. This was first carried out by Despic and Bockris[37] in 1960. The workers chose the silver deposition reaction to establish the effect and, by working at high overpotential and very short

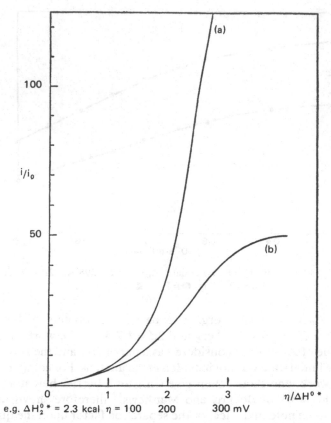

Fig. 10.6. η_a-i relation as predicted by the Butler–Volmer equation (a) with constant β and (b) with β varying with η.

times, avoided transport effects and were able to plot the i/i_0 considering the variation of β with overpotential η (Fig. 10.6).

The theory of the variation of symmetry factor β with overpotential η in terms of the Morse parameters is a more general method than that of the anharmonicity corrections applied by some at a later date. The equations, according to Despic and Bockris, can be derived as follows.

The curves AA and BB (Fig. 10.7) pertaining, respectively, to solvated ions and adions can be represented by the equations

$$U_1 = D_1\{1 - \exp[-a_1 x_0(x/x_0)]\}^2$$
$$U_2 = D_2(1 - \exp\{-a_2 x_0[1 - (x/x_0)]\})^2 + A \qquad (10.64)$$

Then,

$$\tan \theta = (dU_1/dx)_{x'}$$

$$= 2D_1 a_1\{\exp[-a_1 x_0(x'/x_0)] - \exp[-2a_1 x_0(x'/x_0)]\} \qquad (10.65)$$

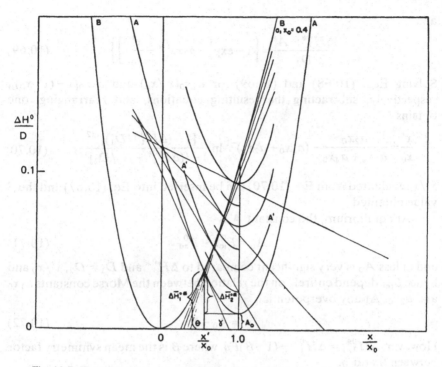

Fig. 10.7. The potential energy–distance relation (Morse function with $a_1x_0 = 0.4$).

and

$$\tan \gamma = (dU_2/dx)_{x'}$$

$$= -2D_2a_2(\exp\{-a_2x_0[1-(x'/x_0)]\}$$
$$- \exp\{-2a_2x_0[1-(x'/x_0)]\}) \qquad (10.66)$$

From Eqs. (10.65) and (10.66),

$$\beta = \frac{-\tan \gamma}{-\tan \gamma + \tan \theta}$$

$$= \frac{c - c^2}{cb^2 + (D_1a_1/D_2a_2)(b - b^2)} \qquad (10.67)$$

where $c = \exp[-a_2x_0(1 - x'/x_0)]$ and $b = \exp[-a_1x_0(1 - x'/x_0)]$. Since at the point of intersection $U_1 = (\Delta H_1^{0*})$ and $U_1 = U_2$ so that $U_2 - A = (\Delta H_2^{0*}) = (\Delta H_1^{0*}) - A$, Eqs. (10.64) can be rewritten as

$$\frac{\Delta H_1^{0*}}{D_1} = \left[1 - \exp\left(-a_1x_0\frac{x'}{x_0}\right)\right]^2 \qquad (10.68)$$

and

$$\frac{\Delta H_1^{0*} - A'}{D_2} = \left\{ 1 - \exp\left[-a_2 x_0 \left(1 - \frac{x'}{x_0} \right) \right] \right\} \qquad (10.69)$$

Solving Eqs. (10.68) and (10.69) for $a_1 x_0 (x'/x_0)$ and $a_2 x_0 [1 - (x'/x_0)]$, respectively, subtracting the resulting equations, and rearranging, one obtains

$$\frac{x'}{x_0} = \frac{a_2 x_0}{a_1 x_0 + a_2 x_0} - (a_1 x_0 + a_2 x_0)^{-1} \ln \frac{[1 - (\Delta H_1^{0*}/D_1)]^{1/2}}{1 - [(\Delta H_1^{0*} - A)/D_2]^{1/2}} \qquad (10.70)$$

x'/x_0 evaluated from Eq. (10.70) can be inserted into Eq. (10.67) and the β value obtained.

At equilibrium, the constant A is

$$A_0 = F_{\text{ion}}^0 - F_{\text{ad}}^0 \qquad (10.71)$$

and unless A_0 is very significant compared to ΔH_1^{0*} and $D_1 \neq D_2$, x'/x_0 and hence β_{rev} depend entirely on the relation between the Morse constants $a_1 x_0$ and $a_2 x_0$. At any overpotential,

$$A = A_0 - F \qquad (10.72)$$

However, $\Delta H_{1,\eta}^{0*} = \Delta H_1^{0*} - (1 - \beta) F \eta$, where β is the mean symmetry factor between 0 and η.

Thus,

$$\frac{x'}{x_0} = \frac{a_2 x_0}{a_1 x_0 + a_2 x_0} - (a_1 x_0 + a_2 x_0)^{-1} \ln \frac{1 - \{[\Delta H_1^{0*} - (1 - \beta) F \eta] / D_1\}^{1/2}}{1 - [(\Delta H_1^{0*} - A_0 + \beta F \eta) / D_2]^{1/2}} \qquad (10.73)$$

and the dependence of β on overpotential is established.

The calculated $i - \eta_a$ relation is shown in Fig. 10.6. The shape of the theoretical relation (b) in Fig. 10.6 reproduces the tendency toward a transfer-controlled limiting current established experimentally (Fig. 10.8). Other work upon the variation of β with potential has been restricted. The work of Parsons and Passeron[73] was interpreted in terms of the continuum electrostatic approximation, according to which, however, the value of β should be continuously changing with overpotential (Fig. 7.6).

10.11. A BEBO Approach to Proton Transfer Calculations

In the potential energy surface work of Bockris and Srinivasan and Bockris and Matthews there is neglect of attractive interaction at the crossing point of the surfaces. The bond energy–bond order method

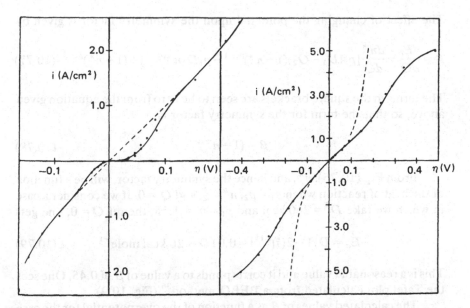

Fig. 10.8. Typical i–η relation in case of (*left*) a nonactivated and (*right*) an activated electrode. (*Left*) 0.0724 N AgClO$_4$ + 1 N HClO$_4$ solution. (*Right*) 0.364 N AgClO$_4$ + 1 N HClO$_4$ solution. Dashed line indicates Butler–Volmer curve with β = constant.

(BEBO) of Johnston[74,75] avoids this difficulty. One can construct a potential energy surface by empirical equations that connect bond length and bond order and bond energy for the reacting system. The potential energy surface is defined by the expression

$$E = D_1 - D_1 n_1^{p_2} - D_2 n_2^{p_2} \qquad (10.74)$$

where E is the potential change along the reaction coordinate above the level of the reactants, the potential energy of which is D_1. D_2 represents the proton–oxygen potential energy in H_3O^+. The bond orders are n_1 and n_2. The p terms are empirical constants, which differ little from unity. An assumption is made that bond order is served over the reaction coordinate, i.e., $n_1 + n_2 = 1$.

The activation energy E_a is given by the condition that $dE/dn = 0$, or when (if n^* is the bond order at the col)

$$p_1 D_1 n^{*p_1/p_2-1} = p_2(D-Q)(1-n^*)^{p_1/p_2-1} \qquad (10.75)$$

$$E_a = D_1 - D_2 n^{*p_1} - (D-Q)(1-n^*)^{p_2} \qquad (10.76)$$

where Q is the potential energy difference between protons and reactants.

The effect of changing the potential upon the symmetry factor is given by

$$\beta = \frac{dE_a}{dQ} = \frac{dn^*}{dQ}[p_2(D_1 - Q_2)(1 - n^*)^{p_1-1} - p_1 D_1 n^{*p_1-1}] + (1 - n^*)^{p_2} \quad (10.77)$$

The terms in the square brackets are seen to be zero from the equation given above, so that the term for the symmetry factor is

$$\beta = (1 - n^*)^{p_2} \quad (10.78)$$

From Eq. (10.75), n^*, and hence the symmetry factor, will be a function of the heat of reaction when $p_1 = p_2$, $n^* = \frac{1}{2}$, and $Q = 0$. If we consider a case in which we take $D^1 = 260$ kcal and $p_1 = p_2 = 1.15$, then if $Q = 0$, one gets

$$E_a = D[1 - 2(\tfrac{1}{2})^{1.15}] \approx 0.01D \sim 26 \text{ kcal mole}^{-1} \quad (10.79)$$

This is a reasonable value and it corresponds to a value of β of 0.45. One sees the Tafel plot calculated from a BEBO method[36] (Fig. 10.9).

The calculated values of β as a function of the overpotential for the case of ΔH^* of 15.4 kcal mole^{-1} are shown in Table 10.6. It is seen that the variation of β with potential is less than that which arises in the continuum theory.

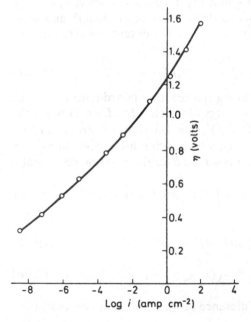

Fig. 10.9. Tafel line obtained by using BEBO method.

Table 10.6

Variation of the Symmetry Factor (β) with Potential

η	$(\Delta H^*)_{calc}$ (kcal/mole)	$(\Delta H^*)_{expt}$ (kcal/mole)	β
0.0	15.54	20	0.561
0.25			0.558
0.5			0.554
0.8			0.551
0.9			0.547
1.0			0.591
1.25			0.531
1.5			0.52
1.63			0.48

10.12. Double-Layer Model and Proton Transfer

Having to meet the three demands of constant b value by a correct absolute value of S, the separation factor, and an ability to interpret the variation of S with potential, all put together, make a powerful weapon in our ability to decide the barrier height and other parameters with which the proton discharge reaction is associated. A table of barrier heights is given in Table 10.7.

However, apart from giving a basis for estimates of the barrier height, the proton transfer work allows us also to make other equations concerning the double-layer structure. It is, of course, necessary to have a consistent picture of the proton transfer work, along with work on the structure of the double layer.[59] In the latter, one of the major points has been to attempt to

Table 10.7

Computed Activation Energies

Barrier height (10^{-12} erg)	2d (Å)	Protium		Deuterium	
		Computed	Classical	Computed	Classical
		(kcal/mole)		(kcal/mole)	
1.6	5.0	17.2	18.8	19.2	19.9
1.6	4.0	15.8	18.8	19.4	19.9
1.6	3.0	10.6	18.8	17.8	19.9
1.8	4.0	18.1	21.7	21.5	22.8
2.0	4.0	20.0	24.5	24.9	25.7

put together the fact that the capacitance of the double layer at mercury offers the almost invariant long section of numerical value in the neighborhood of $16 \ \mu F \ cm^{-2}$ independent of the ion size. This leads one to conclude[77] that there is a probable layer of water molecules attached to the electrode surface and oriented to it.[78] A proof of this structure has recently been given by the statistical and thermodynamic interpretation of the entropy of mixing of the water molecules as a function of charge on the electrode.[79]

If the water molecule structure of the electrode is maintained, then the distance of the H_3O^+ particles from which the molecules must start in proton transfer is in the neighborhood of 5 Å from the metal surface (see Fig. 1.2). This value turns out to be far too great in the vital thickness of the barrier calculation, upon which the value of the separation factor very clearly depends.

In order to avoid the difficulty of inconsistency between the double-layer structure work and the barrier structure required for proton transfer, Bockris and Matthews[17] suggested the situation shown in Fig. 10.10, in which the H_3O^+ in the double layer, transfers a proton first to a water molecule outside the first layer, and then this is the one from which the discharge takes place. Such a structure is consistent with the necessary values of the structure of the double-layer capacitance and other aspects of the structure of the double layer.

10.13. Validity of the WKB Tunneling Expression

Most of the quantum mechanics of proton transfer that have been calculated to this time involve the Gamow equation. It is desirable to make analysis of this equation in respect to its validity. There are two aspects of

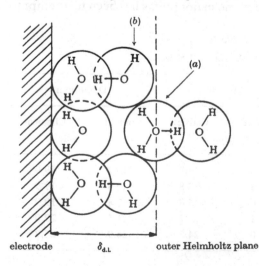

Fig. 10.10. Three-dimensional model of the double layer for proton discharge. The H_3O^+ at (a) transfers a proton via a hydrogen bond to the water at (b). (b) The proton nearest the electrode is the one discharged.

electrode $\delta_{d.L}$ outer Helmholtz plane

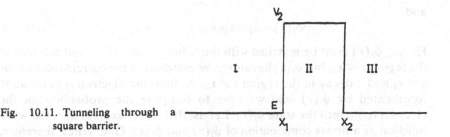

Fig. 10.11. Tunneling through a square barrier.

that. On the one hand, there is the fundamental question of the validity of when the WKB expression can be taken into account. It has been seen from Eq. (8.78) that the WKB transition probability expression applies quantitatively for proton transfer through the double layer.

Another aspect of the situation is the barrier itself, and its fluctuation with time, and this occurs (e.g., in respect to librations or rotations of water molecules) independently of the act of transfer.[†]

The question of the degree of validity of the WKB method has been examined by Sen and Bockris.[26] Consider a square barrier as is shown in Fig. 10.11. The wave functions may be represented by

$$\psi_I = a_{1i} \exp(ik_1 x) + b_{1f} \exp(-ik_1 x)$$

$$\psi_{II} = a_{2i} \exp(-k_2 x) + b_{2f} \exp(k_2 x) \qquad (10.80)$$

$$\psi_{III} = a_{3i} \exp(ik_3 x) + b_{3f} \exp(-ik_3 x)$$

where k_1, k_2, and k_3 are the wave vectors of the particle in regions, I, II, and III, respectively (Fig. 10.11). There is a well-known solution to the tunneling problem provided by these wave functions, and it gives the solution for the transition probability P_T as

$$P_T = \frac{16 k_1^2 k_2^2}{(k_1^2 + k_2^2)(k_3^2 + k_2^2)} \exp(-2k_2 \omega) \qquad (10.81)$$

where $\omega = (x_2 - x_1)$ is the barrier thickness, and $k_2 = [2m(V_2 - E)/h^2]^{1/2}$ and $k_1 = k_3 = (2mE/h^2)^{1/2}$

The following wave functions for the electron were chosen (Fig. 10.11):

$$\psi_i(x) = a_{2i} \exp(-k_2 x), \qquad x \le x_1 \qquad (10.82a)$$

---

† However, Bockris and Habib,[80] in their analysis of the entropy of the double layer, did not detect this variation.

and

$$\psi_f(x) = b_{2f} \exp(k_2 x), \qquad x \geq x_2 \tag{10.82b}$$

Hence, $\psi_f(x)$ must be matched with the solution for $x \geq x_2$, and will fade in the region $x \leq x_1$, but $\psi_i(x)$ has also to be matched to the correct solution for $x \leq x_1$ and a decay in the region $x \geq x_2$. At first, the electron is in the state represented by $\psi_i(x)$, and we have to compute the probability for the electron to go into the state $\psi_f(x)$. Let us, therefore, express the total wave function as a linear combination of $\psi_i(x)$ and $\psi_f(x)$. One writes, therefore,

$$\psi_{II} = C(t)\psi_i(x) \exp(-iE_i t) + d(t)\psi_f(x) \exp(-iE_f t) \tag{10.83}$$

The substitution of Eq. (10.83) into the time-dependent Schrödinger equation gives us

$$C \exp(-iE_i t) H\psi_i + d \exp(-iE_i t) H\psi_f$$

$$= i\dot{C}\psi_i \exp(-iE_i t) + C\psi_i E_i \exp(-iE_i t) + i\dot{d}\psi_i \exp(-E_f t) + d\psi_f E_f \exp(-iE_f t) \tag{10.84}$$

Initially, however, the electron is represented by the wave function $\psi_i(x)$. Hence, we can set $C = 1$, $d = 0$, and $\dot{C} = 0$. The latter condition follows from the normalization condition that $d[CC^* + (dd^*)]/dt = 0$.

What obtains from Eq. (10.84) is

$$i\dot{d}\psi_f \exp(-iE_f t) = (H - E_i)\psi_i \exp(-iE_i t)$$

or

$$i\dot{d}\psi_f = \exp[i(E_f - E_i)t] \int \psi_f^*(H - E_i)\psi_i \, dx \tag{10.85}$$

The effective matrix element for tunneling is therefore

$$T_{if} = \int \psi_f^*(H - E_i)\psi_i \, dx \tag{10.86}$$

The integral can be put into the same symmetric form by subtracting $\psi_i(H - E_f)\psi_f^*$, since this contribution gives no contribution to the range of integration $x \geq x_1$.

$$T_{if} = \int_{x_B}^{\infty} [\psi_f^*(H - E_i)\psi_i - \psi_i(H - E_f)\psi_f^*] \, dx, \qquad x_1 \leq x_B \leq x_2 \tag{10.87}$$

Integrating Eq. (10.87) by parts gives us, therefore,

$$T_{if} = \frac{\hbar^2}{2m}\left(\psi_f^*\frac{d\psi_i}{dx} - \psi_i\frac{d\psi_f^*}{dx}\right)_{x_B} = -i\hbar j_{if} \tag{10.88}$$

where j_{if} is the current operator. Using Eqs. (10.82) and (10.88) gives us

$$T_{if} = (\hbar^2 k_2/m)b_{2f}^* a_{2i} \tag{10.89}$$

One now obtains b_{2f} in terms of b_{3f}, and a_{2i} in terms of a_{1i} by solving the standard matching problem at x_1 and x_2 of Fig. 10.11; one then introduces them into Eq. (10.89) and obtains

$$|T_{if}|^2 = \frac{\hbar^4 k_2^2}{m^2} \frac{16 k_1^2 k_3^2 |a_{1i}|^2 |b_{3f}|^2}{(k_1^2 + k_2^2)(k_3^2 + k_2^2)} \exp(-2k_2\omega) \qquad (10.90)$$

In terms of Fermi's golden rule, the current comes out to be

$$j_f = h^{-1} |T_{if}|^2 \, dn/dE_f \qquad (10.91)$$

where dn/dE_f is the density of states without spin in the transmitted wave and can be written

$$dn/dE_f = |b_{3f}|^{-2} m/2\pi h^2 k_3 \qquad (10.92)$$

The incident current is therefore

$$j_i = (h\mathbf{k}_1/m)|a_{1i}|^2 \qquad (10.93)$$

so that the transition probability becomes

$$P_T = \frac{j_f}{j_i} = \frac{16 k_1^2 k_2^2 k_3^2}{k_1 k_3 (k_1^2 + k_2^2)(k_3^2 + k_2^2)} \exp(-2k_2\omega) \qquad (10.94)$$

using for j_f the expression resulting from the substitution of Eqs. (10.90) and (10.92) in Eq. (10.91).

If one compares Eq. (10.94) with Eq. (10.81), one finds that the exponential part is the same. There is some difference in the preexponential expressions for P_T given by the WKB method and that given by the time-dependent perturbation theory. As k_1 and k_3 have the same order of magnitude, the two factors do not indicate important differences.

Thus, from this point of view, fairly good agreement should be obtained by application of Gamow's equations. Nevertheless, there is advantage in applying time-dependent perturbation as has been done by Khan, Wright, and Bockris,[38] not only because fundamentally it avoids approximations of the WKB method, but also because it is more precise in its quantum mechanical approach.

As pointed out above, there is another argument that could make the Gamow equation less than valid, namely, the fluctuation of the barrier with time. If the tunneling time is *less* than the fluctuation time of the barrier, the usual tunneling expression will indeed be valid. Sen and Bockris[26] indicated this effect by consideration of the situation in Fig. 10.12. Due to the permeability of this barrier, there is a splitting in the energy levels in the two wells. Let this be represented energywise by a δE. Let ψ_{II} and ψ_{IV} represent the wave functions that are applicable in well II and well IV (Fig. 10.12). It can be said, therefore, that $\psi_{II} + \psi_{IV}$ has the energy $E_0 - \delta E$, and $\psi_{II} - \psi_{IV}$ has the energy $E_0 + \delta E$. The time-dependent Schrödinger equation can then be applied, and conserving it, we obtain

$$\psi = \exp(-iE_0 t/\hbar)(\psi_{II} + \psi_{IV})\exp(+i\delta E t/\hbar) + (\psi_{II} - \psi_{IV})\exp(-iEt/\hbar)$$

or

$$\psi = 2\exp(-iE_0 t/\hbar)[\cos(\delta E t/\hbar)\psi_{II} + i\sin(\delta E t/\hbar)\psi_{IV}] \qquad (10.95)$$

If the particle is in well II at $t = 0$, it is in well IV at $t = \hbar\pi/\delta E$. The tunneling time can be defined as $\frac{1}{2}\hbar\pi/\delta E$. The tunneling time must be small compared with the barrier fluctuations. For electron and proton transfer in aqueous solution, the vibrational or librational modes in initial and final states make up the barrier. These modes have frequencies in the realm of 10^{13} sec^{-1} for water. The fluctuation time of the barrier would, then, be in the neighborhood of 10^{-13} sec.

The evaluation of the tunneling time depends upon a calculation of δE, and to do this, Sen and Bockris[26] utilized an expression of Dennison and Uhlenbeck,[81] which is

$$\delta E = \frac{h[2m(E - V)]^{1/2}}{2\pi m l[\exp\{(2\pi l/h)[2m(E - V)]^{1/2}\}]^2} \qquad (10.96)$$

where $(E - V)$ is the barrier height, m is the mass of the tunneling particle, and l is the barrier thickness. The magnitudes of tunneling time can be seen in Table 10.8, and it is seen that in most cases they are smaller than the fluctuation times of the barrier and hence acceptable. The values of $(E - V)$ and l have been taken from Caldin and Kasparian.[83]

In electrochemical proton transfer, the ground vibrational level of the H_3O^+ ion, the height of the barrier for proton transfer in electrochemical hydrogen evolution reaction, and other parameters have been estimated by Matthews and Bockris.[17] The width of the barrier, according to them, is 2.42 Å and 20 kcal mole^{-1} for the height. The tunneling does not occur

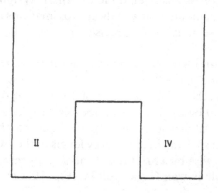

Fig. 10.12. Double-well potential.

Table 10.8
Data Relevant to Calculation of the Splitting Energy in Proton Transfer[26]

Reactions	$(E - V)$ (kcal/mole)	l (Å)	δE (eV)	Tunneling time (sec)	Comments about WKB approximation
1. $CH_2BrC \cdot MePh \cdot H + QEt^-$	22.1	1.59	0.06	1.6×10^{-14}	Valid
2. $RH + D_2O$	13.1	1.26	0.08	1.3×10^{-14}	Valid
3. $RH + F^-$	18.0	1.17	0.09	1.2×10^{-14}	Valid
4. $RH + F^-$ $[H_2O + NaBr(5M)]$	24.2	1.46	0.01	1.1×10^{-13}	Doubtful
5. $C_6H_2(NO_2)_3CH_2^- + HOAc$	10.1	1.66	10.06	1.6×10^{-14}	Valid
6. $H_3O^+ + H_2O$			0.12	8.47×10^{-15}	Valid

principally from the ground state of H_3O^+. The first vibrational level of H_3O^+ is 10 kcal mole^{-1}, and at this point the width is 2 Å. The splitting energy for that case, using Eq. (10.96) is 0.07 eV, and the tunneling time is about 1.3×10^{-14} sec, and consequently the Gamow equation is applicable.

10.14. The Continuum Theory to Proton Transfer

A model for the quantum mechanical aspects of the transfer of protons to and from an electrode in solution was published by Dogonadze, Kuznetsov, and Levich[84] in 1968. Not only the mode of treatment but also the model used and the whole picture of the reaction given by Levich, Dogonadze, and Kuznetsov (LDK) and, more recently, particularly by Dogonadze,[12] differ radically from the model used by Gurney,[3] Gerischer,[51-53] Christov,[8-10] Bockris and Matthews,[17,27,28] and others,[38] during the last two decades.

Apart from the actual physicochemical details of what the model of these workers indicates, the situation concerned with its publication and presentation within the literature has several special features that must be brought out:

(a) The standard of mathematical discussion in such presentations, concerning as it does the well-developed subject of continuum dielectrics, is high.

(b) The presentation of the same material has been made monotonously in many journals in many languages.

Any physicist or applied mathematician coming to the electrode process area finds a few papers giving a relatively elementary treatment in terms of molecular models and a mathematically sophisticated treatment, with many dozens of very similar papers by the Russian theoretical group at the Institute of Electrochemistry, similar to theories earlier published by the respected kineticist R. A. Marcus. It is less often realized that the basic equation of the theory is inconsistent with experiment in numerous ways. The situation is a special one.

10.14.1. Qualitative Discussion of the Continuum Model for Proton Transfer

Although the proton transfer reaction is normally seen as differing from the redox reactions by the existence of metal–hydrogen bonding, the LDK model treats the proton transfer reaction in the same way as it does redox reactions. The reaction takes place as a result of continuum *solvent fluctuation*; no account of any activation or stretching of the O—H bond is made.

Thus (cf. Chapter 6) the mechanism of activation of the H_3O^+ entity to transfer a proton to the electrode is given by fluctuations in the polarization energy of the surrounding solvent and is not connected directly with the thermal heat sink that is supposed to give rise to the activated states for proton transfer in other theories.

Correspondingly, the only levels that are supposed to give rise to any transfer in the LDK view are the lowest levels, i.e., the ground state of H_3O^+, because the other states are higher and thus less activated (since $h\nu \gg kT$). The Russian workers treat proton "tunneling" in a different way from that used by other workers. Their view is that the tunneling does indeed occur, but that the Gamow equation cannot be applied because of fluctuations in the solvent barrier, a view that has been discussed by Sen and Bockris[26] (see Section 10.13).

Thus, one way of showing this quantum mechanical transition in the LDK view is that of Fig. 10.13 when the proton coordinate of the initial state does not change (only the solvent coordinate of the initial state changes). The expression for the time-dependent perturbation theory that would

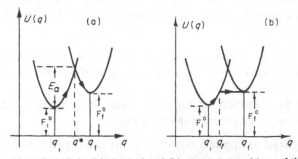

Fig. 10.13. Interpretation of the (a) classical and (b) quantum transitions of the system from initial to final state.

correspond to the alternative to utilizing the Gamow equation and the transition matrix is not explicitly calculated in the LDK view, though its calculation is advocated.

The representation of the energy coordinates for the proton transfer case in the Dogonadze view[12] is given in a plot such as that of Fig. 10.14. The Dogonadze treatment considered that, due to the quantum nature of the proton, only the medium dipoles perform classical fluctuations. Therefore, it assumed the proton coordinate in the initial state to be rigidly fixed $(R = R_{i0})$ and the motion along the electron term to occur only toward q. When, due to fluctuation, the solvent coordinate assumes the value q^*, the system will have the energy $F_i^0 + \frac{1}{2}\hbar\omega_0 (q^* - q_{i0})^2$ (Fig. 10.14a); if the proton in the initial state was at the excited state with the energy $E_p^{(i)} = \hbar\omega_p(l + \frac{1}{2})$ (Fig. 10.14a) the total energy of the system in the activated state q^* will be

$$U_i(q^*, R_{i0}) + E_p^{(i)} = F_i^0 + \frac{1}{2}\hbar\omega_0(q^* - q_{i0})^2 + \frac{1}{2}\hbar\omega_p(2l + 1) \quad (10.97)$$

where F_i^0 is the free energy in the critical state. The solvent being a slow subsystem with respect both to electron and proton, the transition from one

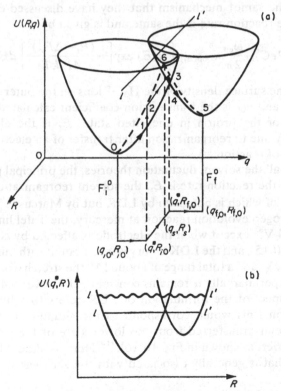

Fig. 10.14. Schematic diagram of the "reaction path" for a proton transition (a) according to the theory of absolute reaction rates (dashed curve) and (b) the true reaction path (wavy line).[12]

energy surface to another occurs with a fixed coordinate q^* (result of the Franck–Condon principle). The transition from the initial to final terms involves, in addition to the change in the electronic state, a change in the proton coordinate (Fig. 10.14a). To illustrate this stage, Fig. 10.14b shows the intersection of the electron terms with the plane ll' parallel to the plane UOR and passing the point (q^*, R_{i0}). Since $U(q^*, R)$ is the proton potential energy, the nature of the proton transition should be explained on this plot (Fig. 10.14b). It can be both subbarrier (straight line in Fig. 10.14b) and above the barrier (wiggly line in Fig. 10.14b). In both cases one has to deal with a purely quantum proton transfer. After the transition from the initial to final energy surface, the system relaxes classically from the point (q^*, R_{a0}) to the equilibrium point of final state (q_{f0}, R_{f0}). In terms of the conservation of energy, the quantum transition (Fig. 10.14a) can occur only if the energies of the system before and after the reaction coincide (see also Fig. 10.14b). It is worth mentioning that the bowllike shape arises in Fig. 10.14a due to a three-dimensional plot, where the energy coordinate U, the solvent coordinate q, and the proton coordinate R, are involved.

A characteristic of the formulation given by LDK is that, as the mechanism is the same† mechanism that they have discussed earlier, the equation for the reaction rate is the same and is given by

$$i_c^0 = 2eC_{H^+}^0 \frac{\omega_{\text{eff}}}{2\pi} \kappa_{00\rho(E_f)} \int n(E) \exp\left[-\frac{(E_s - \Delta F_r^0)^2}{4E_s kT}\right] dE \quad (10.98)$$

where $C_{H^+}^0$ is the surface density of the H_3O^+ ions on the outer side of the Helmoltz layer and κ_{00} is the transmission coefficient calculated using the wave function of the proton in unexcited state. E_s is the electrostatic dielectric energy due to reorganization after transfer of an electron, ΔF_r^0 is the free energy of reaction.

Thus, as in all the solvent fluctuation theories, the principal parameter that determines the reaction rate is E_s, the solvent reorganization energy, the formulation of which is not given by LDK but by Marcus.

In the hydrogen evolution reaction at mercury, the Tafel line is linear over at least 1.4 V[85] except when the electrode is affected by competitive reactions (Fig. 10.15), and the LDK theory is inconsistent with this linearity. Utilizing $E_s = 2$ eV, and a total range of about 2 V, the stretch of α is 0.1 and 0.9, although experimentally α remains constant ± 0.01 over 1.4 V.

Another aspect of the formulation of LDK refers to a hypothetical barrierless region (this would correspond to the situation in which the protons were being transferred from the lower state of their own curve through the barrier, as shown in Fig. 10.16).[91a] Here, α should be 1.

The state that is generally associated with the zero variation of α is

---

† LDK appear to assume that there is one rate-determining step in all electrode kinetics—electron transfer from the metal to the ion.

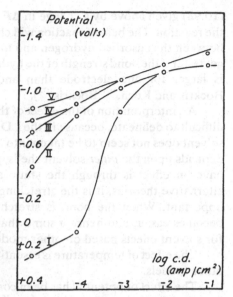

Fig. 10.15. Overpotential in an aqueous acid solution at a mercury cathode in presence of various quantities of oxygen corresponding to I, II, III, IV, and V, as mentioned by Azam, Bockris, Conway, and Rosenberg, *Trans. Faraday Soc.* **46**, 1918 (1950).

called by LDK the normal state, although in fact here the α varies so considerably, if one applies the expressions of their theory.[36] LDK also recognize an activationless region when the heat of activation is equal to the reaction heat (Fig. 10.16). Here, α should be 0. Although other authors recognize that there is a region in which α tends to zero [Despic and Bockris (1960)[37]], there is little experimental evidence for the barrierless region. Thus, work[86] upon which it is based has been carried out in circumstances in which specific adsorption may interfere with the interpretation.

There are several qualitative interpretations that may be made from the LDK model of proton transfer kinetics. The first is catalysis. Here, the physical meaning of the presentation does not differ much from that espoused by Butler[6] in 1936. The essentials arise in the LDK equation

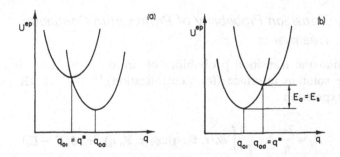

Fig. 10.16. (a) So-called "activationless" barrier. (b) So-called "barrierless" case.

(10.98) given above by a change in ΔF_r^0 which is the free energy change in the reaction. The heat of reaction will change with a change of bond strength between the adsorbed hydrogen and the metal, and in fact the reaction will get faster if the bond strength of the hydrogen atom attached to the electrode is larger for one electrode than another (cf. Conway and Bockris,[100] Bockris and Koch,[101] Bockris[102]).

An interpretation of the effect of the solvent on proton transfer is more difficult to delineate, because in the LDK view the *bond* of the proton to the solvent does not seem to be taken into account, and the effect of the solvent depends upon the *outer* solvent shells, so that the only way the solvent can have an effect is through the static and optical dielectric constants. In alternative theories, it is the stretching of the oxygen–proton bond that is important. When the bond is stretched, the electron transfer to H_3O^+ becomes easier, and hence a somewhat different result would be obtained for solvent effects based on other models.[36,103]

The effect of temperature is essentially the same in the LDK view[36] as in other models.

The effect of potential has been commented upon above, and appears a persistent difficulty in the LDK work.[36] More recently, however, it has been recognized by these workers that their original assumption concerning the square term dependence of the potential energy of the states is an approximation, and they have stressed the necessity of adding considerations of anharmonicity.[12]

One of the more recent additions to the theory is the effect of isotopes.[87] The isotopic effects are well known and have been discussed above in respect to the experimental and theoretical work of Bockris and Matthews[17,27,28] (see Section 10.9). In the LDK view, the isotopic effect should be independent of potential, because the isotopic effect arises from bond stretching, and bond stretching does not have any part to play in the LDK view. The characteristic shape of the separation factor–potential relation is difficult to interpret on the LDK model,[87] and particularly the marked effect of potential upon these quantities.

10.14.2. Transition Probability of Proton from Continuum Treatments

To find the transition probability of proton from H_3O^+ ion to the electrode solution interface (for neutralization),[11,12,88,89] LDK used the general expression

$$P_{if} = \frac{2\pi}{\hbar} \text{Av} \sum_f \left| \int \psi_r(\mathbf{r}, \mathbf{R}, \mathbf{q}) v \psi_i(\mathbf{r}, \mathbf{R}, \mathbf{q}) \, d\tau \right|^2 \delta(E_f - E_i) \quad (10.99)$$

They used the Born–Oppenheimer approximation to express the wave functions of the initial and final states as a product of the electronic, $\psi(\mathbf{r})$, proton, $\chi(\mathbf{R}, \mathbf{q})$, and solvent, $\phi(\mathbf{q})$, wave function, i.e.,

$$\Psi(\mathbf{r}, \mathbf{R}, \mathbf{q}) = \psi(\mathbf{r})\chi(\mathbf{R}, \mathbf{q})\phi(\mathbf{q}) \qquad (10.100)$$

where $\mathbf{r}, \mathbf{R}, \mathbf{q}$ are the electron, proton, and solvent coordinates, respectively. The vibrational frequencies of a proton attached to the hydronium ion in the initial state and that of the hydrogen atom adsorbed on the electrode surface in the final state are orders of magnitude higher than the frequencies of solvent orientational mode. The Franck–Condon approximation is used twice, one for the electronic subsystem and once for the proton subsystem. Levich, Dogonadze, and Kuznetsov[11,12,90,91a] expressed the transition probability of proton using Eqs. (10.99) and (10.100) as

$$P_{if} = \frac{2\pi}{\hbar} \mathrm{Av} \sum_f \left| \int \Psi_f^*(\mathbf{r}, \mathbf{R}, \mathbf{q}) v \Psi_i(\mathbf{r}, \mathbf{R}, \mathbf{q}) \, d\tau \right|^2 \delta(E_f - E_i)$$

$$= \frac{2\pi}{\hbar} \left| \int \psi_f^*(\mathbf{r}) v \psi_i \, d\tau \right|^2$$

$$\times \mathrm{Av} \sum_f \left| \int \chi_{jn}^*(\mathbf{R}, \mathbf{q}) \chi_{im}(\mathbf{R}, \varepsilon) \phi_{fn}^*(\mathbf{q}) \phi(\mathbf{q})_{im} \, d\tau \right|^2 \delta(E_f - E_i) \qquad (10.101)$$

The subscripts *fn* and *im* represent, respectively, the final state in the vibrational level n and the initial state in the vibrational level m.

Now considering the initial state (the electron is at the Fermi level E_f of the metal, the proton in the H_3O^+ at the energy level E_m) and the final state (the electron is in the bound state of hydrogen and the proton in the adsorbed state at the energy level E_m), one can write from Eq. (10.101) the transition probability per unit time as

$$P_{if}(E_f) = \frac{2\pi}{\hbar} |L_{nm}(E_f)|^2 \, \mathrm{Av} \sum_l \left| \int \phi_i^* \phi_l \, d\tau \right|^2 \delta(E_f - E_i) \qquad (10.102)$$

where

$$\delta(E_f - E_i) = \delta(F_f^0 + l'\hbar\omega_0 - F_i^0 - l\hbar\omega_0) = \delta[\Delta F^0 + \hbar\omega_0(l' - l)] \qquad (10.103)$$

with $(l + \frac{1}{2})\hbar\omega_0$ and $(l' + \frac{1}{2})\hbar\omega_0$ the solvent energies in the initial state l and final state l' in harmonic approximation. After substraction, the $\frac{1}{2}$ cancels and does not appear in Eq. (10.103) and

$$L_{mn}(E_f) = \int \psi_f^*(\mathbf{r}) \chi_f^*(\mathbf{R}) V(\mathbf{r}, \mathbf{R}) \psi_i(\mathbf{r}) \chi_m(\mathbf{R}) \, d\mathbf{r} \, d\mathbf{R} \qquad (10.104)$$

The Condon approximation is used, according to which the wave function of the fast subsystems (electron and proton) depend on $\mathbf{q}$ slightly and can be taken at the point corresponding to the activated state of the solvent with $\mathbf{q} = \mathbf{q}^*$.

It has been assumed by DKL[11,12,90,91a] that proton transition must occur from a given level i to a definite level f. For definiteness they have shown in Fig. 10.14a the transition of the system at $\mathbf{R}_0 = m = 0$, and also it is clear from the figure that at first the activation of the solvent occurs from $\mathbf{q}$ to $\mathbf{q}^*$ for a fixed position of proton $\mathbf{R}_0$. At the point $\mathbf{q}^*$ (at fixed coordinate of solvent), a quantum proton transition occurs.

The exchange integral (10.104) is determined by the overlapping both of the electron and the proton wave functions. The contribution of this integral is small due to weak overlapping of the proton wave functions, except for that of electron wave functions as in the case of redox reactions. It is important to mention that Dogonadze, Kuznetsov, and Levich[84] used the harmonic approximation for the wave functions of solvent and proton in their treatment.

The above discussion of transition probability of proton is based on the assumption that the quantum proton states n and m are fixed. To find the total transition probability, one can express

$$P_{if}(E_f) = A V_n \sum_m P_{nm}(E_f) \tag{10.105}$$

where the sum is over all final proton quantum states m and averaging is over all initial proton quantum states n.

After mathematical derivation and simplification, LDK[11,12,86,87,89] gave the following expression for the proton transition probability:

$$P_{if} = \frac{|L_e|^2}{M^2} \frac{\omega_0}{2\pi} \sum_{l=-\infty}^{\infty} I_l(z_p) \exp\left[z_p \coth\left(\frac{\hbar\omega_0}{2kT} - \frac{l\hbar\omega_0}{2kT}\right) \right]$$

$$\times \exp\left[-\frac{(\Delta F^0 - m\hbar\omega_p + E_s)^2}{4E_s kT} \right] \tag{10.106}$$

where

$$M^2 = \frac{\hbar\omega_0}{2} \left(\frac{E_s kT}{\pi^3} \right)^{1/2} \tag{10.107}$$

$$z_p = E_p / \hbar\omega_p \tag{10.108}$$

when E_p is the proton state reorganization energy, ω_p is the proton frequency, $I_l(z_p)$ is the Bessel function of imaginary argument, and ω_0 is the solvent frequency. In the high-temperature approximation i.e., when

$$kT \gg \hbar\omega_p \tag{10.109}$$

one gets a simplified expression for the transition. But this high-temperature condition is unrealistic, since the frequency proton vibration is such that, in fact, within the range of experimental temperature, this condition (10.66) will not be fulfilled.

The low temperature approximation, i.e.,

$$\hbar\omega_p \gg kT \tag{10.110}$$

is realistic. On the basis of this approximation, the expression for the total transition probability (for $\Delta F_0 = 0$) is found as

$$P_{if} = \frac{\omega_0}{2\pi} \frac{|L_e|^2}{|M|^2} \sum_{l=-\infty}^{\infty} \frac{1}{l!} \left(\frac{z_p}{2}\right)^{|l|} \exp\left[-\frac{m^2(\hbar\omega_p)}{4E_s kT}\right] \tag{10.111}$$

It is clear from Eq. (10.111) that the largest term in the sum is the term corresponding to quantum state $l = 0$ (ground state), which corresponds to $n = m = 0$. LDK remarks that this is the only situation present in the case of proton discharge.

We would like to mention that, like the electron transfer case, the transition rate of proton has not also been computed quantitatively by LDK or any other authors until the present time. We think that progress may be made in doing some realistic computations of transition probabilities of protons (without undergoing so much mathematical complication as did LDK) by the procedure used by Khan, Wright, and Bockris,[38] with a simple change in the wave function of the electron in the metal in the initial state and the electron in the adsorbed hydrogen atom in solution in the final state. The latter can be obtained by modifying the hydrogenic wave function to take into account the interaction due to the metal surface and the solvent around it. This can be accomplished by using the usual Schrödinger time-independent perturbation theory.

Thus, from the wave functions and the perturbing potential acting between the metal and proton in the outer Helmholtz plane, one can compute the probability of proton discharge per unit time using Fermi's golden rule (which is an expression in which the sum over all final states is taken into consideration).

If the electron transfer to the proton is not considered only in the vicinity of Fermi's level (as an approximation), one may take the sum over all the initial states and then find the average transition probability.

10.15. Work of Kharkats and Ulstrup[90]

Although the results of this work are still basically within the continuum approach set up by Dogonadze and others, a step toward the molecular approach is taken. In particular, in the work of Appleby *et al.*,[36] the difficulties of the continuum approach are pointed out and the importance of taking into account the difference of structure of the H_3O^+ ion in water and in the gas phase is taken into account. Kharkats and Ulstrup's work,[90] may indeed by termed a "physicochemical" treatment, rather than the "dielectric continuum" treatment of earlier workers. For example, they recognize that the apparent "gaps" in the Tafel line that would seem to arise from Levich's discussion, were activation to take place by thermal modes, do not in fact take place because of the smoothing out of these rotation–vibration modes in solution (Gurney[3]).

Another important advance shown by Kharkats and Ulstrup[90] is that they take into account all the quantum states in $H^+—O$ and no longer restrict their discussion to the ground state. However, their explanation of the discrepancy between the Tafel equation, experimentally observed, and the absence of applicability of the model to this criterion, is still in terms of anharmonicity in the H—O bond.

10.16. Christov's Work on an Oscillator Model for Proton Transfer

Recently Christov[91] put forward an interesting model for proton transfer processes. He defines two frequencies, ν_x and ν_y ($\nu_y > \nu_x$). One (ν_x) is related to some undefined oscillation in the solvent and the other (ν_y) is that of the proton in H_3O^+; or in the case of redox reactions, it represents the frequency of the inner coordination ion–solvent oscillators. Christov[91a] considered the two reactants at a fixed distance apart and represented the potential energies of the reactants–products system in terms of the two frequencies as

$$U_1(x, y) = k_1 x^2 + k_2 y^2 + Q_i \tag{10.112}$$

$$U_2(x, y) = k_1(x - x_0)^2 + k_2(y - y_0)^2 + Q \tag{10.113}$$

where x_0 and y_0 represent the difference between the equilibrium position of the solvent and proton system in the initial and final states, respectively.

It seems that the interpretation of the frequencies lies in a concept in which the first frequency, ν_y, represents the O—H bond energy under a condition of small displacement, when the potential energy is proportional to $(\Delta y)^2$, and the second frequency, ν_x, seems to be a kind of "equivalent frequency" to allow for an influence from the surrounding solvent on the energy of the proton. The equilibrium positions of the two oscillators are chosen at $x = 0$ and $y = 0$ of the coordinate system in the initial (reactant)

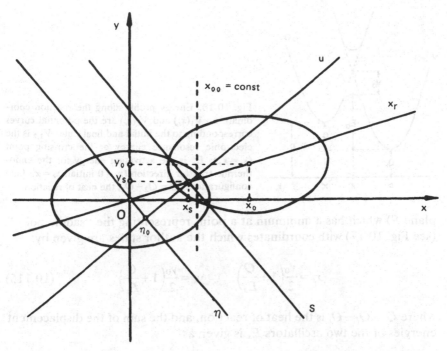

Fig. 10.17. Projection of the potential surfaces on the plane x, y. The two ellipses correspond to a section $V_1(x, y) = V_2(x, y) = \text{const}$; O is $(x = 0, y = 0)$ initial, x_0, y_0 final configuration; S is the intersection plane of the two paraboloids; x_s, y_s is the saddle point; x_r is the reaction coordinate; u is the line normal to S, η is the line parallel to S, η_0 is the crossing point of u and η, $x_{00} = \text{const} = \text{line parallel to } y \ (x_{00} > x_s)$.

state, and x and y are Cartesian coordinates describing two vibrations with frequencies

$$\nu_x = \frac{1}{2\pi}\left(\frac{2k_1}{\mu_x}\right)^{1/2}, \quad \nu_y = \frac{1}{2\pi}\left(\frac{k_2}{\mu_y}\right)^{1/2} \tag{10.114}$$

where μ_x and μ_y are the corresponding effective masses and k_1 and k_2 are the force constants for the solvent and proton system, respectively.†

Equations (10.112) and (10.113) describe two elliptic paraboloids in the three dimensions U, x, and y. Their projections in the x, y plane are shown in Fig. 10.17. The plane of energy U is orthogonal to the plane of the paper. The two paraboloids in three dimensions intersect along a line (in a

---

† It is worth mentioning that following the continuum theories Christov also took the same force constant k_1 and k_2 for both the initial and product states, an unrealistic assumption, though one implicitly made in most continuum approach papers.

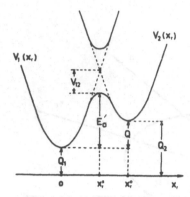

Fig. 10.18. Energy profile along the reaction coordinate x_r, $V_1(x_r)$ and $V_2(x_r)$ are the potential curves corresponding to the initial and final state; $V_{1,2}$ is the electronic resonance energy at the crossing point $(x_r = x_r^*)$; E_a is the activation energy for the endothermic transfer direction $(x_r = 0$ initial, $x_r = x_r^0$ final configuration); $Q = Q_2 - Q_1$, the heat of reaction.

plane S) which has a minimum at a point representing the "saddle point" (see Fig. 10.17) with coordinates which the author states are given by:

$$x_s = \frac{x_0}{2}\left(1 + \frac{Q}{E_s}\right), \qquad y_s = \frac{y_0}{2}\left(1 + \frac{Q}{E_s}\right) \tag{10.115}$$

where $Q = Q_f - Q_i$ is the heat of reaction, and the sum of the displacement energies of the two oscillators E_s is given as†

$$E_s = E_s^x + E_s^y$$
$$= k_1 x_0^2 + k_2 y_0^2 \tag{10.116}$$

where E_s^x is the energy to organize the final state (the surroundings of $M-H-H_2O$) in a way corresponding to its equilibrium configuration from the configuration it had at the equilibrium of the initial state, and E_s^y is the displacement energy due to proton–solvent bond stretching of $M-H$ to a situation it would have were it to have the configuration of the initial state.

The energy of activation is taken at first by Christov conventionally as the energy of the saddle point diminished by that of the initial state, assuming the force constant to be the same in the initial and final states (see Fig. 10.18). (It is noteworthy that Libby identified the Franck–Condon barrier as E_s as defined here.) Correspondingly, Christov finds (see Appleby et al.[36])

$$E_a' = \frac{(E_s + Q)^2}{4 E_s} \tag{10.117}$$

† The physical meaning of Eq. (10.116) is that the energy E_s is that which would occur if the product state were produced but had the same configuration as the initial state. It is apparently not implied that such a change of energy actually takes place. E_s is a construct to calculate E_a.

Due to the electronic interaction between the reactants there is resonance splitting in the intersection region of the potential surface [Eqs. (10.112) and (10.113)] and hence the height of the saddle point is lowered by an amount equal to the resonance energy V_{12} at that point. Thus, the activation energy, in fact, becomes

$$E_a'' = E_a' - V_{12} \tag{10.118}$$

However, Christov then (for unclear reasons) does not consider the value of (10.118) as the activation energy but adds to it the heat content of the initial state Q_1 so that, finally,

$$E_a = E_a' - V_{12} + Q_1$$

In this treatment, Christov derived the rate constants for the proton (also electron) transfer reaction using the general expression

$$k_{12} = \kappa \frac{kT}{h} \frac{f^{\ddagger}}{f} \exp\left(\frac{-E_a}{kT}\right) \tag{10.119}$$

where κ is a "quantum correction factor" (e.g., tunneling) and

$$\kappa = \kappa_t \chi \tag{10.120}$$

where κ_t is the nuclear tunneling correction and is $\geqslant 1$, and χ is the transmission coefficient for the system going from the initial state to the final state and is $\leqslant 1$. In Eq. (10.119), f is the partition function of the reactants (including solvent molecules), and $f^{\ddagger}$ is the partition function of the reacting system in which the motion along the reaction path is excluded.

Using the expression (10.119), Christov derived the expressions for the rate constants at three temperature ranges corresponding to a characteristic temperature, T_K: (1) the semiclassical temperature range $(T > 2T_K)$, (2) the intermediate temperature range $(2T_K > T > T_K/2)$, and (3) the low temperature range $(T < T_K/2)$. The characteristic temperature is expressed as

$$T_K = h\nu/k \tag{10.121}$$

where ν represents the frequency of the proton–solvent bond (for the proton transfer case) oscillation and the ion–solvent bond for the redox reaction. k is the Boltzmann constant.

If one calculates the characteristic temperature for the case of the proton transfer reaction, one finds for the H^+-H_2O bond frequency, $\sim 10^{14}$ sec^{-1}, that $T_K \simeq 1500°K$. For the $Fe^{3+}-H_2O$ bond frequency, $\sim 10^{13}$ sec^{-1}, the characteristic temperature becomes $T_K \simeq 150°K$.

Thus, one notices that the proton transfer case is always under the low-temperature condition, where according to Christov the nuclear (proton) tunneling† becomes important. Thus, for the proton transfer case, the derivation of a rate constant for the three temperature ranges is not necessary. Hence, we consider here the low-temperature $(T < T_K/2)$ treatment of Christov[91a] for the proton transfer case.

Christov[91a] considers the low temperature case where nuclear tunneling occurs mainly along the lines parallel to the y axis (Fig. 10.17). The most probable reaction path can be (somewhat artificially) portrayed as going initially from the origin of the coordinate system $(x = y = 0)$ along x until some point x_{00} (dotted line) near to x_s. The solvent coordinate movement has then become optimized and the movement of the proton coordinate ("inner ion–solvent bond") occurs parallel to the y axis and crosses the saddle point parallel to y until the point $x = x_{00}$, $y = y_0$. It then is imagined as going parallel to x once more until the final state $x = x_0$, $y = y_0$. We, therefore, consider a surface X_r, representing the initial state, which goes through the origin $(0, 0)$. This surface has to be crossed in the direction x. The motion along x is assumed to be a classical one. Equation (10.119) can then be written as

$$k_{12} = \kappa_x \frac{kT}{h} \frac{f_y^\ddagger}{f} \exp\left(\frac{-E_a}{kT}\right) \tag{10.122}$$

where (see Fig. 10.19)[91b]

$$\kappa_x = \int_{-E_c}^{\infty} k(\varepsilon_x) \exp\left(\frac{-\varepsilon_x}{kT}\right) \frac{d\varepsilon_x}{kT}$$

and

$$k(\varepsilon_x) = \sum_n p(\varepsilon_n, T) \sum_{n'} k_{nn'}(\varepsilon_x) \tag{10.123}$$

The rate constant $k(\varepsilon_x)$ is obtained by summing the rate constants for the transition n to n', multiplied at each level by the probability, $p(\varepsilon_n, T)$ of there being a particle in the level n at temperature T. Here

$$p(\varepsilon_n, T) = \frac{\exp(-\varepsilon_n/kT)}{f_y^\ddagger}, \quad f_y^\ddagger = \sum_n \exp\left(\frac{-\varepsilon_n}{kT}\right) \tag{10.124}$$

Here the quantity $\varepsilon_x = E_x - E_c$, $E_x = E_0$, is the initial energy for motion along x; ε_n is the vibration energy along y; $f_y^\ddagger$ is the corresponding partition function; $k_{nn'}$ is the transition probability from the initial state n to the final

---

† Of course, there is no *nuclear* tunneling in case of redox reaction, since redox ions are heavy particles and no measurable tunneling is possible by such particles [e.g., $Fe^{3+}(H_2O)_6$ ion].

state n'; $k(\varepsilon_x)$ is an average value of the total transition probability $k_n = \sum_{n'} k_{nn'}$ over the initial states n.

Christov considers what he calls an adiabatic condition $n = n'$, so that, in (10.123), for $n \neq n'$, $k_{nn'}(\varepsilon_x) = 0$

$$k_{nn'}(\varepsilon_x) = k_{nn}(\varepsilon_x) \neq 0 \qquad \text{for } n = n' \qquad (10.125)$$

Using (10.125) and (10.123) in (10.122), one obtains the rate constant as

$$\kappa_{12} = \kappa_x \frac{kT}{h} \frac{f^{\ddagger}}{f} \exp\left(\frac{-E_a}{kT}\right)$$

$$= \int_{-E_c}^{\infty} k(\varepsilon_x) \exp\left(\frac{-\varepsilon_x}{kT}\right) \frac{d\varepsilon_x}{kT} \frac{kT}{h} \frac{f^{\ddagger}}{f} \exp\left(\frac{-E_a}{kT}\right)$$

$$= \frac{kT}{h} \int_{-E_c}^{\infty} \sum_n \exp\left[\frac{(-\varepsilon_n/kT)}{f''}\right] \sum_n k_{nn}(\varepsilon_x) \exp\left(\frac{-\varepsilon_x}{kT}\right) \frac{d\varepsilon_x}{kT} \frac{f^{\ddagger}}{f} \exp\left(\frac{-E_a}{kT}\right)$$

$$= \frac{kT}{hf} \sum_n \kappa_n \exp\left(\frac{-\varepsilon_n}{kT}\right) \exp\left(\frac{-E_a}{kT}\right)$$

$$= \frac{kT}{hf} \sum_n \kappa_n \exp\left(\frac{-E_a''}{kT}\right) \qquad (10.126)$$

where

$$\kappa_n = \int_{-E_c}^{\infty} k_{nn}(\varepsilon_x) \exp\left(\frac{-\varepsilon_x}{kT}\right) \frac{d\varepsilon_x^n}{kT} \qquad (10.127)$$

and

$$E_a'' = \varepsilon_n + E_a \qquad (10.128)$$

where $\varepsilon_x^n = E_x - E_c^{nn}$, E_c^{nn} is the height of any of the potential barriers, $2V_{nn} = 2u_{nn}(x_{00})$ being the resonance splitting at the corresponding crossing point. The transition probability k_{nn} involves now the nuclear tunneling only along the line $x_{00} = \text{const}$, parallel to y, with a simultaneous rearrangement of the electronic state. This suggests the use of the Landau–Zener theory.[92]

V_{nn} can be expressed in terms exchange integral

$$V_{nn} = \int \psi_n^{(1)}(y) V_{12} \psi_n^{(2)}(y) \, dy \qquad (10.129)$$

and can be used as a matrix element to calculate the transition probability P_{nn} along $x_{00} = \text{const}$ in the framework of first-order perturbation theory and this can be expressed as

$$P_{nn} = P_e(\varepsilon_n) P_{\text{nuc}}(\varepsilon_n) \qquad (10.130)$$

where $P_e(\varepsilon_n)$ and $P_{\text{nuc}}(\varepsilon_n)$ are the probabilities of electron and nuclear rearrangements, respectively.

For $P_{\text{nuc}}(\varepsilon_n) \ll 1$ and $0 \le P_e(\varepsilon_n) \le 1$, Christov used the expressions[91a]

$$P_e(\varepsilon_n) = \frac{2\pi}{\gamma_n \Gamma^2(\gamma_n)} \exp[-2\gamma_n(1 - \ln \gamma_n)] \tag{10.131}$$

$$\gamma_n = \frac{V_{12}^2}{2\pi h \nu_y} \left[\frac{\pi^2}{E_s^y(E_s^y/4 - \varepsilon_n)} \right]^{1/2} \tag{10.132}$$

and according to the WKB tunneling formula

$$P_{\text{nuc}}(\varepsilon_n) = \exp\left[-\frac{4\pi(2\mu_y)^{1/2}}{h} \int_{y_1}^{y_2} [V(x_{00}, y) - \varepsilon_n]^{1/2} \, dy \right] \tag{10.133}$$

More accurately,

$$P_{\text{nuc}}(\varepsilon_n) = \frac{h}{4\mu_y \nu_y} \left(\psi_n^{(2)} \frac{d\psi_n^{(1)}}{dy} - \psi_n^{(1)} \frac{d\psi_n^{(2)}}{dy} \right)_{y=y_0/2} \tag{10.134}$$

where μ_y is the reduced mass of the system, $\psi_n^{(1)}$ and $\psi_n^{(2)}$ are the wave functions of the initial and final states. The more accurate tunneling expression (10.134) is evaluated exactly, using the harmonic oscillator wave functions. The condition $P_e(\varepsilon_n) \ll 1$ is practically fulfilled if $\varepsilon_n \ll E_s^y/4$. This allows one to use (10.132) with the approximation $\gamma_n \approx \gamma_0$, i.e.,

$$P_e(\varepsilon_n) \approx P_e(\varepsilon_0), \qquad \gamma_n \approx \gamma_0 \approx \frac{V_{12}^2}{h\nu_y E_s^y} \tag{10.135}$$

Now evaluating the transition probability expressions from Eqs. (10.130), (10.131), and (10.133) and using κ from Eq. (10.120) and the expression for E_a from Eq. (10.118), and finally, putting the quantities into the expression (10.119), Christov[91a] obtained the following expressions for the rate constants as:

(a) *For nonadiabatic reaction* $(T < T_K/2)$

$$k_{12} = \frac{V_{12}^2}{h} \left(\frac{\pi}{E_s^x kT} \right)^{1/2} \exp\left(-\frac{E_s^y}{h\nu_y} \right) \kappa_t \exp\left(-\frac{Q_1}{kT} \right) \exp\left(-\frac{(E_s^x + Q)}{4E_s^x kT} \right)^2 \tag{10.136}$$

where k_t is expressed as

$$k_t = \frac{(\pi/2)(T_K/T)}{\sin[(\pi/2)(T_K/T)]} \tag{10.137}$$

and ν_y is the frequency of $H^+ - H_2O$ bond.

(b) *For adiabatic reaction* $(T < T_K/2)$

$$k_{12} = \nu_x \kappa_t \exp\left(-\frac{E_s^x/4 - V_{00} + Q_1}{kT}\right) \exp\left[-\frac{Q}{2kT}\left(1 + \frac{\alpha^2 Q}{4E_a}\right)\right] \quad (10.138)$$

where V_{00} is the half of the splitting energy corresponding to the ground electronic state of the systems (Fig. 10.19) and

$$\alpha = 2d_x/x_0 \quad (10.139)$$

where $2d_x$ is the width of the parabolic type of solvent barrier (see Fig. 10.19) and x_0 is the equilibrium solvent coordinate of the product state.

One notices that in the case of nonadiabatic proton transfer reactions, Eq. (10.136) is not in accordance with the Tafel equation of electrode kinetics, which is known to be valid for hydrogen evolution in a very wide range of electrode potential (≥ 1 V). This is because, in Eq. (10.136), the heat of reaction Q, which is the potential dependent term, is not linear in the exponential.

Christov claims that in the adiabatic case Eq. (10.138) does give rise to the Tafel behavior if the factor $\alpha < 1$. This is because if $\alpha < 1$ the Q^2 term can be neglected. One can calculate α from Eq. (10.139) for the typical parabolic barrier width of 2 Å. The value of x_0 is calculated from the relation

$$E_s = k_1 x_0^2 \quad (10.140)$$

For the typical electrochemical reaction, E_s is ~ 0.5 eV[33] and the force constant for the outer solvent system can be taken as its maximum limit as 10^5 dyn cm^{-1}. This is because from the spectroscopic frequency data one

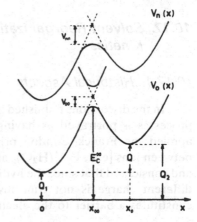

Fig. 10.19. Adiabatic potential curves ($n = 0$ and $n = n$); $x = 0$ initial and $x = x_0$ final configuration. V_{nn} is the resonance energy at the crossing point ($x = x_{00}$) of the unperturbed ($u_{nn} = 0$) potential curves (note that $V_{nn} > V_{00}$). E_c^{00} is the height of the ground-state ($n = 0$) potential barrier for the endothermic transfer direction ($Q \rightarrow x_0$); $Q = Q_2 - Q_1$, the heat of reaction.

gets[33] an inner-sphere ion–solvent bond force constant of the order of 10^5 dyn cm^{-1}, and such a force constant is likely to be greater than the outer-space solvent system force constant. Hence, from Eq. (10.140), one gets $x_0 \simeq 0.3$ Å, and from Eq. (10.139), $\alpha = 6.6$. Thus, it is doubtful that one will get $\alpha < 1$. If α is not less than unity, the expression (10.138) for the adiabatic reaction does not also follow the Tafel behavior. Thus, one notices that the harmonic oscillator model cannot give a theory to explain the fundamental Tafel behavior in electrode kinetics. Christov[91a] also clearly pointed out that though the harmonic oscillator model can be conveniently used, it *greatly overestimates the part played by the solvent in the activation process.* The main imperfection of the harmonic oscillator model is that it neglects the translational and rotational motion of reactants and solvent molecules.

Christov mentioned that the consideration of proton transfer only from the ground vibrational level of the initial state to the ground vibrational level of the final state, as is the normal practice of the LDK[11,12,88,89] school, disagrees with the well-known experimental fact[17,27,94] that the energy of activation is isotope dependent. Hence, the LDK model of no change in the $H^+ - H_2O$ bond length during the H^+ transfer process, and thus the idea that only the continuum solvent reorganization is involved, is erroneous.[91a]

Christov[91a] pointed out that by using two-dimensional models for gas-phase reactions of the type $AH + B \rightarrow A + HB$, it has been shown[95] that there is a considerable stretching of the $A-H$ bond in the transition region. It is reasonable to suppose that this situation takes place in proton transfer in solution, too. We think that at the interface, the stretching situation will be even more pronounced, due to interaction with the electrode surface, than that in the gas phase. Thus, Christov's examination comes nearer to a confirmation of the Gurney-oriented model (activated $M-H$), than that of LDK and is consistent with the model of Bockris and Matthews.[17]

10.17. Solvent Reorganizational Viewpoint in Proton Transfer Kinetics

10.17.1. Historical Aspects

In the discussions published by the LDK group, the history of electrode processes is portrayed as having started with Libby[29] in 1952. Libby[29] applied the Franck–Condon principle to the electron transfer reaction between ions [e.g., $Fe^{2+}(H_2O)_6$ and $Fe^{3+}(H_2O)_6$ ions] in aqueous solution and considered that since the hydration energy between the reactant ions of different charge is not transmissible as rapidly as that of electrons it constitutes a barrier to the electron transfer reaction, called the Franck–

Condon barrier. Libby[29] mentioned that this situation would apply to electron transfer at the electrode–solution interface.

Libby describes something like the radiational transfer from a lower to a higher energy state as in a spectroscopic transition. The electron does not tunnel to a state of equal energy. It is difficult to understand the reason for the electron transfer, i.e., Libby's penetration of the "Franck–Condon barrier," in the absence of an activating photon (and in the absence of a polaronic transfer of fluctuational energy from the surrounding solution).

The necessary energy of activation, in this view, would be determined by the energy of the activated final state (e.g., a ferric ion having the solvent configuration of a ferrous ion) diminished by the equilibrium energy of the ferrous ion. (It is of value to note that Weiss' 1954 paper[31] implies an energy of activation to cause the solvent to orient, i.e., to reorganize, and this would clearly have to occur before the electron transfer.)

Libby[29] calculated, thus, the "energy barrier" for the electron transfer reaction in solution by taking into account the *Born solvation energy* difference when the ion having a smaller charge $+Z$ is changed to $+(Z+1)$ without change in the environment (i.e., without change in its ionic radius r, including the first solvation shell) and given as

$$E_a = \frac{e_0^2}{2r\varepsilon_s}[(Z+1)^2 - Z^2] \tag{10.141}$$

Such considerations given by Libby[29] are partial in many senses. Activation energy is not simply the Born solvation energy difference when a charge of $+Z$ becomes $+(Z+1)$. The Born part of the solvation energy of ions is less than half of the total solvation energy. The solvation energy of the *inner* shells would change with charge independently of a configurational change and therefore should have been counted in the Franck–Condon barrier.

The quotation of Libby's 1952 work as the origin of quantum electrochemistry is often made, in spite of the fact that Gurney introduced quantum mechanics into electrochemistry in 1931, in a molecular model, using certain approximate solutions of Schrödinger's equation. In spite of the Weiss work, the LDK group refer to Platzmann and Franck[30] (1954) as the origin of the concept of a change of Born solvation energy due to surrounding polarized medium on electronic transition.

In the presentation given by LDK, the important work of the Horiuti and Polanyi paper obtains an enormously enhanced status—as an anti-paper. In fact, the Horiuti and Polanyi paper[98] of 1935 was important because it was the first paper to bring out the part played by chemical bonding of chemical radicals on the catalyst and the effect of this upon the heat of activation. This was taken up by Butler,[6] who applied it to the quantum mechanical theory of Gurney[3] in 1936.

In the LDK presentation, the Horiuti and Polanyi paper[98] of 1935 is taken as the archetype of papers that (beginning in 1931) take a molecular viewpoint of the transfer of hydrogen to the electrode. Thus, the thermally activated quantum mechanical viewpoints of Gurney,[3] Christov,[8-10] Gerischer,[51-53] and Bockris and Matthews,[17,27,28] which have a relationship to the work of Bell[20,21] and others in nonelectrochemical proton transfer reactions in solution, are ignored in the LDK presentations and attention is paid to the Horiuti and Polanyi[98] paper alone; it is attacked *because* it involves activation of the O—H bond as a prerequisite to reaction.

In the LDK presentation of quantum aspects of proton discharge, the early work of Bawn and Ogden[5b] and that of Christov,[8-10] a direct continuation of that of Gurney,[3] are also always omitted from mention. The first paper usually quoted is that of Libby.[29]

10.17.2. Role Played by the Solvent

A point of confusion in the literature has to do with the part played by the solvent in the continuum and molecular models. In the LDK theories, it is stated (in respect to the molecular model approach) that "the solvent has been neglected." However, the Horiuti and Polanyi[98] paper stresses the effect of the heat of activation of the M—H bond, and implies an effect of the heat of solvation. Furthermore, hydration energy was taken into account in the original quantum mechanical paper by Gurney.[3] None of the molecular models neglect the solvent. They stress it. Solvational energy, and its variation with distance, during proton vibration is an important factor in these theories. The hydrogen–oxygen bond in H^+—OH_2 is vital. The *position* of the H_3O^+ ion in the solution (with respect to the electrode) is of importance, too. Detailed modelistic discussions of all this were current before the continuum dielectric work of Libby,[29] Marcus,[32] or Levich[88] began. The LDK model does not take into account the detailed model of rotation and vibration of the solvent, which importantly affects electrode kinetics and is indeed an integral part of the theory of isotopic effects.

10.17.3. Isotopic Effects

One aspect in determining which approach, that of the continuum dielectric or the molecular model, is more appropriate in the treatment of proton transfer is the interpretation of isotopic effects. It may be reiterated that there is a strong dependence of the separation factor both for tritium and deuterium upon the nature of the electrode material. Electrode materials divide themselves into three groups, depending upon the factors obtained; these numbers vary greatly (e.g., from 3 to 20). The groups obtained from isotopic separation values correspond to the groups that are obtained from other considerations.[100,103]

Correspondingly, the degree to which the separation factor *depends upon potential* seems to be an important factor in a decision in respect to models. The only interpretation[87] that has been hitherto made upon the solvent reorganization view of the isotopic effect situation is to refer to the unevaluated exchange integral that exists before the exponential term in the solvent reorganization theories, the evaluation of which is described as difficult. There is no direct consideration of the tunneling that is the principal reason for the difference in the velocity of tritium and proton transfer through the double layer.

10.17.4. Tafel Equation

The simplest point is the most important. It is not possible to obtain Tafel's law from the solvent reorganization viewpoint. The small ability of this theory to obtain the Tafel equation by means of additional hypotheses has been described.

A *presentational* aspect of the solvent reorganization view, the reiteration (by LDK) in many articles but different journals of similar theoretical material, has been described. Another aspect is the neglect of corresponding (often prior) work carried out in other countries. To quote three examples:

(a) The founding paper of quantum electrode kinetics, that of Gurney, is never mentioned.

(b) The work of Christov[8-10] is always omitted.

(c) The experimental and theoretical work of Bockris and Matthews[17,27,28] upon the potential dependence of the separation factor predates the later work of Krishtalik.[86] However, it is the work of Krishtalik that Dogonadze describes as "establishing the quantum nature of proton transfer."

(d) LDK describe the partaking of the proton "in the quantum subsystem" as a new thought. However, it was introduced by Gurney nearly fifty years ago.

References

1. J. Tafel, *Z. Physik. Chem.* **50**, 641 (1905).
2. J. Smits, *The Theory of Allotropy*, p. 115, Longmans and Green, London (1922).
3. R. W. Gurney, *Proc. Roy. Soc. (London)* **134A**, 137 (1931).
4. J. A. V. Butler, *Trans. Faraday Soc.* **19**, 729 (1924).
5a. B. Topley and H. Eyring, *J. Electrochem. Soc.* **55**, 508 (1933).
5b. C. E. H. Bawn and G. Ogden, *Trans. Faraday Soc.* **30**, 432 (1934).

6. J. A. V. Butler, *Proc. Roy. Soc. (London)* **A157**, 423 (1936).
7. J. Horiuti and M. Polanyi, *Acta Physicochim. URSS* **16**, 169 (1942).
8. St. G. Christov, *Electrochem. Acta* **4**, 306 (1961).
9. St. G. Christov, *Electrochim. Acta* **9**, 575 (1964).
10. St. G. Christov, *J. Res. Inst. Catal. Hokkaido Univ.* **16**, 169 (1968).
11. V. G. Levich, *Physical Chemistry, An Advanced Treatise* (H. Eyring, ed.), Chapter 12, Academic Press, New York (1970).
12. R. R. Dogonadze, *Reactions of Molecules at Electrodes* (N. S. Hush, ed.), p. 135, Wiley (Interscience) London (1971).
13. J. O'M. Bockris and S. Srinivasan, *J. Electrochem. Soc.* **111**, 844 (1964).
14. J. O'M. Bockris and S. Srinivasan, *J. Electrochem. Soc.* **111**, 853 (1964).
15. J. O'M. Bockris and S. Srinivasan, *J. Electrochem. Soc.* **111**, 858 (1964).
16. J. O'M. Bockris, S. Srinivasan, and D. B. Matthews, *Discuss. Faraday Soc.* **39**, 239 (1965).
17. J. O'M. Bockris and D. B. Matthews, *J. Chem. Phys.* **44**, 298 (1966).
18. J. O'M. Bockris and D. B. Matthews, *Electrochim. Acta* **11**, 143 (1966).
19. T. Keii and T. Kodera, *J. Res. Inst. Catal. Hokkaido Univ.* **5**, 105 (1957).
20. R. P. Bell, *Trans. Faraday Soc.* **55**, 1 (1959).
21. R. P. Bell and D. M. Goodal, *Proc. Roy. Soc. (London)* **A294**, 273 (1953).
22. G. Okamoto, J. Horiuti, and K. Hirota, *Sci. Inst. Phys. Chem. Res. (Tokyo)* **29**, 223 (1936).
23. J. Horiuti, T. Kei and K. Hirota, *J. Res. Inst. Catal. Hokkaido Univ.* **2**, 73 (1951).
24. T. Kodera and T. Saito, *J. Res. Inst. Catal. Hokkaido Univ.* **7**, 5 (1959).
25. B. E. Conway, *Proc. Roy. Soc. (London)* **A247**, 400 (1958).
26. R. K. Sen and J. O'M. Bockris, *Chem. Phys. Lett.* **18**, 166 (1973).
27. J. O'M. Bockris and D. B. Matthews, *J. Electroanal. Chem.* **9**, 325 (1966).
28. J. O'M. Bockris and D. B. Matthews, *Proc. Roy. Soc. (London)* **A292**, 479 (1966).
29. W. Libby, *J. Phys. Chem.* **56**, 863 (1952).
30. R. Platzmann and T. Franck, *Z. Phys.* **138**, 411 (1954).
31. J. Weiss, *Proc. Roy. Soc. (London)* **A222**, 128 (1954).
32. R. A. Marcus, *J. Chem. Phys.* (a) **24**, 966 (1956); (b) **26** 867 (1957); (c) **38**, 1858 (1963); (d) **39**, 1739 (1963); (e) **43**, 679 (1965).
33. J. O'M. Bockris, S. U. M. Khan, and D. B. Matthews, *J. Res. Inst. Catal. Hokkaido Univ.* **22**, 1 (1974).
34. R. Parsons and J. O'M. Bockris, *Trans. Faraday Soc.* **47**, 914 (1951).
35. B. E. Conway and J. O'M. Bockris, *Can. J. Chem.* **35**, 1124 (1957).
36. A. J. Appleby, J. O'M. Bockris, R. K. Sen, and B. E. Conway, *M.T.P. International Review of Sciences* (J. O'M. Bockris, ed.), Vol. 6, *Electrochemistry*, Butterworths, London (1973).
37. A. R. Despic and J. O'M. Bockris, *J. Chem. Phys.* **32**, 389 (1960).
38. S. U. M. Khan, P. Wright, and J. O'M. Bockris, *Electrokhim.* **13**, 914 (1977).
39. T. Ackerman, *Discuss. Faraday Soc.* **24**, 180 (1957).
40. R. A. O'Ferrall, G. W. Koepple, and A. J. Kresge, *J. Amer. Chem. Soc.* **93**, 1 (1971).
41. J. Rudolph and H. Zimmerman, *Z. Phys. Chem.* **43**, 311 (1964).
42. W. Heitler and F. London, *Z. Physik* **44**, 455 (1927).
43. J. A. V. Butler, *Electrocapillarity*, Methuen and Co., London (1940).
44. J. O'M. Bockris, R. J. Mannan, and A. Damjanovic, *J. Chem. Phys.* **48**, 1989 (1968).
45. J. O'M. Bockris, *Chem. Rev.* **3**, 525 (1948).
46. J. O'M. Bockris, *Ann. Rev. Phys. Chem.* **5**, 477 (1954).
47. A. Wiberg, *Chem. Rev.* **55**, 713 (1955).
48. B. E. Conway and J. O'M. Bockris, *Proc. Roy. Soc. (London)* **A248**, 1394 (1958).

49. B. E. Conway and J. O'M. Bockris, *J. Chem. Phys.* **26**, 532 (1957).

49a. J. O'M. Bockris, *Z. Elektrochem.* **55**, 105 (1951).

49b. J. O'M. Bockris and A. M. Azzam, *Experientia.* **4**, 220 (1948).

50. B. E. Conway and J. O'M. Bockris, *Electrochim. Acta* **3**, 340 (1961).

51. H. Gerischer, *Z. Physik. Chem.* **26**, 223 (1960).

52. H. Gerischer, *Z. Physik. Chem.* **29**, 325 (1960).

53. H. Gerischer, *Z. Physik. Chem.* **27**, 48 (1961).

54. M. Enyo, *Modern Aspects of Electrochemistry* (J. O'M. Bockris and B. E. Conway, eds.), Vol. 11, Plenum Press, New York (1976).

55. B. E. Conway, *Proc. Roy. Soc. (London)*, **A247**, 400 (1958).

56. B. E. Conway, *Proc. Roy. Soc. (London)* **A256**, 128 (1960).

57. B. E. Conway and M. Solomon, *J. Chem. Phys.* **68**, 2009 (1969); **41**, 3169 (1964).

58. B. E. Conway and D. J. Mackinnon, *J. Electrochem. Soc.* **116**, 1665 (1969).

59. J. O'M. Bockris, M. A. V. Devanathan, and K. Muller, *Proc. Roy. Soc.* **A274**, 55 (1963).

60. J. O'M. Bockris, *Modern Aspects of Electrochemistry*, Vol. 1, Chapter IV, Butterworths, London (1954).

61. G. Herzberg, *Infrared and Raman Spectra of Polyatomic Molecules*, p. 173, Van Nostrand, New York (1945).

62. L. Melander, *Isotope Effects on Reaction Rates*, Ronald Press, New York (1960).

63. S. Glasstone, K. J. Laidler, and H. Eyring, *The Theory of the Rate Process*, McGraw–Hill, New York (1941).

64. W. S. Benedict, N. Gailer, and E. K. Plyer, *J. Chem. Phys.* **24**, 1139 (1956).

65. W. F. Libby, *J. Electrochem. Soc.* **11**, 101 (1943).

66. M. Ikusima and S. Azakami, *J. Chem. Soc. Japan* **59**, 40 (1938).

67. O. Sepall and S. G. Mason, *Can. J. Chem.* **38**, 2024 (1960).

68. B. E. Conway, *Can. J. Chem.* **37**, 178 (1959).

69. S. G. Christov, *Croatica Chem. Acta* **44**, 67 (1972).

70. S. G. Christov, *J. Res. Inst. Catal. Hokkaido Univ.* **24**, 27 (1976).

71. B. Post and C. F. Hiskey, *J. Amer. Chem. Soc.* **72**, 4203 (1950).

72. J. O'M. Bockris, I. A. Ammar, and A. K. M. S. Huq, *J. Phys. Chem.* **61**, 879 (1957).

73. R. Parsons and P. Passerron, *J. Electroanal. Chem.* **11**, 123 (1967).

74. H. S. Johnston, *Advan. Chem. Phys.* **3**, 131 (1960).

75. H. S. Johnston, *Gas Phase Reaction Rate Theory*, Ronald Press, New York (1966).

76. R. K. Sen, Ph.D. Thesis, University of Pennsylvania, Philadelphia (1973).

77. R. J. Watts-Tobin, *Phil. Mag.* **6**(8), 133 (1961).

78. J. O'M. Bockris and E. C. Potter, *J. Chem. Phys.* **20**, 614 (1952).

79. J. O'M. Bockris and M. A. Habib, *Z. Phys. Chem. N. F.* **98**, 43 (1975).

78. J. O'M. Bockris and E. C. Potter, *J. Chem. Phys.* **20**, 614 (1952).

79. J. O'M. Bockris and M. A. Habib, *Z. Phys. Chem. N. F.* **98**, 43 (1975).

80. J. O'M. Bockris and M. A. Habib, *J. Electroanal. Chem.* **65**, 473 (1975).

81a. D. M. Dennison and G. E. Uhlenbeck, *Phys. Rev.* **41**, 313 (1932).

81b. A. N. Baker, *J. Chem. Phys.* **22**, 1625 (1959).

82. R. L. Somarjai and D. F. Hornig, *J. Chem. Phys.* **36**, 1980 (1962).

83. E. F. Caldin and M. Kaspavian, *Discuss. Faraday Soc.* **39**, 25 (1965).

84. R. Dogonadze, A. Kuznetsov, and V. Levich, *Electrochim. Acta* **13**, 1025 (1968).

85. J. O'M. Bockris and A. M. Azzam, *Nature* **165**, 403 (1950).

86. L. I. Krishtalik, *Electrokhim.* **6**, 507 (1970).

87. E. D. German, R. R. Dogonadze, R. R. Kuznetsov, A. M. Levich, and Yu. I. Kharkats, *Sov. Electrochem.* **6**, 342 (1970).

88. R. R. Dogonadze, A. M. Kuznetsov, and V. G. Levich, *Electrokhim.* **3**, 739 (1967).

89. R. R. Dogonadze, A. M. Kuznetsov, and V. G. Levich, *Electrochim. Acta* **13**, 1025 (1968).
90. Yu. I. Kharkats and J. Ulstrup, *J. Electroanal. Chem.* **65**, 555 (1975).
91*a*. S. G. Christov, *J. Electrochem. Soc.* **124**, 69 (1977).
91*b*. S. G. Christov, *Ber. Bunsenges Phys. Chem.* **79**, 357 (1975).
92. L. Landau and E. Lifshitz, *Quantum Mechanics*, Nauka, Moskow (1974).
93. M. Ovchinnikova, *Dokl. Akad. Nauk SSSR* **161**, 641 (1965); *Opt. Spektrosk.* **17**, 821 (1964).
94. R. P. Bell, *Chem. Soc. Rev.* **3**, 513 (1974).
95. S. G. Christov and Z. L. Gueorguiev, *J. Electrochem. Soc.* **75**, 1748 (1971).
96. J. O'M. Bockris and A. K. N. Reddy, *Modern Electrochemistry*, Vol. 1, Plenum Press, New York (1973).
97. R. A. Marcus, *J. Chem. Phys.* **24**, 966 (1956); **26**, 867 (1957); **39**, 1734 (1963); **43**, 679 (1965).
98. J. Horiuti and M. Polanyi, *Acta Electrochem. USSR* **2**, 505 (1935).
99. P. George and J. Griffith, *The Enzymes* Vol. 1, Chapter 8, p. 347 (P. D. Boyer, H. Latzky, and K. Myrberg, eds.), Academic Press, New York (1959).
100. B. E. Conway and J. O'M. Bockris, *J. Chem. Phys.* **26**, 532 (1957).
101. J. O'M. Bockris and D. F. A. Koch, *J. Phys. Chem.* **65**, 1941 (1961).
102. J. O'M. Bockris, *Z. Elektrochem.* **55**, 105 (1951).
103. J. O'M. Bockris, *J. Chim. Phys.* **49**, 41 (1952).

11

The Hydrated Electron

11.1. Introduction

Solvation of electrons in polar liquids is a well-investigated phenomenon. Advanced and elegant observations upon the properties of such species have become abundant.[1-4] But there is as yet no single model in terms of which the observations may be rationalized and fresh phenomena predicted. Solvated electrons have been observed in various solvent systems; we will deal in this chapter only with the hydrated electron, because it is this that has the greatest relevance to electrochemists interested in discussions of quantal treatments.

In the early sixties, Hart and Boag[5,6] demonstrated the formation of hydrated electrons in aqueous solvent system under the action of radiation. Though electrochemical cathodic generation of ammoniated electrons was known to electrochemists in the late forties, it is only relatively recently that electrochemical generation of hydrated electrons has been investigated.[8-11] There reigns doubt[12] as to whether electrochemical generation of solvated electrons in water is *possible* from the energetic point of view. Hydrated electrons can probably be generated photoelectrochemically.[13-18]

In this chapter, we shall first discuss theoretical treatments of hydrated electrons in terms of continuum, semicontinuum, and structural models. The theory of the homogeneous electron transfer reaction involving a hydrated electron is given. The electrochemical and photoelectrochemical generation of the hydrated electron is discussed in terms of energetic feasibility. In a later section, the role of the hydrated electron in the non-Tafelian behavior of photocurrents is outlined.

11.2. Continuum Approach to the Hydrated Electron

It is assumed[19-21] in the continuum approach that the electron is trapped by the polarization of the solvent medium interacting with the charge on the electron. The solvent is represented by a continuous dielectric medium characterized by macroscopic properties, i.e., the static dielectric constant ε_s and the optical dielectric constant ε_{op}. An additional electron (the hydrated electron) in the solvent medium interacts with the polarization of the dielectric medium† caused by its presence. The interaction energy is considered only with the solvent outside the first layer.

Tachiya *et al.*[22,23] introduced an additional consideration in taking into account the polarization function $P(r)$ as a dipole moment per unit volume at a given point, and the time evolution of polarization, energy level, and transition energy is calculated.[24]

The continuum treatment of the hydrated electron has been given by Jortner.[19–21] He introduced several approximations in the treatment, such as:

(1) An additional electron (the hydrated electron) in a solvent medium is trapped in a cavity of finite radius and spherical shape.

(2) The interaction of this additional electron with the medium is taken to be of long-range interaction only, even up to the center of the cavity. No short-range interaction, e.g., with first-layer dipoles, is taken into account.§

(3) To find the energy of the hydrated electron, excitation energy and heat of solvation, a one-parameter hydrogenlike wave function was used for the electron.

(4) Jortner[19,20] introduced two other approximations to reduce the many-electron system in a solvent medium to a one-electron problem, and hence, a soluble one. These are:

(a) *The electronic adiabatic approximation.* Here, electronic and orientational polarization energy are separable. The excess (additional) electron motion is separated from the motions of the electron present in the solvent molecules of the medium. This separation is made on the basis of the assumption that the average velocity of the additional electron is low compared to electrons in the solvent molecules. The gross assumption in the adiabatic approximation is that, though the motion of the additional electron is taken as slow compared with the motion of the electrons in the solvent molecules, the additional electron is taken as subject only to the potential associated with the orientational polarization of the continuum. The total polarization should be included. The orientational polarization energy is

---

† The authors neglect the variation of dielectric constant with distance, which should be taken into account even in the continuum model. Iguchi[25] considered the dielectric saturation effect near the first layer and its variation outside the first layer.

§ Surprising as this approach seems, it was regarded as reasonable by continuum authors because they regarded the electron and the first water layer as a kind of cohesive negative jellium.

given (Appendix 11.I) by[4]

$$
V(r) = \begin{cases} -\dfrac{e_0^2}{R}\left(\dfrac{1}{\varepsilon_{op}} - \dfrac{1}{\varepsilon_s}\right), & r \leqslant R \\[3mm] -\dfrac{e_0^2}{r}\left(\dfrac{1}{\varepsilon_{op}} - \dfrac{1}{\varepsilon_s}\right), & r \geqslant R \end{cases}
\tag{11.1}
$$

where R is the cavity radius. On this basis, the potential energy is considered constant inside the cavity.

(b) *The self-consistent (SCF) field approximation.* Here, the electronic and orientational polarizations are taken into account. This is called the SCF approximation because the energy of the solvated electron is calculated in the self-consistent manner.[25] The wave function is chosen, called here $\psi_i(\mathbf{r})$. From this wave function, which may be a poor one, it is possible to then calculate the potential created by the electron. When one knows the potential, one can calculate the associated polarization caused by the surrounding solvent. However, once one has a value for the potential of the polarization, i.e., the potential energy term for the Schrödinger equation, it is possible to find a better wave function. Thereafter, one can then get a better self-consistent potential f_i from the Poisson equation

$$
\nabla^2 f_i = 4\pi e_0 |\psi_i|^2
\tag{11.2}
$$

In the SCF treatment, the potential energy of interaction of the hydrated electron with the solvent medium is given by[21]

$$
V(\mathbf{r}) = \begin{cases} \dfrac{e_0}{2}\left(1 - \dfrac{1}{\varepsilon_s}\right) f_i(R), & r \leqslant R_0 \\[3mm] \dfrac{e_0}{2}\left(1 - \dfrac{1}{\varepsilon_s}\right) f_i(\mathbf{r}), & r \geqslant R_0 \end{cases}
\tag{11.3}
$$

where R_0 is the cavity radius in this treatment also. In Eq. (11.3), that in the SCF approach, electronic, nuclear, and orientational polarization, i.e., the total polarization, are each taken into account.

Numerical Results of Continuum Treatment (from the SCF Approach)

(i) Hydration Energy

To find the heat of hydration, the ground-state energy of the hydrated electron is determined from Eq. (11.3) by the variation method. A one-parameter (i.e., λ) 1s hydrogentic wave function is chosen for the ground state of hydrated electron, i.e.,

$$\Psi_{1s} = (\lambda^3/\pi)^{1/2} e^{-\lambda r} \tag{11.4}$$

The electrostatic potential f_{1s}, which depends on the wave function, ψ_{1s}, can be written from the solution of Eq. (11.2) as[21] (Appendix 11.II)

$$f_{1s}(r) = -\frac{e_0}{r} + \frac{e_0(1+\lambda r)\exp(-2\lambda r)}{r} \tag{11.5}$$

This is the Coulomb potential for $r > R$. To treat the electron in a *spherical* cavity, it is assumed that the potential $f(r)$ is continuous up to the cavity boundary and constant within the cavity. This leads from Eq. (11.3) to the following expression for energy E_{1s} of the ground state of hydrated electron:

$$E_{1s} = \int_0^\alpha \Psi_{1s}\left(-\frac{\hbar^2}{2m_e}\nabla^2\right)\Psi_{1s}\, d\mathbf{r} + \frac{e_0}{2}\left(1-\frac{1}{\varepsilon_s}\right) f_{1s}(R) \int_0^R \Psi_{1s}^2\, d\mathbf{r}$$

$$+ \frac{e}{2}\left(1-\frac{1}{\varepsilon_s}\right) \int_R^\infty f_{1s}(\mathbf{r})\Psi_{1s}^2\, d\mathbf{r} \tag{11.6}$$

The first term of Eq. (11.6) takes into account the contribution due to the kinetic energy of the hydrated electron having a mass m_e. The second and third integrals represent the potential energy contributions due to the potential within and outside the cavity, respectively. The value of E_{1s} as a function of the parameter λ is obtained by substituting Ψ_{1s} from Eq. (11.4) and f_{1s} from Eq. (11.5) in Eq. (11.6). The best energy value according to this scheme is obtained, subject to the variation method condition,

$$\frac{dE_{1s}}{d\lambda} = 0 \tag{11.7}$$

The value of λ, determined in accordance with this minimum condition, is then used to obtain E_{1s}.

The limiting case, when $R = 0$, is considered for mathematical convenience (though physically it means that no cavity is considered). One gets from Eq. (11.6) using Eqs. (11.4) and (11.5) that

$$E_{1s} = \frac{\hbar^2\lambda^2}{2m_e} - \frac{5e_0^2\lambda}{16}\left(1-\frac{1}{\varepsilon_s}\right) \tag{11.8}$$

Hence, from Eq. (11.7) it follows that

$$\lambda = \frac{5}{16}\left(1-\frac{1}{\varepsilon_s}\right)a_0 \tag{11.9}$$

and

$$E_{1s} = -\frac{25}{512}\left(1-\frac{1}{\varepsilon_s}\right)^2 \frac{e_0^2}{a_0} \tag{11.10}$$

where $a_0 = 0.53$ Å is the Bohr radius for the hydrogen atom. Thus, one gets the ground-state energy of the hydrated electron from Eq. (11.10) to be -1.3 eV, which is the *hydration energy* of the hydrated electron. This value of -1.3 eV is comparable to the experimental value of solvation energy of the hydrated electron ~ -1.7 eV.[4]†

(ii) Optical Transition Energy of the Hydrated Electron

The absorption band of the hydrated electron[5] is assigned to a 1s → 2p transition. Hence, we need to know energies of both the 1s and the 2p states of the hydrated electron. The energy of the 1s state was calculated in the previous section.

The energy of the 2p excited state, as in the 1s state, is determined by the polarization field of the dielectric medium. The 2p state is not an equilibrium state, so we consider the excited state to exist in the same nuclear configuration as the ground state. This is in accordance with the Franck–Condon principle. Thus, the medium orientational polarization of the optically excited 2p state is determined prior to the transition by the $e_0|\psi_{1s}|^2$ charge distribution of the 1s state. The difference in energy value of the excited state from that of the ground state comes from the difference in charge distribution in the excited 2p state compared to ground 1s state. This nonequilibrium excited state is treated again by the variation method with the one-electron and one-parameter (μ) 2p state wave function

$$\Psi_{2p} = (\mu^5/\pi)^{1/2} r e^{-\mu r} \cos \theta \qquad (11.11)$$

where the angle θ represents the orientation of the orbital from the polar axis of the coordinate system. The total energy of the 2p electronic state with the surrounding solvent molecules in the 1s state is given by[5]

$$E_{2p} = \int \Psi_{2p} \left(-\frac{\hbar^2}{2m_e} \nabla^2 \right) \Psi_{2p} \, d\tau$$
$$+ \frac{e_0}{2} \left(1 - \frac{1}{\varepsilon_{op}} \right) \int \Psi_{2p}^2 f_{2p} \, d\tau + \frac{e_0}{2} \left(\frac{1}{\varepsilon_{op}} - \frac{1}{\varepsilon_s} \right) \int \Psi_{2p}^2 f_{1s} \, d\tau \qquad (11.12)$$

The first integral in Eq. (11.12) represents the mean kinetic energy of the optically excited hydrated electron in the 2p state. The second integral takes into account the potential energy of interaction due to polarization of the medium due to charge distribution of the electron in the 2p excited state. The third integral gives the potential energy of the electron due to the interaction with the orientational polarization of the medium when the electron is in the 2p state and the electrostatic potential f_{1s} corresponds to

† The hydration energy of the hydrated electron was obtained by Baxendale[26] using the thermodynamic cycle for the process $e(aq) + H^+ \rightarrow (\frac{1}{2}H_2)_{aq}$ and the corresponding experimental data involved in the process.

that of 1s state. It seems at first that the orientational polarization effect should not appear in Eq. (11.12) for the excited state. But the lifetime of the excited states is of the order of 10^{-8} to 10^{-10} sec, and the time of orientation is of the order of 10^{-11} sec. Hence, the orientational polarization contributes to the excited state energy. From Eqs. (11.11) and (11.12), the energy E_{2p} is obtained as a function of the variational parameter μ and is minimized with respect to this parameter.

The 1s → 2p transition energy is given by

$$hv = E_{2p} - E_{1s} \tag{11.13}$$

By means of Eqs. (11.5) and (11.12) the result of transition energy can be given as

$$hv = \left[\int \Psi_{2p}\left(-\frac{\hbar^2}{2m_e}\nabla^2\right)\Psi_{2p}\,d\tau - \int \Psi_{1s}\left(-\frac{\hbar^2}{2m_e}\nabla^2\right)\Psi_{1s}\,dr \right]$$
$$+ \left[e_0\left(\frac{1}{\varepsilon_{op}}-\frac{1}{\varepsilon_s}\right)\int \Psi_{2p}^2 f_{1s}\,d\tau - e_0\left(\frac{1}{\varepsilon_{op}}-\frac{1}{\varepsilon_s}\right)\int \Psi_{1s}^2 f_{1s}\,d\tau \right]$$
$$+ \left[\frac{e_0}{2}\left(1-\frac{1}{\varepsilon_{op}}\right)\int \Psi_{2p}^2 f_{2p}\,d\tau - \frac{e_0}{2}\left(1-\frac{1}{\varepsilon_{op}}\right)\int \Psi_{1s}^2 f_{1s}\,d\tau \right] \tag{11.14}$$

In Eq. (11.14), the terms in the first square brackets represent the kinetic energy difference when the hydrated electron is in the excited 2p state and ground 1s state. The terms in the second square brackets represent the difference between the contributions of the electronic polarization energies, each determined by the charge distribution for 2p and 1s states, respectively, but in both cases with the consideration of the electrostatic potential f_{1s} corresponding to the 1s state only. The terms in the third square brackets represent the difference between the contributions of the orientational polarization energy when the electron is in the 2p and 1s states, respectively. The transition energy, calculated from Eq. (11.14) using $R_0 = 0$, is 1.35 eV, only fairly comparable with the experimental value of 1.7 eV.[27] Larger values of R_0 *increase* the discrepancy between the theoretical (e.g., 0.93 eV at $R_0 = 3.3$ Å) and the experimental transition energies (1.7 eV). Thus, *the continuum model does not allow estimation of the cavity radius*; rather, it gives a value comparable to experiment only in the limiting case of a cavity radius of $R_0 = 0$. This is a rather devastating limitation.

11.3. Semicontinuum Model of the Hydrated Electron

In the semicontinuum model,[28-32] the hydrated electron is considered to interact with the specifically oriented water dipoles in the first solvation shell by a *short-range* attractive potential. The solvent molecules beyond the first solvation shell are treated as a continuous dielectric medium with which the electron interacts by means of a *long-range* polarization potential. Most other approximations of continuum treatment, such as the consideration of hydrogenlike one-parameter wave function for the hydrated electron, cavity approximation, and the use of either adiabatic or self-consistent approximations, have been used in most semicontinuum treatments.

The treatments of Fueki *et al.*[28–32] have used the SCF approximation. The energy level of the hydrated electron are calculated as a function of the cavity radius by means of the SCF approximation for particular values of the radius. To find the optimum value of the cavity radius of the hydrated electron, its value is adjusted to fit with some relevant experimental quantity, such as the optical transition energy, and then using this optimum value of the cavity radius, other experimental observations are predicted.

The advantage of the semicontinuum model is that the configurational stability of the hydrated electron can be established, and a particular cavity radius predicted. Fueki *et al.*[29,30] established the configurational stability of the hydrated electron. We will discuss here the treatment of the semicontinuum model based on the treatment of Fueki *et al.*[29,30]

In the semicontinuum model of hydrated electrons by Fueki, Feng, and Kevan,[30] the electron is considered to interact with a certain number N of specifically oriented solvent dipoles in the first solvation shell by means of short-range *attractive* and *repulsive* potentials and with the rest of the solvent molecules beyond the first solvation shell by long-range average polarization potentials. The total energy of the excess electrons in the medium is given by the sum of the energy of the interaction of the electronic energy with the first neighboring dipoles and the rearrangement energy of the medium due to its interaction with the electron.

The dipoles in the first solvation shell are considered to be arranged symmetrically around the electron in a tetrahedral fashion for $N = 4$ and an octahedral fashion for $N = 6$, and these dipoles are assumed to be in thermal equilibrium with the medium.

11.3.1. Short-Range Energies of the Hydrated Electron

For the solvent dipole moment μ_d, and polarizability α, the short-range electronic energy $E_e^s(i)$ of the ith bound state (where $i = 1s$ for the ground state and $2p$ for the excited state and superscript s represents the short range) of the hydrated electron for the short-range attractive interaction is

$$E_e^s(i) = -4\pi \int_0^{R_d} \Psi_i \left(\frac{N e_0 \mu_d \langle \cos \theta \rangle_i}{R_d^2} + \frac{N e_0^2 \alpha C_i}{2 R_d^4} \right) \Psi_i r^2 \, dr \quad (11.15)$$

where Ψ_i is the wave function of the hydrated electron in the state i, R_d is the distance between the center of the hydrated electron charge distribution and the center of the dipole, θ is the angle between the dipole moment vector and the line joining the center of the hydrated electron charge distribution to the dipole, C_i is the total charge density enclosed with the radius R_d, and the 4π term that appears outside the integral sign in the equation appears due to the θ and ϕ integration in the spherical polar coordinate system used in the calculation. To overcome an integrational difficulty, an average value for $\langle \cos \theta \rangle_i$ is taken. The average value of $\cos \theta$ is given using Langevin's relation, $\langle \cos \theta \rangle_i = \coth \chi - 1/\chi$, where $\chi = \mu_d \chi_{loc}/kT$ where the local field $\chi_{loc} = e_0 C_i/R_d^2$.

The short-range *repulsion* energy E_r of the hydrated electron with electrons in the bulk solvent molecules is approximated by the "free electron energy" U_0 times the charge density outside R_d† and is given by

$$E_r^s = U_0(1 - C_i) \tag{11.16}$$

where $(1 - C_i)$ is considered the charge density outside R_d. However, the authors have taken U_0 as an adjustable parameter, and it is chosen so that the calculated $1s \rightarrow 2p$ transition energy fits with the observed optical transition energy. This approach decreases to some degree the implications of the agreement that the model shows with other experimental quantities (e.g., the hydration energy).

There is also a short-range dipole–dipole repulsion energy involved due to the medium rearrangement associated with the dipole orientation for the presence of the hydrated electron, and given by

$$E_m^s(i) = \frac{D_N}{R_d^3}\left(\mu_d\langle\cos\theta\rangle_i + \frac{e_0\alpha C_i}{R_d^2}\right)^2 \tag{11.17}$$

where D_N is a numerical constant calculated from the number and geometrical arrangement of the dipoles. Both the terms in $E_e(i)$ and $E_m(i)$ in Eqs. (11.15) and (11.17) depend on C_i and are determined by Fueki et al.[30] by an SCF approach.

Another short-range medium rearrangement energy arises due to the energy required to form a void or cavity in the medium to accommodate the hydrated electron. This is

$$E_v^s = 4\pi(R_d^2 - R_s^2)\gamma \tag{11.18}$$

where γ is the surface energy and R_s is the radius of the solvent dipole.

---

† The authors, apparently, only repel the electron with part of the solvent. The part that they choose for the solvent is the part *after* R_d. There seems to have been a lack of accounting for the part of the repulsion energy before R_d. As far as U_0 is concerned, it is a provocative use of the adjustable parameter "free electron energy" for electrons taking part in covalent bonds.

11.3.2. Long-Range Energies of the Hydrated Electron

We have considered the short-range attractive and repulsive interaction in the first solvent shell of the hydrated electron. The solvent system outside the first solvation shell, beginning at $R = R_d + R_s$, has been treated by Fueki et al.[30] as a continuous dielectric characterized by the static ε_s and optical ε_{op} dielectric constants. The *long-range* interaction of the hydrated electron with the continuum dielectric consists of an electronic part $E_e^l(i)$ and a medium rearrangement part $E_m^l(i)$, where the superscript l is used to represent the long-range interaction.

The electronic *potential energy* due to the long-range total polarization (electronic and orientational) interaction when the hydrated electron is in the ground 1s state, having a wave function ψ_{1s}, is expressed as[4]

$$E_e^l(1s) = 4\pi e_0\left(1 - \frac{1}{\varepsilon_s}\right)f_{1s}(R)\int_0^R \psi_{1s}^2 r^2\,dr + 4\pi e_0\left(1 - \frac{1}{\varepsilon_s}\right)\int_R^\infty f_{1s}(r)\psi_{1s}^2 r^2\,dr$$

(11.19)

where the first term of Eq. (11.19) represents the long-range electronic energy due to the polarization of the continuum for the electrostatic potential within the distance R, i.e., $f_{1s}(R)$ [see Eq. (11.5)] from the charge center of the hydrated electron, and the second term represents the same when the electrostatic potential is beyond the range R, i.e., $f_{1s}(r)$.

Similarly, the excited 2p state long-range electronic potential energy is given by[4,29]

$$E_e^l(2p) = 4\pi e_0\left(1 - \frac{1}{\varepsilon_{op}}\right)f_{2p}(R)\int \psi_{2p}^2 r^2\,dr + 4\pi e_0\left(1 - \frac{1}{\varepsilon_{op}}\right)\int_R^\infty f_{2p}(r)\psi_{2p}^2 r^2\,dr$$

$$+ 4\pi e_0\left(\frac{1}{\varepsilon_{op}} - \frac{1}{\varepsilon_s}\right)f_{1s}(R)\int_0^R \psi_{2p}^2 r^2\,dr$$

$$+ 4\pi e_0\left(\frac{1}{\varepsilon_{op}} - \frac{1}{\varepsilon_s}\right)\int_R^\alpha f_{1s}(r)\psi_{2p}^2 r^2\,dr \qquad (11.20)$$

where the first two terms give the contribution of the long-range electronic potential energy due to the electronic polarization of the medium and when an electron is in the excited 2p state for the electrostatic potential f_{2p}, both for the range when $r \le R$ and $r \ge R$, respectively. The third and fourth terms of Eq. (11.20) represent the contribution due to orientational polarization only when the electrostatic potential f_{1s} corresponds for the electron in the ground state both for the range $r \le R$ and $r \ge R$, respectively.

Equation (11.20) is a difficulty in the treatment of Fueki et al.[29,30] In the first two terms of Eq. (11.20), the contribution due to the electronic polarization has been taken into account with the idea that when the hydrated electron is excited by a perturbation due to the field in the electromagnetic radiation, only the electronic part of the polarization can respond. The third and fourth terms in the equation concern the orientational polarization, which changes slowly. The 2p wave function, subject to this polarization, "sees" the potential due to the 1s charge distribution.

The authors may have involved themselves in the same difficulty as in the continuum treatment of Jortner *et al.*[19-21] Should not the electronic potential energy in the excited 2p state of the hydrated electron be with the total polarization (electronic and orientational)? The lifetime of the excited state is 10^{-8} to 10^{-10} sec and the time needed for the electronic polarization is of the order of 10^{-15} sec and that needed for orientational polarization is of the order of 10^{-11} sec. Hence, it is correct to consider the total polarization effect for the excited state like the ground state. A more acceptable expression for the electron potential (due to long-range polarization interaction) when the hydrated electron is in the *excited* 2p state would be

$$E_e^l(2p) = 4\pi e_0\left(1 - \frac{1}{\varepsilon_s}\right) f_{2p}(R) \int_0^R \psi_{2p}^2 r^2 \, dr + 4\pi e_0\left(1 - \frac{1}{\varepsilon_s}\right) \int_R^\infty f_{2p}(r)\psi_{2p}^2 r^2 \, dr$$

$$(11.21)$$

The *medium rearrangement* energy $E_m^l(1s)$ (to be distinguished from the electronic potential energy) in the ground 1s state is expressed as[4,29,30]

$$E_m^l(1s) = \frac{4\pi e_0}{2}\left(1 - \frac{1}{\varepsilon_s}\right)\left[f_{1s}(R) \int_0^R \psi_{1s}^2 r^2 \, dr + \int_R^\infty f_{1s}(r)\psi_{1s}^2 r^2 \, dr\right]$$

$$(11.22)$$

The rearrangement energy in the 2p state is[4]

$$E_m^l(2p) = -\frac{e_0}{2}\left(1 - \frac{1}{\varepsilon_{op}}\right)\int f_{2p}\psi_{2p}^2 \, d\tau - \frac{e_0}{2}\left(\frac{1}{\varepsilon_{op}} - \frac{1}{\varepsilon_s}\right)\int f_{1s}\psi_{1s}^2 \, d\tau$$

$$(11.23)$$

Each term of expression (11.23) should be expressed in two terms [like Eq. (11.20)] corresponding to two integration limits, i.e., 0 to R and R to ∞, but for brevity, no separation of the integration range is indicated in Eq. (11.23). In the expression (11.23) for the excited state, a medium rearrangement energy of the electron polarization component has been considered in the first term with the charge distribution of the excited 2p state, but in the second term, the orientational polarization component is considered with the charge distribution of the ground 1s state. However, the consideration seems doubtful, and authors have not given physical reasons for it. Thus, Eq. (11.23) involves a difficulty similar to that of Eq. (11.20). A more acceptable form of (11.23) would be in terms of the total polarization [for the reason explained after Eq. (11.20)], namely,

$$E^l_m(2p) = -\frac{4\pi e_0}{2}\left(1 - \frac{1}{\varepsilon_s}\right)\left[f_{2p}(R)\int_0^R \psi^2_{2p}r^2\,dr + \int_R^\infty f_{2p}(r)\psi^2_{2p}r^2\,dr\right]$$

(11.24)

11.3.3. Total Ground-State Energy of the Hydrated Electron

The total energy of the hydrated electron when it is in the ground state in the solvent medium can be expressed as

$$E_t(1s) = E_k(1s) + E^s_e(1s) + E^s_r(1s) + E^s_m(1s) + E^s_v + E^l_e(1s) + E^l_m(1s) \quad (11.25)$$

where $i = 1s$ has been used and $E_k(1s)$ is the kinetic energy of the hydrated electron when it is in the ground 1s state and is expressed as

$$E_k = \int \psi_{1s}\left(\frac{\hbar^2}{2m_e}\nabla^2\right)\psi_{1s}\,d\tau$$

(11.26)

the other terms from the corresponding expressions given above. For the numerical evaluation of $E_t(1s)$, Fueki *et al.*[30] used the one-parameter hydrogenlike wave function given in Eq. (11.4). The variational procedure is applied to the wave function parameter λ to obtain the minimum energy for the given value of R_d, the distance from the center of the cavity to the center of the first-layer dipole. This procedure is repeated for various values of R_d to construct configuration coordinate curves and establish configurational stability corresponding to the minimum of the curve. The configuration coordinate diagram given by Fueki *et al.*[30] is shown in Fig. 11.1.

11.3.4. Excited-State Energy of the Hydrated Electron

Fueki *et al.*, in their semicontinuum treatment,[30] considered the 2p hydrogenic state as the excited state of the hydrated electron. The total energy of the hydrated electron in the excited 2p state is expressed as

$$E_t(2p) = E_k(2p) + E^s_e(2p) + E^s_r(2p) + E^s_m(2p) + E^s_v + E^{l(2p)}_e + E^l_m(2p)$$

(11.27)

where $i = 2p$ has been used and $E_k(2p)$ is the kinetic energy of the hydrated electron. When it is the excited 2p state, it is expressed as

$$E_k = \int \psi_{2p}\left(\frac{\hbar^2}{2m_e}\nabla^2\right)\psi_{2p}\,d\tau$$

(11.28)

The 2p state wave function ψ_{2p} is used as the one-parameter hydrogenic wave function given in Eq. (11.11). The other terms are obtained from the

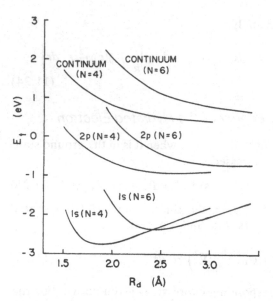

Fig. 11.1. Energy level diagram of a hydrated electron.[30]

expressions given above. In this also, the variational procedure is applied to the wave function parameter μ to get the minimum energy for a given R_d. This process is repeated for different values of R_d to construct configuration coordinated curves as are given in Fig. 11.1.

11.3.5. Numerical Results of the Semicontinuum Treatment (from the SCF Approach)

11.3.5.1. Hydration Energy

The hydration energy of the hydrated electron is given by Fueki et al.[30] as the negative of the ground state energy of the hydrated electron. In their terminology, the hydration energy has been taken (unusually) as a *positive* quantity and the energy of the hydrated electron in the ground state is a negative quantity. Hence, the hydration energy is written as the negative of $E_t(1s)$ as

$$\Delta H = -E_t(1s) \tag{11.29}$$

The total energy $E_t(1s)$ is determined from Eq. (11.25) by finding the value of each term in the equation for its corresponding expression using the ψ_{1s} wave function and electrostatic potential f_{1s} using the SCF treatment from Poisson's equation (11.2). The result obtained by Fueki et al.[30] is given in Table 11.1. The calculated values of ΔH are higher than the experimental

Table 11.1
Physical Properties of the Hydrated Electron

Property	Temperature (°K)	N	Calculated values (eV)	Experimental values (eV)
Hydration energy (ΔH)	298	4	2.75	1.7
		4	2.36	1.7
Cavity radius (R_0)	298	4	1.93	—
		6	2.46	—
Excitation energy ($h\nu$)	298	4	2.15	1.72
		6	1.94	1.72

value, 1.7 eV.[26] This difference arises, according to the authors,[30] because, in their treatment, they neglect the energy required to break the hydrogen bond for electron solvation. If all of the hydrogen bonds in the first layer are broken, the calculated hydration energy ΔH becomes much greater than the experimental value. However, if two hydrogen bonds are broken per molecule in the first solvation shell, the value of ΔH becomes 2.19 eV for the number $N = 4$ of water molecules in the first solvation shell and 1.52 eV for $N = 6$, which is somewhat better.

11.3.5.2. Cavity Radius of the Hydrated Electron

The radius R_0 of the hydrated electron is obtained in the semicontinuum treatment from the value of R_d corresponding to the minimum of the configurational coordinate diagram (Fig. 11.1). The results are given in the Table 11.1. The cavity radius in the excited state increases compared with the unexcited (Fig. 11.1).

11.3.5.3. Transition Energy of the Hydrated Electron

The transition energy is given as[4,30]

$$h\nu = E_t(2p)_{R_0} - E_t(1s)_{R_0} \tag{11.30}$$

where R_0 is the cavity radius, i.e., the value of R_d at the configurational minimum in the 1s and 2p states (Fig. 11.1). The vertical distance from the configurational minimum to that of the excited 2p state curve represents the transition energy $h\nu$ (Fig. 11.1). The values of $h\nu$ are given in Table 11.1 for $N = 4$ and $N = 6$ and compared with the experimental value of 1.72 eV.[27]

Fueki et al.[30] obtained in their variational calculation of the energy of the hydrated electron short-range interactions ($E_e^s + E_m^s = -2.64$ eV for $N = 4$ and -2.21 eV for $N = 6$) and concluded that these are more important than the long-range interactions $(E_e^l + E_m^s = -2.05 \text{ eV}$ for $N = 4$ and -1.77 eV for $N = 6)$. This conclusion supports the calculation of Bockris *et al.*[33] that the contribution of the activation energy from the short-range stretching interaction in electron transfer is more important than the continuum contribution.

Schlick, Narayana, and Kevan[34] have studied the structure of a hydrated electron using ESR technique. Fueki *et al.*[30] obtained the equilibrium energy configuration (corresponding to the configurational minimum) when the first hydration layer dipoles are oriented (typically within 10°) toward the electron, and such orientation is not indicated from the ESR study.[34] The semicontinuum model fails in predicting the structure of hydrated electrons, although it does give good energy-level information.

We have mentioned earlier that Fueki *et al.*,[4,30] in their semicontinuum treatments, used the one-parameter hydrogenic wave function. Recently Feng, Ebbing, and Kevan[32] used the Gaussian wave functions for the electron rather than the hydrogenic one in a semicontinuum model for the hydrated electron. The Gaussian wave functions for 1s and 2p states are expressed as

$$\psi_{1s} = (2\lambda/\pi)^{3/2} \exp(-\lambda r^2) \tag{11.31}$$

and

$$\psi_{2p} = \left[\frac{2}{\pi}\left(\frac{2\mu}{\pi}\right)\right]^{1/2} r \exp(-\mu r^2) \cos\theta \tag{11.32}$$

where λ and μ are the variational parameters in 1s and 2p states, respectively. The other details of this treatment of Feng *et al.*[32] are similar to those of Fueki *et al.*[30] The physical properties of the hydrated electron calculated from Gaussian (G) and hydrogen (H) wave functions show little difference in the total energies and the optical transition energies calculated with the two functions (Table 11.2). The results calculated from Gaussian wave function deviate more from that of experiment than do the earlier calculated ones.

11.4. Structural Models for the Hydrated Electron

In the previous sections we have discussed the continuum and semicontinuum models of the hydrated electron. In this section we will discuss the structural models of the hydrated electron and the quantum mechanical computation of its hydration and excitation energies, etc.

We have observed in the previous section (11.3) that the semicontinuum treatments of the hydrated electron did not consider the specific

Table 11.2

Physical Properties of the Hydrated Electron Calculated from Gaussian and Hydrogenic Wave Functions

Basis	Number of coordinated water dipoles, N	Hydration energy ΔH (eV)	Cavity radius R_0 (Å)	Excitation energy $h\nu$ (eV)
Gaussian	4	2.99	1.89	2.57
	6	2.67	2.40	2.52
Hydrogenic	4	2.75	1.93	2.15
	6	2.36	2.46	1.94
Experimental	—	1.7	—	1.72

structure of individual coordinating water molecules. In fact, the specific structure gives rise to tetrahedral coordination, the four water molecules being interconnected by hydrogen bonding. ESR spectra indicate that electrons are trapped in structural defects in the field due to oriented water molecules.

Natori and Watanabe[35] have given a structural model of the hydrated electron in which they considered the electron to be trapped in a tetrahedron as in Fig. 11.2. As in Fig. 11.2, numbers were assigned to the atoms of the structure according to the following scheme: the inside hydrogen atoms were numbered from 1 to 4, the oxygen atoms from 5 to 8, and the outside hydrogen atoms from 9 to 12. The above structure is analogous to that of an electron trapped in a lattice defect in an ice crystal.

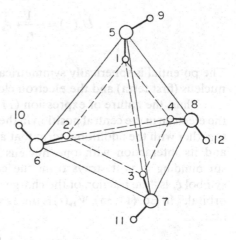

Fig. 11.2. Structural model of the hydrated electron.

⬤ OXYGEN   ○ HYDROGEN

In the vicinity of the defect due to the presence of an extra electron, there exists a potential field that acts on the electron, and if one considers the center of the tetrahedron as the origin of the coordinate system, the potential field can be given in atomic units as[35]

$$U(\mathbf{r}) = \sum_{s=1}^{12} U_s(\mathbf{r} - \mathbf{R}_s) + U_{\text{out}} \tag{11.33}$$

where $\mathbf{r}$ denotes the position vector of the electron, $\mathbf{R}_s$ is the position vector (or coordinate) of the sth nucleus, and U_{out} is the potential acting on the electron from the molecules outside the molecular group under consideration, i.e., the second water layer outside the cavity. $U_s(\mathbf{r} - \mathbf{R}_s)$ is assumed to be spherically symmetrical and determined by the wave function of the sth atom, one of twelve atoms concerned.

The energy state of the trapped hydrated electron can be determined by solving the Schrödinger equation for a one-body problem with this potential (Eq. 11.33). Neglecting U_{out}, it can be written for the three groups of atoms $s = 1$–4, $s = 5$–8, and $s = 9$–12:

Potential Due to the Inside Hydrogen Atoms ($s = 1$–4 of Fig. 11.2). Defining

$$\mathbf{r}_s = \mathbf{r} - \mathbf{R}_s \tag{11.34}$$

[which denotes the coordinate of the electron with respect to the sth nucleus of the H_2O molecules around the electron (Fig. 11.2)], we can write the normalized potential due to the sth hydrogen molecule as[35]

$$U_s(r_s) = -\frac{1}{r_s} + \xi_s \int \frac{|\Psi_{1s}(\mathbf{r})|^2}{|\mathbf{r} - \mathbf{R}|} d\tau \tag{11.35}$$

The potential is spherically symmetrical and due to the contribution of the nucleus (first term) and the electron cloud (second term).

Thus, the nature of expression (11.35) entails the potential energy for the electron in the central position in the cavity of the water molecule, $-1/r_s$, together with the additional term that allows for the energy of the electron, and its interaction with one nucleus among the water molecule nuclei surrounding it as it strays from the central position denoted by r_s. The symbol ξ_s is "the fraction of the charge of the electron that remains in the 1s orbital." In Eq. (11.35), $\Psi_{1s}(r)$ is the 1s wave function of the hydrogen atom, i.e.,

$$\Psi_{1s}(r) = \pi^{-1/2} e^{-r} \tag{11.36}$$

(r is in dimensionless atomic units). From Eqs. (11.35) and (11.36), one obtains[35]

$$U_s(r_s) = -\frac{1-\xi_s}{r_s} - \xi_s\left(1+\frac{1}{r_s}\right)\exp(-2r_s) \tag{11.37}$$

Potential Due to the Oxygen Atoms ($s = 5–8$ of Fig. 11.2). The potential of the oxygen atom, which has eight electrons, is given (summing the nuclear and outer electron contributions to the interaction with the hydrated electron) as

$$U_s(\mathbf{r}_s) = -\frac{8}{r_s} + 2\int\frac{|\Psi_{1s}(\mathbf{r})|^2}{|\mathbf{r}-\mathbf{r}_s|}\,d\tau + 2\int\frac{|\Psi_{2s}(\mathbf{r})|^2}{|\mathbf{r}-\mathbf{r}_s|}\,d\tau + 4g\int\frac{|\Psi_{2p}(\mathbf{r})|}{|\mathbf{r}-\mathbf{r}_s|}\,d\tau \tag{11.38}$$

where the factor g can be determined so that the oxygen atom has the effective charge -2×0.326 in a water molecule. The potential represented by Eq. (11.38) can be determined using the radial wave functions given for the oxygen atom given by Morse *et al.*[36]

Potential from the Outside Hydrogen Atoms ($s = 9–12$ of Fig. 11.2). The outside hydrogen atom is considered to be concentrated at a point charge of 0.326 (in atomic unit) and thus

$$V_s(r_s) = -0.326/r_s \tag{11.39}$$

The Schrödinger equation of the one-electron system, i.e.,

$$H\psi = E\psi \tag{11.40}$$

has been solved by Natori and Watanabe[35] with

$$H = \frac{\hbar^2}{2m}\frac{\partial^2\psi}{\partial r^2} + U \tag{11.41}$$

and

$$U = \sum_{s=1}^{12} U_s(r_s) \tag{11.42}$$

The variational method was introduced to determine the hydrated electron wave function and the corresponding eigenvalue E. An appropriate variational wave function was considered to be a linear combination of four wave functions, each representing the electron bound to one of the inside hydrogen atoms, i.e.,

$$\psi(r) = \sum_{i=1}^{4} C_i\Psi(r_i) \tag{11.43}$$

where $r_i = r - R_i$ as in Eq. (11.34) and $\Psi_i(r_i)$ is a 1s-like wave function having one parameter Z, i.e.,

$$\psi_i(r_i) = (Z^3/\pi)^{1/2} \exp(-Z r_i) \tag{11.44}$$

The resulting wave function was introduced in a molecular orbital method derivation of the hydrated electron energy levels. Natori and Watanabe[35] obtained a secular equation of the fourth order, which can be expressed in the determinental form as

$$\det|H_{ij} - ES_{ij}| = 0 \tag{11.45}$$

where

$$S_{ij} = \int \psi_i \psi_j \, d\tau \tag{11.46}$$

$$H_{ij} = \int \psi_i H \psi_j \, d\tau \tag{11.47}$$

and the overlap integral has the property,

$$S_{ii} = 1 \qquad \text{for all } i$$
$$S_{ij} = S \qquad \text{for all } i \neq j \tag{11.48}$$

Since the potential U is spherically symmetrical, the exchange energy integrals can be written as

$$H_{ii} = \alpha \qquad \text{for all } i$$
$$H_{ij} = \beta \qquad \text{for all } i \neq j \tag{11.49}$$

The two-centered overlap integrals S_{ij} and the exchange integrals α were evaluated using elliptic coordinates, whereas the three-centered integrals β were solved analytically without assumptions.

Natori and Watanabe[35] obtained the ground-state energy E_0 from the single root of Eq. (11.45) as

$$E_0 = \frac{\alpha + 3\beta}{1 + S} \tag{11.50}$$

and the excited state energy E is given by the triply degenerate root, i.e.,

$$E_1 = \frac{\alpha - \beta}{1 - S} \tag{11.51}$$

Now, from the knowledge of E_0 and E_1, one obtains[35] the excitation energy

$$\Delta E = E_1 - E_0$$
$$= 3.58 - (-4.38) = 0.8 \text{ eV} \tag{11.52}$$

This compares rather poorly with the experimental value of 1.72 eV given in Table 11.1.

The total energy of the system E_{tot}, i.e., the hydration energy of the hydrated electrons, can be written in terms of the ground-state energy of the trapped electron and the polarization energy E_{pol} of the medium as

$$E_{tot} = E_0 + E_{pol}$$
$$= -4.38 + 2 = 2.4 \text{ eV} \qquad (11.53)$$

where E_{pol} was estimated to be 2 eV considering the hydrogen bond energy (4–7 kcal mole^{-1}) and the electrostatic energy between atoms.

One finds that the hydration energy is -2.4 eV compared to the experimental value[26] 1.7 eV. The excitation energy of the hydrated electron, 0.8 eV, is compared to the experimental value[27] of 1.72 eV. One observes that this simplified treatment of the structural model gives less good agreement with experiment than the continuum models.

Later, Natori[37] suggested an improved treatment of the structural model of the hydrated electron in which the wave functions of the hydrated electron were refined and orthogonalized to the wave functions of the neighboring water molecules. The excitation energy of the hydrated electron obtained according to this treatment (1.74 eV) agreed better with that of experiment (1.72 eV). The broadness of the optical absorption spectrum and the dependence of the peak of the spectrum on temperature and pressure are explained by the model of the hydrated electron, which is regarded as a hypothetical $(H_2O)_4^-$ ion located in the cavity (Fig. 11.2).

Weissmann and Cohan[38] calculated the energy of the hydrated electron using the molecular orbital method modified according to the CNDO/2 approximation. They considered the following three types of structures:

(a) Five water molecules—the central one and the four nearest neighbors located as in ice.

(b) Four water molecules—the cavity obtained by removing the central molecule and rotating two of the four neighbors so that the four inside hydrogen atoms are symmetrically arranged about the vacancy.[35]

(c) Same as (b) but rotating the four neighbors so that the dipole moment vectors point toward the center of the cavity.

Although the computation overestimated the hydration energy of the electron, the results support the normal icelike structure for the hydrated electron and do not indicate cavity formation.

Later, Weissmann and Cohan[39] studied the structure and the short-range contributions to the binding energy of the hydrated electron with a modified version of the CNDO/2 method. They have calculated the energy of $(H_2O)_n^-$ using the basis set both in terms of ground- and excited-state atomic orbitals. They have considered n water molecules rather than the fixed number of neighboring water molecules. In their calculation, they support the many-electron calculation by Howat and Webster[40] of the system of $(H_2O)_n^-$ using the INDO (intermediate neglect of differential overlap) method that the hydrated electron becomes stable for $n \geq 4$, and it is not bound for $n < 4$.[35,37]

Ray[41] considered a tetrahedral model of the hydrated electron in which it was assumed that the effective dipoles at the corners represent not only the four nearest-neighbor water molecules, but also the accumulated effects on them due to the polar medium of water. This accumulated effect has been taken in terms of the increased dipole moment of water ($3.8D$) in the liquid water from the work of Onsager and Dupuis.[42] Ab initio SCF calculations were carried out for the central H_2O or H_2O^- in the tetrahedral system considered.

The interaction Hamiltonian of H_2O^- in the environment of four assigned dipoles was written for the Ray model as[41]

$$H = H_0 + U \tag{11.54}$$

where H_0 represents the interaction terms of ten electrons and three nuclei for H_2O or interaction terms of eleven electrons and three nuclei when the H_2O^- is being considered. U gives the additional interactions of ten or eleven electrons with the four dipoles and was expressed as

$$U = -\sum_{n=1}^{4} \sum_{i=1}^{N} \frac{e_0 \mu_n \cos \theta_n}{|\mathbf{r}_n - \mathbf{r}_i|^2} \tag{11.55}$$

where the first summation is over the number of dipoles, while the second summation is over the number of electrons N; $\boldsymbol{\mu}_n$ denotes the dipole moment of the nth dipole; $\mathbf{r}_n - \mathbf{r}_i$ is the radial vector of the ith electron from the nth dipole; and θ_n is the angle between the radial vector $\mathbf{r}_n - \mathbf{r}_i$ and dipole moment vector $\boldsymbol{\mu}_n$.

A wave function for water was constructed from five closed-shell (i.e., paired-electron) molecular orbitals, ϕ_i ($i = 1$ to 5), in the case of H_2O, or six molecular orbitals ϕ_i ($i = 1$ to 6) consisting of five closed shells (i.e., paired) and one open shell (i.e., unpaired electron) in the case of H_2O^-.

The molecular orbitals were formed as a linear combination of the atomic wave functions $\chi_{n_\rho l_\rho m_\rho}$ (i.e., the atomic basis set) as follows:

$$\phi_i = \sum_{\rho=1}^{12} C_{i\rho} \chi_{n_\rho l_\rho m_\rho} \tag{11.56}$$

where the summation over ρ corresponds to a basis set of twelve Slater orbital† wave functions. The $C_{i\rho}$ values are orbital coefficients and the basis functions $\chi_{n_\rho l_\rho m_\rho}$, which involve the principal quantum number n, azimuthal quantum number l, and magnetic quantum number m, are expressed as

$$\chi_{n_\rho l_\rho m_\rho} = (2\xi_\rho)^{n_\rho + 1/2} [(2n_\rho)!]^{-1/2} r^n \rho^{-1} \exp(-\xi_\rho r) Y_{l_\rho m_\rho}(\theta, \phi) \quad (11.57)$$

where ξ_ρ are the normal exponents and $Y_{l_\rho m_\rho}(\theta, \phi)$ are spherical harmonics where subscript ρ in n, l, and m and ξ indicates the number in the basis set.

The self-consistent molecular field calculations[41] were carried out to determine $C_{i\rho}$ values subject to the convergency of optimum energies such as

$$E_{H_2O}^0 = \langle \Psi^{(5)} | H_0 | \Psi^{(5)} \rangle \quad (11.58)$$

$$E_{H_2O} = \langle \Psi^{(5)} | H | \Psi^{(5)} \rangle \quad (11.59)$$

$$E_{H_2O^-} = \langle \Psi^{(6)} | H | \Psi^{(6)} \rangle \quad (11.60)$$

Ray[41] computed the expressions (11.58), (11.59), and (11.60), which involve one-centered, two-centered, and three-centered one-electron–two-electron integrals by use of a multicentered integral program.[60]

The energy $E_{H_2O}^0$ corresponds to free H_2O, while E_{H_2O} and $E_{H_2O^-}$ refer to energies for H_2O and H_2O^-, respectively in the complex. The quantity

$$\Delta E = E_{H_2O} - E_{H_2O^-} \quad (11.61)$$

if positive, represents the electron affinity to H_2O in the complex. E_{H_2O}, $E_{H_2O^-}$ and ΔE were calculated for various values of θ_1, θ_2, θ_3, and θ_4.

The variation of the electron affinity in respect to the central H_2O molecule in the tetrahedron depends only on the orientations of the dipoles. If all the dipoles were oriented normally, Ray[41] obtained the maximum value for the electron affinity, i.e., 1.79 eV, which is comparable with the binding energy of the hydrated electron,[26] i.e., 1.7 eV. Thus, the hydrated electron, e^-(aq), can exist in water in the form of a complex, $(H_2O)_n^-$ (where n can be 5 or more), provided that the structure of the water favors the formation of such a complex.

Although the structural and semicontinuum treatments incorporate the spatial arrangements of coordinating water molecules, the semicontinuum models approximate the coordinating molecules by structureless dipoles, whereas the models of Natori and Watanabe[35] and of Ray[41] do acknowledge

---

† The orbitals are the wave functions of the atomic orbitals as given in the general form of Eq. (11.57) by Slater, and he devised a set of general rules to find the parameters involved in this equation on the basis of the variation principle for different orbitals in different atoms.[43]

the structure of the individual water molecule. Nevertheless, the semicontinuum treatment and some structural models agree substantially with the experimental hydrated electron binding energy.

Ishimaru, Kato, Yamabe, and Fukui[44] used the semiempirical unrestricted Hartree–Fock treatment based on the INDO approximation to investigate the structure of the hydrated electron. In their calculation, two models of the hydrated electron were considered. Model I is the tetrahedral model by Natori and Watanabe (Fig. 11.2), which is formed by four water molecules with one OH bond tetrahedrally oriented around the cavity center and one of each four water molecules pointing toward the cavity center.

Model II is the cavity model by Nilson,[45] which is formed by twelve water molecules in the first coordination shell, and of these twelve water molecules, six water molecules are nearer to the cavity center than the other six water molecules. In this case, both hydrogen atoms of each water molecule surround the cavity center.

For the CNDO (*c*omplete *n*eglect of *d*ifferential *o*verlap) calculation, the wave functions are written as a linear combination of *a*tomic *o*rbital (AO) basis sets. Ishimaru *et al.*[44] used the valence AOs of solvent molecules and hydrogenlike AOs for the hydrated electron located at the cavity center. Ishimaru calculated the proton spin density of water according to Model I, which agrees with that obtained from ESR (*e*lectron *s*pin *r*esonance) experiments. Thus, Natori and Watanabe's[35] tetrahedral model is supported by Ishimaru *et al.*'s[44] Hartree–Fock calculation.

11.5. Theory of Electron Transfer Reaction from a Hydrated Electron to an Acceptor in Solution

There are many electron transfer reactions involving hydrated electrons.[1] Marcus[46a,47] formulated the rate theory of electron transfer between a hydrated electron and an acceptor, i.e., $e_{aq}^- + ox \rightarrow red$. In formulating the theory, Marcus used a similar type of treatment and formalism to that which was used for the electron transfer reactions at electrodes, with the inclusion of new features characteristic of the electron transfer reaction involving the hydrated electron. The novel aspects of the hydrated electron transfer process are the following:

(1) The hydrated electron disappears into the other reactant so that there is a change in the *number of reactants*.

(2) The electronic wave function of the hydrated electron, spread over several solvent molecules, is sensitive to orientation fluctuations of these molecules, unlike the electron within the shell of a reactant. This leads to a

quantitative difference between electron transfer reactions involving the hydrated electron and those of conventional electron transfer reactions.

(3) Unlike most of the conventional electron transfer reactions, the reactions of hydrated electrons are diffusion controlled.

In the derivation of the theoretical expression for the rate constant k_a and free energy of activation ΔF^* for the electron transfer reaction involving the hydrated electron, Marcus[46a] includes the following assumptions:

(a) The adiabatic Born–Oppenheimer approximation is used to treat the electronic and nuclear motion.

(b) The solvent polarization (outside that of the inner coordination shell) caused by the two reactants is treated in terms of dielectric continuum theory and *no cavity formation of the electron* is considered† (cf. Jortner[21]).

(c) The vibrations of the second reactant (which reacts with the solvated electron) are treated as harmonic oscillations.

(d) A symmetrical vibrational potential energy function is used for the reactant and products.

The system considered is the reaction between a hydrated ferric ion and a hydrated electron to form the hydrated ferrous ion.

11.5.1. Calculation of the Rate Constant k_r

The rate constant k_r is given for the electron transfer step on the basis of assumption (a) and use of equilibrium statistical mechanics[48,49] (for separation of reactants at a distance R) as

$$k_r = Z\kappa\rho \exp(-\Delta F^*/kT) \qquad (11.62)$$

where ΔF^* is the free energy of activation, and Z is the reactant collision frequency. The κ is taken to be ~ 1 (adiabatic), and ρ is a ratio of certain root mean square displacements§ and is taken as unity.

11.5.2. Calculation of the Free Energy of Activation ΔF^*

The free energy of activation ΔF^* to be used in Eq. (11.62) can be given in terms of its vibrational contribution from the charge–dipole bond ΔF_{vib}^*, the average contribution due to the continuum polarization $\langle F_{pol}^* \rangle$, the

---

† No physical unreality is really implied by neglecting the cavity radius. The latter is a classical device introduced to avoid an absurdly large self energy of the electron. However, in treatments such as the present one, the wave function of the electron takes into account the distribution of electron density around the center and deals with the problem in a more quantum mechanical sense.

§ ρ arises from the separation distance between reactants in the activated complex.

kinetic energy of the hydrated electron E_k and the free energy of the system when the reactants are far apart F_0^r (Fig. 11.3). Hence,

$$\Delta F^{*r} = \Delta F_{vib}^* + \langle F_{pol}^* \rangle + E_k - F_0^r \tag{11.63}$$

where the vibrational contribution is

$$\Delta F_{vib}^* = \tfrac{1}{2} \sum_{i,j} k_{ij}(q_i^* - q_i^r)(q_j^* - q_j^r) \tag{11.64}$$

In considering ΔF_{vib}^*, the author is taking into account the vibrational energy of the inner coordination sphere (in the way introduced first by George and Griffith[46b]), where k_{ij} is the reduced force constant and is expressed as[46a]

$$k_{ij} = \frac{k_{ij}^r k_{ij}^p}{k_{ij}^r + k_{ij}^p} \tag{11.65}$$

where k_{ij}^r and k_{ij}^p are the force constants of the reactant and product species. q_i^r and q_i^* in Eq. (11.64) are the vibrational coordinates for the reactant i when it is in the unconstrained (i.e., relaxed and unactivated) state and constrained (i.e., activated) state, respectively. q_j and q_j^* are the vibrational coordinates for the reactant j when it is in unconstrained and constrained states, respectively.

The contribution of the polarization energy ΔF_{pol}^*, can be obtained by considerations of polaron theory. In this theory, the situation is envisaged of a charge embedded in a dielectric, and the energy of the polaron is the

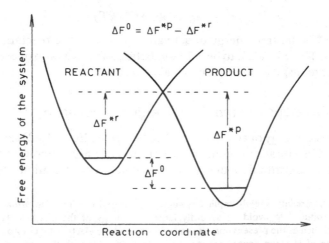

Fig. 11.3. Free energy diagram for the electron transfer reaction between a hydrated electron and a redox ion in solution.

energy of this charge within the surrounding dielectric. The value of F^*_{pol} is given by[46a]

$$F^*_{pol}(\mathbf{r}_1) = -\int \mathbf{P} \cdot \mathbf{X} \, d\mathbf{r} + 2\pi C \int \mathbf{P}^2 \, d\mathbf{r} - \frac{e_2}{2a_2}\left(1 - \frac{1}{\varepsilon_{op}}\right) + \frac{e_1 e_2}{\varepsilon_{op}|\mathbf{r}_1 - \mathbf{r}_2|}$$

(11.66)

and C is a constant given by

$$C = \left(\frac{1}{\varepsilon_{op}} - \frac{1}{\varepsilon_s}\right)$$

(11.67)

ε_{op} and ε_s denote the optical and static dielectric constants of the medium. $\mathbf{X}$ is the field due to the permanent charges and is given by

$$\mathbf{X} = -\nabla\left[\frac{e_1}{|\mathbf{r}_1 - \mathbf{r}_2|} + \frac{e_2}{|\mathbf{r}_1 - \mathbf{r}_2|}\right]$$

(11.68)

where

$$\nabla = \frac{1}{r^2}\frac{\partial}{\partial r}\left(r^2\frac{\partial}{\partial r}\right)$$

(11.69)

In Eq. (11.66), a_2 is the radius of the second reactant, including the inner coordination shell, and e_1 is the charge of the hydrated electron centered at $\mathbf{r}_1$, while e_2 is that of the second reactant centered at $\mathbf{r}_2$ at the activated state.

Here, the first two terms refer to the energy of the electron itself embedded within the dielectric, i.e., the energy of the polaron as given in Chapter 7. The first term in Eq. (11.66) is the interaction energy of the electron with the permanent dipoles in the solvent, and the second term that with the induced poles. The third term, having the optical dielectric constant within it, is the Born term for the energy of the reacting ion (e.g., Fe^{3+}) within the dielectric. *The author has neglected all the other sources of polarization that accrue to the ion* and has expressed the polarization only in terms of the Born energy, about half of the total polarization energy of the ion. In this respect, the work differs markedly from that of Fueki *et al.*[28] The fourth term in Eq. (11.66) represents the attractive Coulombic interaction between the hydrated electron and the positive ferric ion at a distance $|\mathbf{r}_1 - \mathbf{r}_2|$ apart.

The kinetic energy contribution to the free energy of activation [see Eq. (11.63)] of the electron (where the reactants are in the constrained state) E_k is given by

$$E_k = \frac{\hbar^2}{2m_e}\int |\nabla\Psi|^2 \, d\mathbf{r}_1$$

(11.70)

where Ψ is the wave function of the electron and m_e is the mass of the electron.

Now, using Eq. (11.66) as an operator for the free energy of polarization, one can find the free energy contribution to activation by multiplying Eq. (11.66) by the wave function of the electron (when the reactants are in the centered distribution in the activated state). Hence the average polarization energy is

$$\langle F^*_{pol} \rangle = \int F^*_{pol}(\mathbf{r}_1)|\Psi|^2 \, d\mathbf{r}_1 \qquad (11.66a)$$

Now, adding Eqs. (11.64), (11.66a), and (11.70), one gets the free energy of activation as

$$\Delta F^{*r} = \Delta F^*_{vib} + \int F^*_{pol}(\mathbf{r}_1)|\Psi|^2 \, d\mathbf{r}_1 + \frac{\hbar^2}{2m_e} \int |\nabla\Psi|^2 \, d\mathbf{r}_1 - F^r_0 \qquad (11.71)$$

where F^r_0, the free energy of the reactants when they are far apart, is denoted as the sum of the electron kinetic energy [first term of Eq. (11.72)] and the free energy of polarization of the unconstrained system of reactants:

$$F^r_0 = \frac{\hbar^2}{2m_e} \int |\nabla\Psi_0|^2 \, d\mathbf{r}_1 - \frac{C}{8\pi} \int (\mathbf{X}^r_0)^2 \, d\mathbf{r} \qquad (11.72)$$

where

$$\mathbf{X}^r_0 = -\nabla \frac{e_1}{|\mathbf{r} - \mathbf{r}_1|} \qquad (11.73)$$

is the field due to the permanent charges of reactants, and Ψ_0 is the wave function of the electron when the reactants are far apart and unconstrained.

An analogous expression to Eq. (11.71) for the product state can be written as

$$\Delta F^{*p} = \Delta F^*_{vib} + \frac{2\pi}{C} \int \mathbf{P}^2 \, d\mathbf{r} - \int \mathbf{P} \cdot \mathbf{X}^p \, d\mathbf{r} - \frac{e_2}{2a_2}\left(1 - \frac{1}{\varepsilon_{op}}\right) \qquad (11.71a)$$

Expression (11.71a) is simpler than Eq. (11.71). The kinetic energy term is missing, since the electron now resides in the second reactant (e.g., Fe^{3+}), where its kinetic energy is insensitive to solvent polarization fluctuation.†

To determine the polarization contribution F^*_{pol} from Eq. (11.66a) using the operator of Eq. (11.66), one needs to determine the polarization $\mathbf{P}$ of the medium. This can be obtained by minimizing ΔF^{*r} in Eq. (11.71) with respect to $\mathbf{P}$ [since the second term in Eq. (11.71) contains $\mathbf{P}$] under the condition that

$$F^{*r}(R) = F^{*p}(R) \qquad (11.74)$$

---

† However, when the acceptor ion loses its charge after transfer of a hydrated electron (e.g., neutralization of a proton by a hydrated electron), this expression (11.71a) will vanish.

where $F^{*r}(R)$ denotes the free energy of the system of reactants fixed in position a distance R apart and having the constrained equilibrium distribution, and $F^{*p}(R)$ similarly denotes the free energy of the system of products. After minimizing, $\mathbf{P}$ was found by Marcus[49] to be

$$\mathbf{P} = \frac{C}{4\pi}\left[(m+1)\int |\Psi|^2 \mathbf{X}^r \, d\mathbf{r}_1 - m\mathbf{X}^p\right] \qquad (11.75)$$

where $\mathbf{X}^r$ and $\mathbf{X}^p$ are fields due to the permanent charges of the reactants and products, respectively, and m is an undetermined Lagrangian multiplier.

By minimizing Eq. (11.71) with respect to q_i^*, subject to the condition (11.74), Marcus[46a] obtained

$$q_i^* = (m+1)q_i^r - mq_i^p \qquad (11.76)$$

where q_i^p is the vibrational coordinate for the product when it is not in the constrained state. q_j^* will be of similar type of expression as Eq. (11.76) with subscript j only.

Taking the value of q_i^* from Eq. (11.76) and a similar expression with subscript j in Eq. (11.64), one obtains (cf. reference 46b)

$$\Delta F_{\text{vib}}^* = \tfrac{1}{2}m_{i,j}^2 \sum k_{ij}(q_p^i - q_r^i)(q_p^j - q_r^j) \qquad (11.64a)$$

Substituting $\mathbf{P}$ from Eq. (11.75) in the second term of Eq. (11.71) where F_{pol}^{*r} is used from Eq. (11.66), Marcus obtained (Appendix 11.III)

$$\int F_{\text{pol}}^{*r}(\mathbf{r}_1)|\Psi|^2 \, d\mathbf{r}_1 = \frac{m^2 C}{8\pi}\int (\mathbf{X}_\alpha^r - \mathbf{X}^p)^2 \, d\mathbf{r} - \frac{C}{8\pi}\int (\mathbf{X}_\alpha^r)^2 \, d\mathbf{r} - \frac{e_2^2}{2a_2}\left(1 - \frac{1}{\varepsilon_{\text{op}}}\right)$$

$$+ \frac{e_1 e_2}{\varepsilon_{\text{op}}}\int \frac{|\Psi|^2 \, d\mathbf{r}}{|\mathbf{r}_1 - \mathbf{r}_2|} \qquad (11.66b)$$

Now putting the value of ΔF_{vib}^* from Eq. (11.64a) and $\int F_{\text{pol}}^{*r}(\mathbf{r}_1)|\Psi|^2 \, d\mathbf{r}_1$ from Eq. (11.66b) and E_k from Eq. (11.70) into Eq. (11.63) and rearranging, one gets

$$\Delta F^{*r} = \frac{\hbar^2}{2m_e}\int |\nabla\Psi|^2 \, d\mathbf{r}_1 + \frac{m^2 C}{8\pi}\int (\mathbf{X}_\alpha^r - \mathbf{X}^p)^2 \, d\mathbf{r} - \frac{C}{8\pi}\int (\mathbf{X}_\alpha^r)^2 \, d\mathbf{r}$$

$$- \frac{e_2^2}{2a_2}\left(1 - \frac{1}{\varepsilon_{\text{op}}}\right) + \frac{e_1 e_2}{\varepsilon_{\text{op}}}\int \frac{|\Psi|^2 \, d\mathbf{r}}{|\mathbf{r}_1 - \mathbf{r}_2|} + m^2\lambda_i - F_0^r \qquad (11.77)$$

where

$$\lambda_i = \tfrac{1}{2}\sum_{i,j} k_{ij}(q_p^i - q_r^i)(q_p^j - q_r^j) \qquad (11.78)$$

A procedure analogous to that adopted in the derivation of F^{*r} is applied to the corresponding free energy of formation ΔF^{*p} of the activated state with the electronic structure of the product from an initial state. One thereby obtains

$$\Delta F^{*p} = \frac{(m+1)^2 C}{8\pi} \int (\mathbf{X}^r - \mathbf{X}^p)^2 \, d\mathbf{r} - \frac{C}{8\pi} \int (\mathbf{X}^p)^2 \, d\mathbf{r} - \frac{e_2^2}{2a_2}\left(1 - \frac{1}{\varepsilon_{op}}\right)$$
$$+ (m+1)\lambda_i - F_0^p \tag{11.79}$$

The free energy of the reaction ΔF^0 for the elementary electron transfer step can be found from the relation (Fig. 11.3)

$$\Delta F^0 = \Delta F^{*o} - \Delta F^{*r} \tag{11.80}$$

Taking the values of $\mathbf{X}^r$ and $\mathbf{X}^p$ from Eq. (11.68) in Eq. (11.77), Marcus[46a] gave the following expression for the free energy of activation for the electron transfer reaction involving the hydrated electron, i.e.,

$$\Delta F^* = \frac{\hbar^2}{2m_e} \int |\nabla \Psi|^2 \, d\mathbf{r}_1 + \frac{(m^2+1)C}{8\pi} \int (\mathbf{X}_\alpha^r)^2 \, d\mathbf{r} - \frac{e_1 e_2}{\varepsilon_{op}} \int \frac{|\Psi|^2 \, d\mathbf{r}_1}{|\mathbf{r}_1 - \mathbf{r}_2|}$$
$$+ \frac{m^2 C e_0^2}{2a_2} - m^2 C e_0^2 \int \frac{|\Psi|^2 \, d\mathbf{r}_1}{|\mathbf{r}_1 - \mathbf{r}_2|} + m^2 \lambda_i - F_0 \tag{11.81}$$

where

$$F_0 = \frac{\hbar^2}{2m_e} \int |\nabla \Psi_0|^2 \, d\mathbf{r}_1 - \frac{C}{8\pi} \int (\mathbf{X}_c^r)^2 \, d\mathbf{r} \equiv F_0^r \tag{11.82}$$

and $\mathbf{X}_\alpha^r = \int |\Psi_0|^2 \mathbf{X}^r \, d\mathbf{r}$.

The solution of Eq. (11.82) was deduced by Marcus[46a] to be of the form

$$F_0^r = -\frac{A e_0^4 C^2}{\hbar^2} \tag{11.83}$$

where A is a dimensionless constant and is -0.0544 as given by Pakar[50] using the variational wave function $\psi_0 = A(1 + \alpha r + \beta r^2)e^{-2r}$ in Eq. (11.82). A value of $A = -0.0488$ was given by Jortner from the hydrogenic 1s wave function $\psi_0 = A e^{-\gamma r}$

Equation (11.81) is applicable for any finite separation $R = |\mathbf{r}_1 - \mathbf{r}_2|$ between reactants. The third term in Eq. (11.81) involving $|\mathbf{r}_1 - \mathbf{r}_2|^{-1}$ is small in comparison with the electronic energy of the polaron, as is the second term involving $|\mathbf{r}_1 - \mathbf{r}_2|^{-1}$ when m is small (≈ 0.1). Due to the spherical nature of $|\Psi|^2$ in the undistorted ground state, the term $\int |\psi|^2 \, d\mathbf{r}/|\mathbf{r}_1 - \mathbf{r}_2|$ need not be expanded in terms of spherical harmonics but is represented simply by $1/R$.

Hence, Eq. (11.81) then simplifies to:

$$\Delta F^* = -\Delta F_0^r - (m^2-1)2F - \frac{e_0 e_2^r}{\varepsilon_s R} + \frac{m^2 Ce_0^2}{2a_2} - \frac{m^2 Ce_0^2}{R} + m^2 \lambda_i - F_0 \tag{11.84}$$

where $F = -(\hbar^2/2m_e)\int |\nabla \psi|^2 \, d\mathbf{r}$, and $2F = (C/8\pi)\int (\mathbf{X}_\alpha^r)^2 \, d\mathbf{r}$.

In his paper, the author then says [46a] "H [the F used in this book] is obtained from the well-known value of H_0 [F_0 used here] in the literature merely by multiplying H_0 by $(1-m^2)^2$." Putting, then, the kinetic energy $F = (1-m^2)^2 F_0$ in (11.84) and rearranging, one gets

$$\Delta F^* = -2m^2(1-m^2)^2 F_0 + [(1-m^2)^2 - 1]F_0$$

$$+ m^2 \left[e_0^2 C\left(\frac{1}{2a_2} - \frac{1}{R}\right) \right] + m^2 \lambda_i + \frac{e_0^2 e_2^r}{\varepsilon_s R} \tag{11.85}$$

11.5.3. Numerical Results of ΔF^* and k_r

From Eq. (11.85), one finds that if m is small (i.e., $m^2 \ll 1$) the free energy of activation becomes (with the neglect of terms containing factors of m^3 and higher terms)

$$\Delta F^* = \omega' + m^2 \lambda \tag{11.86}$$

where

$$\omega' = -\frac{e_0 e_2^r}{\varepsilon_s R} \tag{11.87}$$

and

$$\lambda = -4F_0 + \lambda_i + e_0^2 [(2a_2)^{-1} - R^{-1}](\varepsilon_{op}^{-1} - \varepsilon_s^{-1}) \tag{11.88}$$

or

$$\lambda = \lambda_e + \lambda_2 \tag{11.89}$$

with

$$\lambda_e = -4F_0 - \frac{e^2}{2R}(\varepsilon_{op}^{-1} - \varepsilon_s^{-1}) \tag{11.90}$$

and

$$\lambda_2 = \lambda_i + \tfrac{1}{2}e_0^2 (\varepsilon_{op}^{-1} - \varepsilon_s^{-1})(a_2^{-1} - R^{-1}) \tag{11.91}$$

For water as a solvent at room temperature, the term $(e_0^2/2R) \times (\varepsilon_{op}^{-1} - \varepsilon_s^{-1})$ in Eq. (11.90) becomes $90/R$ kcal mole^{-1} with R (the distance between the center of the solvated ion and the electron at the activated state) in angstroms. Thus, if R is about 6 Å, this contribution to λ_e in Eq. (11.90) is about 15 kcal mole^{-1}. This choice of R is possible since the radius for e(aq) is about 2.5–3 Å, while a radius of a small hydrated cation is typically of the order of 3.5 Å. However, the value of F_0 is presently uncertain, and Marcus[40] chose a value of 15 kcal mole^{-1} because it fitted the free energy of activation ΔF^* and rate constant k_r for electron transfer to Sm^{3+} from a hydrated electron. In the case of Sm^{3+}, ΔF^* is estimated to be about 1, 1.5, or 7 kcal mole^{-1} according as λ_e is taken to be 15, 20, or 45 kcal mole^{-1} with λ_2 to be 40 kcal mole^{-1}.

The experimental value of $k_a = 9 \times 10^{10}$ mole^{-1} sec^{-1} for Sm^{3+} corresponds to ΔF^* of about 0.5 kcal mole^{-1}. Thus, the value of λ_e around 15 kcal mole^{-1} seems slightly favorable (cf. above estimates). But the many uncertainties in the values of λ for the solvated electron make it difficult to make accurate comparison with experiment. It is important to know what λ_e is in Eq. (11.90) and thus what ΔF^* is in Eq. (11.86). We need to evaluate F_0, which in principle can be done.

However, the author does not do this, but says "the value of F_0 is uncertain" and then, instead of leaving the equation unevaluated because he has not evaluated F_0, awards to it a value, the basis for which is not given. This arbitrary choice of the principal parameter effecting the equation decreases to some extent the integrity of the treatment.[46a] The author finally describes his results as "slightly favorable," although the numerical results are based not on the equations of the theory, but upon their replacement by an arbitrary number.

It may be wondered that we have reproduced this work in this book. However, it is often quoted, and it is therefore important to realize what the author has done; specifically, that he has assumed a continuous polarization model, neglected all structure of the solvent, neglected the hydration energy due to the inner shells of the reactant ions, and based the test on validity of comparison with experiment of the theoretical result, the main quantity in which is a number chosen without a stated basis.

11.6. Hydrated Electron in Photoelectrochemical Processes

The hydrated electron can be generated in electrode processes when a cathode is illuminated by light (photon) of suitable frequency. Several authors[13-18] support the formation of the hydrated electron e$^-$(aq) as the intermediate step in many photoelectrochemical reactions, such as hydrogen evolution under illumination and scavenging processes.

Photoelectrochemical effects involving the hydrated electron have been investigated experimentally by Barker and co-workers,[13-16] Delahay and

Srinivasan,[17] and de Levie *et al.*[18] They suggest that reactions of the following types occur when hydrated electrons are formed due to photoemission of electrons at the electrode–electrolyte interface:

Hydrogen evolution reaction

$$H_3O^+ + e^-(aq) \rightarrow H + H_2O \qquad \text{(acidic solution)} \qquad (11.92)$$

$$H_2O + e^-(aq) \rightarrow H + OH^- \qquad \text{(neutral solution)} \qquad (11.93)$$

$$e^-(aq) + e^-(aq) \rightarrow H_2 + 2OH \qquad \text{(bimolecular recombination)} \qquad (11.94)$$

Scavenging process (e.g., for N_2O as scavenger)

$$e^-(aq) + N_2O \rightarrow N_2O^- \qquad (11.95)$$

followed by

$$N_2O^- + H_2O \rightarrow N_2 + OH + OH^- \qquad (11.96)$$

11.6.1. Energetic Condition for the Production of Hydrated Electrons at an Electrode

The condition of electron transfer from a metal electrode to an acceptor (e.g., a hydrated electron cavity site, H_3O^+, N_2O, etc.) is that of Gurney, in which the electron energy in the metal $E_{e(M)}$ must be less than or equal to that in the acceptor, $E_{e(aq)}$, i.e.,

$$E_{e(M)} \leq E_{e(aq)} \qquad (11.97)$$

with respect to vacuum taken as the zero of energy (Fig. 11.3). Only under condition (11.97) can $e^-(aq)$ be formed in an electrochemical process. Let us see whether the condition of Eq. (11.97) is satisfied for the formation of the hydrated electron at a mercury electrode. Assuming the electron to originate from the Fermi level of the metal, the energy $E_{e(M)}$ with respect to that in a vacuum is given by $-\Phi$, where Φ is the work function of the mercury electrode (4.5 eV) when the electrode is at the potential of zero charge (p.z.c.). The acceptor energy level $E_{e(aq)}$ is negative with respect to the zero of energy represented by e(vacuum) and can be determined from the solvation energy of the hydrated electron $S_{e(aq)}$ (~ -1.7 eV). The difference in energy between the levels $E_{e(M)}$ and $E_{e(aq)}$ is determined according to the following process:

$$
\begin{array}{c}
e(M) \xrightarrow{\Delta E} e^-(aq) \\
-\Phi \searrow \quad \swarrow -S_{e(aq)} \\
e(vacuum)
\end{array}
\qquad (11.98)
$$

Hence, for the assigned transition from $E_{e(M)}$ to $E_{e(aq)}$,

$$E_{e(M)} - E_{e(aq)} = \Delta E = \Phi + S_{e(aq)} = 4.5 - 1.7 = 2.8 \text{ eV} \qquad (11.99)$$

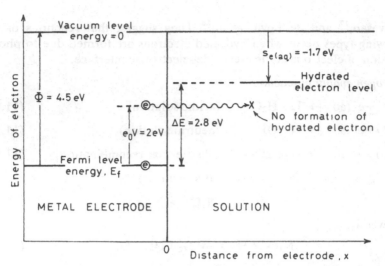

Fig. 11.4. Schematic diagram of energy levels of the electron in the metal electrode and unhydrated electron level in solution at the interface. The diagram indicates that no hydrated electron is formed even at an electrode potential of 2.0 V in the electrochemical process.

which immediately violates the condition (11.97), which implies that $E_M - E_{e(aq)}$ must be negative (for the level E_M must be *above* that of E_e).

An applied cathodic potential of V volts relative to the potential of zero charge is introduced in condition (11.97) as a virtual shift in the electron energy level of the metal (cf. Fig. 11.4), i.e.,

$$\Phi + e_0 V \geqslant E_{e(aq)} \tag{11.100}$$

where $\Phi + e_0 V = E_M$ in this situation. The corresponding value of ΔE is

$$\Delta E = E_M - E_e = \Phi + eV + S_{e(aq)} \tag{11.101}$$

A schematic energy level diagram for a typical cathodic potential of 2 V with respect to the p.z.c. is shown in Fig. 11.4.

The associated value of ΔE (2.8 eV) indicates that even at the cathodic potential of 2 V (with respect to the p.z.c.) the condition (11.97) is not satisfied. *Thus, one concludes that the formation of $e(aq)$ electrochemically is highly improbable, not only at the p.z.c., but at a cathodic potential of as much as 2 V relative to the p.z.c.*† According to Eq. (11.100), it is observed that the

---

† We involve ourselves in the assumption that the p.z.c. is not different sensibly from the absolute potential. But this is an approximation with a decreasing validity, and the numerical estimates need reconsideration in view of the calculations of Argade and Bockris,[56] Trasatti,[57] and Bockris and Habib.[58]

formation e(aq) is theoretically possible when the cathodic potential is [see Eq. (11.93)]

$$e_0 V \geqslant \Delta E = 2.8 \text{ eV} \qquad (11.102)$$

However, such high cathodic potentials with respect to p.z.c. (i.e., with respect to rational potential) are not achieved in normal electrochemical processes.

11.6.2. Can There Be an Electrochemical Production of the Hydrated Electron?

The view that the hydrated electron is an intermediate product in the cathodic evolution of hydrogen and in metal dissolution in *aqueous* solution was originated in the papers of Pyle and Roberts[8] and Walker.[7,9,52-54] Walker obtained experimental evidence for this view by means of an ingenious multiple-reflectance light absorption experiment involving laser light of wavelength 6330 Å. However, Conway[12] suggested that such experiments need further verification. The reason that Walker[52,53] was able apparently (cf. Conway[12]) to detect e(aq) spectrophotometrically may be related to the illumination of the silver electrode surface with light wavelength of 6330 Å. Thus, the incident radiation represents a means of exciting surface electrons of the metal by 1.96 eV relative to the Fermi level. The additional influence of the applied cathodic potential of 1.1 V (p.z.c.) yields a total excitation of 3.06 eV, sufficient to fulfil the condition (11.97). Thus, e^-(aq) formation becomes possible, although doubt remains as to whether the intensity of the illumination was sufficient to produce the observed effects.

An alternative explanation is that the absorption observed at 6330 Å in Walker's experiment was due to potential-modulated reflectance[7] of silver surface itself. Schindelwoff (unpublished experiment[59]) carried out similar studies and found that optical changes induced by cathodic polarization were more likely to be connected with the reflectance spectrum of silver than with the cathodically formed e(aq). Thus, it seems improbable that hydrated electrons are formed electrochemically without the aid of photons. A facile test would be to use a high-speed rotating electrode. If hydrated electrons were the cause of electrode reactions, the active zone would be outside the Prandtl layer and hence subject in reaction velocity to rotation rate.

11.6.3. Photoelectrochemical Production of the Hydrated Electron

A subsection has indicated that the production of hydrated electrons by electrochemical methods is improbable [Fig. 11.4 and Eq. (11.97)].

However, the energetic feasibility of the production of e(aq) by illuminating the cathode with light (photons) of suitable frequency is yet to be considered.

When an electrode is illuminated by light of energy $h\nu$, the energy of the electron (which has absorbed the photon) is increased by $h\nu$, and the effective work function corresponding to the photoelectron becomes $(\Phi - h\nu)$. Hence, for the photoelectrochemical process, condition (11.100) is modified to

$$(\Phi - h\nu) \leqslant E_{e(aq)} \tag{11.103}$$

and in the presence of applied potential V, the condition becomes

$$(\Phi - h\nu + e_0 V) \leqslant E_{e(aq)} \tag{11.104}$$

One notices from Eq. (11.104) that photons of higher frequency are required in the absence of an applied potential for condition (11.97) to be satisfied. Figures 11.5 and 11.6 depict electron emission in the presence of photon energy $h\nu_1$ only and in the presence of photon energy $h\nu_2$ in addition to the cathodic potential of 1 V (with respect to the p.z.c.), respectively.

From Figs. 11.5 and 11.6, $h\nu_1 > h\nu_2$ in order that condition (11.93) be satisfied. Simple calculations for the mercury electrode ($\Phi = 4.5$ eV) show that a minimum photon energy of $h\nu_1 = 2.8$ eV is required for condition (11.97) to be satisfied, whereas a photon energy of $h\nu_2 = 1.8$ eV is required at the cathodic potential of 1 V to satisfy the condition. Thus, the formation

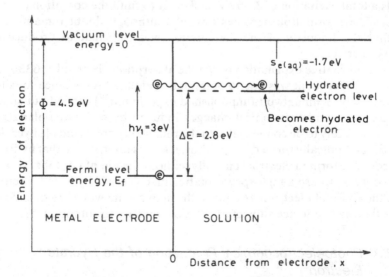

Fig. 11.5. Schematic diagram of energy levels of electron in the metal electrode and the hydrated electron level in solution at the interface. The diagram indicates hydrated electron formation at the photon energy $h\nu_1 = 3$ eV in the photoelectrochemical process.

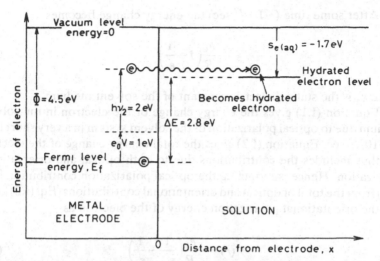

Fig. 11.6. Schematic diagram of energy levels of electron in the metal electrode and the hydrated electron level in solution at the interface. The diagram indicates the formation of a hydrated electron at the photon energy $h\nu_2 = 2$ eV and potential of 1 V in the photoelectrochemical process.

of e(aq) is energetically possible and confirmed in photoelectrochemical experiments by such workers as Barker *et al.*,[13–16] Heyrovsky,[55] and de Levie.[18] Barker *et al.*[13] envisaged the formation of e(aq) at some distance from the electrode in three distinct steps: (a) photoejection of the electron from electrode, (b) thermalization of electron (removal of excess kinetic energy), and finally, (c) its hydration. The last is considered by Barker *et al.*[13] to be the slow process. Heyrovsky[55] interpreted the photoelectrochemical effect in terms of hydrated electron formation subsequent to the photoejection of an electron from the cathode.

Appendix 11.I. Calculation of Orientational Polarization Energy of an Electron in a Solvent

From the viewpoint of continuum theory, when one puts an electron from vacuum into a solvent medium the instantaneous energy change in a very short time ($\sim 10^{-15}$ sec) can be expressed in terms of Born's solvation energy as

$$-\frac{e_0^2}{R}\left(1-\frac{1}{\varepsilon_{op}}\right) \tag{I.1}$$

where R is the cavity radius of the electron in the solvent, ε_{op} is the optical dielectric constant of the solvent medium.

After some time ($\sim 10^{-12}$ sec) the energy change becomes

$$-\frac{e_0^2}{R}\left(1-\frac{1}{\varepsilon_s}\right) \tag{I.2}$$

where ε_s is the static dielectric constant of the solvent medium.

Equation (I.1) gives the energy change of the electron in the solvent medium due to optical polarization of the solvent system in a very short time of $\sim 10^{-15}$ sec. Equation (I.2) gives the total energy change of the electron and thus includes the contributions due to both optical and orientational polarization. Hence, subtracting the optical polarization contribution [Eq. (I.1)] from the total of optical and orientational contributions [Eq. (I.2)], one gets the orientational polarization energy of the electron as

$$V(r) = -\frac{e_0^2}{R}\left(\frac{1}{\varepsilon_{op}}-\frac{1}{\varepsilon_s}\right) \tag{I.3}$$

Appendix 11.II. Solution of Poisson Equation (11.2)

We need to prove that

$$f = -\frac{e_0}{r} + \frac{e_0(1+\lambda r)\,e^{-2\lambda r}}{r} \tag{II.1}$$

is a solution of

$$\nabla^2 f_i = 4\pi e_0 |\psi_i|^2 \tag{II.2}$$

where

$$\psi_i = (\lambda^3/\pi)^{1/2}\, e^{-\lambda r} \tag{II.3}$$

$$\nabla = \frac{1}{r^2}\frac{\partial}{\partial r}\left(r^2\frac{\partial}{\partial r}\right) \tag{II.4}$$

Using the value of ψ_i from Eq. (II.3) and ∇ from Eq. (II.4), we get

$$\frac{1}{r^2}\frac{d}{dr}\left(r^2\frac{df}{dr}\right) = 4e_0\lambda^3\, e^{-2\lambda r}$$

$$= \beta\, e^{-2\lambda r} \tag{II.5}$$

where

$$\beta = 4e_0\lambda^3 \tag{II.6}$$

Let us take a homogeneous solution as

$$\frac{1}{r^2}\frac{d}{dr}\left(r^2\frac{df}{dr}\right)=0 \tag{II.7}$$

or

$$\frac{d}{dr}\left(r^2\frac{df}{dr}\right)=0 \tag{II.8}$$

or

$$r^2\frac{df}{dr}=C_1 \tag{II.9}$$

or

$$\frac{df}{dr}=\frac{C_1}{r^2} \tag{II.10}$$

Hence, the homogeneous solution is

$$f_{HS}=-\frac{C_1}{r}+C_2 \tag{II.11}$$

where C_1 and C_2 are constants.

So we can have two solutions, $1/r$ and 1, for f as a homogeneous solution. The particular solution will be

$$f_{PS}=u_i\frac{1}{r}+(u_2\cdot 1) \tag{II.12}$$

where u_1 and u_2 are variational coefficients.

Now we should get

$$u_1'\frac{1}{r}+(u_2'\cdot 1)=0 \tag{II.13}$$

and

$$u_1'\frac{\partial}{\partial r}\left(\frac{1}{r}\right)+u_2'\frac{\partial}{\partial r}(1)=\beta\,e^{-2\lambda r} \tag{II.14}$$

where the prime indicates differentiation with respect to r. Or,

$$\frac{u_1'}{r}+u_2'=0 \tag{II.15}$$

$$u_1'-\frac{1}{r^2}+0=\beta\,e^{-2\lambda r} \tag{II.16}$$

Hence, from Eq. (II.16) we get

$$-u_1'/r^2 = \beta\, e^{-2\lambda r} \tag{II.17}$$

or

$$u_1' = -\beta r^2\, e^{-2\lambda r} \tag{II.18}$$

Hence,

$$u_1 = -\beta \int r^2\, e^{-2\lambda r}\, dr$$

$$= -\beta\left(-\frac{r^2}{2\lambda}\, e^{-2\lambda r} + \frac{2}{2\lambda}\int r\, e^{-2\lambda r}\, dr\right)$$

$$= -\beta\left[-\frac{r^2}{2\lambda}\, e^{-2\lambda r} + \frac{1}{\lambda}\left(-\frac{r}{2\lambda}\, e^{-2\lambda r} - \int \frac{e^{-2\lambda r}}{-2\lambda}\, dr\right)\right]$$

$$= -\beta\left(-\frac{r^2}{2\lambda}\, e^{-2\lambda r} - \frac{r}{2\lambda^2}\, e^{-2\lambda r} - \frac{e^{-2\lambda r}}{4\lambda^3}\right) \tag{II.19}$$

Putting value of $\beta = 4e_0\lambda^3$, we get,

$$u_1 = \frac{4e_0\lambda^3 r^2}{2\lambda}\, e^{-2\lambda r} + \frac{4e_0\lambda^3 r}{2\lambda^2}\, e^{-2\lambda r} + \frac{4e_0\lambda^3\, e^{-2\lambda r}}{4\lambda^3} \tag{II.20}$$

Therefore,

$$u_1 = 2e_0\lambda^2 r^2\, e^{-2\lambda r} + 2e_0\lambda r\, e^{-2\lambda r} + e_0\, e^{-2\lambda r} \tag{II.21}$$

From Eq. (II.15), we get

$$u_2' = -u_1'/r \tag{II.22}$$

Using u_1' from Eq. (II.18), one gets from Eq. (II.22)

$$u_2' = \beta r\, e^{-2\lambda r} \tag{II.23}$$

$$u_2 = \beta \int r\, e^{-2\lambda r}\, dr$$

$$= \beta\left(-\frac{r}{2\lambda}\, e^{-2\lambda r} - \int \frac{e^{-2\lambda r}}{-2\lambda}\, dr\right)$$

$$= \beta\left(-\frac{r}{2\lambda}\, e^{-2\lambda r} - \frac{1}{4\lambda^2}\, e^{-2\lambda r}\right) \tag{II.24}$$

Putting the value of $\beta = 4e_0\lambda^3$, we get

$$u_2 = -e\frac{4e_0\lambda^3 r}{2\lambda}\, e^{-2\lambda r} - \frac{4e_0\lambda^3}{4\lambda^2}\, e^{-2\lambda r} \tag{II.25}$$

Hence, the particular solution is

$$f_{PS} = u_1/r + u_2$$

$$= 2e_0\lambda^2 r e^{-2\lambda r} + 2e_0\lambda e^{-2\lambda r} + \frac{e_0}{r} e^{-2\lambda r} - ze_0\lambda^2 r e^{-2\lambda r} - e_0\lambda e^{-2\lambda r}$$

$$= e_0\lambda e^{-2\lambda r} + \frac{e_0}{r} e^{-2\lambda r}$$

$$= \frac{1}{r}[e_0\lambda r + e_0] e^{-2\lambda r}$$

$$= \frac{e_0(1+\lambda r) e^{-2\lambda r}}{r} \tag{II.26}$$

We need the general solution, which is the sum of homogeneous and particular solution, i.e.,

$$f_{gen} = f_{HS} + f_{PS}$$

$$= -\frac{C_1}{r} + C_2 + \frac{e_0(1+\lambda r)^{-2\lambda r}}{r} \tag{II.27}$$

To find the C_1 and C_2 in Eq. (II.27), we use the boundary condition that the solution f, which is physically the potential, is zero as $r \to \infty$. Hence,

$$f = C_2 = 0 \tag{II.28}$$

C_1 can be obtained from the definition of potential

$$f = C_1/r = \text{potential}$$

Hence, C_1 should be equal to electric charge. Hence,

$$C_1 = e_0 \tag{II.29}$$

Hence, putting the value of C_1 for Eq. (II.29) and C_2 from Eq. (II.28) in Eq. (II.27), we get

$$f = -\frac{e_0}{r} + \frac{e_0(1-\lambda r) e^{-2\lambda r}}{r} \tag{II.30}$$

which is the result.

Appendix 11.III.   Proof of Equation for Average Polarization Energy [Eq. (11.66b)]

$$\langle F^*_{pol} \rangle = \int F^*_{pol}(\mathbf{r}_1)|\Psi|^2 \, d\mathbf{r}_1$$

$$= \frac{m^2 C}{8\pi} \int (\mathbf{X}^r_\alpha - \mathbf{X}^p)^2 \, d\mathbf{r} - \frac{C}{8\pi} \int (\mathbf{X}^r_\alpha)^2 \, d\mathbf{r}$$

$$- \frac{e_2^2}{2a_2}\left(1 - \frac{1}{\varepsilon_{op}}\right) + \frac{e_1 e_2}{\varepsilon_{op}} \int \frac{|\Psi|^2}{|\mathbf{r}_1 - \mathbf{r}_2|} \, d\mathbf{r}_1 \qquad \text{(III.1)}$$

We know from Eq. (11.66) that the polarization energy operator

$$F_{pol}(\mathbf{r}) = -\int \mathbf{P} \cdot \mathbf{X} \, d\mathbf{r} + \frac{2\pi}{C} \int \mathbf{P}^2 \, d\mathbf{r} - \frac{e_2^2}{2a_2}\left(1 - \frac{1}{\varepsilon_{op}}\right) + \frac{e_1 e_2}{\varepsilon_{op}|\mathbf{r}_1 - \mathbf{r}_2|} \qquad \text{(III.2)}$$

where $\mathbf{P}$ can be written from Eq. (11.75) as

$$\mathbf{P} = \frac{C}{4\pi}[(m+1)\mathbf{X}^r_\alpha - m\mathbf{X}^p] \qquad \text{(III.3)}$$

where

$$\mathbf{X}^r_\alpha = \int |\Psi|^2 \mathbf{X}^r \, d\mathbf{r}_1 \qquad \text{(III.4)}$$

Using Eq. (III.3) we can write

$$\mathbf{P}^2 = \left(\frac{C}{4\pi}\right)^2 m^2 (\mathbf{X}^r_\alpha - \mathbf{X}^p)^2 + 2\left(\frac{C}{4\pi}\right)^2 m(\mathbf{X}^r_\alpha - \mathbf{X}^p)\mathbf{X}^r_\alpha + \left(\frac{C}{4\pi}\right)^2 (\mathbf{X}^r_\alpha)^2 \qquad \text{(III.5)}$$

Using (III.2) in (III.1), we get

$$\langle F^*_{pol} \rangle = -\int \mathbf{P} \cdot \int \mathbf{X}|\Psi|^2 \, d\mathbf{r}_1 \, d\mathbf{r} + \frac{2\pi}{C} \int \mathbf{P}^2 \int |\Psi|^2 \, d\mathbf{r}_1 \, d\mathbf{r}$$

$$- \frac{e_2^2}{2a_2}\left(1 - \frac{1}{\varepsilon_{op}}\right) \int |\Psi|^2 \, d\mathbf{r}_1 + \frac{e_1 e_2}{\varepsilon_{op}} \int \frac{|\Psi|^2}{|\mathbf{r}_1 - \mathbf{r}_2|} \, d\mathbf{r}_2 \qquad \text{(III.6)}$$

The first integral of Eq. (III.6) can be expressed using Eqs. (III.3) and (III.4) as

$$-\int \mathbf{P} \cdot \int \mathbf{X}|\Psi|^2 \, d\mathbf{r}_1 \, d\mathbf{r} = -\int \mathbf{P} \cdot \mathbf{X}^r_\alpha \, d\mathbf{r}$$

$$= -\frac{C}{4\pi} m \int (\mathbf{X}^r_\alpha - \mathbf{X}^p)\mathbf{X}^r_\alpha \, d\mathbf{r} - \frac{C^2}{4\pi}(\mathbf{X}^r_\alpha)^2 \, d\mathbf{r} \qquad \text{(III.7)}$$

The second integral of Eq. (III.6) can be expressed using Eq. (III.5) as

$$\frac{2\pi}{C} \int \mathbf{P}^2 \int |\Psi|^2 \, d\mathbf{r}_1 \, d\mathbf{r} = \frac{m^2 C}{8\pi} \int (\mathbf{X}_\alpha^r - \mathbf{X}^p)^2 \, d\mathbf{r} + \frac{mC}{4\pi} \int (\mathbf{X}_\alpha^r - \mathbf{X}^p)\mathbf{X}_\alpha^r \, d\mathbf{r}$$

$$+ \frac{C}{8\pi} \int (\mathbf{X}_\alpha^r)^2 \, d\mathbf{r} \tag{III.8}$$

The integral $\int |\Psi|^2 \, d\mathbf{r}_1$ in the third term of Eq. (III.6) becomes unity, and we keep the fourth term of Eq. (III.6) as such. Hence, putting the value of integrals from Eqs. (III.7) and (III.8) into Eq. (III.6), we get the value $\langle F_{pol}^* \rangle$ as given in Eq. (III.1).

References

1. E. J. Hart (ed.), *Advances in Chemistry Series*, Vol. 50, American Chemical Society, New York (1965).
2. E. J. Hart and M. Anbar, *The Hydrated Electron*, John Wiley and Sons, New York (1970).
3. G. Scholes, *Ann. Repts. Chem. Soc.* **67A**, 169 (1970).
4. L. Kevan, *Advances in Radiation Chemistry* (M. Burton and J. L. Magee, eds.), John Wiley and Sons, New York (1974).
5. E. J. Hart and J. W. Boag, *J. Amer. Chem. Soc.* **89**, 4090 (1962).
6. J. W. Boag and E. J. Hart, *Nature* **197**, 45 (1963).
7. D. C. Walker, *Anal. Chem.* **39**, 896 (1967).
8. T. Pyle and C. Roberts, *J. Electroanal. Chem.* **115**, 247 (1968).
9. I. A. Kennedy and D. C. Walker, *Electroanalytical Chemistry* (A. J. Bard, ed.), Vol. 5, Marcel Dekker, New York (1971).
10. L. I. Antropov and I. Nauki, *Elektrokhim.* **6**, 5 (1971).
11. L. I. Krishtalik, *Electrochem. Acta* **21**, 693 (1976).
12. B. E. Conway, *Modern Aspects of Electrochemistry* (B. E. Conway and J. O'M. Bockris, eds.), Vol. 7, Chapter 2, Plenum Press, New York (1972).
13. G. C. Barker, A. W. Gardner, and D. C. Sammon, *J. Electrochem. Soc.* **113**, 1183 (1966).
14. G. C. Barker, *Trans. Faraday Soc.* **66**, 1498, 1509 (1970).
15. G. C. Barker and V. Concialini, *J. Electroanal. Chem.* **45**, 320 (1973).
16. G. C. Barker, B. Stringer, and M. J. Williams, *J. Electroanal. Chem.* **51**, 305 (1974).
17. P. Delahay and V. S. Srinivasan, *J. Phys. Chem.* **20**, 520 (1966).
18. R. de Levie and J. E. Krenser, *J. Electroanal. Chem.* **21**, 221 (1969).
19. J. Jortner, *J. Chem. Phys.* **30**, 834 (1959).
20. J. Jortner, *Molec. Phys.* **5**, 257 (1962).
21. J. Jortner, *Radiation Res. Suppl.* **4**, 24 (1964).
21a. M. Tachiya and H. Watanabe, *J. Chem. Phys.* **66**, 3056 (1977).
22. M. Tachiya, Y. Tabata, and K. Oshima, *J. Phys. Chem.* **77**, 263 (1973).
23. M. Tachiya and A. Mozumder, *J. Chem. Phys.* **60**, 3037 (1974); **63**, 1959 (1975).
24. M. Tachiya and H. Watanabe, *J. Chem. Phys.* **66**, 3056 (1977).
25. K. Iguchi, *J. Chem. Phys.* **48**, 1735 (1968).
26. J. H. Baxendale, *Radiation Res. Suppl.* **4**, 139 (1964).
27. W. C. Gottschal and E. J. Hart, *J. Phys. Chem.* **71**, 2102 (1967).
28. K. Fueki, D. F. Feng, and L. Kevan, *J. Phys. Chem.* **74**, 1976 (1970).

29. K. Fueki, D. F. Feng, L. Kevan, and R. E. Christoffersen, *J. Phys. Chem.* **75**, 2297 (1971).
30. K. Fueki, D. F. Feng, and L. Kevan, *J. Amer. Chem. Soc.* **95**, 1398 (1973).
31. K. Fueki, D. F. Feng, and L. Kevan, *J. Phys. Chem.* **80**, 1381 (1976).
32. D. F. Feng, D. Ebbing, and L. Kevan, *J. Chem. Phys.* **61**, 249 (1974).
33. J. O'M. Bockris, S. U. M. Khan, and D. B. Matthews, *J. Res. Inst. Catal. Hokkaido Univ.* **22**, 1 (1974).
34. S. Schlick, P. A. Narayana, and L. Kevan, *J. Chem. Phys.* **64**, 3153 (1976).
35. M. Natori and T. Watanabe, *J. Phys. Soc. Japan* **21**, 1573 (1966).
36. J. Moorse, L. A. Young, and E. S. Huurwitz, *Phys. Rev.* **48**, 948 (1935).
37. M. Natori, *J. Phys. Soc. Japan* **24**, 1735 (1968); **27**, 1309 (1969).
38. M. Weissmann and N. V. Cohan, *Chem. Phys. Lett.* **7**, 445 (1970).
39. M. Weissmann and N. V. Cohan, *J. Chem. Phys.* **59**, 1385 (1973).
40. G. Howat and B. C. Webster, *J. Phys. Chem.* **76**, 3714 (1972).
41. S. Ray, *Chem. Phys. Lett.* **11**, 573 (1971).
42. L. Onsager and M. Dupuis, *Electrolytes* (B. Pasco, ed.), Pergamon Press, London (1972).
43. C. R. Gatz, *Introduction to Quantum Chemistry*, Charles E. Morrill, Columbus, Ohio (1971).
44. S. Ishimaru, H. Kato, T. Yamabe, and K. Fukui, *J. Phys. Chem.* **77**, 1450 (1973).
45. G. Nilson, *J. Chem. Phys.* **56**, 3427 (1972).
46a. R. A. Marcus, *J. Chem. Phys.* **43**, 3477 (1965).
46b. P. George and J. S. Griffith, *The Enzymes*, Vol. 1, Chapter 8, p. 347 (P. D. Boyer, H. Lardy, and K. Myrberg, eds.), Academic Press, New York (1959).
47. R. A. Marcus, *Advan. Chem. Ser.* **50**, 138 (1965).
48. R. A. Marcus, *J. Chem. Phys.* **24**, 966 (1956).
49. R. A. Marcus, *Ann. Rev. Phys. Chem.* **15**, 155 (1964).
50. S. I. Pekar, *Untersuchungen über die Elecktronentheorie der Kristalle*, Academische Verlag, Berlin (1954).
51. K. Kohru and S. Annaka, *J. Cryst. Soc. Japan* **7**, 21 (1965).
52. D. C. Walker, *Can. J. Chem.* **44**, 2226 (1966).
53. D. C. Walker, *Can. J. Chem.* **45**, 807 (1967).
54. D. C. Walker, *Quart. Rev.* **21**, 79 (1967).
55. M. Heyrovsky, *Proc. Roy. Soc.* (*London*) **A301**, 411 (1967).
56. S. Argade and J. O'M. Bockris, *J. Chem. Phys.* **49**, 5133 (1968).
57. S. Trasatti, *J. Electro-analyt. Chem.* **33**, 351 (1971).
58. J. O'M. Bockris and M. A. Habib, *J. Electro-analyt. Chem.* **68**, 367 (1976).
59. P. Schindelwoff, quoted by B. E. Conway.[12]
60. A. C. Wahl, R. H. Land, and F. T. Janis, KRAG (A Computing System for Polyatomic Molecules), Argonne National Laboratory (1971).

Photoelectrochemical Kinetics

12.1. Introduction

In this chapter we will discuss the phenomenon of ejection of electrons from an electrode material (metal and semiconductor) due to the influence of light (photons of electromagnetic radiation) when it is exposed to a light source of appropriate energy.

First, we will discuss the theories that explain the observed facts regarding the photoeffect at an electrode when it is in free space (vacuum). Second, we will focus our attention on an important aspect of the photoeffect at an electrode when it is dipped into a solution and electrochemical phenomena take place. These aspects are of great interest to electrochemists. This is because a quite different current–potential behavior is observed when the electrode is illuminated by light (the well known Tafel law is not followed). Further, by irradiating semiconductors, electrochemists can produce significant amounts of energy in the form of electricity, and, simultaneously, hydrogen from water. Hence, we give in some detail the various theories that clarify the fundamental aspects responsible for the behavior of the photocurrent both at the metal and semiconductor electrodes in solution.

12.2. Rate of Photoemission into a Vacuum

Investigations of the rate of photoemission into a vacuum for different metals were carried out by several authors.[1-4] Such photocurrents at the metal–vacuum interface obey the square law of Fowler.[5] According to this law, the photocurrent I_p at the metal–vacuum interface is proportional to the square of the frequency of the incident light.

Fowler[5] gave a mathematical expression for the photoemission current on the basis that the number of electrons emitted is proportional to the number of electrons per unit volume of the metal the kinetic energy of which

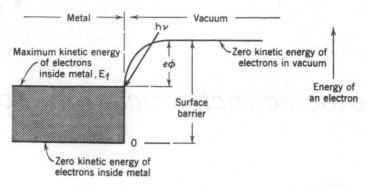

Fig. 12.1. Electron energy relation at the surface of a metal.

(normal to the surface) augmented by light energy $h\nu$ is sufficient to overcome the potential step (work function) at the surface (see Fig. 12.1). Fowler named these electrons the *available electrons*. Thus, he did not calculate the number of electrons reaching the surface per unit time. He considered a free-electron model and neglected internal electron–electron and electron–phonon scattering, etc. It is interesting to note that at the time Fowler wrote the paper[5] (1931), it was not thought necessary to treat specifically the absorption of the photon and the relation between the distance inside the electrode and the intensity of the remaining photons (we shall introduce this later, see Eq. 12.32).

The electron gas[6] in the metal will obey the Fermi–Dirac statistics, so that the number of electrons per unit volume having velocity components in the ranges u to $u + du$, v to $v + dv$, and w to $w + dw$ (u being the velocity component normal to the surface) is expressed by the following formula

$$n(u, v, w)\, du\, dv\, dw = 2\left(\frac{m}{h}\right)^3 \frac{du\, dv\, dw}{\exp\{[\tfrac{1}{2}m(u^2 + v^2 + w^2) - E_f]/kT\} + 1} \quad (12.1)$$

where m is the mass of the electron, h is Planck's constant, and E_f is the electron energy corresponding to the Fermi level in the metal.

However, one needs to determine the number per unit volume $n(u)\, du$ with a velocity component u normal to the surface in the range u, $u + du$. To find this number, $n(u)\, du$, it is convenient to transform the expression (12.1) to an equation in terms of spherical polar coordinates and integrate over the velocity components v and w (corresponding to the y and z axes) but not over the velocity component u (which corresponds to the x axis, perpendicular to electrode surface).[7] Thus,

$$n(u)\, du = 2\left(\frac{m}{h}\right)^3 du \int_0^\infty \int_0^{2\pi} \frac{d\xi\, d\theta}{\exp\{[\tfrac{1}{2}m(u^2 + \xi^2) - E_f]/kT\} + 1} \quad (12.2)$$

where

$$\xi^2 = v^2 + w^2 \tag{12.3}$$

and θ is the angle between the direction of the electron and the axis perpendicular to the surface. After integration with respect to ξ and θ, Eq. (12.2) becomes

$$n(u)\,du = \frac{4\pi m^2 kT}{h^3} \ln\{1 + \exp[(E_f - \tfrac{1}{2}mu^2)/kT]\}\,du \tag{12.4}$$

Since the photocurrent is supposed to be proportional to the available electrons per unit volume, N_a, that have sufficient energy to overcome the surface potential step (work function Φ), Fowler integrated Eq. (12.4) from u_0 to ∞, where u_0 is the minimum velocity of an electron needed to jump over Φ and is obviously given (with an energy zero at the bottom of the conduction band) as

$$u_0 = [2(E_f + \Phi)/m]^{1/2} \tag{12.5}$$

Hence, for photon energy, $h\nu$, one gets N_a:

$$N_a = \frac{4\pi m^2 kT}{h^3} \int_{u_0}^{\infty} \ln\{1 + \exp[(E_f + h\nu - \tfrac{1}{2}mu^2)/kT]\}\,du \tag{12.6}$$

Fowler[5] then wrote the equation for the photocurrent as

$$I_p \propto N_a = \frac{4\pi m^2 kT}{h^3} \int_{u_0}^{\infty} \ln\{1 + \exp[(E_f + h\nu - \tfrac{1}{2}mu^2)/kT]\}\,du \tag{12.7}$$

Expression (12.7) in terms of velocity when converted to the variable of energy using $u = (2E/m)^{1/2}$ becomes

$$I_p \propto \frac{4\pi m^2 kT}{h^3}(2m)^{-1/2} \int_{E_f + \Phi}^{\infty} \ln\{1 + \exp[(E_f + h\nu - E)/kT]\}E^{-1/2}\,dE \tag{12.8}$$

Fowler expanded the logarithm of Eq. (12.7) and integrated term by term to obtain the following analytical expression for photocurrent in the limiting condition of $T \to 0$:

$$I_p = \begin{cases} A\dfrac{(h\nu - \phi)^2}{(E_f + \phi - h\nu)^{1/2}} & \text{for } h\nu > \phi \\[2ex] 0 & \text{for } h\nu < \phi \end{cases} \tag{12.9}$$

where A is a constant, which includes factors that make correct the units of Eq. (12.9), which should be in Coulombs per square centimeter per second.

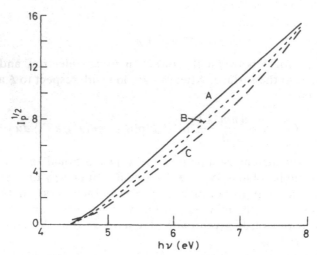

Fig. 12.2. The dependence of the square root of the photocurrent I_p at the metal–vacuum interface on the radiation energy $h\nu$ according to: (A) the corrected Fowler function, Eq. (12.10); (B) the uncorrected Fowler function, Eq. (12.8); (C) the analytical Fowler function, Eq. (12.9). A value of 4.5 eV was used for the vacuum electronic work function Φ.

On the basis of expression (12.9), Fowler stated a square law between photocurrent and frequency of the incident light, though Eq. (12.9) suggests that I_p is not exactly proportional to the square of the light energy $(h\nu)^2$ (Fig. 12.2) (see Matthews and Khan[7]).

However, if one considers the transfer from Eq. (12.6) to Eq. (12.7), the proportionality sign is found to cover a change in dimensions. Thus, N_a is a number per unit volume, and this cannot be a rate, which in the present system is a number per unit area per second. There is no further explicit treatment in Fowler's expression of this aspect, which is terminated by use of the inexplicit proportionality constant A.

Matthews and Khan[7] corrected this aspect of the theory of emission to a vacuum. They multiplied N_a of Eq. (12.7) by u, the component velocity of electrons traveling toward the surface, and thus obtained an expression in terms of number arriving per unit area of the surface per unit time, i.e., the rate of arrival of photoelectrons at the surface from the interior of the metal with energy sufficient to overcome the barrier.† Thus, multiplying by the

---

† The correction of the Fowler expression to take into account the *rate* of electron approach to the surface, rather than merely calculating the number of electrons in appropriate states, is a step in the right direction. Thus, this change is not merely one in which the faulty dimensional properties of the Fowler equation are corrected, but it changes the effect of energy (or frequency) of the incident light because the electrons are going along the x axis toward the surface at a larger transitional velocity after they have been energized by a photon of higher energy.

electronic charge e_0, one obtains

$$I_p = \frac{4\pi m^2 kTe_0}{h^3} \int_{u_0}^{\infty} \ln\{1 + \exp[(E + h\nu - \tfrac{1}{2}mu^2)/kT]\}u\, du \quad (12.10)$$

Transforming expression (12.10) to one in terms of energy, it follows that

$$I_p = \frac{4\pi mkTe_0}{h^3} \int_{E_f+\Phi}^{\infty} \ln\{1 + \exp[(E_f + h\nu - E)/kT]\}\, dE = e_0 \int_{E_f+\Phi}^{\infty} N(E)\, dE$$
$$(12.11)$$

where

$$N(E)\, dE = \frac{4\pi mkT}{h^3} \ln\{1 + \exp[(E_f + h\nu - E)/kT]\}\, dE \quad (12.12)$$

Expressions (12.7) and (12.11) were numerically integrated by Matthews and Khan[7] and the results are shown as in Fig. 12.2. Matthews and Khan's expression (Eq. 12.11) has the observed square law behavior and yields better threshold energy $h\nu_0$ than that of Fowler. This is because expression (12.11) gives a linear $(I_p^{1/2}$ vs. $h\nu)$ plot that is easily extrapolatable to get $h\nu_0$ when $I_p = 0$.

Moreover, after analytical integration of the corrected expression (12.11), Matthews and Khan[7] obtained the following expression for the photocurrent under the condition $h\nu > \phi$:

$$I_p = \frac{2\pi me_0(h\nu - \phi)^2}{h^3} \quad (12.13)$$

Expression (12.13) shows that the square root of the photocurrent in vacuum is linear with frequency, as observed.

12.3. Rate of Photoemission into a Vacuum under an Electric Field

Photoemission into a vacuum in the presence of an applied electric field is different from the simple emission and gives some approach to the electrochemical case, since in the former case, the electron may tunnel through the barrier created by the image potential and the field at the interface. A similar tunneling contribution is expected in the case of electron transfer at the electrode–solution interface.[7,8] In the absence of an applied electric field, the barrier height is equal to the work function of the metal but the barrier width is infinite (Fig. 12.1), and no electron tunneling is possible. All electrons must go over the top of the barrier.

However, in the presence of an applied electric field, the barrier becomes thin enough to permit electron tunneling. Moreover, the applied field in combination with the image potential diminishes the barrier height (Fig. 12.3). Hence, the expression for the photocurrent is modified and incorporates both the number of electrons $N(E)\, dE$ in the metal with an energy between E and $E + dE$ that impinge from within the metal onto unit area of the metal surface at right angles to it in unit time and the probability of tunneling of the electron $P_T(E)$ at an electron energy E. The convention here is to take the energy at the bottom of the conduction as zero and not the energy of an electron in vacuum as zero, the normal convention in electrochemistry.

Thus, the expression for the photoemission current in vacuum in the presence of applied field X can be written as[7]

$$I_{p,x} = e_0 \int P_T(E)N(E)\, dE \tag{12.14}$$

where e_0 is the electronic charge and $P_T(E)$ is the tunneling probability through the barrier of height U_m and width l and is given by

$$P_T(E) = \exp\left[-\frac{\pi^2 l}{h}[2m(U_m - E)]^{1/2}\right] \tag{12.15}$$

which is the WKB tunneling probability expression for a parabolic barrier (the effect of field will be introduced in barrier height U_m and width l). $N(E)\, dE$ is given in Eq. (12.12).

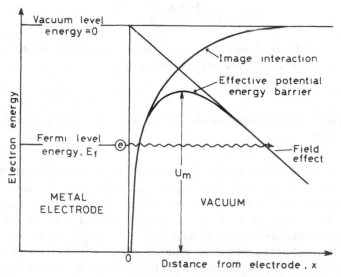

Fig. 12.3. The potential energy barrier for the electron at the electrode (metal)–vacuum interface.

In this case, an electron can escape from the metal surface *both* by tunneling across the barrier or by going over the top of the barrier. Hence, the integral in Eq. (12.14) is expressed as two terms:

$$I_p = e_0 \int_0^{U_m} P(E)N(E) \, dE + e_0 \int_0^\infty N(E) \, dE \qquad (12.16)$$

where the first integral of Eq. (12.16) takes into account the photocurrent due to photoelectrons that tunnel through the barrier, and the second integral takes into account the photocurrent due to photoelectrons that are excited above the barrier.

Using expressions (12.12) and (12.15) in (12.16) and rearranging, one obtains an expression for the photofield emission current as

$$I_p = \frac{4\pi mkTe_0}{h^3} \int_0^{U_m} \ln\left[1+\exp\left(\frac{E_f+h\nu-E}{kT}\right)\right] \exp\left[-\frac{\pi^2 l}{h}[2m(U_m-E)]^{1/2}\right] dE$$

$$+ \frac{4\pi mkTe_0}{h^3} \int_0^{U_m} \ln\left[1+\exp\left(\frac{E_f+h\nu-E}{kT}\right)\right] dE \qquad (12.17)$$

where the barrier height U_m with respect to the bottom of the conduction band in the presence of an applied electric field is easily shown to be

$$U_m = E_f + \Phi - h\nu - e_0^{3/2} X^{1/2}$$

$$= W - e_0^{3/2} X^{1/2} \qquad (12.18)$$

where X is the field in V cm^{-1} and $W = E_f + \Phi$.

The barrier width can be obtained from the relation

$$E = W - \frac{e_0^2}{4x} - e_0 Xx \qquad (12.19)$$

where x is the perpendicular distance from the metal (Fig. 12.3).

Equation (12.19) after rearrangement yields a quadratic equation in x as a function of E that gives two values of x. Thereby, the expression for the barrier width l as a function of energy is

$$l = x_2 - x_1 = [(E-W)^2 - e_0^3 X]^{1/2}/e_0 X \qquad (12.20)$$

The results for the photocurrent as a function of $h\nu$ obtained by numerical integration of expression (12.17) both for medium field strength $(10^5-10^6$ V cm$^{-1})$ and for high fields $(10^7-10^8$ V cm$^{-1})$ are given in Figs. 12.4 and 12.5, respectively. In both the medium- and high-field cases, the square law between photocurrent and frequency is obeyed, but the threshold energy (i.e., the effective work function) differs in both cases. This is due to the difference in the tunneling contribution in medium- and high-field cases.

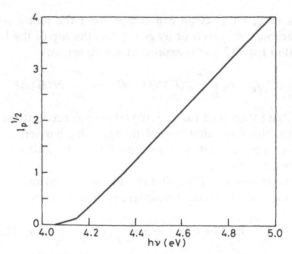

Fig. 12.4. The dependence of the square root of the photocurrent at the metal–vacuum interface on the radiation energy under the influence of an electric field strength of $5 \times 10^5 \, V \, cm^{-1}$ and for $\Phi = 4.5 \, eV$.

In the high-field case, the barrier becomes thin and low, and the tunneling contribution increases.

However, there is a problem that needs attention in the theory of photoemission, and this is the effect of change (increase or decrease) in the number of charge on the electrode upon the occupancy in the conduction band due to a change of charge on the electrode surface.

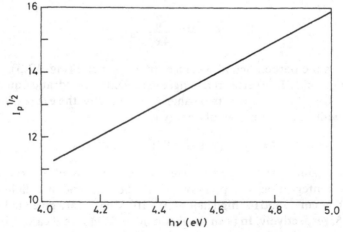

Fig. 12.5. The dependence of the square root of the photocurrent at the metal–vacuum interface on the radiation energy under the influence of an electric field strength of $5 \times 10^7 \, V \, cm^{-1}$ and for $\Phi = 4.5 \, eV$.

Thus, in the treatment given above, the effect of change on charge on the electrode, i.e., the effect of electrode potential, has been taken into account by allowing the field strengths in the double layer to affect the energy barrier. The emitted electrons thus have to penetrate a barrier that is higher or lower, according to the field. For this reason, their rate of emission would alter with electrode potential or charge on the electrode.

However, the conduction band is filled to different degrees when charges enter to or are withdrawn from a metal under applied potential. One of the definitions of the Fermi level is the energy at which the probability of filling the electron levels in a metal is one-half. If we fill the conduction band to a greater or lesser extent by charging the metal, then it is obvious that the work function will change with potential (see page 441). This has not been taken into account in the treatments we have been giving so far and needs attention in the future treatments.

12.4. Quantum Mechanical Theory of the Photoemission Rate into an Electrolytic Solution

12.4.1. Models

Photoemission of electrons at the metal (electrode)–electrolyte interface have been studied from early times.[9–11] However, most of the work before about 1950 was complicated by a lack of definition in the experimental sense of the state of the surface of the electrode. Since the mid-1960s, reliable work on photoemission into solutions involving better definition of electrode surfaces have appeared.[12–23]

Two models of photoemission into an electrolyte exist. In one, the effect of light upon the electrode reaction originates in the absorption of photons by adsorbed species that are hypothesized to be taking part in the rate-determining step.[11,14] Heyrovsky[14] associated the photoeffect with decomposition under the action of light of surface-charge transfer complexes formed between the electrode and reactants. The absorption of light quanta breaks a bond; the bonding electrons are transferred to the electrode (anodic photocurrent) or to the adsorbed molecules, which migrate away from the electrode surface (cathodic photocurrent).

The advantage of this type of approach to photoelectrochemistry is that it can explain why there are both anodic and cathodic photoeffects (as established by Hillson and Rideal). The alternative model, in which the photocurrent is due to an acceleration of cathodic electron emission,[66,67] can obviously only explain (on metals) cathodic photoelectrochemical kinetics, concerning which much more data exist than with anodic kinetics stimulated by illumination. We have given little attention here to the view that it is the surface-active complex that absorbs light. Thus, at certain wavelengths in semiconductors, nearly all the light is used up in causing electron emission. However, in fact, light is absorbed in a metal in the first 100 Å or so. It seems unlikely that the one atom reactant layer on the surface would be able to absorb so much. Again, theories (Heyrovsky,[14] Barker et al.[16,17]) in which this activation occurs, and in which the light effects are supposed to have to do with some reaction other than that of electron emission, have hitherto been presented only qualitatively, so that a rigorous test is less easy.

Hillson and Rideal[11] used this surface absorption model in 1949 as the basis to a phenomenological expression for the rate of a photoelectrochemical reaction. In their consideration of the hydrogen evolution reaction in the presence of light, it was assumed that illumination leads to the activation of some of the adsorbed hydrogen atoms. Thus, surface combination of such atoms (or dissociation of H_2 molecules in the anodic direction) has been considered as rate determining in hydrogen evolution, but this is now regarded as a tenable hypothesis only on noble metals. On the chosen model, the number of atoms activated per second is proportional to $I[H]$, where I is the intensity of illumination and $[H]$ is the concentration of adsorbed atoms. With this basis, the expression of the photocurrent at constant electrode potential given by Hillson and Rideal[11] is

$$i_p = 2e_0 \frac{kT}{h}[H^*][H_D^+]\exp\left(-\frac{\Delta G^{\ddagger}}{kT}\right)\exp\left(\frac{\beta h\nu}{kT}\right) \qquad (12.21)$$

where $[H^*]$ is the number of hydrogen atoms activated by light per second, $[H_D^+]$ is the concentration of hydrogen ion in the Helmholtz double layer, $\Delta G^{\ddagger}$ is the free energy of activation for the reaction, $h\nu$ is the energy of the light of frequency ν, and β is the fraction of the energy acquired by the atoms from the photon energy. The expression for the quantum efficiency[11] γ is

$$\gamma = i_p/I \qquad (12.22)$$

Substitution of i_p from Eq. (12.21) gives an equation which was found by the authors to accord well with their experiments.[11]

One advantage of the radical activation model is that it can explain why (as in the hydrogen evolution reaction) there are light effects both on the cathodic and anodic reaction mechanism. Both combination and dissociation of H and H_2, respectively, could be photoactivated. However, the second model, applied below to metals, is only able to explain cathodic photoeffects, for which, indeed, there is far more evidence than for anodic effects.

The second mechanism† concerns light-induced electron emission from the electrode. The first equations for such a model were given by Bockris[24] in 1954 and were the basis of later developments by Barker *et al.*,[16,17] Brodskii and Gurevich *et al.*,[25-27] and the theoretical investigation of photoemission at the metal–solution interface.

Brodskii *et al.*[25-27] developed a quantum mechanical theory of photo-emission at metal–electrolyte interfaces. They have given an expression for the photocurrent that involved the following assumptions:

(a) Photoemission of electrons from the electrode surface to an electrolyte is the rate-determining step.

(b) The surface of the metal electrode is illuminated by a monochromatic light of frequency ω.

(c) The general expression used for the photocurrent is

$$I_p = \int j_x(E_i, p_{\parallel}, A) f(E_i, E_f) \rho(E_i, p_{\parallel}) \, dE_i \, p_{\parallel} \qquad (12.23)$$

where $j_x(E_i, p_{\parallel}, A)$ is the partial current due to electrons having an initial energy E_i and momentum $p_{\parallel}$ and influenced by the electromagnetic field A. The Fermi distribution function of the initial electrons inside the metal is $f(E_i, E_f)$, and $\rho(E_i, p_{\parallel})$ is the density of state function of electrons in the metal electrode.

(d) The partial current (in the general sense—not limited to electron flow) j_x is, e.g., that due to the electron flow along the x axis normal to the electrode surface and is obtained by using the quantum mechanical expression for flow or current, i.e.,

$$j_x = \frac{e_0 i \hbar}{2 m_e} \left(\psi_e \frac{\partial \psi_e^*}{\partial x} - \psi_e^* \frac{\partial \psi_e}{\partial x} \right) \qquad (12.24)$$

where e_0 is the electronic charge, m_e is the electron mass, and ψ_e is the wave function of the excited electron in the interface, and ψ_e^* is its complex conjugate.

(e) The wave function of the electron ψ_e^* is obtained by solving the Schrödinger equation using the image interaction term only as the potential energy of the electron.

(f) Using the wave function of the electron thus obtained in (12.24), the expression for the partial current was found by Brodskii *et al.*[25-27] to be

$$j_x = \frac{P_\alpha}{m_e} |\Lambda|^2 [1 - \exp(-P_\alpha/P)]^{-1} \qquad (12.25)$$

† This mechanism was first suggested qualitatively by Bowden[64] and was worked upon by Price.[65]

where

$$P_\alpha = \frac{2\pi e_0^2 m_e}{4h\varepsilon_s} \qquad (12.26)$$

where ε_s is the dielectric constant of the electrolyte medium into which photoemission of the electron takes place, P is the momentum of the electron, and Λ is a dimensionless constant that depends on the properties of the metal and the interface.

(g) Using the value of j_x from Eq. (12.25) in Eq. (12.23), and after introducing the density of states and the Fermi distribution law in Eq. (12.23), Brodskii and Gurevich[25-27] obtained an expression for the photo-current as

$$i_p = \tfrac{4}{15}A_0\chi|\Lambda|^2 \frac{h^2}{k^2\omega_0^{1/2}}(n\omega_p - \omega_0)^{5/2} \qquad (12.27)$$

where n is the number of photons absorbed by one electron, k is the Boltzmann constant and ω_p is the actual photon frequency, and ω_0 is the threshold frequency.

Equation (12.27) is obtained under the condition

$$\frac{\hbar\omega_p - \hbar\omega_0}{p_\alpha^2/2m_e} \gg 1 \qquad (12.28)$$

However, condition (12.28) is satisfied only when the dielectric constant of the medium is >10, an unsatisfying condition for the optical interaction with water.

In Eq. (12.27), $A_0 = 4\pi e_0 k^2 m_0/(2\pi h)^2$, where k is the Boltzmann constant and χ is a dimensionless function that distinguishes between the behavior of electrons in the metal and that described by the ideal Fermi gas model, from which $\chi = 1$. Equation (12.27) represents a linear $\frac{2}{5}$ law of the photocurrent as a function of ω and is called Brodskii's law.

12.4.2. Limitations of Brodskii's Quantum Mechanical Treatment of Photoemission

Brodskii et al.[25-27] determined the partial current using the quantum mechanical expression for flux (Eq. 12.24), but the wave function used was obtained by solving a Schrödinger equation in which it was assumed that the potential energy experienced by the escaping electron at the electrode-solution interface was only the image potential of the electron in the metal. Thus, Brodskii's model neglects the Coulombic interaction of the electron with the ions in the outer Helmholtz plane (OHP) of the double layer, the field effect at the interface, and the solvating effect of the solvent. It is, in fact, a treatment of emission into a vacuum.

Correspondingly, Brodskii's model [Eq. (10.23)] does not involve a factor that takes into account the distribution law for the solvent acceptor states. His model could not thus give the dark current, which must be Tafelian and which only comes out thus in the presence of exponentially distributed acceptor states in solution.

Last, Brodskii's treatment neglects any account of the interaction of photons with the metal, i.e., it avoids the possibility of understanding the spectral property of metal as a function of frequency. Scattering in the metal is also neglected.[25-27]

12.5. Theory of Photoelectrochemical Kinetics

12.5.1. Introduction

In the previous section, the quantum mechanical theory of photoemission into an electrolyte was discussed in general. This theory gave an expression for the photocurrent that neglected all details of the electrode material and the properties of the electrode–solution interface. In this section we consider a theory of photoelectrochemical kinetics by Bockris, Khan, and Uosaki[28] (BKU) that gives an expression for the photocurrent that can be computed in absolute terms. This theory takes into account the reflective properties of the electrode material, the excitation probability from photon–electron interactions, electron–phonon and electron–electron scattering, the probability of barrier penetration, and the presence of acceptor states and their distributions in solution. The effect of the double layer is taken into account in constructing the barrier at the electrode–solution interface.

12.5.2. The BKU Theory

In the BKU theory,[28] the photoemission of an electron from the metal to acceptor states in the electrolyte across the potential energy barrier at the metal–solution interface is taken to be the rate-determining step. The following aspects were considered in deriving an expression of the photo-emission current at the electrode–solution interface:

(a) The potential energy barrier for the electron at the interface.

(b) The number of photoexcited electrons in the metal electrode that finally reach the surface, taking into account electron–electron and electron–phonon interaction, after absorbing a quantum of light energy $h\nu$.

(c) The availability of acceptor states in solution.

(d) The number of photoelectrons per unit area of electrode surface that tunnel through or go over the potential energy barrier at the interface.

12.5.2.1. Potential Energy Barrier at the Interface

The potential energy barrier was constructed by considering the path of a hypothetical electron that would move classically over the top of the barrier to a defined acceptor state. The following considerations are made.

Interaction Energy of an Electron with the Water Molecules Absorbed on the Electrode. During transfer of an electron across the electrode–electrolyte interface, an important interaction will be that due to the positive and negative centers of adsorbed water molecules on the electrode surface[29] (Fig. 12.6).

The classical interaction energy can be calculated as a function of distance from the metal by considering the Coulomb interaction of the electron with the positive and negative centers of the adsorbed water molecules during its transit across them. This Coulomb interaction in the first adsorbed water layer, U_{H_2O}, can be obtained using the relation[30]

$$U_{H_2O} = \begin{cases} \dfrac{3Ze_0^2}{2R_2} - \dfrac{1}{2}\dfrac{r^2}{R^2}, & 0 < r < R \\[3mm] -\dfrac{Ze_0}{r}, & r > R \end{cases} \tag{12.29}$$

where Z is the charge on the position and/or the negative center (i.e., on the hydrogen or oxygen atom), respectively, R is the radius of the sphere (i.e., of hydrogen or oxygen atoms), which is considered to have a uniform distribution of charge Ze_0, and r is the distance from the center of the sphere (Fig. 12.6).

Coulomb Interaction. When a photoejected electron leaves the metal surface, it experiences Coulombic attraction and repulsion from all ions in the outer Helmholtz plane (OHP) and their images in the metal at an optical frequency, so the optical dielectric constant ε_{op} is used. Equation (12.30) represents the Coulomb energy for these interactions.†

$$U_{Coulomb} = U_1 + U_2 + U_3 + U_4$$

$$= -\frac{e_0^2}{(d-x)\varepsilon_{op}} + \frac{e_0^2}{(d+x)\varepsilon_{op}}$$

$$- \frac{2e_0^2}{\varepsilon_{op}} \sum_{n=1}^{\infty} \left\{ \frac{1}{[(d-x)^2 + n^2 R_i^2]^{1/2}} - \frac{1}{[(d+x)^2 + n^2 R_i^2]^{1/2}} \right\} \tag{12.30}$$

where U_1 is the Coulomb interaction energy between the emitting electron and the central ion in the OHP, U_2 is the Coulomb interaction energy between the image of the central ion in the metal electrode and the emitting

---

† This is deduced in a similar way as for Eq. (7.139) in Bockris and Reddy.[31]

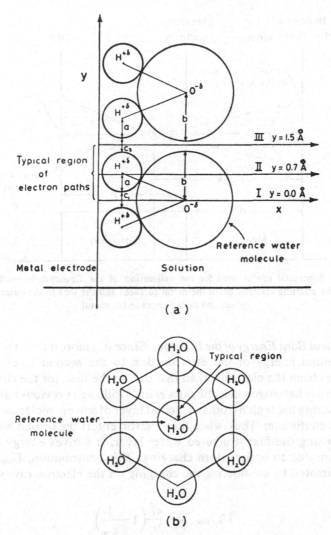

Fig. 12.6. Schematic model used for the estimation of the Coulomb interaction energy of electron with the positive and negative centers of the water molecules adsorbed on the electrode surface. (a) Typical region of electron paths using two water molecules. (b) Hexagonal arrays of six water molecules.

electron, U_3 is the Coulomb interaction energy between the emitting electron and the ions in the rings around the central ion in the OHP, and U_4 is the corresponding interaction energy with images of the ions of the rings in the metal electrode; R_i is the average distance between the ions in the OHP depending on its coverage and is determined using $R_i = 4r_i/(\pi\theta)^{1/2}$, where θ is the coverage of ions in the OHP, and r_i is the radius of the ions (Fig. 12.7).

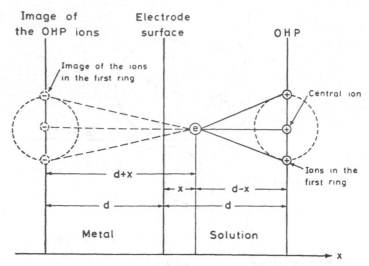

Fig. 12.7. Schematic model used for the estimation of the Coulomb interaction energy between the emitting electron (from the metal surface) and the ions in the outer Helmholtz plane and their images in the metal.

Optical Born Energy of the Electron. Since it is more difficult to evaluate the potential energy of the electron due to the second layer of water molecules from the electrode than that due to the first, for the structure of water here is heterogeneous, Bockris *et al.*[28] followed previous authors[32-34] in considering the region from the second layer of solvent molecules into the bulk as a continuum. Thus, when the electron enters the second water layer after crossing the first structured water layer, it derives energy from the continuum due to optical Born charging. This contribution, $U_{\text{op Born}}$, has been estimated by considering the charging on the electron cavity[35,36] and gives us

$$U_{\text{op Born}} = -\frac{e_0^2}{2r_c}\left(1 - \frac{1}{\varepsilon_{\text{op}}}\right) \qquad (12.31)$$

where r_c is the radius of the electron cavity. The radius r_c was taken here in the numerical computations of BKU from the SCF calculations of Fueki *et al.*[35]

12.5.2.2. Number of Photoexcited Electrons

Let a monochromatic beam of photons, each having an energy $h\nu$, be incident upon a metal electrode surface in the x direction of a rectangular coordinate system. Some light will be reflected, some adsorbed within the metal, and some will excite electrons into the metal conduction band. Thus, one can calculate the following:

(a) The number of electrons per unit area and time that are excited by an incident photon of intensity I_0 at depths between x and $(x + dx)$ inside the metal from the surface that can be expressed as

$$N(x, x + dx) = I_0(1 - R_f)\alpha_p \exp(-\alpha_p x) \, dx \qquad (12.32)$$

where R_f is the reflection coefficient of the metal and α_p is the adsorption coefficient of the metal for a photon of frequency ν.

(b) The probability $P_1(E, h\nu)$ of excitation of an electron in the metal electrode to an energy state E from an energy state $(E - h\nu)$. This is given by[37]

$$P_1(E, h\nu) \alpha \rho_E(E)[1 - f(E)]\rho(E - h\nu)f(E - h\nu) \, dE \qquad (12.33)$$

Since P_1 is a probability, it must be normalized to unity. Thus,

$$P_1(E, h\nu) = \frac{\rho(E)[1 - f(E)]\rho(E - h\nu)f(E - h\nu) \, dE}{\int_{h\nu}^{\infty} \rho(E)[1 - f(E)]\rho(E - h\nu)f(E - h\nu)f \, dE}$$

$$= \frac{1}{Q}\rho(E)[1 - f(E)]\rho(E - h\nu)f(E - h\nu) \, dE \qquad (12.33a)$$

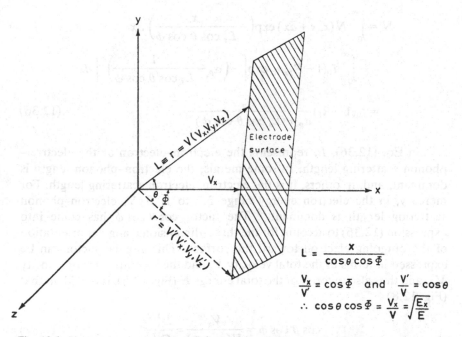

Fig. 12.8. Vectorial representation of the angular distribution of the electrons that move toward the metal surface from within the bulk when stimulated by light.

i.e.,

$$Q = \int_{h\nu}^{\infty} \rho(E)[1 - f(E)]\rho(E - h\nu) f(E - h\nu) \, dE \qquad (12.34)$$

where Q is the normalization factor, $\rho(E)$ is the density of electron states, $[1 - f(E)]\rho_E$ is the number of vacant states at energy E, $\rho(E - h\nu)$ is the density states corresponding to an energy $(E - h\nu)$, $f(E - h\nu)$ is the Fermi distribution of electrons in the metal corresponding to the energy $(E - h\nu)$.

(c) The probabilities of photoexcited electrons from within the bulk of the metal reaching the surface at different angles without suffering an inelastic electron–phonon or electron–electron scattering event. These probabilities are given (Fig. 12.8) by

$$P_2 = \exp\left(-\frac{x}{L_s \cos \theta \cos \phi}\right) \qquad (12.35)$$

where L_s is the electron–phonon or electron–electron scattering length. Thus, the number of electrons, N ($\text{cm}^{-2} \text{sec}^{-1}$), which absorb photons at a depth between x and $(x + dx)$ and which reach the surface without undergoing inelastic scattering, is given from Eqs. (12.32) and (12.35) by

$$N = \int_0^{\infty} N(x, x + dx) \exp\left(-\frac{x}{L_s \cos \theta \cos \phi}\right) dx$$

$$- \int_0^{\infty} I_0(1 - R_f)\alpha_p \exp\left[-\left(\alpha_p - \frac{1}{L_s \cos \theta \cos \phi}\right)^x\right] dx$$

$$= I_0(1 - R_f)\frac{\alpha_p}{\alpha_p + (L_s \cos \theta \cos \phi)^{-1}} \qquad (12.36)$$

In Eq. (12.36), L_s represents the electron–electron *or* the electron–phonon scattering length. In some metals, the electron–phonon length is dominant, and, in others, it is the electron–electron scattering length. For mercury, in the electron energy range 2.5 to 10.0 eV, electron–phonon scattering length is dominant.[38] The factor $\cos \theta \cos \phi$ has come into expression (12.36) to account for the three-dimensional angular orientation of the outgoing electron toward the surface. This angular factor can be expressed in terms of the total velocity V, and the x component of velocity V_x, and thus also in terms of the total energy E (Fig. 12.8). It is evident that (Fig. 12.8)

$$\cos \theta \cos \phi = \frac{V_x}{V(x, y, z)} = \frac{E_x^{1/2}}{E^{1/2}} \qquad (12.37)$$

From Eqs. (12.36) and (12.37), one gets

$$N = I_0(1 - R_f)\left[1 + \frac{\alpha_s}{\alpha_p}\left(\frac{E}{E_x}\right)^{1/2}\right]^{-1} \tag{12.38}$$

where $\alpha_s = (L_s)^{-1}$.

Since $\alpha_s/\alpha_p \gg 1$, one can write

$$N \simeq I_0(1 - R_f)\frac{\alpha_p}{\alpha_s}\left(\frac{E_x}{E}\right)^{1/2} \tag{12.39}$$

which is the number of electrons having a velocity component perpendicular to the surface (i.e., parallel to the x axis) corresponding to energy E_x and $(E_x + dE_x)$. This number can be obtained from Eqs. (12.33) and (12.39) as follows:

The number (per unit area and time) of unscattered excited electrons at any energy E_x is given from Eqs. (12.33a) and (12.39) (considering the x component of energy E_x) as

$$NP_1(E_x, h\nu) = \frac{I_0(1 - R_f)}{Q}\frac{\alpha_p}{\alpha_s}\frac{E_x^{1/2}}{E^{1/2}}\rho(E_x)[1 - f(E_x)]\rho(E_x - h\nu)f(E_x - h\nu)\, dE_x \tag{12.40}$$

Putting the standard values of $\rho(E)$ and $1 - f(E)$ in (12.40), one gets:

$$NP_1(E_x, h\nu) = \frac{B'I_0(1 - R_f)\alpha_s}{Q}E_x^{1/2}\left[\exp\left(\frac{E_f - E_x}{kT}\right) + 1\right]^{-1}$$

$$\times[\rho(E_x - h\nu)f(E_x - h\nu)\, dE_x] \tag{12.41}$$

where $B' = 4\pi(2m)^{3/2}/h^3$.

In Eq. (12.41), we have considered the probability of having vacant states of fraction $[1 - f(E_x)]$, corresponding to x components only of the velocity in velocity space, i.e., corresponding to an energy E_x without finding them in the proper way by integrating with respect to E_y and E_z. This is a good approximation, since $[1 - f(E)]$ is unity in most situations when $(E_x - E_f) \gg kT$.

The factor $\rho(E_x - h\nu)f(E_x - h\nu)\, dE_x$ of Eq. (12.41) is obtained by integrating $\rho(E - h\nu)f(E - h\nu)\, dE$ with respect to the y and z components of energy E, and this gives the number of electrons having their x components of energy E_x directed normal to the surface. Since energy E is a scalar quantity, one prefers to do this integration in terms of velocity. The equivalent expression for $\rho(E - h\nu)f(E - h\nu)\, dE$ in terms of momentum and then velocity is given by

$$\rho(E-h\nu)f(E-h\nu)\,dE \equiv \frac{2}{h^3}\frac{dp_x\,dp_y\,dp_z}{\exp[(E-h\nu-E_f)/kT]+1}$$

$$=2\left(\frac{m}{h}\right)^3\frac{dv_x\,dv_y\,dv_z}{\exp\{[\frac{1}{2}m(v_x^2+v_y^2+v_z^2)-h\nu-E_f]/kT\}+1}$$

(12.42)

Transforming in terms of a polar coordinate system, the integral of Eq. (12.42) can be written as[5]

$$\int \rho(E-h\nu)f(E-h\nu)\,dE \equiv 2\left(\frac{m}{h}\right)^3 dv_x$$

$$\times\int_0^\infty\int_0^{2\pi}\frac{\xi\,d\xi\,d\theta}{\exp\{[\frac{1}{2}m(v_x^2+\xi^2)-h\nu-E_f]/kT\}+1}$$

(12.43)

where $\xi = (v_y^2+v_z^2)^{1/2}$.

After integration with respect to the y and z components, i.e., with respect to ξ and θ, one gets

$$\rho(E_x-h\nu)f(E_x-h\nu)\,dE_x \equiv \frac{4\pi kT}{m}\left(\frac{m}{h}\right)^3\ln\left[1+\exp\left(\frac{E_f+h\nu-\frac{1}{2}mv_x^2}{kT}\right)\right]dv_x$$

(12.44)

Rewriting Eq. (12.44) in terms of energy, one obtains

$$\rho(E_x-h\nu)f(E_x-h\nu)\,dE_x = \frac{\pi(2m)^{3/2}}{h^3}kT\ln\left[1+\exp\left(\frac{E_f+h\nu-E_x}{kT}\right)\right]$$

$$\times E_x^{-1/2}\,dE_x$$

(12.45)

Thus, Eq. (12.45) gives the number of electrons with a velocity component between v_x and v_x+dv_x and, in terms of energy, the number of electrons in the direction x toward the surface that have an energy between E_x and E_x+dE_x. Substitution of $\rho(E_x-h\nu)f(E_x-h\nu)\,dE_x$ from Eq. (12.45) in Eq. (12.41) followed by a transformation to allow for the effect of the electrode potential V (without considering sign, i.e., negative for cathodic polarization and positive for the anodic polarization) gives the number of photoelectrons per unit area and time that strike the surface and have energies between E_x and E_x+dE_x. This (with V as the rational potential) is[28]

$$N_p(E_x, E_x + dE) = NP_1(E_x, h\nu)$$

$$= \frac{I_0(1 - R_f)\alpha_p KTB}{4Q'B\alpha_s}\left[\exp\left(\frac{E_f - e_0 V - E_x}{kT}\right) + 1\right]^{-1}$$

$$\times \ln\left[1 + \exp\left(\frac{E_f + h\nu - e_0 V - E_x}{kT}\right)\right] dE_x \qquad (12.46)$$

where

$$B = [4\pi(2m)^{3/2}/h^3]^2 \qquad (12.47)$$

and [see Eq. (12.34)]

$$Q'B = Q \qquad (12.48)$$

Equation (12.46) represents the number of photoelectrons at any energy E_x corresponding to velocity components v_x normal to the surface that reach the surface without scattering. The quantity Q' defined in Eq. (12.48) is obtained from Eq. (12.34) with the use of standard values of $\rho(E)$, $\rho(E - h\nu)$, $f(E)$ and $f(E - h\nu)$ as

$$Q' = \int_{h\nu}^{\infty} E^{1/2}(E - h\nu)^{1/2}\left[\exp\left(\frac{E_f - E}{kT}\right) + 1\right]^{-1}\left[\exp\left(\frac{E - h\nu - E_f}{kT}\right) + 1\right]^{-1} dE_x$$

$$(12.49)$$

12.5.2.3. Acceptor States in Solution

The availability of acceptor states for electrons that cross the interface is given by the probability of energy states in the acceptor equal in energy to that of the ejecting electrons. There are two kinds of acceptor states available for photoelectrons, depending on their energy. If the energy of the photoelectrons is more than the electron energy corresponding to the ground vibrational–rotational energy of the acceptor ion in solution, electrons go to the solvent states (Fig. 12.9). They lose their kinetic energy after collision with solvent molecules in the bulk and ultimately get accepted by an acceptor that is in its ground state or reacts with H_2O. The probability of having such solvent states is unity.

If the photoelectron energy is less than that corresponding to the ground-state energy of the acceptor ion they go to the distributed energy states of the acceptor. The probability of having such distributed acceptor states is given by the classical Boltzmann distribution (Fig. 12.9), according

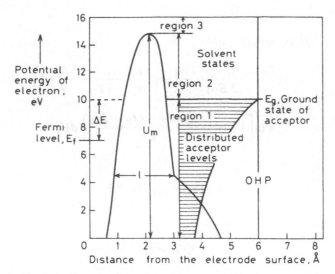

Fig. 12.9. A schematic diagram of the potential energy barrier for electron transfer from a mercury electrode ($\Phi = 4.5$ eV and $E_f = 7.0$ eV) to acceptors in the electrolyte at the outer Helmholtz plane (6 Å from the electrode surface).

to which the distribution of acceptor state $G(E_x)$ is given as[7,28]

$$G(E_x) = \exp[-\beta(E_g - E_x)/kT] \qquad (12.50)$$

where E_g is the ground-state energy of the acceptor, E_x is the energy of the photoelectron corresponding to the x component of velocity of an electron, and β is the electrochemical symmetry factor.

Why do electrons having an energy less than the ground state energy E_g of an acceptor go only to acceptor states and not to free solvent states as well? Low-energy electrons ($E_x < E_g$) may go to free solvent states and lose their energy by collision with the surrounding solvent molecules, but after losing their energy, they do not find any acceptor states (unlike the high-energy electrons above the ground-state energy levels of acceptor) in the energy region below E_g due to low probability of population acceptors in such regions arising from the Boltzmann distribution (Eq. 12.50). Thus, the low-energy region for the electron in the metal represents the high-energy region for the acceptor states in solution. Electrons that go to free solvent states in the energy range $E_x < E_g$ are not ultimately accepted by the acceptor and thus form a local space charge near the electrode. Hence, no contribution to the photocurrent is expected from such electrons. Moreover, to accede to the free state of the solvent, electrons need a relatively high energy. Low-energy electrons cannot form such states. To determine $G(E_x)$ of Eq. (12.50), one needs to find E_g, the ground-state energy of the acceptor.

Ground-State Electron Energy of an H_3O^+ Acceptor. To find the energy of an electron in the ground state of H_3O^+ near an electrode in solution, one considers the transfer of an electron from vacuum to H_3O^+ in solution forming H_2O-H near the electrode. The energy change in this process can be written as (see Section 8.5)

$$-L_0 - J + R + A$$

where L_0 is the solvation energy of H^+ in water, J is the ionization energy of the hydrogen atom, R is the repulsive interaction for the $H \cdots H_2O$ at the moment of arrival of the electron, and A is the attractive $M \cdots H$ interaction at that moment.

One uses values[39,40] $L_0 = -11.4$ eV, $J = 13.6$ eV, $R = 1.0$ eV, and $A = -0.3$ eV. Thus, $-L_0 - J + R + A = -1.5$ eV. Hence, a difference in potential energy between the electron in the ground state of H_3O^+ and that at the Fermi level of the electrode (mercury) is (Fig. 12.9):

$$\Delta E = -1.5 \text{ eV} + \Phi$$

$$= -1.5 + 4.5 = 3.0 \text{ eV} \tag{12.51}$$

where the value of the work function Φ is taken (e.g.) as that of the mercury electrode.

The ΔE value indicates that the energy of the electron in the ground state of the acceptor (H_3O^+ ion) is 3.0 eV above the Fermi level of the mercury electrode, and the ground-state energy of the acceptor E_g is

$$E_g = E_f + \Delta E$$

$$= 7.0 \text{ eV} + 3.0 \text{ eV} = 10.0 \text{ eV} \tag{12.52}$$

relative to the bottom of the conduction band of mercury (Fig. 12.9).

In the estimation of the ground-state energy of the acceptor (H_3O^+ ion), we have considered that the Fermi level of the mercury electrode is not changed from its vacuum value when it is dipped into the solution, disconnected from an electric circuit, inasmuch as the Fermi energy is a bulk property of the metal, and hence, $E_f(\text{vacuum}) = E_f$ (at the p.z.c., in contact with the solution).†

---

† In taking the potential contribution to the energy of the electron as zero at the p.z.c., namely, as identifying the Galvani potential difference to be zero, there is of course, the error of neglecting some of the dipole potential differences and the electron overlap potential differences. In respect to the mercury solution interface,[41] an error of about 0.2 eV arises because of this.

12.5.3. Photocurrent from the BKU Theory

The photocurrent is proportional to the number of photoelectrons that reach the surface without scattering, $N_p(E_x, E_x + dE_x)$ [the probability of tunneling of these electrons through the potential barrier at the interface $P_T(E_x)$ and the probability of the presence of acceptor states $G(E_x)$ of the same energy as that of an emitted electron]. Hence, using Eq. (12.46), the photocurrent (amperes per square centimeter) can be expressed as[28]

$$I_p = e_0 \frac{C_A}{C_T} \int_{h\nu}^{\infty} N_p(E_x, E_x + dE_x) P_T(E_x) G(E_x) \, dE_x \qquad (12.53)$$

where C_A and C_T are respectively the total number of acceptors per unit area of the outer Helmholtz plane (OHP) and the total number of sites per unit area of OHP; $P_T(E_x)$ is the WKB tunneling probability of an electron across the potential barrier of height U_m and width l, i.e.,

$$P_T(E_x) = \exp\left\{ -\frac{\pi^2 l}{h} [2m(U_m - E_x)]^{1/2} \right\} \qquad (12.54)$$

and $G(E_x)$ represents the population of vibrational–rotational states of the acceptor in solution at different energies E_x (Eq. 12.50).

Using the result of the right side of (12.46) in Eq. (12.54), one obtains

$$I_p = \frac{e_0 A'}{Q'} \frac{C_A}{C_T} \int_{h\nu}^{\infty} \left[\exp\left(\frac{E_f - e_0 V - E_x}{kT} \right) + 1 \right]^{-1}$$

$$\times \ln\left[1 + \exp\left(\frac{E_f + h\nu - e_0 V - E_x}{kT} \right) \right] P_T(E_x) G(E_x) \, dE_x$$

$$(12.55)$$

where

$$A' = \frac{I_0(1 - R_f)\alpha_p kT}{4\alpha_s} \qquad (12.56)$$

For the practical evaluation of the photocurrent, one can write the integral in Eq. (12.55) in three parts for the three regions of the barrier (Fig. 12.9), i.e.,

$$I_p = \frac{e_0 A'}{Q'} \frac{C_A}{C_T} \left\{ \int_{h\nu}^{E_g} \left[\exp\left(\frac{E_f - e_0 V - E_x}{kT} \right) + 1 \right]^{-1} \right.$$

$$\times \ln\left[1 + \exp\left(\frac{E_f + h\nu - e_0 V - E_x}{kT} \right) \right] P_T(E_x) G(E_x) \, dE_x$$

$$+ \int_{E}^{U_m} \left[\exp\left(\frac{E_f - e_0 V - E_x}{kT} \right) + 1 \right]^{-1}$$

$$\times \ln\left[1+\exp\left(\frac{E_f+h\nu-e_0V-E_x}{kT}\right)\right]P_T(E_x)\,dE_x$$

$$+\int_{U_m}^{\infty}\left[\exp\left(\frac{E_f-e_0V-E_x}{kT}\right)+1\right]^{-1}$$

$$\times \ln\left[1+\exp\left(\frac{E_f+h\nu-e_0V-E_x}{kT}\right)\right]dE_x\Bigg\} \tag{12.57}$$

The first integral of Eq. (12.57) gives the contribution of the photocurrent from photoelectrons that tunnel through the barrier to distributed acceptor states corresponding to the vibrational–rotational energy of an acceptor, H_3O^+.

The second integral of Eq. (12.57) gives the contribution to the photocurrent from the photoelectrons that tunnel through the barrier but are above E_g (see Bockris and Matthews[8]) and hence go to the bulk solvent states. The major contribution for photon energies greater than 3.0 eV and less than about 10.0 eV comes from such electrons. After emission into the solution, these electrons lose their kinetic energy by inelastic collision with the bulk water molecules and become accepted by the solvent molecules, forming hydrated electrons, or accepted by the bulk H_3O^+ in the ground state.

The third integral of Eq. (12.57) gives the contribution to the photocurrent from the photoelectrons that can go over the barrier.

Later, Uosaki[42] gave an expression for the photocurrent which imvolves double integral to avoid the approximation that only $[1-f(x)]$ is considered in Eq. (12.57). This is

$$I_p = e_0 A'' \frac{C_A}{C_T}\Bigg[\int_{h\nu}^{E_g}\int_{E_x}^{\infty}\frac{1}{\alpha_p+\frac{1}{2}(E/E_x)^{1/2}}\frac{1}{2EQ}\rho(E)$$

$$\times[1-f(E)]\rho(E-h\nu)f(E-h\nu)\,dE\,P_T(E_x)G(E_x)\,dE_x$$

$$+\int_{E_g}^{U_m}\int_{E_x}^{\infty}\frac{1}{\alpha_{ph}+\frac{1}{2}(E/E_x)^{1/2}}\frac{1}{2EQ}\rho(E)$$

$$\times[1-f(E)]\rho(E-h\nu)f(E-h\nu)\,dE\,P_T(E_x)\,dE_x$$

$$+\int_{U_m}^{\infty}\int_{E_x}^{\infty}\frac{1}{\alpha_{ph}+\frac{1}{2}(E/E_x)^{1/2}}\frac{1}{2E}\rho(E)$$

$$\times[1-f(E)]\rho(E-h\nu)\,d(E-h\nu)\,dE\,dE_x\Bigg] \tag{12.58}$$

where

$$A'' = I_0(1-R_f)\alpha_{ph} \tag{12.59}$$

and

$$Q = \int_{h\nu}^{\infty} \rho(E)[1 - f(E)]\rho(E - h\nu)f(E - h\nu) \, dE \qquad (12.60)$$

12.5.4. Computation of the Photocurrent from the BKU Expression

To compute the photocurrent from Eq. (12.57) one needs to estimate barrier height and width. Bockris *et al.*[28] plotted the barrier for the interface finding potential energies of interactions from Eqs. (12.29), (12.30), and (12.31). The barrier height is 14.8 eV with respect to the bottom of the conduction band of the electrode and the barrier width 2.2 Å (Fig. 12.9). Photocurrents at the metal–solution interface were computed at rational potentials V by numerical integration of expression (12.57) using the above-mentioned barrier parameters. Figure 12.10 shows the calculated photocurrents for different incident light energies $h\nu$.

The photocurrent I_p is given in absolute units[28] (nanoamperes per square centimeter). The incident light intensity[11] I_0 used in this calculation is 10^{15} quanta per square centimeter per second.

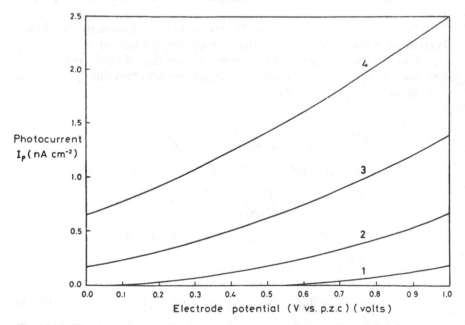

Fig. 12.10. The dependence of the photocurrent on electrode potential for electron transfer from a mercury electrode to acceptors in the electrolyte for different incident photon energies, $h\nu$; (1) $h\nu = 2.5$ eV, (2) $h\nu = 3.0$ eV, (3) $h\nu = 3.5$ eV, (4) $h\nu = 4.0$ eV.

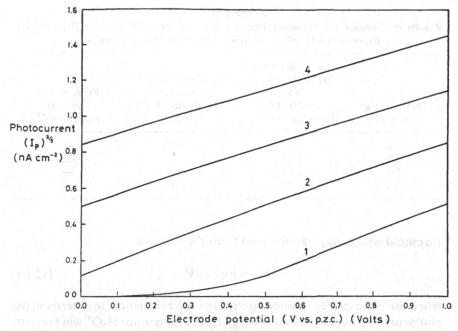

Fig. 12.11. The dependence of the two-fifths power of the photocurrent on electrode potential for electron transfer from a mercury electrode to acceptors in the electrolyte for different incident photon energies $h\nu$: (1) $h\nu = 2.5$ eV, (2) $h\nu = 3.0$ eV, (3) $h\nu = 3.5$ eV, (4) $h\nu = 4.0$ eV.

The constant A' of Eq. (12.57) was computed using a reflection coefficient R_f (values for mercury electrodes) and values of the α_p for different radiation energies $h\nu$. The absorption coefficient of the photon α_p was calculated using $\alpha_p = 4\pi\kappa_p/\lambda_p$, where κ_p is the extinction coefficient of the photon[43] and λ_p is the photon wavelength. The parameter α_s has been calculated for the energy of electrons corresponding to the Fermi energy of mercury using $\alpha_s = (V_F\tau)^{-1}$, where V_F is the electron velocity corresponding to the Fermi energy and τ is the relaxation time of the Fermi level electrons. The C_A/C_T value of Eq. (12.57) was taken as unity for the part of Eq. (12.57) in which electrons are being donated to the solvent and 0.1 when the acceptor is H_3O^+.

The photocurrent $(I_p)^{2/5}$ against potential V was plotted (Fig. 12.11) for different incident light energies $h\nu$. There is linearity for a potential change from the p.z.c. for about one decade, in consistence with the experimental results. The threshold energy $h\nu_0$ for photoemission into the electrolyte was determined by plotting $(I_p)^{2/5}$ against potentials V (Fig. 12.11) and extrapolating to $(I_p)^{2/5} = 0$. The extrapolated potential corresponding to $(I_p)^{2/5} = 0$ is the threshold (or critical) potential V_0. The

Table 12.1

Results to Compare the Threshold Energies $h\nu_0$ (Extrapolated) with the Expected and Experimental Threshold Energies at Different Photon Energies $h\nu$

Photon energy $h\nu$ (eV)	Threshold energy $h\nu_0$ (extrapolated) (eV) from Fig. 12.11 [from Eq. (12.57)]	Theoretically expected threshold $h\nu$ (eV)	Experimental threshold energy (eV)[44]
2.5	2.9	3.0	—
3.0	2.85	3.0	3.15
3.5	2.85	3.0	—
4.0	2.85	3.0	—

threshold energy $h\nu_0$ was obtained from the relation

$$h\nu_0 = h\nu - e_0 V_0 \tag{12.61}$$

where $h\nu$ is the incident photon energy. In each calculation, to determine the photocurrent, the ground-state energy E_g of the acceptor H_3O^+ was taken to be 10.0 eV with respect to the bottom of the conduction band and 3.0 eV above the Fermi level of the mercury electrode. The results of the extrapolated value of the threshold energy from the plots of Fig. 12.11 are in good agreement with the theoretically expected threshold† (Table 12.1). The linear behavior of the plot of $(I_p)^{2/5}$ against the potential V and agreement of the extrapolated threshold light energy $h\nu_0$ with the experimental one[44] and also with the theoretically expected threshold energy confirm the consistence of the model with experiment and with the $\frac{5}{2}$ law for the photocurrent.

The present model takes into account the variation of the depth of penetration of light with energy, the effect of absorption and reflectivity of light by different electrode materials, and the effect of electron–electron and electron–phonon scattering length of electrons in metals on photocurrents.

---

† Figure 12.11 shows that only electrons that have an energy equal or greater than E_g will be emitted to acceptors of electrons in the continuum solvent states and not in the distribution stated in H_3O^+. Detailed computations of the relative contributions of the various terms in the expression for the photocurrent [see Eq. (12.57)] show that the second term predominates. As this term represents the rate of electron transfer through the barrier to solvent states in the solution, it is possible to identify the threshold energy of the photons as that which increases the energy of electrons in the Fermi level to values equal to E_g. This theoretically expected threshold value is 3.0 eV above the Fermi level of mercury electrode (Fig. 12.9). The experimental value of threshold energy[44] is 3.15 eV for mercury electrode.

It gives some account of the angular distribution of nonscattered photo-excited electrons from within the metal bulk.

12.5.5. Non-Tafel Behavior of the Photocurrent

Electrochemical dark currents follow Tafel's Law when free from diffusion control, but one observes that the photocurrent does not follow it (Fig. 12.12) when computed from Eq. (12.57). There is a Boltzmann type of distribution of acceptor states in solution. For the hydrogen evolution at the cathode, one finds that the ground state of the acceptor is at 3.0 eV above the Fermi level of the mercury electrode (Fig. 12.9). The higher energy states of ions in solution are closer to the Fermi level because of the distribution, and their number is reduced exponentially from the ground state and are given by[7,28]

$$N(E) = N_0 \exp\{-\beta(E_g - E)/kT\} \tag{12.62}$$

where N_0 is the number of acceptors in the ground state and E_g is the energy of the acceptor ion corresponding to the ground vibrational–rotational state of the ion solvent bond. E is the energy of the electron.

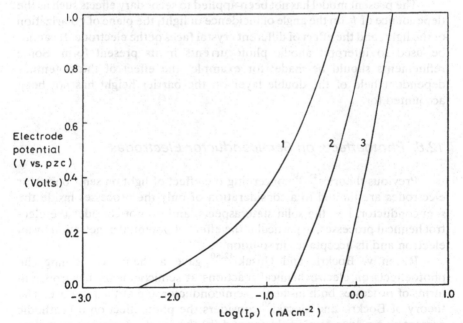

Fig. 12.12. The dependence of the logarithm of the photocurrent I_p on the electrode potential (shows non-Tafelian behavior): (1) $h\nu = 3.0$ eV, (2) $h\nu = 3.5$ eV, (3) $h\nu = 4.0$ eV.

Correspondingly, dark current electrons come from near the Fermi level, because, even if the potential is as high as 2 V (versus the p.z.c.), the Fermi level remains below the ground state of the acceptor, i.e., of H_3O^+ ions, so that the electrons go to distributed states of H_3O^+, and not to the solvent state. As the acceptor H_3O^+ ion states are classically distributed in solution, one obtains an exponential behavior of the dark current with potential, and therefore, the logarithmic Tafel law is followed.

For photocurrents, the photon energy is such that Fermi electrons can be excited above the ground state of the acceptor (Fig. 12.9). The major contribution to the photocurrent is then due to the photoelectrons that tunnel through the potential barrier and go to solvent states above the energy of the H_3O^+ states. The number of these states per unit area is not of Boltzmann distribution, because after the electrons produced have lost sufficient energy, they will *all* be accepted by acceptors in the ground state or by the solvent molecules. Thus, the photocurrents no longer depend on the availability of exponentially distributed vibrational–rotational states, and accordingly do not follow Tafel's law for the exponential dependence of current on potential.

12.5.6. Limitations of the BKU Theory

The present model has not been applied to secondary effects such as the dependence of I_p on the angle of incidence of light, the plane of polarization of the light, and the effect of different crystal faces of the electrode. It cannot be used to interpret anodic photocurrents in its present form. Some refinements should be made; for example, the effect of the potential-dependent field of the double layer on the barrier height has not been accounted for.

12.6. Photoeffects on Semiconductor Electrodes

Previous theories[45–49] concerning the effect of light on semiconductor electrodes are limited to a consideration of only the processes inside the semiconductor, i.e., the solid state aspects, and do not consider the electrochemical processes, in particular the effect of potential penetration by an electron and its acceptance in solution.

Recently, Bockris and Uosaki[42,50] gave a theory concerning the photoeffects on electrochemical reactions at semiconductor electrodes in terms of processes both inside the semiconductor and at the interface. The theory of Bockris and Uosaki[50] considers the photoeffect on (a) cathodic current as a p-type semiconductor and (b) the anodic current at an n-type semiconductor.

12.6.1. Photoeffect on a Cathodic Current at a p-Type Semiconductor–Solution Interface

There are many holes in a valence band and few electrons in the conduction band of p-type semiconductors. Illumination with light, the energy of which is larger than the energy gap of the semiconductor, produces electrons in the conduction band and holes in the valence band (Fig. 12.13). Since there are many holes in the valence band in the absence of illumination and, usually, the number of holes created by light is small compared with the number of holes without light, the values of the anodic current via the valence band with and without illumination are nearly the same. On the other hand, there are almost zero electrons in the conduction band and therefore zero cathodic current via the conduction band without illumination. Consequently, a photoeffect would only be expected via the conduction band and would be cathodic.

Bockris and Uosaki[50] considered a typical reaction to be the hydrogen evolution reaction in acid solutions. The following step was assumed to be the rate-determining step on the semiconductor concerned:

$$p\text{-SC(e)} + H_3O^+ \rightarrow p\text{-SC}\cdots H\cdots H_2O \qquad (12.63)$$

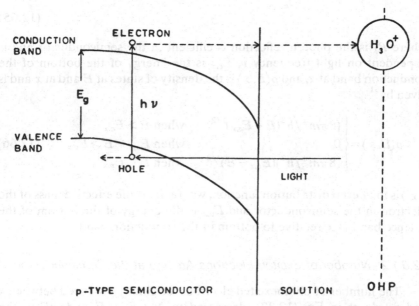

Fig. 12.13. The schematic diagram of the photoeffect on a cathodic current at a p-type semiconductor–solution interface.

where p-SC(e) represents an electron in a p-type semiconductor. The photocurrent corresponding to this reaction i_p is given by

$$i_p = e_0 \frac{C_A}{C_T} \int_0^\infty N_p(E)G(E)P_T(E) \, dE \tag{12.64}$$

The definitions of terms are given in Section 12.5 after Eq. (12.53).

12.6.1.1. Creation of Excited Electrons

The semiconductor electrode was considered to be illuminated normal to the surface with a light source of intensity I_0. Some of the photons are adsorbed and some are reflected, depending on the reflectivity R of the semiconductor. The number of excited electrons due to absorption of photons (with energy greater than the band gap), the energy of which is between E and $E + dE$ on the conduction band between x and $x + dx$, i.e., $N_e(E, x) \, dx \, dE$, is given by[28]

$$N_e(E, x) \, dx \, dE = I_0(1 - R_f)\alpha_{ph} \exp(-\alpha_{ph}x)$$

$$\times \frac{\rho(E, x)(1 - f(E)\rho(E - h\nu, x)f(E - h\nu)}{\int_{E_{c,x}}^\infty \rho(E, x)[1 - f(E)]\rho(E - h\nu, x)f(E - h\nu) \, dE}$$

$$\tag{12.65}$$

where R_f is the photon reflection coefficient in the semiconductor and is dependent on light frequency ν, $E_{c,x}$ is the energy of the bottom of the conduction band at x, and $\rho(E, x)$ is the density of states at E and at x and is given by[51]

$$\rho(E, x) = \begin{cases} (8\pi m^*/h^3(E - E_{c,x})^{1/2} & \text{when } E > E_{c,x} \\ 0 & \text{when } E_{c,x} > E > E_{v,x} \\ (8\pi m^*/h^3)(E_{v,x} - E)^{1/2} & \text{when } E < E_{v,x} \end{cases} \tag{12.66}$$

$f(E)$ is the Fermi distribution function, where m^* is the effective mass of the electron in the semiconductor and $E_{v,x}$ is the energy of the bottom of the valence band at x relative to bottom of the conduction band.

12.6.1.2. Number of Excited Electrons Arriving at the Semiconductor

The number of photoexcited electrons, originally produced between x and $x + dx$ as in Eq. (12.32), decreased to $N_{e,x = x - dx}(E, x) \, dx \, dE$ at the electrode surface due to the electron–electron or electron–phonon scattering and is given as

$$N_{e,x=x-dx}(E, x)\, dx\, dE = N_e(E, x) \exp\left[-\frac{1}{L(x)}\right] dx\, dE \qquad (12.67)$$

where $L(x)$ is the mean free path of electrons at x and is given by[52]

$$L(x) = \frac{2l_D^2}{(l_E^2 + 4l_D^2)^{1/2} - l_E} \qquad (12.68)$$

where l_D is the diffusion length and L_E is the drift length. The terms l_D and l_E are given by

$$l_D = (D\tau_e)^{1/2} = (300\mu_e\tau_e kT/e_0)^{1/2} \qquad (12.69)$$

$$l_E = \tau_e\mu_e V'(x) \qquad (12.70)$$

where μ_e is the mobility of the electron in the semiconductor (square centimeter per volt second), τ_0 is the lifetime of the electron, and $V'(x)$ is the potential drop inside the semiconductor at $x[=(dV/dx)_x$; volts per centimeter].

When $V'(x) = 0$, $L(x)$ becomes l_D, and when $V'(x)$ is sufficiently high, $l_E \gg l_D$, and $L(x)$ becomes l_E [see Eq. (12.68)]. Here, $L(x)$ changes with l_E corresponding to a change of $V'(x)$, since $L(x)$ is a function of l_E and l_E is a function of $V'(x)$ [see Eq. (12.70)]. Hence for N steps ($N = x/dx$), one gets

$$N_{e,x=0}(E, x)\, dx\, dE = N_{p,x=dx}(E, x) \exp\left[-\frac{1}{L(dx)}\, dx\right] dx\, dE$$

$$= N_p(e, x) \exp\left\{-\left[\frac{1}{L(x)} + \frac{1}{L(x-dx)} + \cdots \right.\right.$$

$$\left.\left. + \frac{1}{L(dx)}\right] dx\right\} dx\, dE \qquad (12.71)$$

$N_{p,x=0}(E, x)\, dx\, dE$ gives the number of photoelectrons arriving at the surface per unit time that were excited between x and $x + dx$ and the energy of which is between E and $E + dE$. Therefore, the total number of electrons $N_e(E)\, dE$ that have an energy between E and $E + dE$ at the surface is given by the integral

$$N_e(E)\, dE = \int_0^\infty N_{l,x=0}(E, x)\, dx\, dE \qquad (12.72)$$

12.6.1.3. Electron Transfer Process—Construction of the Potential Energy Barrier

The potential energy barrier for electron transfer from the semiconductor to an acceptor in the OHP was constructed considering the following interactions.

Interaction of an Electron with the Dipole Potential of Adsorbed Water. The surface potential due to adsorbed water dipoles U_{H_2O} at the potential of zero charge was taken as 0.03–0.04 V at mercury, cadmium, and zinc electrodes.[41] The diameter of the water molecules (i.e., 2.76 Å) was taken as the barrier width.

Interaction with Ions in the OHP and Their Images. When a photo-excited electron leaves the semiconductor surface, it interacts with the ions in the OHP and their electrical images in the semiconductor. The Coulombic force F_x between this electron (x from electrode) and all the ions in the OHP and their images is given by

$$F(x) = -\frac{e_0^2}{(d-x)^2 \varepsilon_{op}} + \frac{e_0^2}{(d+x)^2 \varepsilon_{op}} \frac{\varepsilon_s - \varepsilon_{op}}{\varepsilon_s \varepsilon_{op}}$$
$$-\frac{2\pi e_0^2}{\varepsilon_{op}} \sum_{n=1}^{\infty} n\left[\frac{1}{(d-x)^2 + n^2 R_i^2} - \frac{1}{(d+x)^2 + n^2 R_i^2} \frac{\varepsilon_s - \varepsilon_{op}}{\varepsilon_s \varepsilon_{op}}\right] \quad (12.73)$$

The definitions of terms in Eq. (12.73) are given in Section 12.5 after Eq. (12.30).

The potential energy of an electron at x due to this force $U_{ion}(x)$ is given by

$$U_{ion}(x) = \int_0^x F(x) \, dx \quad (12.74)$$

and this value can be obtained by numerical integration using Eq. (12.73).

Optical Born Energy of the Electron. This is given in Eq. (12.31).

Calculation of the Barrier Maximum. According to these considerations, the potential energy of the electron with respect to the bottom of the conduction band is given by

$$U = E_a + U_{H_2O} + U_{ion} + U_{op\,Born} \quad (12.75)$$

where E_a is the electron affinity of the semiconductor electrode. A potential energy profile based on these considerations is given as a function of distance in Fig. 12.14 assuming d is 6.5 Å in Eq. (12.73). The dimensions of the barrier can also be seen in this figure. The maximum value of the barrier height U_m depends on the electron affinity of the semiconductor E_a and is given (in electron volts) by

$$U_m = E_a - 0.23 \text{ V} \quad (12.76)$$

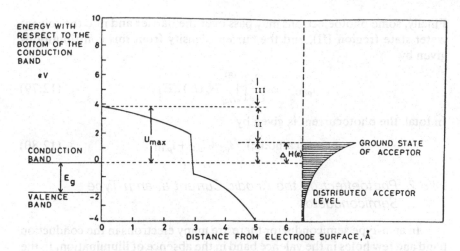

Fig. 12.14. The schematic diagram of the potential energy barrier for electron transfer from a p-type semiconductor to an acceptor in solution.

Equation for the Cathodic Photocurrent. The equation for the current density may be divided into three parts (Fig. 12.14). In region I of this figure, photoelectrons pass through the barrier and are accepted by H_3O^+ ions in solution. The equation for photocurrent for this region $i_{p,I}$ is given by

$$i_{p,I} = e_0 \frac{C_A}{C_T} \int_0^{\Delta H(e)} N_e(E) G_A(E) P_T(E) \, dE$$

$$= e_0 \frac{C_A}{C_T} \int_0^{\Delta H(e)} N_e(E) \exp\{-\beta[\Delta H(e) - E]/kT\}$$

$$\times \exp\left\{-\frac{\pi^2 l}{h}[2m(U_m - E)]^{1/2}\right\} dE \qquad (12.77)$$

where $\Delta H(e)$ is the enthalpy for the electron transfer from semiconductor to H_3O^+ and can be found using the thermodynamic cycle[42,50] for the process given in Eq. (12.63). $N_e(E) \, dE$ is given in Eq. (12.72), and P_T from Eq. (12.15) is used in Eq. (12.77). In region II, the photoelectrons do not find an acceptor state in H_3O^+ but are accepted by water molecules and become solvated electrons. The current density for region II is represented by

$$i_{p,II} = e_0 \frac{C_A}{C_T} \int_{\Delta H(e)}^{U_{max}} N_e(E) \exp\left[-\frac{\pi^2 l}{h}[2m(U_m - E)]^{1/2}\right] dE \qquad (12.78)$$

Finally, some photoelectrons may pass over the barrier and into the solvent water state (region III), and the current density from this contribution is given by

$$i_{p,\text{III}} = e_0 \frac{C_A}{C_T} \left\{ \int_{U_m}^{\infty} N_e(E)\, dE \right\} \tag{12.79}$$

In total, the photocurrent is given by

$$i_p(\text{cathodic}) = i_{p,\text{I}} + i_{p,\text{II}} + i_{p,\text{III}} \tag{12.80}$$

12.6.2. Photoeffect on the Anodic Current at an n-Type Semiconductor

In an n-type semiconductor there are many electrons in the conduction band and few holes in the valence band in the absence of illumination. In the presence of illumination by photons, the energy of which is larger than the energy gap of the semiconductor, holes are created in the valence band and electrons are activated into the conduction band.

However, since there are many electrons in the conduction band without illumination and usually the number of electrons created by light is small compared with the number of electrons without illumination, the value of the cathodic current via the conduction band with illumination and without illumination are almost the same. On the other hand, there are almost zero holes in the valence band, and hence, the anodic current via the valence band is zero without illumination. Consequently, the photoeffect on the anodic current via the valence band will be the significant one.

The mechanism of anodic reactions is less well known than that of cathodic reactions. One complication is the anodic dissolution of the semiconductor itself. Let it be supposed that the anodic current corresponds to the oxidation of OH^- and water and that the rate-determining steps are

$$\text{n-sc(hole)} + OH_{aq}^- \rightarrow \text{n-sc} \cdots OH \cdots H_2O \tag{12.81}$$

$$\text{n-sc(hole)} + H_2O \rightarrow \text{n-sc} \cdots OH \cdots H_3O^+ \tag{12.82}$$

where n-sc(hole) indicates a hole in an n-type semiconductor.

12.6.2.1. Creation of Holes

Each photon, the energy of which is greater than the energy gap of the semiconductor, creates an electron in the conduction band and a hole in the valence band. The number of created holes of energy between $(E - h\nu)$ and $(E - h\nu) + dE$ in the valence band at distances from the electrode surface between x and $x + dx$, i.e., $N_h(E - h\nu, x)\, dx\, dE$, is equal to the number of photoexcited electrons, the energies of which are between E and $E + dE$ in the conduction band and at a distance between x and $x + dx$, i.e., $N_e(E, x)\, dx\, dE$. Hence, using Eq. (12.65), $N_h(E, x)\, dx\, dE$ is given by

$$N_h(E, x)\, dx\, dE = I_0(1 - R)\alpha_{ph} \exp(-\alpha_{ph} x)$$

$$\times \frac{\rho(E + h\nu, x)[1 - f(E + h\nu)]\rho(E, x)f(E)}{\displaystyle\int_\infty^{E_{v,x}} \rho(E + h\nu, x)[1 - f(E + h\nu)]\rho(E, x)f(E)\, dE}\, dx\, dE$$

$$(12.83)$$

where $E_{v,x}$ is the energy of the top of the valence band relative to bottom of the conduction band at x from the electrode surface.

12.6.2.2. Number of Holes Arriving at the Surface

The number of holes at the surface, the energies of which are between E and $E + dE$, can be given in a similar way to that of electrons discussed in Section 12.6.1, with minor changes in the diffusion and drift lengths, as the diffusion length l_D and drift length l_E in this case become

$$l_D = (D\tau_h)^{1/2} = (300\mu_h\tau_h kT/e_0)^{1/2} \tag{12.84}$$

$$l_E = -\tau_h\mu_h V'(x) \tag{12.85}$$

where μ_h is the mobility of the hole in the semiconductor and τ_h is the lifetime of the hole.

12.6.2.3. Electron Transfer Process

In anodic reactions at semiconductor–solution interfaces activated by light, electrons are transferred from a donor in solution to a hole in the valence band. Since $N_D(E)$ is an inverted Boltzmann distribution for donors,[53] a photocurrent corresponding to a direct electron transfer from a donor to a hole is expected only when the energy of the ground states of the donor is lower than the top of the valence band at the surface (Fig. 12.15), i.e. $\Delta H' < 0$.[†]

12.6.2.4. Construction of the Potential Energy Barrier

The potential energy barrier for the electron from OH^- is constructed by considering the optical Born energy, Coulombic interaction, and

---

† $\Delta H'$ is the standard enthalpy change for electron transfer from the donor to a hole in the valence band and can be found using the thermodynamic cycle[42] for the process (12.81).

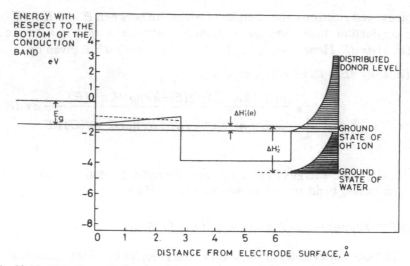

Fig. 12.15. The schematic diagram of the potential energy barrier for electron transfer from donors (OH$^-$ ion and water) in the solution to an n-type semiconductor.

interaction energy between electron and adsorbed water analogously to the p-type semiconductor case.

The potential energy of the electron with respect to its value in the ground state of OH$^-$, U', is given by

$$U' = E_a + U_{\text{op Born}} + U'_{\text{ion}} + U_{\text{H}_2\text{O}} \tag{12.86}$$

where

$$U'_{\text{ion}} = \int_0^x -\frac{e_0^2}{(2d-x)^2 \varepsilon_{\text{op}}} \frac{\varepsilon - \varepsilon_{\text{op}}}{\varepsilon + \varepsilon_{\text{op}}} + \frac{e_0^2}{x^2 \varepsilon_{\text{op}}}$$
$$+ \frac{2\pi e_0^2}{\varepsilon_{\text{op}}} \sum_{n=1}^{\infty} n \left(\frac{1}{(2dx-x)^2 + n^2 R_i^2} \frac{\varepsilon - \varepsilon_{\text{op}}}{\varepsilon + \varepsilon_{\text{op}}} - \frac{1}{x^2 + n^2 R_i^2} \right) dx \tag{12.87}$$

Similarly, the potential energy of the electron from water with respect to its value in the ground state of water U'' is given by

$$U'' = E_{\text{diss}} + J + U_{\text{op Born}} + U'_{\text{ion}} + U_{\text{H}_2\text{O}} \tag{12.88}$$

where E_{diss} is the dissociation energy of water in gas phase and J is the ionization potential of the hydrogen atom. A potential energy profile based on these considerations is shown as a function of distance in Fig. 12.15 where it is assumed that d is 6.5 Å. Other parameters taken in this calculation are $E_g = 1.5$ eV and $E_a = 3.5$ eV.

12.6.2.5. Equation for the Photocurrent

Similarly to Eq. (12.64), the photocurrent corresponding to an electron transfer from donor OH^- to holes in semiconductor is given in general form as Fig. 12.15:

$$i_{p,OH} = e_0 \frac{C_D}{C_T} \int_{\infty}^{-E_g} N_h(E) G_D(E) P_T(E) \, dE \qquad (12.89)$$

where C_D is the number of donors in the OHP and $G_D(E)$ gives the distribution of the donor.

Since $-E_g > E_g - \Delta H_I'(e) > E_g - \Delta H_2'(e)$,† the photocurrent is given by

$$
\begin{aligned}
i_p'(\text{anodic}) = e_0 \Bigg[&\frac{C_{H_2O}}{C_T} \int_{-E_g-\Delta H_2'(e)}^{-E_g} N_h(E) \\
&\times \exp\{\beta[-E_g - \Delta H_2'(e) - E]/kT\} \\
&\times \exp\left(-\frac{\pi^2 l}{h}\{2m[-E_g - \Delta H_2'(e) + U_{\max}'' - E]\}^{1/2}\right) dE \\
+ &\frac{C_{OH^-}}{C_T} \int_{-E_g-\Delta H_1'(e)}^{-E_g} N_h(E) \exp\{\beta[-E_g - \Delta H_1'(e) - E]/kT\} \\
&\times \exp\left(-\frac{\pi^2 l}{h}\{2m[-E_g - \Delta H_1'(e) + U_{\max}' - E]\}^{1/2}\right) dE \Bigg]
\end{aligned}
$$
$$(12.90)$$

where C_{H_2O} and C_{OH^-} are the number of water molecules and OH^- ions per unit area in the OHP, respectively; $U_{\max}''$ and $U_{\max}'$ are the barrier maxima for the electrons from water and OH^- ion, respectively, and $N_h(E)$ is given in Eq. (12.83).

12.6.3. Results of the Calculation of Photocurrent for p- and n-Type Semiconductors

The photocurrents for p- and n-type semiconductors were computed by Bockris *et al.*[50] from Eqs. (12.80) and (12.90) and the results were given in terms of quantum efficiencies. The quantum efficiency is defined as

$$\gamma = I_p/e_0 I_0 \qquad (12.91)$$

where I_p is the photocurrent and I_0 is the intensity of illumination.

---

† $\Delta H_2'(e)$ is the standard enthalpy change for the process (12.82) and can be found from a thermodynamic cycle.[42,45]

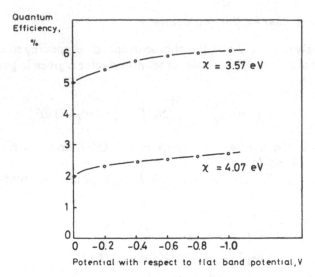

Fig. 12.16. The dependence of the quantum efficiency of a p-type semiconductor on the electrode potential for several values of electron affinities.

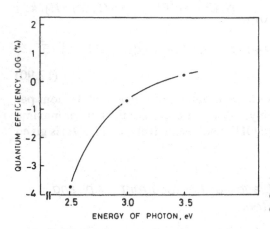

Fig. 12.17. The dependence of the quantum efficiency of a p-type semiconductor on the energy of a photon.

The effects of potential and photon energy, energy gap, and electron mobility on quantum efficiency were computed for the p-type semiconductor GaAs and are given in Figs. (12.16), (12.17), (12.18), and (12.19), respectively. The parameters for GaAs are

$$R_f = 0.47^{54}, \quad \alpha_{ph} = 4.4 \times 10^5 \, cm^{-1}, \quad E_g = 1.4 \, eV$$

$$E_a = 4.07 \, eV^{55}, \quad \mu_e = 8600 \, cm^2/V \, sec^{56}, \quad \tau = 10^{-12} \, sec$$

Similarly, the effects of potential, electron affinity, and energy gap for n-type semiconductor TiO_2 were computed.[50] The results are given in Table 12.2. The parameters used for computation were the same as those of GaAs except for $\mu_h = 400$ cm^2/V sec and $E_a = 3.5$ eV.

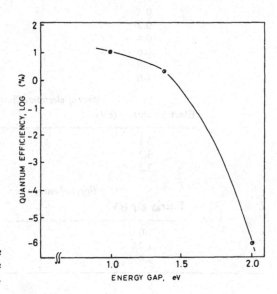

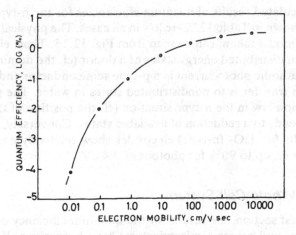

Fig. 12.18. The dependence of the quantum efficiency of a p-type semiconductor on the energy gap.

Fig. 12.19. The dependence of the quantum efficiency of a p-type semiconductor on the electron mobility.

Table 12.2
Computed Results for n-Type Semiconductor

Effect of potential	
Potential (V with respect to the flat-band potential)	Quantum efficiency (%)
0	1.0×10^{-2}
0.2	1.8×10^{-2}
0.4	2.2×10^{-2}
0.6	2.7×10^{-2}
0.8	3.0×10^{-2}
1.0	3.2×10^{-2}

Effect of electron affinity	
Electron affinity (eV)	Quantum efficiency (%)
3.1	4.5×10^{-2}
3.3	4.1×10^{-2}
3.5	3.2×10^{-2}

Effect of energy gap	
Energy gap (eV)	Quantum efficiency (%)
1.0	4.0×10^{-2}
1.2	3.5×10^{-2}
1.4	3.0×10^{-2}

12.6.3.1. Comparison with the Experimental Results of n-Type Semiconductors

The calculated results of quantum efficiencies for the n-type semiconductors, as shown in Table 12.2, are low in all cases. The physical reasons for this, for the model taken, can be seen from Fig. 12.15. Thus, electrons are donated from distributed energy states of a donor (cf. the dominant contribution the cathodic photocurrent in a p-type semiconductor and in metals). The electron transfer is to nondistributed states in water. The presence of the distribution law in the n-type situation (cf. the position of the donor in Fig. 12.15) leads to a reduction of available states. Conversely, the experimental results for TiO_2 (n-type) electrodes show relatively high quantum efficiencies, i.e., up to 90% for photons of 3.4 eV.

12.7. The Whole-Cell System

In the last section, the results for the quantum efficiency of individual electrodes (n- and p-type semiconductors) have been given. Knowing the number of photons of the incident light, one can get the photocurrents by multiplying the quantum efficiency by the number of photons. However, the photocurrent of the entire electrochemical *cell*, which consists of a cathode and an anode without other external potential source, has a more important practical meaning than that of a certain electrode. Thus, in the evaluation of the cell carried out below, the photocurrent–energy relation at chosen electrode potential has been integrated over the whole solar spectrum.

12.7.1. Relation between the Potential of an Electrode and a Cell

The potential V_{cell} of a self-driving cell at a current I, is given by[57]

$$V_{cell} = E_{s0} - E_{si} - IR_c \qquad (12.92)$$

where E_{s0} is the potential of the cathode with respect to a reference electrode, E_{si} is the potential of the anode with respect to the same reference electrode, and R_c is the inner cell resistance. Also, the potential V_{ext} (the external potential to the flow of a current) of a driven cell at a current I is given by[57]

$$V_{ext} = E_{si} - E_{s0} + IR_c \qquad (12.93)$$

Let us define V as

$$V = E_{s0} - E_{si} - IR_c \qquad (12.94)$$

If V has a positive value at a current I, the cell is a self-driving cell, and if V has a negative value at a current I, the cell is a driven cell and the cell potential $-V$ is required to cause a current I to flow. In Fig. 12.20 this relation is shown. Figure 12.20a represents a current–potential relationship of the system H_2O/O_2 and H^+/H_2 for individual electrodes, respectively, both for the cathodic and the anodic reactions. If one dips two electrodes into an acidic solution and one of them is in contact with oxygen gas and another is in contact with hydrogen gas, the system works as a self-driving cell. The oxygen electrode works as a cathode ($\frac{1}{4}O_2 + H^+ + e \rightarrow \frac{1}{2}H_2O$) and the hydrogen electrode works as anode ($\frac{1}{2}H_2 \rightarrow H^+ + e$) until I reaches I_1 (at I_1, $V = E_{s0} - E_{si} - IR_c = 0$). To obtain a current I_2, one needs an extra external potential $I_2R_c(= -V)$. Correspondingly, to get hydrogen from one electrode and oxygen from another in the normal dark case, one needs to supply an external potential $V_{ext}[= -V = -(E_{s0} - E_{si} - IR_c)]$.

12.7.2. Calculated Hydrogen Production Rate from Solar Energy Using TiO₂ Photodriven Cells

Photocurrents from individual electrodes stimulated by solar energy at a certain potential have been calculated by integrating the photocurrent of

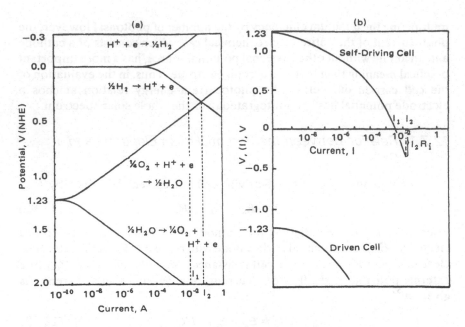

Fig. 12.20. (a) The potential–current relationships of individual electrodes. (b) The potential–current relationships of a self-driving cell and a driven cell.

the monochromatic light through the whole solar energy range. Thus, one can get a current–potential relationship of an individual electrode by carrying out this process over some potential range. Then, one can get a cell potential–current relationship if one knows the value of the flat-band potential of the semiconductor electrode with respect to a reference electrode.

As examples, Eq. (12.94) has been applied to cell 1 [TiO$_2$/solution (pH = 14)/solution (pH = 0, 7, 14)/Pt] and cell 2 [TiO$_2$/solution (pH = 14)/GaAs] with the assumption that all holes arriving at the surface react, i.e., contribute to electricity. The parameters used were as follows: the flat band potential of TiO$_2$ at pH = 14 and at pH = 0 and of GaAs at pH = 0 are −0.8 V(NHE),[58] −0.05 V(NHE),[58] and 0.43 V(NHE)[59]; R is 100 Ω; i_0 of the platinum electrode is 10^{-4} A cm^{-2}; and optical constants of TiO$_2$ were taken from Möllers *et al.*[60] The results are shown in Figs. 12.21 and 12.22. As seen in Fig. 12.21, the maximum cell current of cell 1 is a function of the pH gradient. When pH of the cathodic compartment is 0, the maximum cell current is 0.14 mA cm^{-2} and the maximum hydrogen production rate is 0.06 cm^3 hr^{-1} cm^{-2}. This means that the efficiency of this cell is 0.3% of the total solar energy (0.07 W cm^{-2}). When the pH of the cathodic compartment is 7, the maximum cell current is 0.12 mA cm^{-2}, and when the pH of the

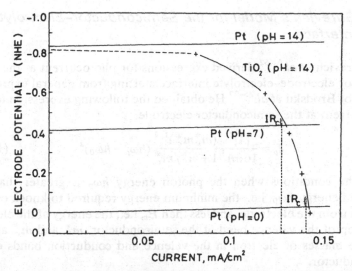

Fig. 12.21. The potential–current relationships of the cell [TiO$_2$/solution (pH = 14)/solution (pH = 0, 7, 14)/Pt].

cathodic compartment is 14, no current is expected. On the other hand, in cell 2 of which both electrodes are illuminated, current flows even when pH gradient is zero. In this case, maximum current is determined by cathodic reaction and is 0.028 mA cm^{-2} (Fig. 12.22).

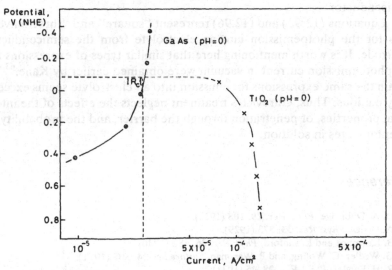

Fig. 12.22. The theoretical potential–current relationship of the cell [TiO$_2$/solution (pH = 0)/GaAs].

12.8. Gurevich's Model for the Semiconductor–Electrolyte Interface

Gurevich et al.[61,62] derived expressions for photocurrent at the semi-conductor electrode–electrolyte interface starting from general expression (12.23) of Brodskii et al.[25-27] He obtained the following expression for the photocurrent at the semiconductor electrode:

$$I_p = \frac{|\Lambda|^2}{16\pi h^3} \frac{(m_0^* m_c^*)^{1/2}}{1 + m_0^*/m_c^*} (h\omega_p - h\omega_0)^2 \qquad (12.95)$$

under the conditions when the photon energy $h\omega_p$ is greater than the threshold energy $h\omega_0$, i.e., the minimum energy required to knock out the electron from the electrode, and less than E_v, i.e., the energy of the electron in the top of the valence band of the semiconductor. m_0^* and m_c^* are the effective masses of electron in the valence and conduction bands of the semiconductor.

When the incident photon energy is further increased (i.e., when $h\omega_p > E_v$) direct photoemission (i.e., without scattering in the volume or at the surface) occurs. In this situation, the following expression for the photocurrent is given by Gurevich et al.[62]:

$$I_p = C(\hbar\omega_p - \hbar\omega_v)^{3/2} \qquad (12.96)$$

where C is a constant independent of different $(h\omega_p - h\omega_0)$ where $h\omega_v$ is the energy corresponding to the electron energy in top of the valence bond in the semiconductor.

Equations (12.95) and (12.96) represent "square" and "three halves" laws for the photoemission into an electrolyte from the semiconductor electrode. It is worth mentioning here that similar types of expressions for the photoemission current in vacuum were obtained earlier by Kane.[63] To obtain the same expressions for emission into an electrolyte seems exceedingly dubious. Thus, Gurevich's treatment neglects the effects of the interfacial properties, of penetration through the barrier, and the probability of acceptor states in solution.

References

1. D. A. Dubridge, *Phys. Rev.* **29**, 108 (1932).
2. A. Goetz, *Phys. Rev.* **33**, 373 (1929).
3. B. Lawrence and L. Linford, *Phys. Rev.* **36**, 482 (1930).
4. G. Wedler, C. Wofing, and P. Wassman, *Surface Sci.* **24**, 302 (1971).
5. R. H. Fowler, *Phys. Rev.* **38**, 45 (1931).
6. A. Summerfeld, *Z. Physik* **47**, 1 (1928).

7. D. B. Matthews and S. U. M. Khan, *Austral. J. Chem.* **28**, 253 (1975).
8. J. O'M. Bockris and D. B. Matthews, *Proc. Roy. Soc.*, *(London)* **A292**, 479 (1966).
9. P. E. Clark and A. B. Garrett, *J. Amer. Chem. Soc.* **61**, 1805 (1939).
10. A. W. Copeland, O. D. Black, and A. D. Garrett, *Chem. Rev.* **31**, 177 (1942).
11. P. J. Hillson and E. K. Rideal, *Proc. Roy. Soc. (London)* **199**, 295 (1949).
12. H. Berg and H. Schweiss, *Electrochim. Acta.* **9**, 425 (1964).
13. H. Berg and H. Schweiss, *J. Electroanal. Chem.* **15**, 415 (1967).
14. M. Heyrovsky, *Proc. Roy. Soc. (London)* **301**, 411 (1967).
15. M. Heyrovsky and F. Pucciarelli, *J. Electroanal. Chem.* **75**, 353 (1977).
16. G. C. Barker, A. W. Gardner, and D. C. Sammon, *J. Electrochem. Soc.* **113**, 1183 (1966).
17. G. C. Barker, *Electrochim. Acta.* **13**, 1221 (1968).
18. G. C. Barker and A. W. Gardner, *Electroanal. Chem.* **47**, 205 (1973).
19. G. C. Barker, D. McKeown, M. J. Williams, G. Bottara, and V. Concialini, *Faraday Discuss. Chem. Soc.* **56**, 41 (1974).
20. R. de Levie and J. E. Kreuser, *J. Electroanal. Chem.* **21**, 221 (1969).
21. Yu. V. Pleskov and Z. A. Rotenberg, *J. Electroanal. Chem.* **20**, 1 (1969).
22. Yu. V. Pleskov, Z. A. Rotenberg, V. V. Eletsky, and V. E. Lakomov, *Faraday Disscuss. Chem. Soc.* **56**, 53 (1974).
23. V. I. Vesselovskii, *Zhur. Fiz. Khim.* **20**, 1493 (1946).
24. J. O'M. Bockris, *Modern Aspects of Electrochemistry* (J. O'M. Bockris and B. E. Conway, eds.), Vol. 1, Plenum Press, New York, (1954).
25. A. M. Brodskii and Yu. Ya. Gurevich, *Zhur. Eksp. Teor. Fiz.* **54**, 213 (1968).
26. A. M. Brodskii, Yu. Ya. Gurevich, and V. G. Levich, *Phys. Status Solidi.* **40**, 139 (1970).
27. A. M. Brodskii and Yu. V. Pleskov, *Progr. Surface Sci.* **2**, 1 (1972).
28. J. O'M. Bockris, S. U. M. Khan, and K. Uosaki, *J. Res. Inst. Catal. Hokkaido Univ.* **24**, 1 (1976).
29. J. O'M. Bockris, M. A. V. Devanathan, and J. Muller, *Proc. Roy. Soc. (London)* **A274**, 55 (1963).
30. F. Constantinescu and E. Magyari, *Problems in Quantum Mechanics* Pergamon Press, Oxford (1971).
31. J. O'M. Bockris and A. K. N. Reddy, *Modern Electrochemistry*, Vol. 2, Plenum Press, New York (1973).
32. R. A. Marcus, (a) *J. Chem. Phys.* **24**, 966 (1956); (b) *J. Chem. Phys.* **43**, 679 (1965).
33. V. G. Levich, *Physical Chemistry, An Advanced Treatise* (H. Eyring, D. Henderson, and W. Jost, ed.), Vol. IXB, Academic Press, New York (1971).
34. P. P. Schmidt, *J. Phys. Chem.* **77**, 488 (1973).
35. K. Fueki, D. F. Feng, and L. Kevan, *J. Phys. Chem.* **74**, 1976 (1970).
36. L. Kevan, *Advances in Radiation Chemistry* (M. Burton and J. L. Magee, eds.), John Wiley & Sons, New York (1974).
37. P. Krolikowski and W. E. Spicer, *Phys. Rev.* **185**, 882 (1969).
38. C. Kittel, *Introduction to Solid State Physics* (4th ed.), John Wiley & Sons, New York (1971).
39. J. O'M. Bockris and D. B. Matthews, *Proc. Roy. Soc. (London)* **A292**, 479 (1966).
40. S. Trasatti, *J. Electroanal. Chem.* **52**, 313 (1974)
41. J. O'M. Bockris and M. A. Habib, *J. Electroanal. Chem.* **68**, 479 (1976).
42. K. Uosaki, Ph.D. Thesis, Flinders University of South Australia, Adelaide (1976).
43. J. Valasek, *International Critical Tables*, Vol. VI, p. 248 (1929).
44. L. I. Korshanov, Ja. M. Zolotovitskii, and V. A. Benderskii, *Russ. Chem. Rev.* **40**, 699 (1971).
45. H. Gerischer, *Physical Chemistry: An Advanced Treatise* (H. Eyring, ed.), Vol. 9A, Chapter 5, Academic Press, New York (1970).

46. V. A. Myamlin and Yu. V. Pleskov, *Electrochemistry of Semiconductors*, Plenum Press, New York (1967); Yu. Ya. Gurevich, M. D. Krotova, and Yu. V. Pleskov, *J. Electroanal. Chem.* **75**, 339 (1977).
47. C. G. B. Garrett and W. H. Brattain, *Phys. Rev.* **99**, 376 (1955).
48. E. O. Johnson, *Phys. Rev.* **111**, 153 (1958).
49. D. Laser and A. J. Bard, *J. Electrochem. Soc.* **123**, 1828 (1976).
50. J. O'M. Bockris and K. Uosaki, *Int. J. Hydrogen Energy* **3**, 157 (1977).
51. L. V. Azaroff and J. J. Brophy, *Electronic Processes in Materials*, Chapter 7, McGraw-Hill, New York (1963).
52. S. M. Ryvkin, *Photoelectric Effect in Semiconductors* (English edition), Chapter XIII, Consultants Bureau, New York (1964).
53. R. W. Gurney, *Proc. Roy. Soc. (London)* **34A**, 137 (1932).
54. B. I. Seraphin and H. E. Bennett, *Semiconductors and Semimetals* (R. K. Willardson and A. C. Beer, eds.), Vol. 3, Chapter 12, Academic Press, New York (1967).
55. G. W. Gobeli and F. G. Allen, *Semiconductors and Semimetals* (R. K. Willardson and A. C. Beer, eds.), Vol. 2, Chapter 11, Academic Press, New York (1966).
56. J. I. Pankove, *Optical Processes in Semiconductors*, Prentice-Hall, Englewood Cliffs, New Jersey (1971).
57. J. O'M. Bockris and R. K. N. Reddy, *Modern Electrochemistry*, Plenum Press, New York (1973).
58. T. Watanabe, A. Fujishima and K. Honda, *Chem. Lett.* p. 897 (1974).
59. D. J. Benard and P. Handler, *Surface Sci.* **40**, 141 (1973).
60. F. Möllers, H. J. Tolle, and R. Memming, *J. Electrochem. Soc.* **121**, 1160 (1974).
61. Yu. Ya. Gurevich, *Electrokhim.*, **8**, 1564 (1972).
62. Yu. Ya. Gurevich, M. D. Krotova, and Yu. V. Pleskov, *J. Electroanal. Chem.* **75**, 339 (1977).
63. E. O. Kane, *Phys. Rev.* **127**, 131 (1962).
64. F. P. Bowden, *Trans. Faraday Soc.* **27**, 505 (1931).
65. L. E. Price, Ph.D. Thesis, Cambridge University, Cambridge, England (1938).
66. C. Durst and D. Taylor, *J. Res. Natl. Bur. Standards* **69a**, 517, (1965).
67. C. Crow and D. Ling, *J. Chem. Soc. Dalton*, p. 698 (1972).

13

Quantum Electrode Kinetics

13.1. Introduction

Electrode kinetics basically deals with charge transfer processes such as electron and proton transfer at the electrode–solution interface. Since electrons and protons are quantum particles, the relevant theory of quantum mechanics can be applied to better understanding the fundamentals of electrode processes.

Ronald Gurney[1] was the pioneer in applying quantum mechanical theory to electrode kinetics. The further use of quantum mechanics and development of the Gurney-based model was not made until the 1960s. Butler,[2] Christov,[3] Bockris and Matthews,[4,5] and Matthews and Khan[6] used the Gurney model to find the electrochemical current at the electrode–solution interface. Gurney's quantum mechanical theory of electron transfer reactions was mainly concerned with radiationless electron transfer (tunneling) to distributed energy states of an acceptor at the interface. He introduced the concept of radiationless electron transfer at the interface and made use of Gamow's tunneling expression.[7]

The Levich group[8,9] applied the quantum theory to electrochemical process in a different manner from Gurney. Their theory, as we have seen in Chapter 9, was concerned with activation process in a polar solvent system with a harmonic approximation for the solvent wave function and a model of interaction in solution in terms of polarons.[9] The transition process at the interface was little discussed, since their treatment was silent regarding the evaluation of the electronic matrix element $L_c = \langle \psi_f | V | \psi_i \rangle$ where the electronic wave function ψ_i is the initial state in the metal (for a cathodic reaction) and ψ_f is the final bound state wave function for the acceptor in solution; V is the perturbing potential at the interface. Absence of the evaluation of the electronic matrix element of transition is characteristic of such work. Thus, a comparative discussion of the Gurney-based expression for the electrochemical current density and related Tafel behavior and the treatments of the Levich group will help our understanding of the basic differences between the two approaches.

Electrochemical kinetics concerns the rate of transfer of an electron between two electronic levels of equal energy and can be described by time-dependent perturbation theory. One electronic level is in the electrode and the other is in the acceptor ion at the interface. This process of transfer of an electron between equal energy levels is termed radiationless transfer, and time-dependent perturbation theory can be introduced to find the transition probability of the electron at the interface. Time-dependent perturbation theory has previously been widely used in finding the transition probabilities and rates of spectroscopic processes such as the spectroscopic dipole transition, electron excitation, and deexcitation processes. All such processes are radiative. The application of time-dependent perturbation theory to electrochemical nonradiative electron transfer processes indicates some analogy between spectroscopic photon and electrochemical processes. A correlation between spectroscopic and electrochemical transitions has been pointed out[10] (see Section 13.3). In Section 13.5, we shall give details of the use of time-dependent perturbation theory in electrode kinetics as developed by Khan, Wright, and Bockris,[11] which suggests the nonadiabaticity of electron transfer to redox system. In the concluding section, we shall discuss electron transfer processes at the semiconductor electrode–solution interface.

13.2. Quantum Theory of the Electrochemical Current Density and Tafel Behavior

The most general expression for the electrochemical current density (considering the interfacial electron transfer step as rate determining) is given (for the cathodic reaction) by[1]

$$i = e_0 \iint c(x)G(x, E)W(x, E)f(E)\rho(E) \, dE \, dx \qquad (13.1)$$

where e_0 is the electronic charge, $c(x)$ is the concentration of acceptor or donor atoms or ions at any distance x from the metal electrode, $G(x, E)$ is the probability of having the acceptor (for the cathodic case) or donor (for the anodic case) at any energy level E and at a distance x from the electrode, $W(x, E)$ is the transition probability of an electron from the electron level in a metal to an acceptor ion in the solution (for the cathodic case) or vice versa, $f(E)$ is the Fermi distribution of *occupied* electron states in the metal electrode and is used to find the cathodic current density and $[1 - f(E)]$, which gives the distribution of *vacant* electronic state in the electrode and is used to find the anodic current density. Finally, $\rho(E)$ is the density of electronic states in the metal electrode. The first integral in Eq. (13.1) is carried out to take into account the total contribution of the current density from all energy states in the metal (considering the bottom of the conduction band as zero) up to infinity (infinity in this case can be roughly taken as the $E_f + \Phi$, because after an energy range $E_f + \Phi$, there are virtually no electrons in the metal), where E_f is the Fermi energy and Φ is the work function in the metal. The second integral in Eq. (13.1) determines the contribution of current density from the acceptance (for the cathodic case) or donation (for the anodic case) of electrons from the solution acceptor or donor, respectively, for all the distances from $x = 0$, at the metal surface to $x \to \infty$, in the bulk of the solution.

Since most of the acceptor and donor ions are situated in the outer Helmholtz plane (OHP), one may remove the integral over distance and write Eq. (13.1) in the simplified form as

$$i = e_0 C \int G(E) W(E) f(E) \rho(E) \, dE \qquad (13.2)$$

What terms of this Eq. (13.2) can be determined quantum mechanically? That which is most widely treated by quantum mechanics is $W(E)$, the transition probability of electrons across the interface. $G(E)$, the probability of having the acceptor or donor state with energy equal to that of the electron in the metal electrode, has not yet been fully determined quantum mechanically. So far, this has been done by all workers in terms of a classical distribution function. Thus, the Levich–Dogonadze group[8,9] has not considered this term separately as they have included it in the term $W(E)$. A quantum mechanical treatment using a time-dependent perturbation theory approach to find the quantity $G(E)$ in the form of a probability of activation P_{act} has been outlined in Section 6.11.

The Fermi distribution term $f(E)$ of Eq. (13.2) is obtained using quantum statistics considering the electron as a fermion that possesses one of two possible opposite spins ($\pm \frac{1}{2}$). The density of electron states $\rho(E)$ is obtained using the quantum mechanical treatment of the free electron theory.[12] Thus, the electrochemical current density can be derived quantum mechanically, but the question remains as to what extent we can use the quantum mechanically derived expression to determine quantitative estimates of the current density and hence the Tafel behavior. In the next subsection we will give the derivation of the current density in the form of a Tafel-like equation for the thermal treatment, originated by Gurney, and the continuum treatment, originated by Libby, and discuss their respective validities.

Rate-Overpotential Relation According to Both the Thermal and Continuum Models

Both models derive the rate of an electron transfer reaction from the basic Eq. (13.2). The main difference in these two treatments is in the calculation of the transition probability $W(E)$ and the probability of attaining a suitable acceptor state (for the cathodic reaction) and donor state (for the anodic reaction). Here, we will outline the essence of each model for a cathodic reaction using a simplified comparative analysis given by Bockris *et al.*[13,14]

(a) Current Density According to the Thermal Approach

One may initially assume for simplicity that electron transfer reactions are adiabatic. For the hydrogen evolution reaction, the rate-determining step may be chosen for the sake of discussion as that corresponding to

$$(H_3O^+)_{dl} + e^-(M) \rightarrow (M—H \cdots OH_2)_{dl} \tag{13.3}$$

where the subscript *dl* means "double layer." The potential energy profile for the process (13.3) is shown in Fig. 13.1.

The vertical transition (Fig. 13.1) $\Delta H_0(e)$ corresponds to the process:

$$e^-(M) + (H_3O^+)_{dl} \rightarrow (M—H \cdots OH_2)_{dl} \tag{13.4}$$

i.e., transferring an electron from the Fermi level of the metal to an H_3O^+ ion in its ground rotation–vibration state with no change in the proton coordinate. Such a transition for an H_3O^+ ion not in its ground state will be accompanied by an energy change $\Delta H_0(e)$. At the intersection point X of curves a and b in Fig. 13.1, $\Delta H_0(e) = 0$, and radiationless electron tunneling from the metal to the proton becomes possible.

Since the thermal model assumes that a Boltzmann distribution exists between the various smoothed-out vibrational–rotational energy states of the $H^+—OH_2$ ion, the probability of finding the latter sufficiently activated

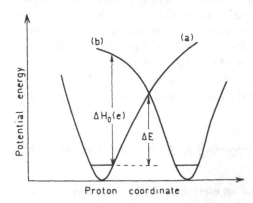

Fig. 13.1. Potential energy–distance profile of hydrogen evolution reaction; curves (a) and (b) show the variation of potential energy with internuclear separation for $M(e) \cdots H^+—H_2O$ and $M—H \cdots OH_2$, respectively.

so that its energy will correspond to the intersection point (Fig. 13.1) is given by

$$G(E) = \exp\left(-\frac{\Delta E}{KT}\right) \tag{13.5}$$

ΔE is a fraction of $\Delta H_0(e)$ (Fig. 13.1), i.e.,

$$\Delta E = \beta \Delta H_0(e) \tag{13.6}$$

where $0 < \beta < 1$.[†] Hence,

$$G(E) = \exp\left[-\frac{\beta \Delta H_0(e)}{kT}\right] \tag{13.7}$$

Since, for a potential $\Delta\phi$ at the electrode solution interface

$$[\Delta H_0(e)]_{\Delta\phi} = \Delta H_0(e) + e_0\Delta\phi \tag{13.8}$$

we have

$$G(E) = \exp\left(-\frac{\beta \Delta H_0(e)}{kT}\right) \exp\left(-\frac{\beta e_0\Delta\phi}{kT}\right) \tag{13.9}$$

The electron tunneling probability, assuming a square barrier,[15] is

$$P_T = A \exp\left\{-\frac{4\pi l}{h}[2m(U_m - E_F)]^{1/2}\right\} \tag{13.10}$$

according to the Gamow equation, where U_m is the barrier height at the interface, E_F is the Fermi level energy of the electron in electrode metal, and l is the barrier width (Fig. 13.2). Hence, using Eqs. (13.9) and (13.10) in Eq. (13.2), one gets the current density at potential $\Delta\phi$ as[§]

$$i = 2eC \frac{4\pi m (kT)^2}{h^2} \exp\left\{-\frac{4\pi l}{h}[2m(U_m - E_F)]^{1/2}\right\}$$

$$\times \exp\left\{-\frac{\beta[\Delta H_0(e) + e\Delta\phi_e]}{kT}\right\} \exp\left(-\frac{\beta e\eta}{kT}\right)$$

$$= i_0 \exp\left(-\frac{\beta e\eta}{kT}\right) \tag{13.11}$$

where the overpotential $\eta = \Delta\phi - \Delta\phi_e$, where $\Delta\phi_e$ is the interfacial potential difference at the equilibrium of the process (Eq. 13.3). This is Tafel's law.

[†] Note that β is in general a function of $\Delta\phi$. It turns out that the dependence on $\Delta\phi$ is small if the potential energy relations are Morse-like.

[§] U_m will be dependent upon the double-layer field and therefore upon η. However, Gurney showed that such an effect was small compared with the effect in the second exponential term.

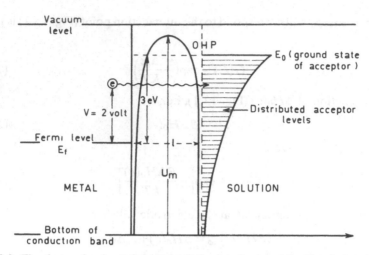

Fig. 13.2. The electron barrier at the metal–solution interface, depicting the electron transfer from metal electrode to distributed acceptor levels in solution even at a cathodic electrode potential as high as 2 V.

The expression is obtained under the approximation that all electrons come from near the Fermi level and the contribution from electrons of other energy levels in the electrode is neglected. Moreover, no consideration has been made about the directional characteristics of electron emission from the metal electrode.

Matthews and Khan[6] derived a more accurate expression of electrochemical current density for the cathodic reaction using the general expression (13.2) and taking into account the component of the motion of an electron-directed normal to the electrode surface with contributions from electrons from all energy states in the metal (i.e., not from only Fermi level electrons). The expression is

$$i = e_0 D \int_0^{E_0} \ln\left[1 + \exp\left(\frac{E_f - e_0 V - E_x}{kT}\right)\right]$$

$$\times \exp\left\{-\frac{\pi^2 l}{h}[2m(U_m - E_x)]^{1/2}\right\} \exp\left[-\frac{\beta(E_0 - E_x)}{kT}\right] dE \quad (13.12)$$

where D is a constant and E_g is the ground state energy of the acceptor in the solution. The first factor in the integral takes into account the number of electrons in the electrode that have a velocity component normal to the metal electrode surface along the x axis and is represented in terms of the component of energy E_x (see Section 12.3). V is the absolute potential. The second factor is the tunneling probability expression for the parabolic type of barrier having width l and height U_m measured with respect to the bottom of the conduction band, and the third factor determines the probability of having the acceptor state at any suitable energy E_x on the solution side of the interface that will correspond the energy of the electron in the electrode. The integration in Eq. (13.12) is done from zero (bottom of the conduction band of the electrode) to the energy state E_0 (with respect to the bottom of the conduction band). Since the major contribution comes from the electrons having this range of energy, the contribution from the energy levels above E_0 has not been considered. This is because, even if one applies an electrode potential of 2 V, an electron energy in the electrode corresponding to Fermi level does not exceed the energy corresponding to E_0, since the energy difference between the Fermi level and the ground acceptor (H_2O^+) level is typically 3.0 eV (Fig. 13.2).

One cannot integrate (13.12) analytically so that it does not have the look of a Tafel equation as does Eq. (13.11). Equation (13.12) has been numerically solved by Matthews and Khan.[6] Log i plotted as a function of V shows Tafel behavior in agreement with experiment.

(b) Current Density According to the Continuum Approach

In the continuum approach,[8,9] the determination of current density commences with the Eq. (13.1) and ultimately uses the following simplified expression:

$$i = e_0 C_n(\delta)\delta \int_0^\infty f(E)\rho(E)W(\delta, E)\, dE \tag{13.13}$$

when δ is some distance from the electrode at which the probability of electron transfer is greatest, C_n is the concentration of the ions in the reaction plane $x = \delta$, and $W(\delta, E)$ includes the transition probability of an electron at a distance δ from the electrode and the probability of producing the activated states in the reaction plane, due to solvent fluctuation; $f(E)$ is the Fermi function and $\rho(E)$ the density of states.

To allow an analytical expression to be evolved, Eq. (13.13) was further simplified by taking an average value of the density ρ of electronic states in the electrode and taking it outside the integral. Then, by using the expression for the transition probability $W(E)$ of Eq. (7.47), one obtains

$$i = e_0 C_n(\delta)\delta\bar{\rho}\int_0^\infty W(\delta, E)f(E)\, dE$$

$$= e_0 C_n(\delta)\bar{\rho}\delta|L_e|^2\left(\frac{\pi}{h^2 kTE_s}\right)^{1/2}\int_0^\infty f(E)\exp\left[-\frac{(\Delta F_r + E_s)^2}{4E_s kT}\right] dE \tag{13.14}$$

where $|L_e|^2$ is the transition moment matrix that has come from $W(\delta_1 E)$, ΔF_r is the free energy of the reaction, and E_s is the organization energy.

Equation (13.14) can be written in a simple form†

$$i = A \exp\left[-\frac{E_s}{4kT} - \frac{\Delta F_r}{2kT} - \frac{\Delta F_r^2}{4E_skT}\right] \int_0^{\infty} f(E)\,dE \qquad (13.15)$$

where

$$A = e_0 C_n(\delta)\delta\bar{\rho}|L_e|^2 \left(\frac{\pi}{h^2 kTE_s}\right)^{1/2} \qquad (13.16)$$

(c) The Tafel Equation from the Continuum Model

Putting the Fermi distribution function for $f(E)$, one can express Eq. (13.15) as

$$i = A \exp\left(-\frac{E_s}{4kT} - \frac{\Delta F_r}{2kT} - \frac{\Delta F_r^2}{4E_skT}\right) \int_0^{E_F + \Phi} \frac{1}{\exp[(E - E_F)/kT] + 1}\,dE$$
$$(13.17)$$

where the upper limit of integration is taken as $E_F + \Phi$ in place of ∞. This is acceptable, since there are virtually no electrons in the metallic states in the energy range beyond $E_F + \Phi$, where E_F is the Fermi level energy of electron and Φ is the work function of the electrode.

$$\Delta F_r = \Delta F^0 + e_0\eta \qquad (13.18)$$

where ΔF^0 is the free energy of reaction when overpotential $\eta = 0$.

Introducing Eq. (13.18) into Eq. (13.17), one gets

$$i = AB \exp\left\{-\left[\frac{E_s}{4} + \frac{1}{2}(\Delta F^0 + e_0\eta)\left(1 + \frac{\Delta F^0 + e_0\eta}{2E_s}\right)\right]\middle/ kT\right\} \quad (13.19)$$

It is observed from Eq. (13.19) that, in general, no Tafel relation is obtained from the continuum model. However, Levich[8] expressed the Tafel-type expression as

$$i = AB\left[-\left(\frac{E_s}{4} + \frac{\Delta F^0}{2} + \frac{e_0\eta}{2}\right)\middle/ kT\right] \qquad (13.20)$$

$$= A' \exp\left(-\frac{e_0\eta}{2kT}\right)$$

$$= A' \exp\left(-\beta\frac{e_0\eta}{kT}\right) \qquad (13.21)$$

where $A' = AB \exp\{-[(E_s + 2\,\Delta F^0)/4kT]\}$

---

† ΔF_r is the free energy of transferring an electron of energy E. It can only be taken outside the integral if electrons are considered to come from a given level E, e.g., the Fermi level.

Equation (13.21) represents the Tafel equation from the continuum model *only under the condition that* in the exponential term of Eq. (13.20) the absolute value

$$\left|\frac{\Delta F^0 + e_0 \eta}{2E_s}\right| \ll 1 \quad \text{or} \quad \left|\frac{\Delta F_r}{2E_s}\right| \ll 1 \tag{13.22}$$

We have made calculations for some typical reactions for the over-potential of 0.5 V and for the typical reorganization energy 0.5 eV. The results (see Table 13.1) is that condition (13.22) is seldom fulfilled for redox reactions. Hence, one can never get the Tafel behavior over a large η of range from the continuum model, and thus this model can not explain the most observed kind of behavior of this electrochemical current density as a function of potential.[†]

It is argued[8,9] that constraints introduced by the harmonic approximation [which is the basis of Eq. (13.14)] are the reason for nonlinearity of Tafel line in the continuum approach. To get the Tafel behavior, it is essential to have an exponential term where the energy of the electron appears as a single power. This is observed in the thermal approach in which the *classical Boltzmann distribution* is introduced [e.g., as in (13.12)] to find the probability of having the acceptor in the required energy state. The proposed model given in Section 6.10 may yield such a relation by a quantum mechanical approach.

13.3. Relations between Spectroscopic and Electrochemical Transitions

The object of this section is to point out correlations (and differences) between the quantum mechanical aspects of fundamental processes in spectroscopy and similar aspects of the emission and acceptance of the electrons across the interfaces between ionic solutions and metals or semi-conductors.

Phenomenologically, the method by which electrode kinetic results are portrayed is only superficially different from that by which spectroscopic results are shown. The current density used in current–potential diagrams is the analog of the intensity of absorption or emission of photons as plotted in spectra. For transfer of an electron across an interface, the electrostatic part of the work done is the potential difference multiplied by the electronic

---

[†] The better behavior in respect to the relative constancy of β with potential, calculated on molecular model theories of electrode kinetics,[14] is a matter of degree. It arises because the molecular models involve bond stretching ("activation of inner levels") and use some Morse-like potential energy–distance relation.

Table 13.1
Experimental Data to Verify the Applicability of Condition (13.22)

Electrode reactions	Free energy[a] of reaction overpotential $\eta = 0$ (ΔF^0 at 25°C) (eV)	Over-potential η (V)	Free energy of reaction at overpotential $\eta (\Delta F_r = \Delta F + \eta)$ (eV) at 25°C	Reorganization energy (E_s at 25°C) (eV)	$\left|\dfrac{\Delta F_r}{2E_s}\right|$	$\left|\dfrac{\Delta F^0}{2E_s}\right| \ll 1$ satisfied?	Can continuum model give Tafel equation?
$Fe^{3+}(H_2O)_6 \rightarrow Fe^{2+}(H_2O)_6$	-0.77	-0.5	-1.27	0.5	1.27	No	No
$CO^{3+}(H_2O)_6 \rightarrow CO^{2+}(H_2O_6)$	-1.82	-0.5	-2.32	0.5	2.32	No	No
$V^{3+}(H_2O)_6 \rightarrow V^{2+}(H_2O)_6$	0.25	-0.5	-0.25	0.5	0.25	Yes	Yes
$Cr^{3+}(H_2O)_6 \rightarrow Cr^{2+}(H_2O)$	0.41	-0.5	-0.09	0.5	0.09	Yes	Yes
$Ti^{3+}(H_2O)_6 \rightarrow T^{1+}(H_2O)_6$	-2.5	-0.5	-3.0	0.5	3	No	No
$[MnO_4(H_2)_2]^{-1} \rightarrow [MnO_4(H_2O)_2]^{-2}$	-0.81	-0.5	-1.31	0.5	1.31	No	No
$Cu^{2+}(H_2O)_4 \rightarrow Cu^{1+}(H_2O)_4$	-0.15	-0.5	-0.65	0.5	0.65	No	No
$U^{4+}(H_2O)_6 \rightarrow U^{3+}(H_2O)_6$	0.6	-0.5	0.1	0.5	0.1	Yes	Yes
$CO(NH_3)_6{}^{3+} \rightarrow CO(NH_3)_6{}^{2+}$	-0.11	-0.5	-0.61	0.5	0.6	No	No
$Ag^{2+}(aq) \rightarrow Ag^{1+}(aq)$	-1.98	-0.5	-2.48	0.5	2.48	No	No
$Au^{3+}(aq) \rightarrow Au^{1+}(aq)$	-2.8	-0.5	-3.3	0.5	3.3	No	No

[a] The values of ΔF^0 are determined from the thermodynamic data of free energy of formation of individual ions in solution and taken from R. E. Dickerson, *Molecular Thermodynamics*, W. A. Benjamin, New York (1969).

charge, and this is the analog of the energy of the photon adsorbed in spectroscopic processes. Thus, the intensity–frequency relation in absorption spectroscopy (Fig. 13.3a) is the analog of the current–potential diagram in cathodic electrochemical processes (Fig. 13.3b). Qualitatively, after a critical energy has been reached, further increase in the photon energy suddenly increases the intensity of absorption of photons by particles in solution (or in the gas phase). Correspondingly, increase in the potential of an electrode beyond a critical value in the negative direction (i.e., increase in electron energy in the electrode) suddenly increases the rate of acceptance of electrons from the electrode by acceptors in solution (or their emission from the metal surface).

An increase of potential in the positive direction causes an increase in the emission of electrons to the electrode from a donor in solution in anodic electrode processes and corresponds to emission of photons from excited particles. Thus, qualitatively, there are operational and presentational similarities between the individual steps in the cathodic electrode process and absorption spectroscopy, and in anodic electrode processes and emission spectroscopy, respectively.

Correspondingly, there are differences in the intensity–energy aspects between spectroscopic and electrochemical phenomena; the spectroscopic situation shows an increase in the rate of transition with respect to energy until a certain maximum followed by a symmetrical fall of transition rate at higher energies (Fig. 13.4a); the electrochemical transition rate increases with electron energy until a certain current density, after which, at higher energies, it does not undergo a diminution but tends toward a limiting value[15,16] (Fig. 13.4b).

We have mentioned in Section 1.10 that in the spectroscopic case, as long as the transition matrix is nonzero (according to selection rules), the probability of photon transition passes as a maximum at $\omega \approx \omega_{ml}$ i.e., when $E_m - E_l = h\omega$, which is the Bohr condition [Eq. (4.85)].

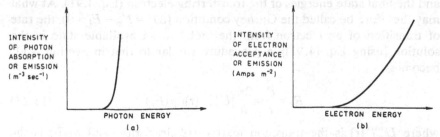

Fig. 13.3. The initial trend of increase in absorption and acceptance of photons and electrons as a function of their respective energies. (a) Increase in intensity of photon absorption or emission with photon energy. (b) Increase in intensity of electron acceptance or emission with electron energy.

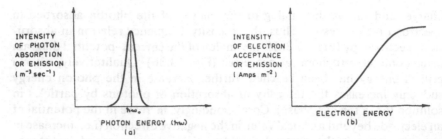

Fig. 13.4. The intensity of absorption and acceptance of photons and electrons as a function of their respective energies. (a) Intensity of photon absorption or emission with photon energy. (b) Intensity of electron acceptance or emission with electron energy.

The intensity–energy relationship for absorption in spectroscopy can be expressed[10,17-20] as

$$I = \frac{|U_{ml}^0(x)|^2 t \sin^2\{[E_m - E_l - \hbar\omega)/2\hbar]/t\}}{\hbar^2\{[(E_m - E_l - \hbar\omega)/2h]\}^2 t^2} N_0 \exp\left(\frac{-v\hbar\omega}{kT}\right) \quad (13.23)$$

where $U_{ml}^0(x)$ is the transition moment matrix $= \langle\psi_m|e_0 F_0 x|\psi_l\rangle$, with F_0 the field of electromagnetic radiation; t is the time of perturbation of the process; and v is the vibrational quantum number of the acceptor ion-solvent bond. N_0 is the number of the acceptor ion–solvent bonds in their ground vibrational quantum state per unit area of electrode, and ω_0 is the frequency of vibration of the photon absorbing bonds. A plot of Eq. (13.23) against I indicates the nature of the spectroscopic intensity–energy relation of the ideal case in the gas phase (Fig. 13.4a), since the transition moment matrix $U_m^0(x)$ is independent of the frequency.

It was shown in Section 4.11 that, in contrast to the spectroscopic case, the transition probability of an electron in the electrochemical case is a maximum when $E_m - E_l \simeq 0$, where E_l and E_m are respectively the initial and the final state energies of the transferring electron (Eq. 4.91). At what may therefore be called the Gurney condition ($\Delta E = E_m - E_l \simeq 0$), the rate of transition of an electron from the metal to an available state in the solution [using Eq. (4.91)], a procedure similar to that in Section 4.11, becomes

$$R_t = \frac{P_t}{t} = \frac{2\pi}{\hbar}|U_{ml}^{\text{pert}}(\mathbf{r})|^2 \rho(E_f) \quad (13.24)$$

where $U_{ml}^{\text{pert}}(\mathbf{r})$ is the transition matrix $= \langle\psi_m|e_0\mathbf{X}_0\mathbf{r}|\psi_l\rangle$ and $\rho(E_f)$ is the density of final electronic states.

The dependence of current density on the energy of the electron in the electrochemical case arises from the Boltzmannian population of acceptor molecules in different vibrational–rotational levels in solution, and this is expressed as,

$$N(E) = N(E_g) \exp[-\beta(E_g - E)/kT] \qquad (13.25)$$

where E_g is the ground-state energy of the acceptor and E is the electron energy (see Fig. 13.4) and $N(E_g)$ is the number of acceptors or donors at energy E_g.

Thus, the intensity–energy relation for electron acceptance in the electrochemical case is obtained approximately from Eqs. (13.24) and (13.25) as

$$i = R_t N(E)$$
$$= \frac{2\pi e_0}{h} |U_{fi}^0(\mathbf{r})|^2 \rho(E_F) N(E_g) \exp\left[\frac{-\beta(E_g - E)}{kT}\right] \qquad (13.26)$$

The spectroscopic and electrochemical intensity–energy diagrams (Figs. 13.4a and 13.4b), in which the spectroscopic curve peaks and the electrochemical one does not, arise from differences in expressions (13.23) and (13.26), which in turn depend on the fact that spectroscopic transitions (e.g., in the infrared region) involve excitation from a lower to a higher vibrational quantum state. Such excitations are governed by the transition moment matrix, which becomes numerically significant only for transitions between certain vibrational states. In the electrochemical case, there is no *excitation* following the process in which the emitted electrons from the metal electrode are accepted in solution.

The selection rule for the spectroscopic (excitation) of a vibrational transition, $v = \pm 1$, arises from the zero and nonzero conditions of the transition moment matrix $U_{ml}^0(x)$. In the electrochemical case, however, the transition matrix $U_{ml}^{\text{pert}}(\mathbf{r})$ is nonzero for the transfer of an electron from the electrode surface to an available acceptor state in solution, or vice versa, and hence no selection rule is involved for the transition of particles between states of equal energy.

As there is no selection rule involved in electrochemical transitions (in contrast to those in spectroscopic systems), electrons from the metal electrode can undergo transition to any energy states of an acceptor in solution so long as it is equal to the energy they have when emitted from the metal electrode, i.e., it obeys the Gurney condition.[1] The population of the corresponding states in solution is given by Eq. (13.25).

In solution, collisional interactions cause a lessening of the sharpness of definition of the energy of the vibrational–rotational states, so that there is, in effect, an overlap of the energies of these states that amounts to the existence of quasicontinuous states.[1] This fact, together with the absence of a selection rule, allows the Gurney condition to be fulfilled at any electron energy, i.e., at any electrode potential, whereas the Bohr condition can exist only for energy transfer for which the transition moment matrix is nonzero, i.e., at an energy at which the selection rules allow transition. Thus, the intensity–energy relation does not peak out in electrochemical transitions, only again starting to increase when a new energy region is reached. With increasing energy of the electronic states in the metal, there will be an exponential increase in the rate of transition [Eq. (13.26)]. Barring the imposition of factors outside this discussion,† this increase of the rate of transitions with increasingly negative electrode potential will continue until the Fermi level in the metal has been shifted to a value equal to that of the ground state of the acceptor E_g. If the electron energy is made still more negative, the electrons have no simple acceptor states in the double layer (e.g., those in an H_3O^+ ion) to which to transfer, but it is found experimentally that the current does not peak; it tends to slow its increase and the intensity–energy (i.e., the current–potential) relation flattens out (Fig. 13.4b). From theoretical computations made of the current–potential relation in photoemission, when incident photons lift electrons in metals above the ground state of H_3O^+ ions, it seems likely that the emitted electrons become solvated, diffuse about in the solution using up energy, until they have lost enough to become capturable by water molecules, with which they react to generate H or OH^-, or by ground state H_3O^+.

13.4. Adiabaticity and Nonadiabaticity in Electron Transfer Processes

A reaction is called adiabatic when the reacting system moves during the whole course of the reaction along the same potential energy curve corresponding to the given electronic state (Fig. 13.5, arrow a). On the other hand, a nonadiabatic reaction is one in which the reacting system undergoes transition during the course of the reaction from one potential energy curve to another corresponding to a higher energy electronic state (as shown in Fig. 13.5, arrow b).

These two possibilities of reaction processes make much difference in calculating the rate of any chemical reaction. According to the transition state theory formalism, these two types of reactions (i.e., adiabatic and nonadiabatic) influence the value of the transmission coefficient κ, which is a preexponential term in the rate expression. The value of κ is unity for the adiabatic reaction and less than unity for nonadiabatic reactions. For the

† For example, the provision by mass transport of an insufficient number of states to an interface to accept the electrons at a rate at which the perturbation due to local electric fields is producing them.

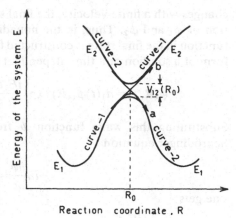

Fig. 13.5. The plot of energy of the reacting system as a function of reaction coordinate. Arrow a represents adiabatic and b represents nonadiabatic motion in the region of closest approach of two potential curves.

electron transfer case, where the quantum aspects (e.g., tunneling) are involved analogously, the transmission coefficient can be formulated in a quantum mechanical sense as the tunneling probability factor (according to semiclassical theory) or the transition probability factor (according to time-dependent perturbation theory).

Physically, in the adiabatic case, the reactants interact strongly at their closest approach when the probability of reaction is a maximum and the system remains in a lower energy curve. This is because when the system interacts strongly, the gap between the lower and higher potential energy curves (Fig. 13.5) increases and the transition to the upper curve becomes unfavorable. In the nonadiabatic case, the system interacts weakly and the gap between the lower and upper electronic state curves become small, and the probability of going to the upper electronic state curve increases and the rate of reaction is reduced, thus indicating a value of transmission coefficient κ lower than unity.

One needs to find out the conditions under which a reaction is adiabatic or nonadiabatic. Landau[21] and Zener[22] made the first attempts to find the probability of transition of the reacting system from a lower electronic state to an upper one, and thus to find the adiabaticity and nonadiabaticity of a reaction. They determined under what conditions the reactions are adiabatic. We give here the essence of their treatments.

Landau–Zener Formulation

Let $\phi_1(\mathbf{R})$ and $\phi_2(\mathbf{R})$ be two electronic states of a reacting system and E_1 and E_2 are the corresponding energies of the states (Fig. 13.5). According to the adiabatic theorem,[21,22] if the molecule is initially in the state ϕ_2 and the reaction coordinate $\mathbf{R}$ changes infinitely slowly from $R \ll R_0$ (reactant state) to $R \gg R_0$ (product state), then the molecule will remain in a state ϕ_1. If $\mathbf{R}$

changes with a finite velocity, the final state $\psi(\mathbf{R})$ will be a linear combination of ϕ_1 and ϕ_2. Thus, in the nonadiabatic case, the total system wave function in the final state is constructed from ϕ_1 and ϕ_2 and expressed in the form of a solution of a time-dependent Schrödinger equation as follows

$$\psi(\mathbf{R}, t) = a_1(t)\phi_1(\mathbf{R}) \exp\left[\frac{iE_1 t}{\hbar}\right] + a_2(t)\phi_2(\mathbf{R}) \exp\left[\frac{iE_2 t}{\hbar}\right] \quad (13.27)$$

Substituting this wave function ψ from Eq. (13.27) in the following Schrödinger equation

$$-i\hbar\frac{\partial \psi}{\partial t} = H\psi \quad (13.28)$$

one gets

$$a_1(t) \exp\left(\frac{iE_1 t}{\hbar}\right) H\phi_1 - \phi_1 E_1 a_1(t) \exp\left(\frac{iE_1 t}{\hbar}\right) + i\hbar\dot{a}_1(t) \exp\left(\frac{iE_1 t}{\hbar}\right)\phi_1$$

$$+ i\hbar a_2(t) \exp\left(\frac{iE_2 t}{\hbar}\right) H\phi_2$$

$$- \phi_2 E_2 a_2(t) \exp\left(\frac{iE_2 t}{\hbar}\right) + i\hbar\dot{a}_2(t) \exp\left(\frac{iE_2 t}{\hbar}\right)\phi_2 = 0 \quad (13.29)$$

where $\dot{a}_1$ and $\dot{a}_2$ are the first differential of the coefficients a_1 and a_2 with respect to time.

It was assumed by Zener that if ϕ_1 and ϕ_2 are made orthogonal, it is tantamount to the removal of an interaction energy V_{12} near the crossing point of Fig. 13.5.[†] Then,

$$H\phi_1 = E_1\phi_1 + V_{12}\phi_2 \quad (13.30)$$

$$H\phi_2 = E_2\phi_2 - V_{12}\phi_1 \quad (13.30a)$$

where V_{12} is the matrix element of perturbation between the two electronic states, represented by E_1 and E_2 of the curves in Fig. 13.5. Thus,

$$V_{12} = \langle\phi_1|V|\phi_2\rangle \quad (13.31)$$

Putting the value of $H\phi_1$ and $H\phi_2$ from Eqs. (13.30) and (13.30a) into Eq. (13.29), one obtains the simplified expression as

$$i\hbar\dot{a}_1(t) \exp\left(\frac{iE_1 t}{\hbar}\right)\phi_1 + i\hbar\dot{a}_2(t) \exp\left(\frac{iE_2 t}{\hbar}\right)\phi_2 - V_{12}a_1(t) \exp\left(\frac{iE_1 t}{\hbar}\right)\phi_2$$

$$- V_{12}a_2(t) \exp\left(\frac{iE_2 t}{\hbar}\right)\phi_1 = 0 \quad (13.32)$$

[†] In Fig. 13.5, the electronic level represented by E_2 is in fact, for the example of H_3O^+ + e → MH, the vibrational set of states for the *second* electronic level of the activated complex.

If the system on a dynamic passage through R_0 of Fig. (13.5) remains in the single surface (i.e., on the curve E_1), the reaction is adiabatic, as indicated before, and if not, the reaction is nonadiabatic (cf. Fig. 13.5, arrow b). Now the problem is to determine the probability that it will remain on the single (lower) surface (E_1) when the system passes through the transition region. It is convenient to specify two states of the system. Hence one considers initially:

$$a_1(-\infty) = 1$$

$$a_2(-\infty) = 0 \tag{13.33}$$

Since one is interested in knowing the system when it passes through the region R_0 (Fig. 13.5) to the right, i.e., to $R > R_0$, it is mathematically convenient to consider a region where $R = \infty$. Thus, ϕ_1 and ϕ_2 may be regarded as finite and constant at $R = \infty$, for here, there is no interaction between states 1 and 2 (Fig. 13.5). Under these conditions, Eq. (13.32) is satisfied only when the coefficients of ϕ_1 and ϕ_2 in the equation vanish, for the time dependence will vanish and hence time-dependent terms must be zero. Thus, equating the coefficient of ϕ_1 and ϕ_2 of Eq. (13.32) to zero, one gets two coupled first-order differential equations as

$$i\hbar \dot{a}_1(t) = V_{12}a_2 \exp\left[\frac{-i(E_1 - E_2)t}{\hbar}\right] \tag{13.34a}$$

$$i\hbar \dot{a}_2(t) = V_{12}a_1 \exp\left[\frac{i(E_1 - E_2)t}{\hbar}\right] \tag{13.34b}$$

From these two coupled equations, one can determine the value of $a_2(\infty)$, the modulus square of which gives the probability that the system remains in the lower potential energy surface in the final state, i.e., remains on the E_1 curve (Fig. 13.5) and transition occurs adiabatically. Hence, the probability of an adiabatic transition is

$$P_{ad} = |a_2(\infty)|^2 = 1 - |a_1(\infty)|^2 \tag{13.35}$$

Similarly, the probability that system jumps from the curve E_1 to the E_2 is given by

$$P_{nonad} = 1 - |a_2(\infty)|^2 = |a_1(\infty)|^2 \tag{13.36}$$

Elimination of $a_1(t)$ from Eq. (13.34b), using Eq. (13.34a), leads to the single equation in terms of $a_2(t)$ as

$$\frac{\partial^2 a_2(t)}{\partial t^2} + \left[\frac{i}{\hbar}(E_1 - E_2) + \frac{1}{V_{12}}\frac{\partial V_{12}}{\partial t}\right]\frac{\partial a_2(t)}{\partial t} + \left(\frac{V_{12}}{\hbar}\right)^2 a_2(t) = 0 \tag{13.37}$$

Since V_{12} is independent of time, one gets

$$\frac{\partial V_{12}}{\partial t} = 0 \tag{13.38}$$

Thus, using Eq. (13.38) in Eq. (13.37), one gets

$$\frac{\partial^2 a_2(t)}{\partial t^2} + \frac{i}{\hbar}(E_1 - E_2)\frac{\partial a_2(t)}{\partial t} + \left(\frac{V_{12}}{\hbar}\right)^2 a_2(t) = 0 \tag{13.39}$$

Zener[22] solved this equation for $a_2(t)$ under two assumptions such as (a) $V_{12}(R_0)$ (the splitting energy) $\ll$ the relative kinetic energy of the two particles in the interacting systems. (b) The transition region between the electronic levels E_1 and E_2, near R_0, is so small that in it one may regard $(E_1 - E_2)$ as a linear function of time. This condition is satisfied provided $V_{12}(R)$ is sufficiently small compared with the relative kinetic energy of the motion of the system to the transition region near R_0 (Fig. 13.5). Hence, it is expressed that

$$\frac{E_1 - E_2}{\hbar} = \alpha t \tag{13.40}$$

where α is a constant that must have the dimensions of (time)$^{-2}$, and

$$\frac{\partial \Phi_1}{\partial t} = \frac{\partial \Phi_2}{\partial t} = 0 \tag{13.41}$$

since ϕ_1 and ϕ_2 are considered independent of time near the crossing point.
    To simplify Eq. (13.39), one can define

$$\beta = \frac{V_{12}}{\hbar} \tag{13.42}$$

and put a value of a_2 as

$$a_2 = F(t) \exp\left[\frac{-i(E_1 - E_2)t}{\hbar}\right] \tag{13.43}$$

where the function $F(t)$ needs to be determined.
    Now, using Eqs. (13.40), (13.42), and (13.43) in Eq. (13.39), one obtains

$$\frac{\partial^2 F}{\partial t^2} + \left(\beta^2 + \frac{i\alpha}{2} + \frac{\alpha^2 t^2}{4}\right)F = 0 \tag{13.44}$$

This is a standard Weber differential equation[23] and its solution can be obtained by means of the Laplace transform method with the application of boundary condition as given in Eq. (13.33). The solution of Eq. (13.44) when put into Eq. (13.43), taking the modulus square, is

$$|a_2(\infty)|^2 = 1 - \exp\left[-\frac{2\pi V_{12}^2}{h|(\partial/\partial t)(E_1 - E_2)|}\right]$$ (13.45)

Now,

$$\frac{\partial}{\partial t}(E_1 - E_2) = v_s|P_1 - P_2|$$ (13.46)

where v_s is the relative velocity of approach of the one reactant to another during reaction, $|P_1 - P_2|$ is the net force exerted on the system tending to restore to its original state or take it to a final state. P_1 is the force exerted on the system to restore to the original state, and P_2 is that for the final state. Thus P_1 and P_2 represent the negative slopes of the potential energy curves E_1 and E_2 (Fig. 13.5), i.e., $P = -\partial E/\partial R$. Using Eq. (13.46) in Eq. (13.45) and putting the result in Eq. (13.35), one gets, respectively, the probability of an adiabatic and a nonadiabatic transition as,

$$P_{ad} = 1 - \exp\left[-\left(\frac{2\pi V_{12}^2}{h v_s|P_1 - P_2|}\right)\right]$$ (13.47)

and

$$P_{nonad} = \exp\left[-\frac{2\pi V_{12}^2}{h v_s|P_1 - P_2|}\right]$$ (13.48)

Equations (13.47) and (13.48) indicate under what conditions the probability of the reaction would be adiabatic or nonadiabatic. Thus, one finds that:

(1) If the perturbation interaction near R_0, i.e., V_{12}, is large, the probability of an adiabatic transition is dominant. If V_{12} is small, the probability of nonadiabatic transition is dominant [Eqs. (13.47) and (13.48)]. If the relative kinetic energy corresponding to the relative velocity of approach of the reacting systems v_s is large compared to V_{12}, the reaction becomes nonadiabatic.

(2) If there exists a large difference in slopes of the energy–distance curves, that is between two restoring forces in the initial and final states, the probability of nonadiabaticity increases. This indicates that, due to higher restoring forces in the initial state than in the final state, the probability of transition of the system to the final state becomes weak and the reaction becomes nonadiabatic.

In electrochemical terminology, there is some relation between these conditions, and the Tafel slopes observed for electrochemical reactions. In the most common case, the Tafel slopes tend to $\frac{1}{2}$, and it is well known that this signifies equal slopes of the potential energy–distance relations, i.e., equal values of P_1 and P_2. Equation (13.48) shows that $P_1 = P_2$ is equivalent to "a zero *non*-nonadiabatic," i.e., a completely adiabatic reaction. Hence, the Tafel slope may be a measure of adiabaticity.

Later, Bates[24] criticised the Landau–Zener[21,22] treatment for finding the probability of transition. The assumption by Landau–Zener that the transition occurs only in the zone of crossing of two potential energy surfaces is not necessarily correct; transition may occur away from the crossing, and Bates calculated the width of the transition zone as

$$\Delta R = \left(\frac{4\pi v_s h}{P}\right)^{1/2} \tag{13.49}$$

where it is observed that the width of the transition zone is proportional to the square root of the velocity of the relative motion of the system v_s. P is a constant and is equal to e_0^2/R_0^2, where e_0 is the electronic charge.

Serious error is likely to arise in applying the treatment of Landau[21] and Zener[22] unless the wave functions Φ_1 and Φ_2 are spherically symmetrical and unless the velocity of relative motion is low. For example, in a charge transfer process that involves electron transfer between p and d states, the Landau–Zener result will not be applicable where contribution of angular orientations, and even the rotational aspects of interacting molecules, have not been taken into account.

Recently, Schmidt[25] favored the use of the Landau–Zener[21,22] expression for the probability of transition [Eqs. (13.47) and (13.48)] and its use to find the adiabatic and nonadiabatic character of electron transfer reaction rather than that from time-dependent perturbation theory expression (e.g., Fermi's golden rule). According to him, Fermi's expression

$$P = \frac{2\pi}{\hbar} |\langle \psi_f | V | \psi_i \rangle|^2 \delta(E_f - E_i) \tag{13.50}$$

is applicable only for a nonadiabatic transition because it involves only a first-order contribution of the transition-inducing perturbation, and a condition of its applicability is that the magnitude of perturbation cannot be large. But, for a *small* perturbation, the separation between the lower and upper energy surfaces will be small, and hence the system can easily jump from the lower to the upper surface, and the reaction will become nonadiabatic.

However, expression (13.50) is applicable for a reasonably *strong perturbation* for a *very short time*.[52] Such a short time perturbation may be also sufficient to allow the reaction to go adiabatically, i.e., that it should have a value of the transition probability close to unity. Schmidt's criticism[25] that Eq. (13.50) cannot be applied to test adiabaticity need not be applicable to electron transfer between phases, when the time interval of the perturbation is about 10^{-16} sec and the perturbation less than 1 eV.

On the other hand, if one notices the derivation of the Landau–Zener[21,22] formula, one observes that one of the conditions for the applicability of the adiabatic condition lies in the fact that the relative velocity of motion of the reacting system must be small and the perturbation V_{12} much less than the relative kinetic energy of the system. Thus, a *weak* perturbation is also assumed in the Landau–Zener treatment, although it is applied both to the adiabatic and nonadiabatic processes. One important point to notice here is that, for the electron or proton transfer reactions from an electrode to an ion, the relative motion of electron and the ion during reaction is in fact high, and the Landau–Zener adiabatic formula will tend to be less applicable in such cases.

13.5. Time-Dependent Perturbation Theory of Electron Transfer at Electrodes

In this section, we present the use of the time-dependent perturbation theory to calculate the transition probability of electron from the electrode to redox systems at a metal–solution interface. The discussion will be on the basis of the recent Khan, Wright, and Bockris[11] (KWB) treatment. In the KWB treatment, redox reactions are chosen, since they constitute a simple electron transfer situation (non-bond – breaking), which has some analogy[10,26] to electronic transitions that occur in the photoexcitation of electron (i.e., transition of electron from lower orbital to higher orbital) in an atom or in a molecule.

The first discussion of the quantal treatment of redox reactions was given on the basis of ideas originated by Weiss,[27] Marcus,[28] Hush,[29] and Levich[8,30] and developed by Dogonadze[9] and co-workers[31] in terms of continuum theory. Recently, Dogonadze et al.,[9,31] Kestner et al.,[32] and Schmickler[33] have attempted to advance the theoretical position in terms of molecular models. Most of such theories (not references 9, 32, and 33) involve the adiabatic assumption regarding the electron transfer reaction. Dogonadze[9] was the first to focus attention on the fact that electrode reactions are not necessarily adiabatic.

The validity of the Landau–Zener formula for the calculation of the degree of nonadiabaticity rests on a sufficiently small splitting energy, but also on spherical symmetry of the wave functions involved. It is not satisfactory for the interacting systems *that contain orbitals other than s orbitals*, as is the case for redox systems.[24]

13.5.1. Transition Probability of Electrons from Electrodes

The transition probability of an electron P_t at a certain energy level in the electrode to an acceptor state of about the same energy (within the limit ΔE, since energy is conserved to within the limit of the uncertainty principle[34]) in the double layer can be calculated by using an expression from time-dependent perturbation theory,[34] i.e.,

$$P_t = \frac{2\pi\tau}{\hbar}|\langle\psi_f|V(\mathbf{r})|\psi_i\rangle|^2\rho(E_F) \tag{13.51}$$

where ψ_i and ψ_f are respectively the initial and final state wave functions of the electron. The initial state of the electron is that in the metal surface, corresponding to energy equal to that of the Fermi level; and the final state is that of the electron in the reduced ion. τ is the time during which the electron in the electrode surface experiences a perturbation[34,35] before it joins the acceptor ion after crossing the interface.† $V(\mathbf{r})$ is the perturbing potential energy which causes the transition. $\rho(E_F)$ is the density of electronic states in the final state.

13.5.2. KWB Model for the Calculation of the Transition Probability

To calculate the transition probability P_t, it was considered that an electron transfers from the electrode (P_t) to an available $Fe^{3+}(H_2O)_6$ ion in solution (Fig. 13.6).

The *initial state* wave function ψ_i of a conducting electron in the electrode (metal) surface is not taken as Bloch wave function because the

† P_t/τ is, then, the transition rate.

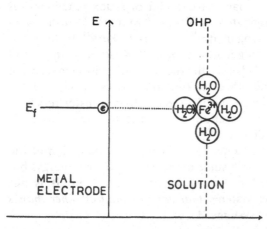

Fig. 13.6. A schematic model of the electron transfer process from a metal electrode to a redox system at the interface. Fe^{3+} is coordinated with six water molecules, but only four are shown on the plane of the paper, since the other two are out of plane of the paper.

periodicity of the metal is no longer effective at the surface. The wave function ψ_i is taken as that of a free particle (cf. Dogonadze[9]), but allowance is made for the interaction in the metal by using an effective mass of the conducting electron. Thus, ψ_I is expressed[36] as

$$\psi_i(\mathbf{r}) = L_0^{-3/2} e^{i\mathbf{k}\cdot\mathbf{r}} \tag{13.52}$$

where L_0 is the normalization constant of the free particle wave function and $\mathbf{k}$ is the wave vector of the electron with magnitude given by

$$|\mathbf{k}| = (2m_e^* E/\hbar^2)^{1/2} \tag{13.53}$$

where m_e^* is the effective mass of the electron. E is the energy of the electron in the metal electrode and is taken to be the Fermi energy of the electron E_F. At the reversible potential on the rational scale V_R, the electron energy becomes $E = E_f - e_0 V_R$.

Expressing Eq. (13.52) in terms of spherical waves, one gets[36]

$$\psi_i(\mathbf{r}) = L_0^{-3/2} 4\pi \sum_{l=0}^{\infty} \sum_{m=-l}^{l} i^l j_l(\mathbf{k}r) Y_{lm}^*(\hat{\mathbf{k}}) Y_{lm}(\hat{\mathbf{r}}) \tag{13.54}$$

where l is the angular momentum quantum number and m is the magnetic quantum number; $j_l(\mathbf{k}r)$ is the spherical Bessel function of order l and argument $\mathbf{k}r$ and $Y_{lm}(\hat{\mathbf{k}})$ and $Y_{lm}(\hat{\mathbf{r}})$ are the spherical harmonics of order l and projection m.

The *final state* wave function ψ_f is expressed as

$$\psi_f = \psi_d^{(1)}(\mathbf{r}) Y_{lm}(\hat{\mathbf{r}}) \tag{13.55}$$

where $\psi_d^{(1)}(\mathbf{r})$ is the *perturbed* radial wave function of the 3d state electron in the $Fe^{2+}(H_2O)_6$ ion in solution (see Section 13.5.4). $Y_{lm}(\hat{\mathbf{r}})$ is the spherical harmonic that represents the angular part of the wave function ψ_f.

The perturbing potential $V(r)$ which acts between the surface electrons and the $Fe^{3+}(H_2O)_6$ ion is taken in terms of the electrostatic work done by the electrode on the electron and the potential between an acceptor ion and a surface electron, i.e.,

$$V(r) = e_0 X r - \frac{z_i e_0^2}{\varepsilon_s r} \tag{13.56}$$

where X is the field across the double layer and ε_s is the static dielectric constant of the medium at the interface, and the other symbols have their usual meaning.

13.5.3. Unperturbed 3d State Wave Function of the Fe^{2+} Ion

The unperturbed 3d state wave function of the Fe^{2+} ion, i.e., $\psi_d(r)$, is obtained by solving the radial part of the Schrödinger equation,

$$\frac{1}{r^2}\frac{d}{dr}\left[r^2\frac{d\psi_d(r)}{dr}\right]+\left[-\frac{l(l+1)}{r^2}+2E_d^0-2V_0(r)\right]\psi_d(r)=0 \quad (13.57)$$

where E_d^0 is the binding energy of the 3d electron in the Fe^{2+} ion and $V_0(r)$ is the corresponding potential energy. Equation (13.57) can be solved numerically using the self-consistent analytical potential given by Yunta *et al.*,[37,38] i.e.,

$$V_0(r)=(N_e-1)\left(\kappa-\frac{1}{r}\right)e^{-\alpha\gamma}-\frac{N_p-N_e+1}{r} \quad (13.58)$$

where N_e is the number of the electrons in the atom or ion, N_p is the number of protons in the atom, and α, κ, and γ are the parameters for the atom or the ion concerned and are given by Yunta *et al.*[37,38]

13.5.4. Perturbed 3d State Wave Function of the $Fe^{2+}(H_2O)_6$ Ion

One can use the theory to generate the wave function $\psi_d^{(1)}(r)$ from the unperturbed 3d state wave function $\psi_d(r)$ of Fe^{2+} ion in the gas phase. Thus, the appropriate wave function of the Fe^{2+} ion in solution can be regarded as the (time-independent) perturbed value of the Yunta wave function. $\psi_d(r)$ is the eigenfunction of the unperturbed Hamiltonian of Eq. (13.57). For the Fe^{2+} ion in aqueous solution, the perturbed part of the Hamiltonian can be represented in terms of the electrostatic potential between an ion and the first-coordination sphere of water molecules as[39]

$$U=\sum_{n=1}^{6}\frac{z_ie_0\mu\cos\theta}{|R-r|^2} \quad (13.59)$$

where the summation is over the number of dipoles (here six in number) in the solvation sheath of the ion; μ is the dipole moment of water molecule, $|R-r|$ is the radial vector of the 3d electron from the dipoles, and θ is the angle between the vector $|R-r|$ and the dipole moment vector μ. Considering that the field is uniformly distributed around the Fe^{3+} ion, one can write Eq. (13.59) as†

$$U=\left(\frac{\xi}{|R-r|^2}\right)\cos\theta \quad (13.60)$$

where

$$\xi=6z_ie_0\mu \quad (13.61)$$

† The structure of the complex is not accounted for in this approximation, i.e., it is assumed that six water molecules surround Fe^{3+} at equal distances.

The first-order perturbed 3d state wave function $\psi_d^{(1)}$ can be obtained as

$$\psi_d^{(1)} = \psi_d + C_{dm}\psi_m \tag{13.62}$$

$$C_{dm} = \frac{\langle \psi_m | V | \psi_d \rangle}{E_d^0 - E_m^0} \tag{13.63}$$

where $m \neq d$.

Now, using the expression for U from Eq. (13.60), one can write C_{dm} as

$$C_{dm} = \frac{\xi}{E_d^0 - E_m^0} \left\langle \psi_m \left| \frac{\cos\theta}{|\mathbf{R} - \mathbf{r}|^2} \right| \psi_d \right\rangle \tag{13.64}$$

The unperturbed 3d state wave function, ψ_d, can be expressed in term of its radial and angular parts as

$$\psi_d = \psi_d(r) Y_{20}(\hat{\mathbf{r}}) \tag{13.65}$$

where $\psi_d(r)$ is the normalized radial wave function of the 3d electron in Fe^{2+} ion. Hence, using Eqs. (13.65) and (13.64), one obtains

$$C_{dm} = \frac{\xi}{(E_d^0 - E_m^0)} \int \psi_m^* \frac{1}{|\mathbf{R} - \mathbf{r}|^2} \psi_d(r) \cos\theta \, Y_{20}(\hat{\mathbf{r}}) \, dr \tag{13.66}$$

One knows that[36,40]

$$\cos\theta \, Y_{lm}(\hat{\mathbf{r}}) = \left[\frac{(l+1+m)(l+1-m)}{(2l+1)(2l+3)} \right]^{1/2} Y_{(l+1)m}(\hat{\mathbf{r}})$$
$$+ \left[\frac{(l+m)(l-m)}{(2l+1)(2l-1)} \right]^{1/2} Y_{(l-1)m}(\hat{\mathbf{r}}) \tag{13.67}$$

Hence, for a d orbital, i.e., $l = 2$ and for $m = 0$,

$$\cos\theta \, Y_{20}(\hat{\mathbf{r}}) = \frac{3}{35^{1/2}} Y_{30}(\hat{\mathbf{r}}) + \frac{2}{15^{1/2}} Y_{10}(\hat{\mathbf{r}}) \tag{13.68}$$

Using Eq. (13.68) the integral of Eq. (13.66) can be expressed as

$$\int \psi_m^* \frac{1}{|\mathbf{R} - \mathbf{r}|^2} \psi_d(r) \left[\frac{3}{35^{1/2}} Y_{30}(\hat{\mathbf{r}}) + \frac{2}{15^{1/2}} Y_{10}(\hat{\mathbf{r}}) \right] dr \tag{13.69}$$

ψ_m^* of Eq. (13.69) can be written as

$$\psi_m^* = \psi_m^*(r) Y_{l0}^*(\hat{\mathbf{r}}) \tag{13.70}$$

When Eq. (13.70) is used in Eq. (13.69), due to orthonormal conditions of spherical harmonics, ψ_m^* becomes

$$\psi_m^* = \begin{cases} \psi_f^*(r)Y_{30}^*(\hat{r}) & \text{for } l = 3 \\ \psi_p^*(r)Y_{10}^*(\hat{r}) & \text{for } l = 1 \end{cases} \tag{13.71}$$

where $\psi_f^*(r)$ and $\psi_p^*(r)$ are the radial part of the 4f and 3p state wave functions of the electron in the Fe^{2+} ion.

Using these values of ψ_m^* from Eq. (13.71) in Eq. (13.69) and putting the result in Eq. (13.66), one obtains

$$C_{dm} = \frac{\xi}{(E_d^0 - E_f^0)} \left[\frac{3}{35^{1/2}} \int \psi_f^*(r)Y_{30}(\hat{r}) \frac{1}{|\mathbf{R}-\mathbf{r}|^2} \psi_d(r)Y_{30}(\hat{r}) \, dr \right]$$

$$+ \frac{\xi}{(E_d^0 - E_p^0)} \left[\frac{2}{15^{1/2}} \int \psi_p^*(r)Y_{10}(\hat{r}) \frac{1}{|\mathbf{R}-\mathbf{r}|^2} \psi_d(r)Y_{10}(\hat{r}) \, dr \right] \tag{13.72}$$

Using C_{dm} from Eq. (13.72) in Eq. (13.63), one gets the perturbed wave function as

$$\psi_d^{(1)} = \psi_d(r)Y_{20}(\hat{r}) + \frac{3\xi}{35^{1/2}(E_d^0 - E_f^0)} \left\langle \psi_f(r)Y_{30}(\hat{r}) \right.$$

$$\times \left[\frac{1}{|\mathbf{R}-\mathbf{r}|^2} \right] \psi_d(r)Y_{30}(\hat{r}) \right\rangle \psi_f(r)Y_{30}(\hat{r})$$

$$+ \frac{2\xi}{15^{1/2}(E_d^0 - E_p^0)} \left\langle \psi_p(r)Y_{10}(\hat{r}) \left[\frac{1}{|\mathbf{R}-\mathbf{r}|^2} \right] \psi_d(r)Y_{10}(\hat{r}) \right\rangle \psi_p(r)Y_{10}(\hat{r})$$

$$= \psi_d(r)Y_{20}(\mathbf{r}) + \frac{3\xi}{35^{1/2}(E_d^0 - E_f^0)} |M_1| \psi_f(r)Y_{30}(\hat{r})$$

$$+ \frac{2\xi}{15^{1/2}(E_d^0 - E_p^0)} |M_2| \psi_p(r)Y_{10}(\hat{r}) \tag{13.73}$$

where E_p^0 and E_f^0 are, respectively, the binding energies of 3p and 4f state electrons in an unperturbed Fe^{2+} ion, and M_1 and M_2 are the matrix elements.

To find the integrals in Eq. (13.73), one needs to expand the vector operator $1/|\mathbf{R}-\mathbf{r}|^2$ in terms of spherical harmonics which in turn takes into account the angular orientations of electron orbital with respect to the dipoles. This expansion is given as[23]

$$\frac{1}{|\mathbf{R}-\mathbf{r}|^2} = \left[2Rr \left(\frac{R^2+r^2}{2Rr} - \cos\theta \right) \right]^{-1} \tag{13.74}$$

Following Heins' expansion,[23] one can show that

$$\frac{1}{|\mathbf{R}-\mathbf{r}|^2} = \sum_{l=0}^{\infty} \frac{2l+1}{2Rr} Q_l\left(\frac{R^2+r^2}{2Rr}\right) P_l(\cos\theta) \tag{13.75}$$

where $P_l(\cos\theta)$ is an associated Legendre function and $Q_l(R^2+r^2/2Rr)$ is a Legendre function of the second kind.

Using the addition theorem of spherical harmonics, Eq. (13.75) can be expressed in terms of spherical harmonics, i.e.,

$$\frac{1}{|\mathbf{R}-\mathbf{r}|^2} = \sum_{l,m} \frac{4\pi}{2Rr} Q_l\left(\frac{R^2+r^2}{2Rr}\right) Y_{lm}^*(\hat{\mathbf{R}}) Y_{lm}(\hat{\mathbf{r}}) \tag{13.76}$$

Hence, using Eq. (13.76), the first matrix element of Eq. (13.73) can be expressed as

$$M_1 = \sum_{l,m} \frac{4\pi}{2R} Y_{lm}^*(\hat{\mathbf{R}}) \left[\int_0^\infty r^2\, dr\, \psi_f(r) \frac{1}{r} \psi_d(r) Q_l\left(\frac{R^2+r^2}{2Rr}\right) \right] \langle Y_{30}(\hat{\mathbf{r}})| Y_{lm}(\hat{\mathbf{r}})| Y_{30}(\hat{\mathbf{r}}) \rangle \tag{13.77}$$

The factor $\langle Y_{30}(\hat{\mathbf{r}})| Y_{lm}(\hat{\mathbf{r}})| Y_{30}(\hat{\mathbf{r}}) \rangle$ vanishes unless $m=0$, and this factor can be expressed as[41]

$$\langle Y_{l_1 m_1}(\hat{\mathbf{r}})| Y_{l0}(\hat{\mathbf{r}})| Y_{l_2 m_2}(\hat{\mathbf{r}}) \rangle = \left| \frac{(2l_1+1)(2l+1)(2l_2+1)}{4\pi} \right|^{1/2} \begin{pmatrix} l_1 & l & l_2 \\ 0 & 0 & 0 \end{pmatrix}^2 \tag{13.78}$$

where $l_1 = l_2 = 3$ and

$$\begin{pmatrix} l_1 & l & l_2 \\ 0 & 0 & 0 \end{pmatrix}$$

is the Wigner $3j$ symbol for $m_1 = m = m_2 = 0$.

Using Eq. (13.78) in Eq. (13.77), one gets the matrix element as

$$M_1 = \sum_l \frac{4\pi}{2R} Y_{l0}^*(\hat{\mathbf{R}}) \int_0^\infty r^2\, dr\, \psi_f(r) \frac{1}{r} \psi_d(r) Q_l\left(\frac{R^2+r^2}{2Rr}\right)$$

$$\times (-1)^m \left(\frac{2l+1}{4\pi}\right)^{1/2} (2l_1+1) \begin{pmatrix} 3 & l & 3 \\ 0 & 0 & 0 \end{pmatrix}^2 \tag{13.79}$$

One can express the spherical harmonics $Y_{l0}^*(\mathbf{R})$ of Eq. (13.79) in terms of an associated Legendre function[36] as

$$Y_{l0}^*(\hat{\mathbf{R}}) = \left(\frac{2l+1}{4\pi}\right)^{1/2} P_l(\cos\theta) \tag{13.80}$$

where θ is the angle between the vector $\mathbf{R}$ and the polar axis.

Using Eq. (13.80) in Eq. (13.79), one obtains the value of M_1 as

$$M_1 = \frac{2l+1}{2R} \sum_l (2l_1+1) \begin{pmatrix} 3 & l & 3 \\ 0 & 0 & 0 \end{pmatrix}^2 P_l(\cos\theta) \int_0^\infty r^2 \, dr \, \psi_f(r) \frac{1}{r} \psi_d(r) Q_l\left(\frac{R^2+r^2}{2Rr}\right)$$

(13.81)

Similarly, the second matrix element of Eq. (13.73) becomes

$$M_2 = \frac{2l+1}{2R} \sum_l (2l+1) \begin{pmatrix} 1 & l & 1 \\ 0 & 0 & 0 \end{pmatrix}^2 P_l(\cos\theta) \int_0^\infty r^2 \, dr \, \psi_p(r) \frac{1}{r} \psi_d(r) Q_l\left(\frac{R^2+r^2}{2Rr}\right)$$

(13.82)

Wigner's $3j$ symbol of Eqs. (13.81) and (13.82) can be determined from the general formula[41]

$$\begin{pmatrix} j_1 & j_2 & j_3 \\ 0 & 0 & 0 \end{pmatrix} = (-1)^{J/2} \left[\frac{(J-2j_1)!(J-2j_2)!(J-2j_3)!}{(J+1)!} \right]^{1/2}$$

$$+ \left[\frac{(J/2)!}{(J/2-j_1)!(J/2-j_2)!(J/2-j_2)!} \right]$$

(13.83)

for the even values of J and where $J = (j_1 + j_2 + j_3)$, and is zero for odd values of J.

Now, using Eqs. (13.81) and (13.82) in Eq. (13.73), the perturbed 3d state wave function of $Fe^{2+}(H_2O)_6$ ion can be expressed as

$$\psi_d^{(1)} = \psi_d(r) Y_{20}(\mathbf{r}) + A\psi_f(\mathbf{r}) Y_{30}(\mathbf{r}) + B\psi_p(r) Y_{10}(\mathbf{r})$$

$$= \psi_F$$

(13.84)

where

$$A = \frac{3\xi |M_1|}{35^{1/2}(E_d^0 - E_f^0)}$$

(13.85)

and

$$B = \frac{2\xi |M_2|}{15^{1/2}(E_d^0 - E_p^0)}$$

(13.86)

13.5.5. Transition Probability Using the Perturbed Wave Function $\psi_d^{(1)}$ of the $Fe^{2+}(H_2O)_6$ Ion

Putting the perturbed 3d state wave function $\psi_d^{(1)}$ from Eq. (13.84) in Eq. (13.51) in place of ψ_f, one gets the matrix element as

$$M = \langle \psi_f | V(r) | \psi_i \rangle$$

$$= \langle \psi_d(r) Y_{20}(\hat{\mathbf{r}}) | V(r) | \psi_i \rangle + A \langle \psi_f(r) Y_{30}(\hat{\mathbf{r}}) | V(r) | \psi_i \rangle$$

$$+ B \langle \psi_p(r) Y_{10}(\hat{\mathbf{r}}) | V(r) | \psi_i \rangle$$

(13.87)

Putting the initial state wave function ψ_i from Eq. (13.54) in Eq. (13.87) and using the orthonormal condition of spherical harmonics one obtains

$$M = \frac{4\pi}{L_0^{3/2}} \sum_{m=-2}^{2} Y_{2m}(\hat{\mathbf{k}})(-1) \int_0^\infty r^2\, dr\, j_2(kr) V(r)\psi_d(r)$$

$$+ \frac{4\pi A}{L_0^{3/2}} \sum_{m=-3}^{3} Y_{3m}(\hat{\mathbf{k}})(-i) \int_0^\infty r^2\, dr\, j_3(kr) V(r)\psi_f(r)$$

$$+ \frac{4\pi B}{L_0^{3/2}} \sum_{m=-1}^{1} Y_{1m}(\hat{\mathbf{k}})(i) \int_0^\infty r^2\, dr\, j_1(kr) V(r)\psi_p(r)$$

$$= -C_1 I_1 - iC_2 I_2 + iC_3 I_3 \qquad (13.88)$$

where the real quantities I_1, I_2, and I_3 are the first, second, and third integrals of Eq. (13.88).

Using the wave vector $\mathbf{k}$ parallel to the z axis, which is considered perpendicular to the metal surface, and considering the $\mathbf{k}$ normalization,[36] L_0 is to be taken as 2π.

$$C_1 = \frac{4\pi}{L_0^{3/2}} \sum_{m=-2}^{2} Y_{2m}(\hat{\mathbf{k}}) = \frac{1}{(2\pi)}\left(\frac{5}{\pi}\right)^{1/2} \qquad (13.89)$$

$$C_2 = \frac{4\pi}{L_0^{3/2}} A \sum_{m=-3}^{3} Y_{3m}(\hat{\mathbf{k}}) = \frac{1}{(2\pi)}\left(\frac{7}{\pi}\right)^{1/2} A \qquad (13.90)$$

$$C_3 = \frac{4\pi}{L_0^{3/2}} B \sum_{m=-1}^{1} Y_{1m}(\hat{\mathbf{k}}) = \frac{1}{(2\pi)}\left(\frac{3}{\pi}\right)^{1/2} B \qquad (13.91)$$

Hence, the square of the matrix element M becomes

$$|M|^2 = |-C_1 I_1 - iC_2 I_2 + iC_3 I_3|^2$$

$$= C_1^2 I_1^2 + C_2^2 I_2^2 + C_3^2 I_3^2 - 2C_2 I_2 C_3 I_3 \qquad (13.92)$$

Then, the transition probability can be written as

$$P_t = \frac{2\pi\tau}{\hbar} |M|^2 \rho(E_F) \qquad (13.93)$$

13.5.6. Transition Probability: Quantitative

The quantitative calculation of the probability of transition of an electron from the platinum electrode to $Fe^{3+}(H_2O)_6$ to form $Fe^{2+}(H_2O)_6$ has been carried out from Eq. (13.93) using $|M|^2$ from Eq. (13.92).

The values of A and B have been determined respectively from Eqs. (13.85) and (13.86) with Eq. (13.61) to find ξ using $\mu = 2.3\, D$ for water dipoles.[42] The radial vector $\mathbf{R}$ between the only 3d electron in the Fe^{3+} ion and the center of the water dipole is taken as $3.81a_0$, where a_0 is the Bohr radius. We have in Eq. (13.80) $P_l(1) = 1$ for all values of l, since $\theta = 0$ for **R** parallel to polar axis (Fig. 13.7).

The radial wave functions $\psi_p(r)$, $\psi_d(r)$, and $\psi_f(r)$ corresponding to 3p, 3d, and 4f electron states in the Fe^{2+} ion [which are needed to find M_1 and M_2 from Eqs. (13.81) and (13.82), respectively] were calculated numerically by solving the radial Schrödinger equation with the use of self consistent potential (13.58) and the computer code ORBITAL with values of $\alpha = 2.4624$, $\kappa = 0.001$, and $\gamma = 0.75$ from Yunta et al.[37,38] The plots of the unperturbed wave functions 3p, 3d, and 4f are given in Fig. (13.8). The values of $3j$ symbols in M_1 and M_2 have been calculated from the general relation (13.83) under the rules governing the evaluation of Eq. (13.83).

The integrals in Eqs. (13.81) and (13.82) were calculated numerically for the different values of the Legendre function $Q_l(R^2 + r^2/2R^2)$ using the standard relations.[43] The plots of the perturbed 3d state wave function $3d^{(1)}$ is given in Fig. 13.9. The integrals of Eq. (13.88) were determined by numerical integration using $V(r)$ from Eq. (13.56) with the field $X = 2.25 \times 10^7$ V cm^{-1} at the reversible potential $V_R = 0.75$ (p.z.c.)[44] for the Fe^{3+}/Fe^{2+} process. In these integrals, the standard forms of the spherical Bessel function $j_1(kr)$, $j_2(kr)$, and $j_3(kr)$ for $l = 1, 2$, and 3 have been used.[43] The values of the electron wave vector **k** were calculated from Eq. (13.53) taking the effective mass of electron $m_e^* = 1.15m_e$ (free electron mass) in the case of platinum metal and the electron energy at the reversible potential V_R as $E = (5.9 - 0.75)$ eV $= 5.15$ eV. The quantities M_1, M_2, A, B, C_1, C_2, C_3, I_1, I_2, and I_3 are given in Table 13.2.

The perturbation time can be found from

$$\tau = h/\Delta E \qquad (13.94)$$

where ΔE is the difference between the initial and the final state energies and can be taken as the uncertainty in the energy of the electron, i.e., as

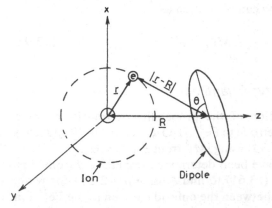

Fig. 13.7. The schematic diagram of an ion–dipole interaction and the coordinate representation for the quantum mechanical computation of energy of such interaction.

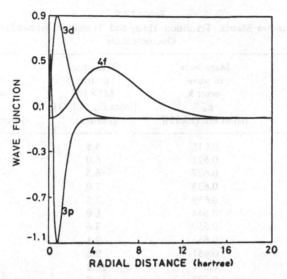

Fig. 13.8. The radial wave functions of 3p, 3d, and 4f state electrons in an Fe^{2+} ion.

$$\Delta E = \frac{(\Delta p)^2}{2m_e} = 6 \times 10^{-12} \text{ ergs} \qquad (13.95)$$

where $\Delta p = h/\Delta x$ and Δx is the uncertainty in the position of an electron across the interface (~ 6 Å). Hence, one gets from Eqs. (13.94) and (13.95) that

$$\tau = 1.66 \times 10^{-16} \text{ sec} \qquad (13.96)$$

Table 13.2

Computed Values of Different Terms Used in Evaluating the Matrix Element M at Reversible Potential of Fe^{3+}/Fe^{2+} Reaction

Terms	Values
Matrix element M_1	0.02551 hartree
Matrix element M_2	−0.070875 hartree
A	−0.018943
B	−0.032655
C_1	0.50329
C_2	−0.011281
C_3	−0.012731
I_1	0.12854×10^{-2} hartree
I_2	−0.13559 hartree
I_3	0.13010 hartree

Table 13.3

Values of Transition Matrix, Transition Time, and Transition Probability at a Series of Overpotentials

Over-potential (V)	Magnitude of wave vector $\mathbf{k}$, a_0^{-1} [from Eq. (13.53)]	Transition matrix $M^2 \times 10^6$ [from Eq. (13.92)] (hartree)^2	Transition probability $P_t \times 10^3$ [from Eq. (13.98)]
0	0.615	5.5	2.4
0.1	0.621	6.0	2.6
0.2	0.627	6.5	2.8
0.3	0.633	7.0	3.0
0.4	0.639	7.5	3.3
0.5	0.644	8.0	3.5
0.6	0.650	8.6	3.8
0.7	0.656	9.1	4.0
0.8	0.661	9.6	4.2
0.9	0.664	10.2	4.5
1.0	0.672	10.7	4.7

There must thus be three states in the acceptor ion identifiable with states having the "same" energy as that of the Fermi energy in the metal, and these states will have the energy, $E_F + \Delta E$, E_F, and $E_F - \Delta E$, so that there are three states per $2\,\Delta E$. Hence, the *maximum* value of the density of states is

$$\rho(E_F) = \frac{\partial N}{\partial E} = \frac{3}{2\,\Delta E} = 2.2 \times 10^{11} \text{ per unit energy (in ergs)} \quad (13.97)$$

13.5.7. Results: Transition Probability

The transition probability was calculated from Eq. (13.93) with the use of computed values of the square of the matrix element M and perturbation time τ [see Eq. (13.96) and Table 13.3]. The result of the transition probability at the reversible potential of Fe^{3+}/Fe^{2+} reaction for which $|M|^2 = 5.5 \times 10^{-6}$ (hartree)^2 and $\tau = 1.66 \times 10^{-16}$ sec becomes

$$P_t = \frac{(6.28)^2 \times 1.66 \times 10^{-16}}{1.054 \times 10^{-27}} [5.5 \times 10^{-6} \times (27.2)^2 \times (1.6)^2 \times 10^{-24}] \times 2.3 \times 10^{11}$$

$$= 2.4 \times 10^{-3} \quad (13.98)$$

This value of the transition probability P_t indicates the nonadiabatic character of the typical redox reaction that has been considered here.†

---

† However, this reaction could be an entrance into some novel aspects of electrocatalysis.

The value of the transition probability P_t given in Eq. (13.98) is obtained considering the dielectric constant $\varepsilon_s = 6$ in the second term of the perturbing potential of Eq. (13.56). For higher values of the dielectric constant one gets smaller values of P_t (e.g., $P_t = 0.6 \times 10^{-3}$ for $\varepsilon_s = 10$ and $P_t = 0.1 \times 10^{-3}$ for $\varepsilon_s = 20$).

13.5.8. Transition Probability from Gamow's Equation

Gamow's tunneling probability expression has often been used for finding the transition probability of an electron between two states across a barrier. Levich[8] claimed that Gamow's expression was not applicable for the electron transfer reactions. Gamow's expression involves the WKB approximation. This is an approximate method for solving the quantum mechanical problems and is applicable when $P(x) \gg (\lambda/2\pi)[dp(x)/dx]$, where $P(x)$ is the momentum of the particle and λ is the wavelength of the particle. We have shown in Section 8.3 that in the electrochemical interfacial situation this condition is approximately satisfied.

However, Gamow's tunneling expression considers just the probability of tunneling of particles (electrons) from one side of the interfacial barrier (e.g., from the electrode surface) to the other side of it. It does not consider whether or not the tunneling electron becomes accepted by an acceptor to form a product having specific wave function in the initial (in the surface) and the final (product) states. Thus, one expects that Gamow's expression will give a higher value of transition probability than that from the time-dependent perturbation theory expression where the consideration of forming a final product state from the initial state has been made in terms of respective wave functions [Eq. (13.51)].†

We give here an estimate of transition probability from Gamow's equation and compare it with that of time-dependent perturbation theory. Gamow's tunneling probability expression for the parabolic type of barrier is

$$P_t = \exp\left\{-\frac{\pi^2 l}{h}[2m_e(U_m - E)]^{1/2}\right\} \tag{13.99}$$

where the barrier height U_m and barrier width l are obtained from the modelistic consideration[6,46] (Fig. 13.10) of the barrier, and the electron energy at the reversible potential is obtained from

$$E = E_{F,vac} - e_0 V_R \tag{13.100}$$

---

† Levich[8] suggested that the oscillation of the dipoles near the electrode would change the barrier so that tunneling took place under changing conditions and so the Gamow expression would be fundamentally inapplicable. Bockris and Sen[53] calculated the relaxation times in the double layer and found them slow, at least for electron tunneling.

where $E_{F,vac}$ is the Fermi energy of the metal electrode in vacuum and V_R is the reversible potential expressed on the rational scale of the reaction concerned. This value of $E_{F,rev}$ for the platinum electrode at the reversible potential of Fe^{3+}/Fe^{2+} reaction becomes $E_{F,rev} = 5.9 - 0.75 = 5.15$ eV.

The barrier model at the interface is constructed by considering the interaction of an outgoing surface electron with the first layer of water molecule on the electrode surface, Coulomb interaction of the electron with the acceptor $[Fe^{3+}(H_2O)_6]$ ions on the outer Helmholtz plane (OHP) and their images in the metal, and the Born interaction of an electron in the second layer of solvent molecules. Detail of such barrier calculation has been given in Section 12.5 with H_3O^+ as an acceptor at the mercury electrode. The schematic diagram of the barrier model is given in Fig. 13.10 for the $Fe^{3+}(H_2O)_6$ ion as an acceptor at the platinum electrode. From the barrier model (Fig. 13.10), $U_m = 9.78$ eV and $l = 1.55$ Å. Using the values of $E_{F,rev}$, U_m, l, and electron mass m_e in Eq. (13.99), one obtains the value of transition probability $P_t = 69 \times 10^{-3}$ at the reversible potential.

Thus, one finds that the value of the transition probability from Gamow's equation is an order of magnitude greater than that of the time–dependent perturbation theory expression [Eq. (13.98)], and this is expected, since Gamow's equation does not consider the *specifics* of the acceptance of the electron on the solution side and should give, hence, a higher result.

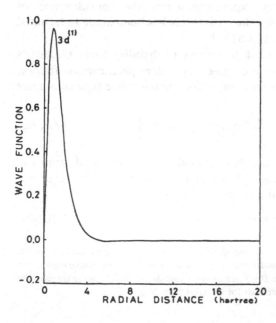

Fig. 13.9. The radial perturbed $3d^{(1)}$ state wave function of electrons in an Fe^{2+} ion in solution.

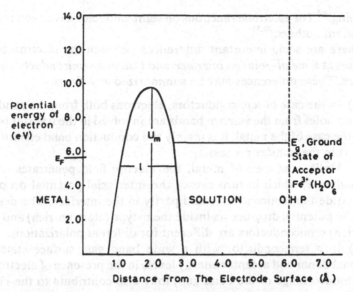

Fig. 13.10. A schematic diagram of potential energy barrier for electron transfer from a platinum electrode to acceptor $Fe^{3+}(H_2O)_6$ ions at the interface.

13.5.9. Potential Dependence of the Transition Probability

When the electrode is subject to an overpotential, the electron energy at the reversible potential changes. The perturbing potential $V(r)$ [Eq. (13.56)] also changes with overpotential, since the field across the interface changes with it. The values of k, matrix element square $|M|^2$, and the transition probability P_t at series of overpotentials are given in Table 13.3.

Over the range of about 1.0 V, the transition probability changes about two times. Simple computations show that this is equivalent to a change in symmetry factor β of about 0.05, which is not inconsistent with most experimental observations.

13.6. Quantum Theory of Electrochemical Processes at Semiconductor Electrodes

There has been little work on the quantum mechanical aspects of electrochemical processes at the semiconductor–solution interfaces. Gerischer[47,48] first put forward a theoretical treatment of semiconductor electrode reactions on the basis of Gurney's quantum mechanical model.[1] Other treatments were made by Levich,[8] Dogonadze,[9] and recently Gleria and

Memming.[49] The electrode reactions on semiconductor have been reviewed by different authors.[50,51]

There are some important differences between the electron transfer processes at a *metal–solution interface* and that at a *semiconductor–solution interface*. These differences may be summarized as follows:

(a) In the case of semiconductors, electrons both from the conduction band and holes from the valence bands are involved in the transfer process. But in the case of the metal, it is mainly the conduction band electrons that take part in the transfer process.

(b) Unlike the case of metal, the electric field penetrates into the semiconductor, which in turn causes the interfacial potential drop to be partly inside the semiconductor and partly in the interface. The degree to which the potential drop occurs inside the n-type (electron-rich) and p-type (hole-rich) semiconductors are different for different polarizations.

(c) In a semiconductor with a wide band gap, surface states (i.e., additional donor and acceptor energy levels in the presence of electrolyte in the forbidden range of the band gap) may also contribute to the electron transfer process.

Due to the above-mentioned differences between the metal and the semiconductors, one will observe some differences in the kinetic expressions for the electron transfer process at the semiconductor–solution interface. Here, in this subsection, we will discuss the electron transfer process from a semiconductor electrode to a redox system or vice versa mainly on the lines of the treatment of Gerischer.[47,48]

The following electron exchange reaction at the semiconductor electrode was considered

$$ox + sc(e) \rightarrow red \qquad (13.101)$$

where "ox" and "red" refer to the oxidized, (e.g., Fe^{3+}) and reduced (e.g., Fe^{2+}) species, respectively in the solution and sc(e) represents the electron in the semiconductor. The "ox" species represent an empty level of an electron, since these species are deficient in an electron, whereas the "red" species represent an occupied electron level in the species in the solution.

The rate of electron transfer to the redox system in a cathodic process occurring over all energy ranges is proportional to the number of occupied states in the semiconductor electrode and the number of empty states in redox electrolyte, and this can be represented by the expression

$$i(\text{cathodic}) = e_0 \int_0^\infty \nu^c(E) \rho_{sc}(E) f^c(E - E_{F,sc}) G_{ox}(E) \, dE \qquad (13.102)$$

where the bottom of the valence band of the semiconductor has been taken as the zero of energy. $\nu^c(E)$ is the proportionality factor that contains the frequency with which electrons in the interface move in an appropriate direction from the interior of the semiconductor electrode to an oxidized ion having the same energy as that of the electron and contains the *tunneling probability factor*.

In Eq. (13.102), $\rho_{sc}^c(E)$ is the density of electron states in the conduction band of semiconductor electrode and is given by

$$\rho_{sc}^c(E) = 4\pi(2m_e^*/h^2)^{3/2}(E - E_c)^{1/2} \quad \text{for } E > E_c \quad (13.103)$$

where m_e^* is the effective mass of an electron in the conduction band. E_c is the energy of the electron corresponding to the bottom of the conduction band.

In Eq. (13.102), $f^c(E - E_{F,sc})$ is the Fermi distribution of electron in the conduction band of semiconductor where $E_{F,sc}$ is the Fermi energy of the semiconductor and given approximately (for the intrinsic semiconductor when number of electrons is equal to the number of holes) as (for an intrinsic substance)

$$E_{f,sc} \approx \tfrac{1}{2}(E_c + E_v) \quad (13.104)$$

where E_v is the energy at the top of the valence band. $G_{ox}(E)$ is the distribution of oxidized species in solution, which has been expressed[48,49] as

$$G_{ox}(E) \simeq C_{ox} \exp[-(E_{F,redox} - E)/kT] \quad (13.105)$$

where C_{ox} is the concentration of the oxidized species, $E_{F,redox}$ is the Fermi energy of the redox electrolyte[47,48] (Fig. 13.11), and this is considered to be equal to the Fermi energy in a semiconductor at equilibrium.

The Fermi energy of the redox electrolyte arises from the analogy of the Fermi energy of the semiconductor, since the level $E_{F,redox}$ coincides with the Fermi level of the semiconductor $E_{f,sc}$ (Figs. 13.11 and 13.12). This means that $E_{F,redox}$ is the energy level where the probability of having the "reduced" species and "oxidized" species are equal.

The corresponding expression for the anodic current density due to transfer of electrons from the reduced species in the solution to the empty electron levels and holes in the semiconductor is given as

$$i(\text{anodic}) = e_0 \int_0^\infty \nu^a(E)\rho_{sc}^a(E)f(E_{F,sc} - E)G_{red}(E)\,dE \quad (13.106)$$

where $\nu^a(E)$ is the frequency factor for the electron transfer in the anodic case and $\rho_{sc}^a(E)$ is the density of hole states in the valence band and is expressed as

$$\rho_{sc}^a(E) = 4\pi(2m_h^*/h^2)^{3/2}(E_v - E)^{1/2} \quad \text{for } E < E_v \quad (13.107)$$

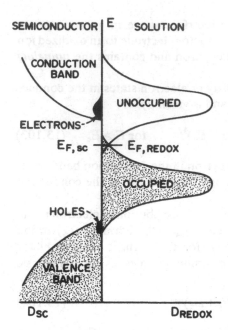

Fig. 13.11. The schematic diagram of the relative distribution of energy states for a semiconductor in contact with an electrolytic solution containing a redox couple (according to Gerischer[47,48]).

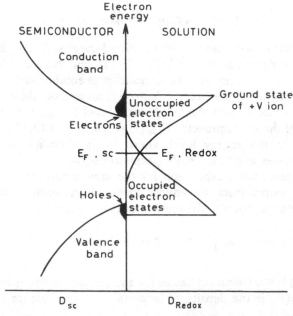

Fig. 13.12. The schematic diagram of the relative distribution of energy states for a semiconductor in contact with an electrolytic solution (form as that of Gurney[1]).

and

$$f^a(E_{F,sc} - E) = 1 - f^c(E - E_{F,sc}) \qquad (13.108)$$

and

$$G_{red}(E) = C_{red} \exp[-(E - E_{F,redox})/kT] \qquad (13.109)$$

where C_{red} is the concentration of the reduced species.

References

1. R. W. Gurney, *Proc. Roy. Soc. (London)* **A134**, 137 (1931).
2. A. V. Butler, *Electrocapillary*, Methuen and Co., London (1940).
3. St. G. Christov, *Electrochim. Acta* **9**, 575 (1964); *J. Res. Inst. Catal. Hokkaido Univ.* **24**, 27 (1976).
4. J. O'M. Bockris and D. B. Matthews, *Proc. Roy. Soc. (London)* **A292**, 479 (1966).
5. J. O'M. Bockris and D. B. Matthews, *J. Chem. Phys.* **44**, 298 (1966).
6. D. B. Matthews and S. U. M. Khan, *Austral. J. Chem.* **28**, 253 (1975).
7. G. Gamow, *Z. Phys.* **51**, 204 (1928).
8. V. G. Levich, *Physical Chemistry: An Advanced Treatise*, (H. Eyring, D. Henderson, and Y. Jost, eds.), Vol. 9B, Chapter 12, Academic Press, New York (1970).
9. R. R. Dogonadze, *Reactions of Molecules at Electrodes* (N. S. Hugh, ed.), Wiley (Interscience), New York (1971).
10. J. O'M. Bockris and S. U. M. Khan, *J. Res. Inst. Catal. Hokkaido Univ.* **25**, 63 (1977).
11. S. U. M. Khan, P. Wright, and J. O'M. Bockris, *Electrokhim.* **13**, 914 (1977).
12. J. S. Blakemore, *Solid State Physics*, W. B. Saunders, London (1970).
13. J. O'M. Bockris, R. R. Sen, and K. L. Mittal, *J. Res. Inst. Catal. Hokkaido Univ.* **20**, 153 (1972).
14. J. Appleby, J. O'M. Bockris, B. E. Conway, and R. K. Sen, *M.T.P. International Review of Science* (J. O'M. Bockris, ed.), Vol. 6, Butterworths, London (1973).
15. J. O'M. Bockris and A. K. N. Reddy, *Modern Electrochemistry*, Plenum Press, New York (1973).
16. A. R. Despic and J. O'M. Bockris, *J. Chem. Phys.* **32**, 389 (1960).
17. S. Gasiorowicz, *Quantum Physics*, John Wiley and Sons, New York (1974).
18. C. R. Gatz, *Introduction to Quantum Chemistry*, Charles E. Merrill Pub. Co., Columbus, Ohio (1971).
19. I. N. Levine, *Quantum Chemistry*, Vol. II, Allyn and Bacon, Boston (1970).
20. P. W. Atkins, *Molecular Quantum Mechanics*, Clarendon Press, Oxford (1970).
21. L. Landau, *Z. Phys. Sowjet.* **2**, 46 (1932).
22. C. Zener, *Proc. Roy. Soc. (London)* **A137**, 696 (1932).
23. E. T. Whitaker and G. N. Watson, *A Course of Modern Analysis*, Cambridge University Press, London (1963).
24. D. R. Bates, *Proc. Roy. Soc. (London)* **A257**, 22 (1960).
25. P. P. Schmidt, *Electrochem.* **5**, 27 (1975).
26. J. Albery, *Electrode Kinetics*, Clarendon Press, Oxford (1975).
27. J. Weiss, (a) *J. Chem. Phys.* **19**, 1066 (1951); (b) *Proc. Roy. Soc. (London)* **A222**, 128 (1959).
28. R. A. Marcus, (a) *J. Chem. Phys.* **43**, 679 (1965); (b) *Electrochim. Acta* **13**, 995 (1968).
29. N. S. Hush, (a) *J. Chem. Phys.* **28**, 962 (1958); (b) *Z. Elektrochem.* **61**, 734 (1957).

30. V. G. Levich, *Advances in Electrochemistry and Electrochemical Engineering* (P. Delahay, ed.), Vol. 4, Wiley (Interscience), New York (1964).
31. R. R. Dogonadze, J. Ulstrup, and Yu. I. Kharkats, *J. Chem. Soc. Faraday Trans.* II(5), 744 (1972).
32. N. R. Kestner, J. Logan, and J. Jortner, *J. Phys. Chem.* **78**, 2148 (1974).
33. W. Schmickler, *Electrochim. Acta* **21**, 161 (1976).
34. E. E. Henderson, *Modern Physics and Quantum Mechanics*, W. B. Saunders, London (1971).
35. H. L. Strauss, *Quantum Mechanics: An Introduction*, Prentice-Hall, Englewood Cliffs, New Jersey (1968).
36. E. Merzbacker, *Quantum Mechanics*, John Wiley and Sons, New York (1970).
37. J. Yunta, E. R. Mayquez, and C. S. Rio, *Phys. Rev.* **9**, 1483 (1974).
38. J. Yunta, *Junta De Energia Nuclear, Report 264*, Madrid, Spain (1974).
39. S. Ray, *Chem. Phys. Lett.* **11**, 573 (1974).
40. F. Constantinescu and E. Magyari, *Problems in Quantum Mechanics*, Pergamon Press, Oxford (1971).
41. A. R. Edmonds, *Angular Momentum in Quantum Mechanics*, Princeton University Press, Princeton, New Jersey (1957).
42. D. Eisenberg and W. Kauzmann, *The Structure and Properties of Water*, Oxford University Press, Oxford, (1969).
43. M. Abramowitz and I. A. Stegun, *Handbook of Mathematical Functions*, Dover Publications, New York (1968).
44. W. M. Latimer, *Oxidation Potentials*, Prentice–Hall, Englewood Cliffs, New Jersey (1952).
45. C. Kittel, *Introduction to Solid State Physics*, Chapter 7, John Wiley and Sons, New York (1971).
46. J. O'M. Bockris, S. U. M. Khan, and K. Uosaki, *J. Res. Inst. Catal. Hokkaido Univ.* **24**, 1 (1976).
47. H. Gerischer, *Advances in Electrochemistry and Electrochemical Engineering* (P. Delahay, ed.), Vol. 1, Wiley (Interscience), New York (1961).
48. H. Gerischer, *Physical Chemistry: An Advanced Treatise* (H. Eyring, ed.), Vol. 9A, Academic Press, New York (1970).
49. M. Gleria and R. Memming, *J. Electroanal. Chem.* **65**, 163 (1975).
50. V. A. Myamlin and Y. V. Pleskov, *Electrochemistry of Semiconductors*, Plenum Press, New York (1967).
51. A. K. Vijh, *Electrochemistry of Metals and Semiconductors*, Marcel Dekker, New York (1973).
52. L. D. Landau and E. M. Lifshitz, *Quantum Mechanics*, p. 142, Pergamon Press, London, (1965).
53. J. O'M. Bockris and R. J. Sen, *Chem. Phys. Lett.* **18**, 166 (1973).

Index